D0919472

Magnetism and Ligand-Field Analysis

For my wife, Annette
and daughters, Sarah and Joanna
who put up with it all.

MAGNETISM AND LIGAND-FIELD ANALYSIS

M. GERLOCH

Université d'Ottawa
BIBLIOTHÈQUES
LIBRARIES
University of Ottawa

CAMBRIDGE UNIVERSITY PRESS

Cambridge

London New York New Rochelle

Melbourne Sydney

Published by the Press Syndicate of the University of Cambridge
The Pitt Building, Trumpington Street, Cambridge CB2 1RP
32 East 57th Street, New York, NY 10022, USA
296 Beaconsfield Parade, Middle Park, Melbourne 3206, Australia

101255

© Cambridge University Press 1983

First published 1983

Printed in Great Britain at the University Press, Cambridge

Library of Congress catalogue card number: 83–1820

British Library cataloguing in publication data
Gerloch, M
Magnetism and ligand-field analysis.
1. Ligand field theory
I. Title
541.2′242   QD475

ISBN 0 521 24939 2

QD
475
.G44
1983

TM

Contents

Preface

By ligand-field analysis, I mean the study of a variety of ligand-field properties of transition-metal and lanthanide complexes leading to a quantified *chemical* insight into the individual metal–ligand interactions in these molecular species. I take the view that we are engaged in chemistry, that our interest is ultimately focussed on molecules, individually and as types, and that we use magnetism, spectroscopy and ligand-field theories as parts of a technique. Earlier exploitation of magnetism and ligand-field theory was largely exemplary and full of promises: a mandate was established but for years the manifesto was not carried out. It now has been: and this book seeks to tell the story of how we start with real – rather than idealized – molecules and end up with answers which are useful and readily comprehensible to the mainstream inorganic chemist. In between are the nuts and bolts of the technique. The mathematical and technical nature of the bulk of this book arise out of a desire to review and firmly establish the foundations upon which the methods of magnetochemistry and ligand-field analysis rest. The book is not, therefore, about magnetism and also about ligand-field analysis – as two separate enterprises – although a fairly clean separation has been possible on purely technical grounds: rather is it about analysis, with magnetism playing one single but important role in it.

The notion of the functional group has long played a ubiquitous role in chemical thinking but, until very recently, has failed to enrich – and, one might argue, give purpose to – all that we call ligand-field theory. As chemists we are especially interested in the separate donor or acceptor functionality of individual ligands in their discrete σ- and π-bonding roles

xi

within a complex. Inevitably, the parametrization schemes of traditional ligand-field models are unable to provide much valid insight into such properties for, being framed too strictly within a group-theoretical straightjacket, they refer only to global quantities. It is correlation with local regions of electron density in the molecule that we seek and in most cases that concept has more validity than an arid recourse to the Wigner–Eckart theorem alone. Our chemical goals are achieved here through the angular overlap model, but in the form of a cellular decomposition of ligand-field theory that resolves the spatially local, chemical functionality without taking on board the premises and approximations of semi-empirical molecular-orbital theory. A common practical objection to the AOM has been the high degree of parametrization required by this factorization of the ligand field. Our response is to argue that an increase in the number of variables must be matched by an increase in the experimental data base; an attitude directly contrary to some earlier views that more unknowns require more assumptions. All this leads to the position that the most successful ligand-field analyses are generally likely to be those based simultaneously upon a variety of experimental techniques – paramagnetic susceptibilities, electron spin resonance, and optical spectroscopy – preferably exploiting samples in the form of single crystals. While it can be more difficult to work with such systems rather than powders or solutions, observations of anisotropic properties adds enormously to the data base. However, anisotropy is manifested only in molecules with lower symmetry. Apparently there is a catch here, for lower molecular symmetry generally begets more independent system variables, so a compromise might be indicated. On the other hand, the lower numbers of independent variables defined by a more symmetrical molecule refer increasingly to overall, global molecular properties rather than to individual, local ligand interactions. Thus, studies of higher-symmetry systems are frequently rather bare of essentially chemical content: indeed, I would argue that such has been especially responsible for a decreasing interest in ligand-field studies at the research level. And the teaching of ligand-field theory relies almost exclusively upon the octahedron and tetrahedron because of the various group-theoretical rules that flow therefrom in a readily communicable way. Often, the mysteries of elementary group theory are allowed to obscure, and even totally replace, the essential reasons for developing the whole enterprise in the first place.

So this book begins with an introductory chapter, in Part I, which summarizes the position of magnetochemistry and ligand-field analysis as it was, say a decade ago; and it identifies the need for a fresh approach to the whole subject. The last section of the book, Part IV, illustrates the results and significance of several recent ligand-field analyses. Both sections involve minimal mathematical technique, the last chapter in particular aiming to show how the conclusions, if not their attainment, are readily discussable by the non-specialist. In between are two large sections: Part II, describing the foundations of paramagnetism from both classical and quantum-mechanical viewpoints, and its exploitation with respect to anisotropic crystals; and Part III, dealing with technical, computational and theoretical aspects of ligand-field theory. The level of presentation varies across the fairly wide range of topics covered in these chapters, in a way which hopefully reflects the typical background of an honours student or young researcher with a chemist's training. Thus, while some of the quantum mechanics is treated with a modicum of briskness, a more elementary path is taken for much of the basic physics. It is, of course, impossible to please everyone and I do apologize to those who find the text either irritating here or baffling there. Some examples of the sorts of questions that are discussed and answered throughout the book are; How are crystal magnetic properties measured? What is the difference between the fields **B** and **H**? Why does the magnetic moment operator take the form $\mu = (1 + 2s)$? Why is spin-orbit coupling called a relativistic correction? How does an explanation for magnetism require both quantum theory and the special theory of relativity? All these points and many others arise in Part II. In Part III we consider questions like: What is the nature of the ligand-field approximation? What is the difference between molecular orbitals and ligand-field orbitals? Are the procedures of ligand-field theory legitimate and why should one doubt it? What is required to construct a complete, experimental, computational and theoretical framework for ligand-field analysis? And many more.

The bonding energies that ligand-field analysis probes are generally much greater than those involved in the processes of exchange coupling, even though these may give rise to dramatic magnetic consequences. It can be argued, but I do not do so here, that the purely chemical content of exchange studies is frequently, though not inevitably, meagre. Certainly, such reliable chemical insight as may be evinced from magnetically concentrated systems often comes only after very thorough and technically difficult analysis. In any case, the subject does not fall naturally into that chosen for the present book and so no attempt has been made to cover it here. Even diamagnetism is really only relevant as a correction so far as ligand-field analysis is concerned and therefore our review of this aspect of the subject is cursory. Electron spin resonance spectroscopy, on the other hand, is only too relevant for our field. However, several splendid texts already exist in this area and so the subject is barely covered in the present book. I do include, nevertheless, discussions on how to calculate esr g values from our quantum models and also procedures for fitting such calculations to experimental results. For rather similar reasons, I present no detailed description of the methods of optical spectroscopy as applied to ligand-field systems: these are covered in many standard books and reviews or are of such a speciality as to require separate professional coverage. Of course, the fitting of spectral transition energies, and the assignment of bands by comparison with theory, *are* topics discussed here.

I consider myself fortunate to have worked with so many able and amiable colleagues and students over the years. In one way and another they have all contributed to this book and their work is cited throughout the text. Three have helped me here explicitly. Dr J.E. Davies provided the material for Appendix A and made many valuable suggestions concerning my approach to chapter 10. Dr R.F. McMeeking has not only checked several of my wilder follies but has also provided me with instruction on, and access to, his recent unpublished work: summaries of this appear, with explicit acknowledgement, in chapters 6 and 10. Finally, Dr R.G. Woolley has guided me on various theoretical matters throughout the preparation of the book and spent much time reading first drafts: above all, he was my co-author on the original material on which chapter 11 is based. Our collaboration, then and since, has afforded me a much greater understanding of both his and my own subjects and given me enormous pleasure. I sincerely thank them all.

Several chapters were written while on sabbatical leave in Australia and I would like to record my appreciation to Professor Hans Freeman in Sydney and to Professor Bruce West at Monash University in Melbourne for making those visits possible; and to them and their colleagues, and to all our many friends in Perth, Canberra, Brisbane, Townsville, Hobart and Adelaide for their company and hospitality. As promised, special thanks are also due to those at the University of Sydney for explaining that King's Cross is not a railway station.

Finally, I thank Mrs G. Neal-Freeman for preparing the lengthy typescript so accurately and quickly.

MG
October 1982

PART I

—

CHEMICAL AIMS IN LIGAND-FIELD STUDIES

While ligand-field theory grew up in the physics literature, it is probably within the discipline of chemistry that the subject is more widely taught and applied. In large measure, the chemist's approach to ligand-field models has been to exploit the purely group-theoretical aspects of the subject in order to establish 'finger-printing' rules by which molecules may be classified according to structure and bonding. Used in conjunction with the wide range of other physical techniques available to chemistry over the years, this approach has enjoyed some measure of success and, in earlier days at least, inspired fresh appreciation of theories of chemical bonding. However, a marked decline of interest in the ligand-field properties of single-centre systems has taken place in the past 15 years. Lest it seem perverse, therefore, to present yet another book in this area, it must be said at the outset that many of the efforts of the past decade or so can now be seen as having paved the way towards a contemporary exploitation of ligand-field theory which begins at last, and almost for the first time, to fulfil the explicitly *chemical* promise, hoped and claimed for the subject as long ago as the 1930s. In order to appreciate the sea change brought about by the contemporary synthesis, it is necessary to understand the reasons for the chemists' disenchantment with ligand-field theory in the recent past. In chapter 1, therefore, we begin with a little history.

1

Decline and recovery

___

1.1 The promise

The developments of the theories of paramagnetism and ligand fields have been inextricably linked from the beginning.[1,2] Understanding magnetic moments was one of the first aims and successes of quantum theory in the late 1920s and, by 1935, Howard had published[3] one of the earliest studies of single-crystal magnetic anisotropy – on potassium ferricyanide. On the other hand, it was probably the delayed development of sufficiently sensitive spectrophotometers that held up recognition and assignment of spin-allowed bands in optical d–d spectra till the early 1950s, despite the group-theoretical basis for ligand-field theory as a whole having begun with Bethe[4] in 1929. Probably for this reason as much as any other, the first and, for a long time, only, exploitation of the subject matter of ligand-field theory by a chemist was that by Pauling in 1932 in his seminal papers,[5] 'The nature of the chemical bond'. We must refer here to the 'subject matter of ligand-field theory' – specifically, paramagnetism – for, in several papers throughout the early 1930s, Pauling was at pains to disassociate his views of the bonding and magnetism of transition-metal complexes from the crystal-field theories of Van Vleck and of Penney & Schlapp. It was Van Vleck, however, who rationalized[6] Pauling's hybrid-bonding orbitals and crystal-field theory under the umbrella of Mulliken's molecular-orbital theory.[7] That synthesis still provides the basis for so much discussion of ligand-field theory and transition-metal complex bonding that we take it as our point of departure.

Figure 1.1 provides a schematic review of the σ bonding in octahedral symmetry in terms of molecular-orbital theory for a first-row transition-metal ML_6 complex. The bonding a_{1g}, t_{1u} and e_g orbitals are completely

3

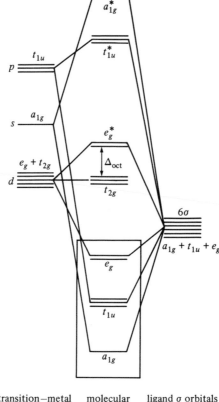

transition–metal   molecular   ligand σ orbitals
   orbitals   orbitals

Fig. 1.1. Metal-ligand σ bonding in an octahedral complex: but see also figure 11.1.

occupied by the six ligand σ-donor electron pairs, so that their relative energies are unimportant in the present qualitative description. Pauling's d^2sp^3 valence hybrids can be related to the block of bonding molecular-orbitals, shown in the box of figure 1.1, when we take note of the metal orbital parentage as involving two d atomic functions (d_{z^2} and $d_{x^2-y^2}$); the valence electrons, formally belonging to the metal atom, are distributed between the t_{2g} and e_g^* molecular orbitals exactly as prescribed in elementary crystal-field theory. The whole scheme brings together the positive aspects of Pauling's hybrids and crystal-field theory while simultaneously avoiding the more problematic features of these one-sided models. The primary property of the hybrid-bond approach was its concern with chemical bond formation. The less successful aspects of Pauling's formalism – being the loss of d orbitals available after hybridization in

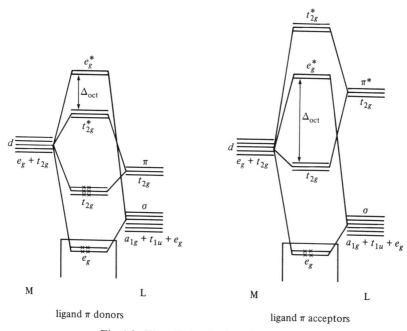

ligand π donors   ligand π acceptors

Fig. 1.2. The effects of π bonding on Δ_{oct}.

which to place the 'metal' electrons, and its general inability to describe excited-state properties and hence spectral transitions – are all adequately dealt with by reference to the non-bonding and antibonding MOs of the ligand-field region of the diagram. Similarly, the original introduction[4] of crystal-field theory in terms of the relative *repulsive* energies between metal orbitals and ligand charges is now much better represented in terms of the antibonding energies of the e_g^* molecular orbitals. The magnitude of Δ_{oct} thus refers to the antibonding energy shift of the metal e_g orbitals and, as this might be expected to change with ionization energies, overlap and resonance integrals somewhat similarly to that of the bonding counterparts, a qualitative association of Δ_{oct} values with bond 'strength' is apparent. The approach thus relieves crystal-field theory of the embarrassment of having to rationalize Δ_{oct} values in terms of the magnitudes of actual ligand charges.

As it stands, however, the model represented in figure 1.1 does not provide even a qualitative rationalization of the phenomenological variation of Δ_{oct} values we call the spectrochemical series. That can be remedied though, in outline at least, by the introduction of ligand π function into the MO scheme. The well-known diagrams in figure 1.2

illustrate the situation for the cases of ligands acting as π donors or π acceptors towards the central metal atom. On the left, the low-lying, filled π orbitals of the ligand donate three electron pairs to the metal, filling up the t_{2g} molecular orbitals: Δ_{oct} here refers to the energy splitting between t_{2g}^* and e_g^* MOs. On the right, the high-lying, empty π^* orbitals of the ligand acceptors interact with the metal t_{2g} orbitals, leaving the t_{2g} and e_g^* molecular orbitals to hold the 'metal' electrons of the ligand-field model. There follows the idea that ligand π donors tend to reduce the magnitude of Δ_{oct} relative to the value deriving from σ bonding alone, while ligand π acceptors tend to increase Δ_{oct}. If we write the contributions to Δ_{oct} from σ and π bonding as Δ_σ and Δ_π, taking the convention that positive values correspond to an upward (antibonding) energy shift in the appropriate metal orbitals, we have the relationship

$$\Delta_{oct} = \Delta_\sigma - \Delta_\pi. \tag{1.1}$$

The effects of σ and π donors as monitored by Δ_{oct} are opposed, therefore. The high position of the carbonyl ligand in the spectrochemical series[8] is thus rationalized in terms of a strong π acceptor or acid property of this ligand. Foreshadowing much later discussion, particularly that in chapter 11, it is worth commenting here on the relationship between the description we have outlined and the synergic, 'back-bonding' model[9,43,44] of Dewar & Chatt. It is commonly argued that the $-$ CO ligand might not be considered an especially good σ donor or π acceptor, *a priori*, in view of its rather modest role as a Lewis base with non-π-bonding acids. The adjective 'synergic' was coined to express the mutual augmentation of σ-donor and π-acceptor properties of this and similar ligands when provided with the appropriate orbital pathways by a metal, a basic theme in the model being the operation of the electroneutrality principle. Incorporation of these ideas within a molecular-orbital calculation means that the process of bond formation requires changes in the relative valence-orbital, ionization potentials for the metal and ligand orbitals, which in turn alter the degree of overlap and electron distribution. We take the diagrams in figures 1.1 and 1.2 to represent the *final situation* in a self-consistent molecular-orbital calculation: this means, not only that metal and ligand energies are different from those of the unbonded species, but also that appropriate modification of the basis orbitals has been made to describe electron polarization effects. For a first-row transition-metal ion then, we avoid labelling the metal orbitals as $3d, 4s, 4p$, since the identification of the energy splitting $t_{2g} - e_g^*$ with the ligand-field parameter Δ_{oct} requires those orbitals to conform with the complex molecule *as it is*, rather than to refer to the free metal ion from which it was formed. We note here that the interpretation of MO diagrams like these is a matter requiring great care, no more so than when applied to the meaning of Δ_{oct} and $10Dq$: we deal with this question in detail in chapter 11, observing for the moment that the *description* given so far is only adequate for complexes possessing a d^1 configuration for the metal ion.

One feature of Pauling's original model is not reproduced by the molecular-orbital approach. His 'magnetic criterion for bond type' ascribed certain differences between high- and low-spin species to the use of inner ($3d$) or outer ($4d$) orbitals for hybrid formation. The approach implied a discontinuity of bond type between what he considered as 'mostly ionic' and 'mostly covalent'. The MO and ligand-field models, on the other hand, recognize no such discontinuity, the change from a high- to a low-spin state reflecting a smooth progression from ligands associated with small Δ_{oct} values to those with large ones. While we can no longer support Pauling's simple criterion we might reasonably identify the birth of 'magnetochemistry' with his early papers[5], for the idea that *chemically* interesting information might be derived from a study of paramagnetism began there. Over the years, and in no small measure due to the influence of Nyholm[10] and his school, magnetochemistry grew to the status of a routine tool within inorganic chemistry research laboratories and, indeed, one may consider it as 'the senior partner' relative to spectroscopy within the broader, chemical, ligand-field discipline up to the early 1960s. One important reason for this interest lay in the models described by Figgis[11–14] at that time and their exploitation by Figgis, Lewis and their colleagues.[12–14]

1.1.1 Magnetic properties of ions with formal, orbital-triplet ground states

Bringing together the work of Kotani, Bleaney and his group, and Stevens, Figgis described an approach to the interpretation of the paramagnetism of ions with formally, triply-degenerate ground terms. Although a number of systems had been studied and interpreted in a somewhat similar fashion previously, Figgis' model was the first to have rather wide application throughout the transition-metal block, with respect to variation in metal, coordination number, and ligand. The average, powder susceptibilities of some 60 complexes and their variation with temperature were measured and 'fitted' within the range 80–300 K by Figgis, Lewis and their students.

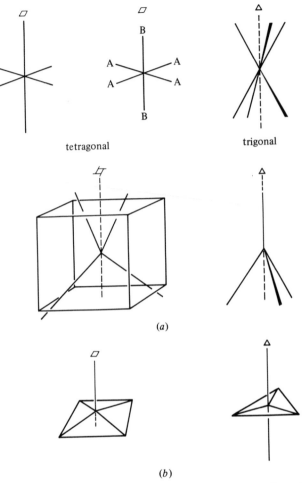

tetragonal   trigonal

(a)

(b)

Fig. 1.3. Types of axial field: (a) tetragonal and trigonal distortions from octahedral and tetrahedral precursors, and (b) examples of four-fold and three-fold geometries which do not approximate cubic precursors.

The approach may be illustrated by its application to ions with formal $^2T_2$ ground terms.

The model involves the simultaneous perturbation of the six-fold degenerate $^2T_2$ basis by the effects of spin-orbit coupling and an axially symmetric, ligand-field distortion, using the effective Hamiltonian

$$\mathcal{H}' = \lambda \mathbf{L} \cdot \mathbf{S} + V_{\text{axial}}. \tag{1.2}$$

The exact nature of the axial field is irrelevant in this model and could

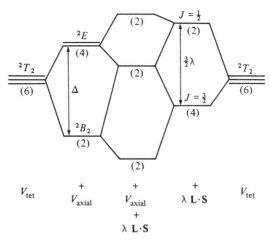

Fig. 1.4. The separate and simultaneous perturbations of a cubic-field $^2T_2$ term by an axial distortion and spin-orbit coupling: total degeneracies are given in parentheses.

represent any of the departures from octahedral or tetrahedral precursors shown throughout figure 1.3(*a*) or even, perhaps, the intrinsically lower symmetries of the types of geometry shown in figure 1.3(*b*). In all cases, the axial field is simply represented by a splitting Δ of the orbital-doublet and -singlet components of the T term, being defined by the energy difference,

$$\Delta = \varepsilon(^2E) - \varepsilon(^2B_2), \tag{1.3}$$

taking, here symmetry labels appropriate for the tetragonal distortion of the tetrahedron in figure 1.3(*a*). The model had emerged from the recognition that low-symmetry splittings can be comparable with those from spin-orbit coupling previously studied by Kotani[15] and, in the middle of figure 1.4, are shown the three Kramers' doublets which arise from these two effects (1.2) together. The energies and composition of these doublets are functions of the two parameters Δ and λ. The model involves one further variable; namely, Stevens' orbital reduction factor[16] k appearing in the magnetic moment operator,

$$\mu_\alpha = \beta_0(kL_\alpha + 2S_\alpha); \quad \alpha = \parallel, \perp. \tag{1.4}$$

Full descriptions of the technical and numerical features of Figgis' approach, which was applied to transition-metal complexes with formal $^2T_{2(g)}$, $^3T_{1(g)}$, $^4T_{1(g)}$ and $^5T_{2(g)}$ ground terms, have been presented several times and will not be repeated here. The importance of the models lay in their promise to illuminate the nature of the bonding in the complexes to which they were applied. Numerical values for the three parameters Δ, k, λ (together with ones for a further parameter A associated with the degree of cubic-field mixing between T_1 terms arising from the free-ion P and F terms) were determined by reproduction of the absolute values and temperature dependences of experimental, mean magnetic moments. These parameter values were to be correlated with broader chemical concepts along the following lines. Firstly the sign and magnitude of the axial-field parameter Δ should comment upon the nature of the ligand-field asymmetry with respect to the unique geometric axis. Something of the nature of the different roles played by A- and B-type ligands in a *trans*-octahedral MA_4B_2 complex, for example, should be discernible from a series of such Δ values. Secondly, some measure of the degree of covalency in the M—L bonds should derive from the values determined for λ and k. At the time these models were first proposed and applied, Stevens' orbital reduction factor was also called[17] an 'electron delocalization factor' or 'covalency factor', being associated with the degree of overlap between metal and ligand π orbitals. The extent by which k was reduced from unity in a given complex should, it was hoped, provide a measure of the delocalization of unpaired metal electrons onto the ligands and, at the same time, comment upon the relative degrees of σ and π bonding in the complex. The effective spin-orbit coupling coefficient λ in (1.2) should also reflect the covalent nature of the bonding by providing some measure of the degree of orbital expansion which had taken place on complex formation.[18]

Should we be less than convinced by these aspirations today, it is well to remember that Figgis' models allowed both absolute values and temperature variations of the magnetic moments of a wide variety of transition-metal complexes to be reproduced theoretically, virtually exactly and almost for the first time since the principles of the subject had been laid down in the late 1920s. This success appeared to signify that magnetochemistry had come of age and promised a rich haul, in terms of our understanding of chemical bonding, to be had from increasingly detailed and sophisticated theory and experiments. At the beginning of the 1960s the mood was certainly optimistic. It didn't last.

1.2 Larger bases

Despite its successful reproduction of the detailed paramagnetism of so many transition-metal complexes, Figgis' approach failed to yield generally acceptable interpretations of the model parameters. Actually, several

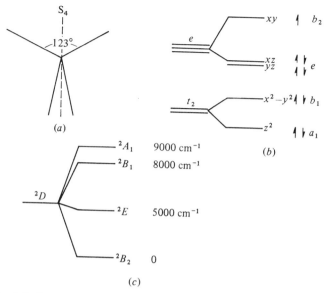

Fig. 1.5. Cs_2CuCl_4. (a) Approximate axial symmetry of $CuCl_4^{2-}$ ions: (b) predicted orbital ordering from compressed geometry, and (c) summary of term splitting from the polarized spectrum.[42]

difficulties were recognized[11–14] by Figgis and Lewis from the beginning. Orbital reduction factors frequently failed to correlate with accepted notions of π bonding and covalency and in due course came to be regarded as 'fudge' factors. Since then, arguments have been presented[19] to show why direct correlations of the sort originally claimed may not obtain and in chapter 11 we shall see more of the origins of the confused state of the interpretation of orbital reduction factors. Similarly, relationships between empirical λ values and covalency or ligand-field strength were not transparent. The most serious problems, which began to emerge a little later, were to do with the fact that various parameter values estimated by the procedures above were seriously in error due to deficiencies in the model itself. The ligand-field properties[20] of some nominally tetrahedral copper(II) complexes illustrate the point.

Crystals of Cs_2CuCl_4 contain nominally tetrahedral $CuCl_4^{2-}$ ions, shown by X-ray crystallographic analysis to suffer the tetragonally compressed, approximate D_{2d} geometry shown in figure 1.5(a). We expect, therefore, the orbitals involving the z label to be stabilized with respect to the tetrahedral energies, as shown in figure 1.5(b) and to give rise to a $^2B_2$ ground term. Application of the $^2T_2$ model, described above, to the

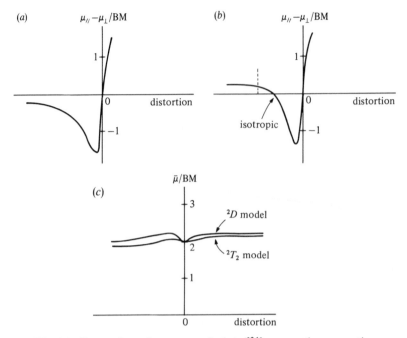

Fig. 1.6. Comparison between calculated[21] magnetic properties within a $^2T_2$ or ^2D basis. (a) $^2T_2$ model for a typical anisotropy; (b) using similar parameters within the ^2D basis, and (c) average, powder susceptibilities for typical parameter values.

powder susceptibilities of Cs_2CuCl_4 leads to a value for Δ, the tetragonal-field splitting of the term, of about $5000\,cm^{-1}$ and of a sign which agrees with the orbital-singlet ground term. Further, the single-crystal polarized spectrum of this complex, recorded and analysed by Ferguson,[42] confirms the size and magnitude of Δ as shown schematically in figure 1.5(c). At this stage, therefore, the crystal structure, crystal spectrum and powder susceptibilities are compatible within the $^2T_2$ model. However, that model involves axially distorted systems and the predicted mean susceptibilities were calculated from initially computed values of the principal molecular susceptibilities parallel and perpendicular to the axis of distortion. While reproducing the powder susceptibilities above, the model simultaneously predicts that the parallel moment should be less than that perpendicular to the S_4 axis. Experimentally, measurements[20] of the crystal magnetic anisotropy yield the opposite result.

The explanation[20] of this contradiction is that the energetically distant $^2B_1$ and $^2A_1$ components of the cubic-field ^2E precursor must be included in the calculation: they contribute through spin-orbit coupling and in second-order Zeeman terms. Their effect, though small, is important[21] and is illustrated in figure 1.6. In part (a) is shown a typical plot of anisotropy, $\mu_\parallel - \mu_\perp$, versus distortion, calculated within the restricted $^2T_2$ basis. The curve lies in two diametrically opposite quadrants, indicating a direct relationship between the sign of the magnetic anisotropy and the sense of the distortion: for the compressed geometry of the $CuCl_4^{2-}$ 'tetrahedron', the model uniquely predicts $\mu_\parallel < \mu_\perp$. Without distortion, the system is isotropic, of course, as required by symmetry. The inclusion of the higher-lying $^2A_1$ and $^2B_1$ terms, corresponding to a diagonalization of the complete ^2D free-ion term, shifts the curve by a small, but important, amount, as shown in figure 1.6(b). No longer does the curve lie only in two quadrants: no longer do we observe an unambiguous relationship between geometry and the sign of magnetic anisotropy. Indeed, at one point on the abscissa the anisotropy vanishes. The 'accidental' isotropy does not indicate structural isotropy, of course; rather, in the present example, it occurs when the ground-state splitting Δ is about $3500\,cm^{-1}$. The double reversal in the sign of $\mu_\parallel - \mu_\perp$ permits the reproduction of the experimental anisotropy in the $CuCl_4^{2-}$ ions near the broken line in figure 1.6(b). All the properties – molecular structure, crystal spectrum and crystal magnetism – are then accounted for simultaneously. Rather different estimates of the orbital reduction factors occur, however, in the later model within the ^2D basis as compared with those restricted to the $^2T_2$ set. While the enlargement of the basis produced important and significant changes in calculated magnetic anisotropies, the same is not true for powder or average magnetic moments. Thus we observe in figure 1.6(c), for the present example, that the difference between the two models is almost negligible in terms of mean moments alone.

Two important generalizations can be made at this stage; they do not derive from this study only but are confirmed[1,8] time and again from studies involving other metals, geometries and ligands. The first is that calculations must include a sufficiently complete basis within the ligand-field d- or (f-) orbital set. Today, we use the complete set of terms with maximum spin multiplicity (at least within the d block) as a matter of course: occasionally, even this may be insufficient and in any case, the convergence of the calculation should be checked. The second is that we conclude that it is impossible to rely on average magnetic moments alone to provide unambiguous fitting parameters, so that a complete set of principal susceptibilities from a single-crystal study constitutes a basic, minimal data set for any useful study of this kind.

1.3 The axial-field symmetry

The way in which various higher-lying excited states may contribute to the 'ground-state' magnetism depends upon the geometry of the system. Consider the average powder moments at room temperature of ferrous Tutton salt, $FeSO_4.(NH_4)_2SO_4.12H_2O$ and ferrous fluorosilicate, $FeSiF_6.6H_2O$: the values are $5.5\mu_B$ and $5.2\mu_B$ respectively. An early fitting[13] of these magnetic moments and their temperature variations within the cubic-field $^5T_{2g}$ ground term had yielded values for Stevens' orbital reduction factor k of about 1.0 and 0.7 respectively. The contemporary understanding[17] of Stevens' factor implied that the M—L bonds in the fluorosilicate complex were much more covalent and involved in π bonding than those in the Tutton salt. However, the coordination in both complexes involves the formally octahedral hexaquo ions $[Fe(H_2O)_6]^{2+}$, the essential difference between the two compounds being concerned with the counterions and the different crystal lattices. While an obvious contribution to understanding this problem, in view of the study[20] of the tetrahedral copper systems, was to work within the complete spin-quintet manifold, namely ^5D, it was still difficult to see why the magnetic moments of two such similar ions should be different. The key lies, however, in the different point-group symmetries adopted in the two lattices. The iron complex possesses 3 symmetry in the fluorosilicate crystals, but (rather approximate) four-fold symmetry in the Tutton salt. The term splitting diagrams for the two geometries are shown in figure 1.7.

The eventual successful interpretation[22] of the paramagnetism of these two salts originates in this difference in molecular geometry. In the tetragonal case, the higher-lying components of the cubic-field ^5E term contribute only through spin-orbit coupling and the second-order Zeeman terms, while in the fluorosilicate symmetry they are also effective by mixing with the lower $^5E(^5T_{2g})$ term via the trigonal component of the ligand field. The final result of the analysis was to yield approximately equal values of 0.7 for the orbital reduction factors in the two systems, once this low-symmetry-field mixing had been taken into account. The importance of this result for the present review of developments in magnetochemistry in recent years is that the exact symmetry of an axial distortion is relevant for a proper calculation of magnetic properties, a description in terms of a splitting like that in (1.3) being quite inadequate. The splitting Δ of (1.3) represents a difference between diagonal energies in the low-symmetry matrix: in the trigonal system there is also an off-diagonal element connecting the two ^5E terms.

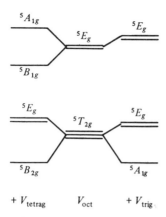

Fig. 1.7. Splittings of ^5D term for iron (II) in octahedral, tetragonal and trigonal fields.

1.4 Rhombic symmetry

Actually, contributions from higher-energy states may not always be important. A dominance by a few low-lying states can yield surprising results, as the magnetism of *trans*-dimesitylbis(diethylphenylphosphine) cobalt(II) shows. In this nominally square-planar, low-spin d^7 complex, shown in figure 1.8(*a*), the α-methyl groups of the *trans*mesitylene ligands, which are oriented almost exactly perpendicular to the coordination plane, effectively block the fifth and sixth octahedral sites.[23,24] Although the symmetry of the first coordination shell closely approaches rhombic D_{2h}, it might be expected that a reasonable approximation for analysis of the ligand-field properties would be tetragonal with the out-of-plane direction as essentially unique. The esr g values,[25] however, tell a different story. While the g tensor *is* approximately axial, in that g_z (1.74) is similar to g_y (1.97) but very different from the g_x (3.72), the 'unique' direction lies parallel to the P...P vector and not perpendicular to the coordination plane. Some details of the ligand-field analysis of this system are shown in figure 1.8(*b*).

While the primary departure from octahedral symmetry *is* represented by a tetragonal field with respect to the coordination plane, the in-equivalence of the phosphine and mesityl ligands superimposes a rhombic component. It transpires that the magnetic properties – both g values and susceptibilities[26] – are dominated by the interaction of the configurations $(xz)^2(xy)^2(yz)^1(z^2)^2$ and $(xz)^1(xy)^2(yz)^2(z^2)^2$ with the ground $(xz)^2(xy)^2(yz)^2(z^2)^1$; and the close proximity of the yz orbital to the z^2

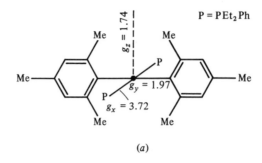

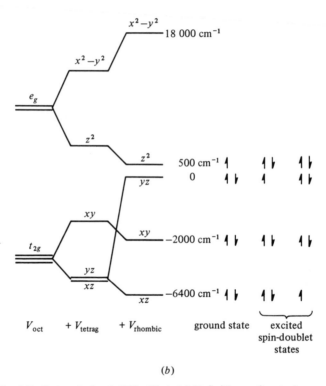

Fig. 1.8. Co(mesitylene)$_2$(PEt$_2$Ph$_2$). (a) Definitions of cartesian axes and principal g values (b) Summary of ligand-field analyses[25, 26] in terms of a rhombic field superimposed upon a tetragonal one.

function is ultimately responsible for the large paramagnetism parallel to the P...P axis. These configurations carry much more information relating to the rhombic component of the low-symmetry ligand field than to the overall tetragonal 'distortion' and so illustrate an important generalization. It is that we should not expect that different ligand-field properties, being more sensitive to different subsets of the eigenvector manifold, should in general equally reflect the pseudosymmetry and electronic structure of the system *as a whole*. We note, further, that small departures from axial symmetry, which may be considered as insignificant from a structural or general chemical point of view, may utterly dominate the magnetic properties for which they must not therefore be ignored nor unduly approximated.

1.5 Low-symmetry fields

By far the majority of real molecules of interest in ligand-field studies possess little or no symmetry. This obvious fact, which is reiterated several times throughout this book, is one which has been almost universally ignored in standard texts on ligand-field theory. While idealizations of real molecular geometries to various high-symmetry types may occasionally form an adequate basis for approximate studies of optical spectra, they are almost always totally unacceptable for interpretations of paramagnetism, of *any* quality. We illustrate this point with a discussion of the relationship between crystal and molecular susceptibilities.

In magnetically dilute systems we associate with each paramagnetic centre a molecular magnetic tensor **K**, defined, for example, by three orthogonal principal molecular susceptibilities and their orientation; and a *crystal* susceptibility tensor χ, which is given by an appropriate tensorial sum of all *molecular* susceptibility tensors in the unit cell. Consider the case for the monoclinic system, illustrated in figure 1.9(a). The monoclinic crystal class is defined by there being a unique symmetry axis, conventionally taken as b, parallel to which is a two-fold rotation or screw axis and/or perpendicular to which is a mirror or glide plane. For clarity, the situation is illustrated for the case of two molecules related by a screw axis. The crystal magnetic property must have at least the point-group symmetry of the crystal, plus, for centrosymmetric susceptibility tensors, a centre of inversion. The tensor thus possesses $2/m$ symmetry with one principal susceptibility lying parallel to the crystal symmetry axis: by convention, χ_3 is defined to lie parallel to b. As shown in figure 1.9(b), the remaining principal susceptibilities χ_1 and χ_2 lie in the crystal ac plane in directions not defined by symmetry. Experiment yields values for χ_1, χ_2 and χ_3 (and possibly ϕ). The tensor addition which relates the crystal and molecular tensors is given by

$$\chi = \sum_{\mathbf{T}}^{\substack{\text{equivalent} \\ \text{molecules}}} \mathbf{T}\mathbf{K}\mathbf{T}^{\dagger}, \qquad (1.5)$$

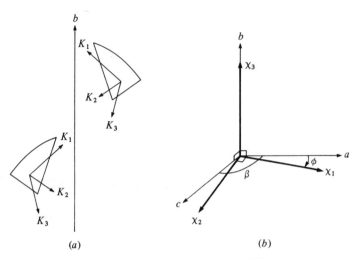

Fig. 1.9. Molecular and crystal principal susceptibilities in the monoclinic system. (*a*) Principal molecular magnetic axes related by a two-
fold screw axis parallel to the crystal symmetry axis, *b*: (*b*) one principal
crystal susceptibility is constrained to lie along *b*, by symmetry.

where **T** is the transformation matrix expressing the orientation of the
molecular tensor in the crystal. Earlier studies were concerned with axially
distorted octahedral or tetrahedral species where the orientation of the
principal molecular susceptibilities could be confidently inferred by
reference to the known crystal structure. Accordingly, the practice grew
of calculating the magnitudes of the molecular susceptibilities from
the experimentally observed crystal ones using (1.5) and the presumed
orientation of the molecular property in the crystal. The procedure
was extended to include molecules which possessed only approximate
symmetry where the orientation of the molecular susceptibilities were
considered 'obvious' by reason of pseudosymmetry and 'common sense'.
Such a system was cobalt Tutton salt, $CoSO_4.(NH_4)_2SO_4.12H_2O$, containing nominally octahedral $Co(H_2O)_6^{2+}$ ions.

The detailed coordination shell in that system approximates that of
a slightly elongated (tetragonal) octahedron. A presumption that the
principal molecular susceptibilities lie parallel, or nearly parallel, to
the Co—O bond vectors yielded,[27] via (1.5), a negative value for one
of the molecular susceptibilities – surely an inherently unlikely situation. An alternative hypothesis, in which the two principal molecular
susceptibilities were deemed to lie parallel to the bisectors of the shorter
Co—O bonds, produced very different magnitudes for the calculated

molecular susceptibilities. Clearly, the presumed orientation of the molecular tensor was untenable. It could be argued that this system was bound to be difficult since approximate structural isotropy (that is, near O_h symmetry) should yield a near-isotropic tensor property; and hence small distortions, perhaps arising from crystal-lattice effects, might dominate the tensor orientation. However, more-or-less the same behaviour is found for the more obviously tetragonally distorted molecule *trans*-Co(pyridine)$_4$(NCS)$_2$, for which some similarly calculated[28] molecular susceptibilities have very low and unreasonable values.

The problem here, of course, is to know what criterion exists by which we may judge the calculated molecular susceptibilities to be 'reasonable'. There is none. Recognizing that we cannot generally calculate six quantities – the magnitudes and orientations of the three molecular susceptibilities – from the three or four observables, we must refrain from the earlier practice of computing the molecular property in this way. Unless the orientation of the molecular magnetism is defined *rigorously* by symmetry, such calculations based on pseudosymmetry – however reasonable – can lead to *grossly* incorrect results: magnetism is frequently extremely sensitive to small structural details. Instead, comparison between observed and theoretically calculated magnetic properties must be made in the *crystal* frame. That principle aside, this discussion emphasizes the importance of low-symmetry fields in the interpretation of magnetic properties and identifies the need for a suitable quantum-mechanical model in which to accommodate them.

1.6 Symmetry-based, ligand-field parameters

It should not need saying that the aim of any magnetochemical study lies beyond the mere reproduction of observed magnetic phenomena. One could, after all, achieve that modest goal simply by fitting the temperature dependence of some experimental susceptibility to a polynomial expansion. Within chemistry, at least, we are interested in any illumination of structure and bonding which may derive from the reproduction of paramagnetism. As such conclusions are almost invariably arrived at via a ligand-field model, magnetochemistry and ligand-field theory are inextricably linked: as pointed out above, this has always been so. It is this author's view that a description of paramagnetic phenomena in chemistry which does not dwell equally on ligand-field theory fails to identify the whole purpose of the discipline. For the purposes of the present introduction, let us briefly review some of the ways in which ligand-field theory has been used in the recent past. Throughout this summary we should bear in mind the importance of low-symmetry in *real* chemical systems, especially with regard to magnetic properties.

It is well known that we may express a general ligand-field potential as a superposition of spherical harmonics, the expressions for the octahedral potential V_{oct} in tetragonal and trigonal quantization being,

$$\text{tetragonal}: \quad V_{oct}^{(4)} = Y_0^4 + \sqrt{(\tfrac{5}{14})}(Y_4^4 + Y_{-4}^4), \tag{1.6}$$

$$\text{trigonal}: \quad V_{oct}^{(3)} = Y_0^4 + \sqrt{(\tfrac{10}{7})}(Y_3^4 - Y_{-3}^4). \tag{1.7}$$

The expansion coefficients $\sqrt{(5/14)}$ and $\sqrt{(10/7)}$ are determined by the O_h symmetry and the normalized forms of the spherical harmonics. Energy calculations involve only one parameter, Δ_{oct} or $10Dq$ (but see chapter 11 concerning the definitions of these quantities), and we summarize the dependence by the expression

$$\langle d|V_{oct}|d \rangle \Rightarrow Dq. \tag{1.8}$$

Axially distorted octahedra, having lower symmetry, leave three expansion coefficients undetermined, and we write

$$V_{D_{4h}} = aY_0^2 + bY_0^4 + c(Y_4^4 + Y_{-4}^4), \tag{1.9}$$

$$\equiv aY_0^2 + b'Y_0^4 + c'V_{oct}^{(4)}, \tag{1.10}$$

for tetragonal distortions, and

$$V_{D_{3d}} = \alpha Y_0^2 + \beta Y_0^4 + \gamma(Y_3^4 - Y_{-3}^4), \tag{1.11}$$

$$\equiv \alpha Y_0^2 + \beta' Y_0^4 + \gamma' V_{oct}^{(3)}, \tag{1.12}$$

for trigonal symmetry. Matrix elements of these ligand-field potentials may be parametrized by the coefficients a, b, c, etc., but if so, it is obvious that little chemical insight will be gained. Some workers prefer the 'descent-in-symmetry' approach implied by (1.10) and (1.12) in order to maintain an apparently constant definition for Dq, and parametrize matrix elements for tetragonal and trigonal symmetry with the quantities,

	tetragonal	trigonal
$\langle d|Y_0^2|d \rangle \Rightarrow$	Ds	$D\sigma ,$
$\langle d|Y_0^4|d \rangle \Rightarrow$	Dt	$D\tau$

$$\left. \right\} \tag{1.13}$$

These parametrization schemes have been reviewed extensively.[1,8,29]

1.7 The point-charge model

A common characteristic of schemes like those involving Dq, Ds, Dt, $D\sigma$, $D\tau$ is that matrix elements or potentials are parametrized at the *global* level; that is, the variables refer to the molecule as a whole. Although such procedures are entirely proper, being well founded group-theoretically, they offer little by way of chemical insight. It is far from obvious, for example, how the magnitudes and signs of Dt and Ds in a tetragonally distorted octahedral complex, say MA_4B_2, reflect the differing characteristics of the various metal–ligand interactions; or, in angularly distorted systems, how $D\sigma$ and $D\tau$ relate, on one hand, to bonding factors and, on the other, to details of the coordination geometry. The point-charge model is one attempt to localize our perception of the ligand field.

In this, as in the closely related point-dipole approach, ligands are represented as point charges (or dipoles) located at their donor atoms. The model was much used[8] for angularly distorted systems and is very briefly reviewed here by reference to trigonally distorted octahedra and tetragonally distorted tetrahedra, as in figure 1.10. In both geometries, we consider cases where all ligands are identical, the departure from cubic symmetry being represented by the angle θ. The point-charge model provides a degree of separation between the angular parameter θ and the radial parameters Cp and Dq, defined as

$$\left.\begin{aligned}
Cp &= \tfrac{2}{7}ze^2 \int R(d)\frac{r^2}{a^3}R(d)r^2dr, \\
Dq &= \tfrac{1}{6}ze^2 \int R(d)\frac{r^4}{a^5}R(d)r^2dr.
\end{aligned}\right\} \tag{1.14}$$

The second- and fourth-order radial integrals (Cp and Dq respectively) do not involve the angular coordinates of the ligands and so hopefully refer directly to features of the M—L bonding. The angle θ must be treated as a parameter (although it should take some value near that defined by the actual molecular geometry) because of the gross approximation involved in representing ligands as point charges. It was also hoped that sensible relationships between Dq and Cp values for different geometries, like those in figure 1.10, should emerge, as the 'chemically unimportant' angular or distortion features are factorized off from the radial parameters in this model. The precise nature of these relationships is not clear, however, when we recognize that the definition of Dq in (1.14), for example, only coincides with that in (1.6) for exact octahedral or tetrahedral symmetry and, in any case, the whole question of the definition

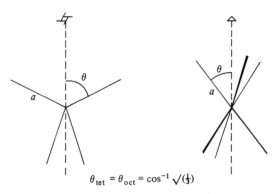

$$\theta_{\mathrm{tet}} = \theta_{\mathrm{oct}} = \cos^{-1}\sqrt{(\tfrac{1}{3})}$$

Fig. 1.10. Angular parameters in tetragonally distorted tetrahedra and trigonally distorted octahedra.

of Dq and its relationship to Δ_{oct} outside of pure *crystal*-field theory is a vexed one, to which we return in chapter 11.

1.8 A nadir in magnetochemistry

Neither the global parameters of §1.6 nor the point-charge model offer a satisfactory route for the description of the ligand-field properties of low-symmetry molecules. This is immediately obvious in the former approach because parameters like Dt and Ds result from a breakdown of the ligand-field potential or matrix elements into a rather small number of parts on the basis of molecular symmetry. The potential for a totally unsymmetrical molecule (C_1) is written as

$$V_{C_1} = aY_0^2 + bY_{\pm1}^2 + cY_{\pm2}^2 + dY_0^4 + eY_{\pm1}^4 + fY_{\pm2}^4 + gY_{\pm3}^4 + hY_{\pm4}^4 \tag{1.15}$$

for d electrons, in which the coefficients of Y_m^l for $m \neq 0$ are complex. The 14 degrees of freedom so engendered are daunting, to say the least, but even if such an approach were pursued, the chemical significance of such expansion coefficients can never be clear. If, instead, we adopt the point-charge model, assigning different Dq and Cp parameters to each ligand *and* parametrizing each unique coordinate-bond angle in the molecule, there would result an equally intractable problem. But the need for some useful model for low-symmetry molecules is evident from our earlier discussion of paramagnetism. This is but one of the reasons why magnetochemistry and ligand-field studies in general had ceased to capture the imagination of mainstream chemists by, say, the end of the 1960s. With all our preceding remarks in mind, together with a little hindsight, let us summarize the problems and failings of ligand-field methods at that time.

Throughout the 1960s ligand-field studies of optical spectra, paramagnetism and esr became more detailed and sophisticated with respect to both experimental and theoretical techniques: one thinks here particularly of the measurement of single-crystal properties and the models for axial distortions that went with them. In large measure, these studies appeared to concentrate far too much on the purely technical parts of the analysis and on the ability of a selected set of approximations to reproduce the observed data. Even when some attempt was made to assess the significance of the model variables so determined, the opportunity for correlation with generally understood chemical concepts was slight. This was not just a question of poor communication between specialist and mainstream chemist, however. The original and still-relevant aims of magnetism and ligand-field theory within chemistry had been largely forgotten in the process of mastering a different level of theoretical and computational detail than had been common before. Thus we observe:

(i) There were too many ligand-field parameters without obvious chemical significance: it is difficult to see how global parameters can provide a ready commentary on ligand chemical functionality, for example.

(ii) Little relationship was evident between the various models described at that time or between the parameters used for similar molecules of different geometry. The diverse functional bases and parametrization schemes appearing in the literature frequently prevented meaningful comparison between otherwise chemically related systems.

(iii) The inability to cope with molecules possessing little or no symmetry: in many cases it had been possible to reproduce observed magnetic data, especially those derived only from experiments on powders, with models which idealized real molecular geometry to axial symmetry. But experiments on single crystals, in particular, had demonstrated that such fits could be false and that analysis of some systems could not progress without the development of a suitable ligand-field model which recognized the true coordination geometry. Historically, one can see why small distortions had been neglected for so long. When crystal-field theory was first exploited, physicists were concerned to see whether small details of a crystal lattice might cause energy splittings which could affect magnetic properties sufficiently to negate the utility of the simple, symmetry-based rules which had been developed from the beginning. However, Van Vleck proved[30] a theorem showing that, provided such splittings are not larger than about kT, their effects on *average* susceptibilities are very small. Accordingly, physicists neglected low-symmetry components for many years and the chemists followed suit. *Now* that we recognize that useful chemical information may frequently not emerge from mean magnetic moments alone, but from measurements of crystal anisotropies, Van Vleck's theorem loses some relevance and the original concerns come to the fore once more. As we shall see, the solution of these problems is not merely an exercise in quantum mechanics, but actually provides magnetochemistry and ligand-field theory with one of its more powerful, and chemically useful, tools. In the early 1970s, however, a general solution had not been developed.

(iv) At that time, there still remained some practical problems associated with the experimental measurement of paramagnetism. The first of these concerned the general lack of *routine* apparatus for the measurement of susceptibilities of either powdered or single-crystal samples at temperatures below that of liquid nitrogen. As susceptibilities vary most rapidly at temperatures lower than this, studies performed within the 80–300 K range often failed to provide a sufficiently exacting data set with which to test quantum models. The second arose in connection with triclinic crystals. While the theoretical basis for the measurement of single-crystal susceptibilities dated back many years, the difficulties in measuring those properties for the completely unsymmetrical, triclinic crystal class had not been fully overcome.

Altogether, therefore, a great deal was required to be understood and solved at that time, relating to apparatus design, measurement technique and theoretical interpretation. Above all, the original aims of magnetochemistry and ligand-field theory – of being directed to the better understanding of structure and bonding in real chemical systems – had to be readdressed.

1.9 The angular overlap model

In the early 1960s a rather different approach to the understanding of ligand-field theory began to emerge. The angular overlap model (AOM) was introduced as a means of assessing the contributions of separate ligands to splitting patterns observed in $d–d$ and $f–f$ electronic spectra. Its original *purpose*, therefore, lay firmly within the area of ligand-field theory. The *manner* of its introduction, however, was as a molecular-orbital scheme based upon the contemporaneously popular Wolfsberg–Helmholz model.[37] It is on this basis that the AOM is expounded in teaching texts[31-33] today: it is equally because of these historical beginnings that some hold the AOM in bad odour despite its demonstrable success in providing a path by which the problems summarized in §1.8 can be solved and the essential chemistry within ligand-field studies recovered. Models come and go, like fashions: this is particularly so of the history of ligand-field theory and magnetochemistry. It is therefore doubly important at this juncture to assure the reader that the development and exploitation of the AOM which forms a large part of the present book do *not* represent yet another ephemeral phase of the ligand-field story. Firstly, the procedures of the AOM as used in ligand-field studies can be totally divorced from the molecular-orbital scheme which gave it birth: in chapter 11, a detailed theoretical structure for ligand-field theory in general, and the AOM in particular, is provided, in terms of well-established, quantum-mechanical principles. Secondly, the chemical relevance and success of the AOM in practical applications to ligand-field analysis are described in chapter 12. Before looking ahead that far, however, let us review the basic method of the AOM roughly as it was first presented.[34-36]

The AOM was introduced, then, as a molecular-orbital scheme. We consider first an individual metal–ligand interaction within some general ML_N complex and initially restrict our attention to the overlap between one metal d orbital and one ligand orbital. In the LCAO approximation we write, as usual, molecular orbitals in the form

$$\psi_b = c_{bM}\chi_M + c_{bL}\chi_L, \qquad (1.16)$$

$$\psi_a = c_{aM}\chi_M + c_{aL}\chi_L, \qquad (1.17)$$

for bonding and antibonding MOs, respectively, in terms of the metal and ligand orbitals χ_M and χ_L. Application of the variation method to optimize the coefficients c leads to the secular determinant

$$\begin{vmatrix} H_M - E & H_{ML} - ES_{ML} \\ H_{ML} - ES_{ML} & H_L - E \end{vmatrix} = 0, \qquad (1.18)$$

in which H_M and H_L represent metal and ligand orbital energies (say VSIEs); H_{ML} is the Hückel 'resonance integral'; and S_{ML} the appropriate M—L overlap integral. The roots of this determinantal equation are given by the roots of the quadratic equation in E, namely

$$(H_M - E)(H_L - E) - (H_{ML} - ES_{ML})^2 = 0. \qquad (1.19)$$

It is assumed that the involvement of transition-metal d orbitals in bond formation is small (relative to that of s and p functions), so that

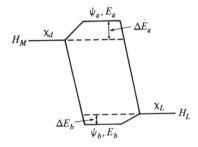

Fig. 1.11. Interaction between metal and ligand orbitals in a simple molecular-orbital model.

energies of the bonding and antibonding MOs should not be too different from the ligand and metal orbital energies, respectively, as sketched in figure 1.11, (ligand-field theory similarly recognizes the small role of the metal d orbitals in M—L bonding which is reflected, for example, in the remarkable relevance of the Laporte rule for spectral intensities). This assumption permits a useful simplification of (1.19), as follows. Consider the root E_a corresponding to the antibonding MO ψ_a. By assumption, $E_a \approx H_M$, so we make this equality in all parts of (1.19) except that which would vanish identically if we did:

For E_a:
$$(H_M - E_a)(H_L - H_M) - (H_{ML} - H_M S_{ML})^2 = 0 \qquad (1.20)$$

and hence,

$$E_a = H_M + \frac{(H_{ML} - H_M S_{ML})^2}{H_M - H_L}. \qquad (1.21)$$

similarly:

for E_b:
$$(H_M - H_L)(H_L - E_b) - (H_{ML} - H_L S_{ML})^2 = 0 \qquad (1.22)$$

and so,

$$E_b = H_L - \frac{(H_{ML} - H_L S_{ML})^2}{H_M - H_L}. \qquad (1.23)$$

During the 1950s and 1960s, the Wolfsberg–Helmholz[8,37] (WH) extension of Hückel theory for transition-metal complex bonding enjoyed considerable popularity. A recipe in the Wolfsberg–Helmholz scheme, inspired by an earlier approximation of Mulliken regarding the partition of 'overlap charge density', concerned the expression of the Hückel resonance integrals as proportional to the average of the Coulomb integrals, namely

$$H_{ML} \approx F . S_{ML} \left(\frac{H_M + H_L}{2} \right). \qquad (1.24)$$

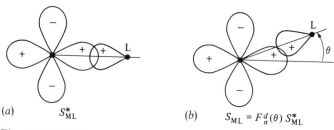

(a) S_{ML}^* (b) $S_{ML} = F_\sigma^d(\theta)\, S_{ML}^*$

Fig. 1.12. Defining the radial and angular overlap integrals involved in a single metal–ligand σ bond.

If this expression, which is in effect a definition of F, is incorporated into equations (1.21) and (1.23), there arise equations for the antibonding and bonding energy shifts ($E_a - H_M$ and $H_L - E_b$ respectively),

$$\Delta E_a = \left(\frac{x}{H_M - H_L}\right) S_{ML}^2, \tag{1.25}$$

$$\Delta E_b = \left(\frac{y}{H_M - H_L}\right) S_{ML}^2, \tag{1.26}$$

where the numerators x and y are functions of F, H_M and H_L; and which describe the *energy shifts of the precursor metal and ligand orbitals resulting from bond formation as proportional to the squares of the metal–ligand overlap integrals.*

Now a general M—L overlap integral can be factorized into two parts – angular and radial – and we write

$$S_{ML} = F_\lambda^l \cdot S_{ML}^*, \tag{1.27}$$

where S_{ML}^* represents the maximum value for S_{ML} corresponding to the situation for 'best-aligned overlap' shown in figure 1.12(a). The angular factor F_λ^l describes the dependence of the total overlap integral upon the 'misdirected' nature of the interaction illustrated in figure 1.12(b): l labels the azimuthal quantum number (here $l = 2$ for metal d orbitals) and λ labels the symmetry of the M—L bond classified with respect to the M—L axis–thus λ stands for σ, π or δ bonding. In ligand-field problems we are interested in the antibonding energy shifts of the metal d orbitals (or f for lanthanide complexes, of course) and so from (1.27) and (1.25) we have

$$\Delta E_a = e_\lambda (F_\lambda^l)^2, \tag{1.28}$$

where

$$e_\lambda = \left(\frac{x}{H_M - H_L}\right) S_{ML}^{*2}. \tag{1.29}$$

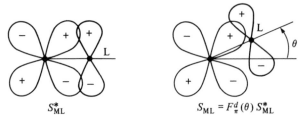

$$S_{ML}^* \qquad\qquad S_{ML} = F_\pi^d(\theta)\, S_{ML}^*$$

Fig. 1.13. Defining the radial and angular overlap integrals in a metal–ligand π bond.

The angular part of the overlap integral is defined entirely by the orientation of the ligand axis with respect to the quantization frame of the metal orbital, together with the specified angular form of the metal d (f) orbital. The radial part, dependent upon all other aspects of the metal–ligand bonding, is sequestered together with VSIEs (or equivalent quantities) into the energy *parameter* e_λ.

To be concrete; suppose we are concerned with a complex ML_N comprising N equivalent ligands arranged *arbitrarily* around the central metal. The approach describes the various antibonding energy shifts of he metal d orbitals as proportional to the *angular* overlap integrals squared. The proportionality constants are given by the e_λ of (1.29), with values to be fitted by comparison with experimental data. This simple scheme originally defined the *angular overlap model*. Our description needs filling out a little, of course, to include the effects of π and δ bonding. For example, the diagram corresponding to that in figure 1.12 but for a π interaction between metal and ligand, is shown in figure 1.13. Here the proportionality constant appearing in (1.28) is e_π which, for the same metal, ligand and bond length is expected to be much smaller than e_σ simply on an overlap criterion for the two S_{ML}^* values. The detailed model is a little more complicated than explicitly described thus far, of course, for the 'misorientation' of a given M—L interaction with respect to the global, metal coordinate frame generally involves three dimensions. Accordingly, the angular factors F_λ^l are not, in general, simply functions of only one angular variable. Further, we can include in the approach, ligands whose π- (and δ-) bonding functionality may depend on the direction perpendicular to the M—L vector. A metal–pyridine interaction, for example, is expected to involve a σ bond along the M—N vector and π bonding perpendicular[††] to the plane of the pyridine ligand: no π bonding is

---

[††] We adopt the convention throughout this book that π bonding perpendicular to a ligand plane involves π orbitals whose nodal plane *is* that ligand plane.

anticipated *within* this plane. An AOM scheme might assign non-zero, parametric values e_σ and $e_{\pi\perp}$ therefore, but a fixed value of zero for $e_{\pi\parallel}$. Accordingly, calculation of F_λ^l values in these, and all general, circumstances requires a knowledge of the the mutual orientation of the fixed, but arbitrary, global frame and the ligand cartesian axes, defined by the local M—L pseudosymmetry. The construction of F_λ^l values is thus relegated to an exercise in coordinate rotations: it is discussed thoroughly in chapters 9 and 10. Finally, if we assume that ligand–ligand overlap and interactions are negligible, we may consider the total antibonding effects of a complete set of ligands within a complex as given simply by the sum of the separate effects from each individual ligand.

The original formulation of the AOM thus involved three assumptions:

(i) that the metal orbital antibonding energy shifts are proportional to the squares of the appropriate overlap integrals;

(ii) that individual M—L interactions may be classified with respect to local pseudosymmetry. This 'overlap criterion' separates the effects of σ, π and δ bonding; and

(iii) that antibonding contributions from different ligands are additive.

This is the first of several definitions of the AOM that we discuss. Although it provides the most transparent description of the approach for the non-mathematical chemist, it also attracts the greatest criticism from a theoretical point of view. While we shall discuss this issue thoroughly throughout this book, we first move on to appraise the advantages offered by the AOM: we further note that the manner in which the AOM is *used* will be demonstrated to be entirely sound, even though the historical formulation we have reviewed is quite irrelevant.

1.10 The chemical appeal of the AOM

The AOM expresses metal orbital antibonding energies within a transition- or lanthanide-metal complex in terms of the squares of the angular parts of metal–ligand overlap integrals, and energy parameters which refer separately to individual ligands in the complex and to their discrete bonding modes. If values for these various parameters can be determined by comparison with spectra and other ligand-field properties, we shall obtain some measure of, and differentiation between, σ, π (and perhaps δ) bonding for each ligand in the complex. The appeal for chemistry is obvious. Instead of parametrizing ligand-field properties in terms of globally defined quantities, the AOM focusses on *local* M—L interactions and, in so doing, brings to ligand-field theory the notion of the functional group. At the same time, the technical procedures of the approach are such that molecules without any symmetry at all can be modelled on the same footing, and with the same facility, as those with high symmetry. In principle, therefore, the AOM offers solutions to the first three problems identified in §1.8. In practice, of course, nothing is that simple. However, let the reader be assured that the promise *has* become a reality, the present book being an attempt to clarify and detail the principles and technicalities involved in making it so.

1.11 The AOM as a perturbation model

While the Wolfsberg–Helmholz method enjoyed some popularity in the recent past as a means of calculating energy levels in transition-metal complexes, its quantitative success was very limited and, in common with many other semi-empirical molecular-orbital methods, it has been the subject of considerable criticism for its theoretical inconsistencies.[38–40] In an attempt to distance the AOM from these problems, Schäffer subsequently suggested[41] a 'perturbation' formalism for the AOM which does not refer to the Wolfsberg–Helmholz scheme. He set out three assumptions for his revised model:

(A) that the ligand-field potential acts as a first-order perturbation on the metal orbital basis, in the sense that admixtures of ligand functions into the basis need not be considered;
(B) that the perturbation is diagonal within the local metal–ligand coordinate frame; and
(C) that the perturbations from different ligands are additive.

In attempting to draw this perturbation approach and the overlap-oriented model together, Schäffer asserted that the first of these assumptions (A) is equivalent to the idea that the metal orbital energy shift is proportional to the square of the relevant overlap integral (assumption (I) in §1.9). He commented, too, that the overlap criterion (II) of the earlier model 'serves to clarify' assumption (B). This is obviously correct, for assumption (B) is a strong assumption otherwise: assumptions (B) and (II) thus appear somewhat tautological. One may be left with the impression that the 'equivalence' of the perturbation and overlap approaches implies that the AOM still relies upon energy shifts being proportional to the squares of overlap integrals, even if the Wolfsberg–Helmholz foundation is discarded. This is not so, however, for the only equivalence required–and one which is trivially true – is that the angular parts of the overlap integrals transform identically to the angular parts of the metal basis wave functions: indeed the name of the AOM follows from this. The Wolfsberg–Helmholz model carries a much stronger assumption of simple proportionality between resonance and overlap integrals which is irrelevant for the AOM.

1.12 The magnitudes of AOM parameters

A number of 'ground rules' concerning the magnitudes of the AOM e parameters have emerged over the years, whose origins lie in the earlier molecular-orbital formulation of the model. They are all based on considerations of the magnitudes of metal–ligand overlap integrals. Most immediate of these is the idea that $e_\sigma > e_\pi \gg e_\delta$ and, so far as it was possible, early applications of the AOM did indeed support this. Had they not, belief in the 'reality' of the approach, and in its relevance to chemistry in general, would have withered on the vine. Further, in the interests of reducing the number of parameters spawned by the AOM, a subject we shall discuss at length in §9.8, effects of M—L δ bonding were usually ignored altogether by setting e_δ to zero. In the same way, presumptions about chemical bonding were frequently built into early AOM analyses: thus e_π values for non-π-bonding ligands like NH_3, or for those like pyridine in the plane of the heterocycle ($e_{\pi\|}$), were fixed at zero by assumption. In all cases, these simplifying assumptions were required in order to reduce the degree of parametrization to a level which was tractable within the amount of experimental data obtainable: this latter almost invariably derived from optical spectroscopy in the earlier AOM studies. The same exigencies also led to a further 'refinement' in which the ratio e_π/e_σ was predetermined in some analyses by reference to a calculated ratio of the squares of associated overlap integrals. The rationale for this approach derives from (1.29) in which e parameters are given as proportional to the squares of S_{ML}^*. On the other hand, in several analyses in which AOM e_σ and e_π parameters were allowed to very freely in the process of reproducing ligand-field properties, the empirical e_π/e_σ ratios so derived were frequently very different indeed from the idealized 'theoretical' values.

We are fast approaching the position of having commended the angular overlap model as the 'chemical saviour' of ligand-field theory, so to speak, only to be seen amassing a new list of objections to it. In order to maintain the superficial level of the present review, we reserve a more detailed critique of the AOM, together with a sound and proper reformulation of the approach, to the later chapters. For the moment, then, we look only in outline at the approach by which the deficiencies of the model can be made good and by which we can now claim that contemporary ligand-field studies genuinely do provide information that relates directly to chemical bonding and the electron distribution in complexes in a way that is transparently obvious to the non-specialist.

1.13 Spectroscopy and magnetism together

Consider a transition-metal complex ML_6 with exact octahedral O_h symmetry and involving ligands which interact with the central metal in cylindrical symmetry with respect to each local M—L frame: these so-called 'linear ligators' are thus parametrized in the AOM by e_σ and $e_{\pi x} = e_{\pi y}$. There results the well-known[8] equation for the energy separation between the basis t_{2g} and e_g d-orbital sets,

$$\Delta_{\text{oct}} = \varepsilon(e_g^*) - \varepsilon(t_{2g}^*) = 3e_\sigma - 4e_\pi. \tag{1.30}$$

Note that this relationship has the same form as (1.1). In the original AOM scheme this is only to be expected, for both (1.1) and (1.30) are based upon the same molecular-orbital formalism: the octahedral-field splitting arises from antibonding energy differences due to ligand σ-donor function minus those due to ligand π-donor function. The molecular-orbital version of the AOM clearly correlates the sign of an e_π parameter directly with the π-donor or acceptor functions of the ligand. Actually, with a high degree of confidence, the same conclusion emerges from an analysis of the AOM within ligand-field theory, as we shall see in chapter 11, but the result is not at all direct.

However, we note also that the measurement of any ligand-field property can only yield a value for the combination of e_σ and e_π parameters in (1.30) which is Δ_{oct} so that, while the relationship may have theoretical interest, the high molecular symmetry denies us the possibility of differentiating σ- and π-bonding functions in this system. The AOM is clearly overparametrized in this situation. Indeed, similar circumstances occur for many molecules which possess relatively high symmetry, when an AOM analysis may only retain any utility if supplemented with further assumptions about the presumed bonding function of several of the ligands, as discussed in §1.12. Many of the earliest AOM studies were of the optical spectra of relatively high-symmetry species and it should not be surprising therefore that the angular overlap model failed to win universal appeal in these circumstances. For it is with *low*-symmetry molecules that the

AOM acquires utility – both absolutely and relatively. By this is meant the fact that a lack of global molecular symmetry will increasingly remove constraints like (1.30) which link different e parameters in fixed ratios and, on the other hand, that the degree of parametrization and lack of relevance increase with lower symmetry within the point-charge or symmetry-based models, as described in §§ 1.6–1.8. Consider, by way of illustration, a molecule MA_4BC_2, possessing no overall symmetry at all but in which the four A-type ligands may be considered sufficiently similar to share common e parameters; and similarly for the two C-type ligands: let us further suppose one of the ligand types, say B, is a linear ligator and that another, C, may be deemed not to enter into π bonding in one direction. Should these conditions seem unduly restrictive, note that they are actually very common: a realistic system conforming to them would be $M(\text{bidentate-carboxylate})_2\text{Br}(\text{quinoline})_2^{n-}$, for example, illustrating incidentally that the donor group A can represent one end of a chelating ligand. The set of AOM parameters comprises, therefore, $e_\sigma(A)$, $e_{\pi x}(A)$, $e_{\pi y}(A)$, $e_\sigma(B)$, $e_\pi(B)$, $e_\sigma(C)$, $e_{\pi x}(C)$, a total of seven degrees of freedom. As discussed in §1.8, a general multipole expansion of the ligand-field potential involves 15 variables (if Y_0^0 is included). In general, the number of AOM e parameters in a given study can be less than, equal to, or greater than the number of expansion coefficients in a symmetry-based, multipole expansion. It is also generally true that low or non-existent symmetry increases the chances that the various parameters in an AOM analysis will be resolvable. And, of course, we must recognize that the great majority of interesting and real, complex species actually do possess little or no symmetry. All these aspects of the degree of parametrization within the AOM – and the problems associated with them must not be minimized – are discussed more formally, and in greater detail, in §9.8. For the purposes of the present discussion, however, we point to the very large number of parameters which frequently remain in AOM studies.

No doubt that was why so many early ligand-field studies within the AOM scheme concentrated on molecules with relatively high symmetry – tetragonally distorted octahedra and *trans* MA_4B_2 type systems being popular systems – for these analyses were performed almost exclusively with data from spectroscopy alone. Of course, the problem here, paradoxically exacerbated by the high symmetry, is the small number of optical transitions which are observed. Accordingly, many of these early studies relied rather heavily upon the notion of transferability of AOM parameter values between chemically related systems, a tactic of which the present author strongly disapproves, for reasons set forth in chapter 9. The way forward in AOM studies involving a large number of parameters is surely to *increase the experimental data base* rather than the number of *ad hoc* assumptions. Measurements of paramagnetic susceptibilities and esr *g* values, particularly from single crystals, can provide a powerful way of doing this. Given the appropriate theory, these magnetic properties may be calculated from the same model used to reproduce an electronic spectrum and so a joint attack on a ligand-field problem, using a variety of ligand-field properties simultaneously, can be sufficiently definitive to yield values for most, if not all, of the AOM system variables. The theoretical and procedural package required for such analysis is lengthy and detailed, but has been produced: it is described in some depth in the present book. For the moment, we might note two particulars.

The first concerns the problem of calculating the magnetic suscepti-bilities and *g* values of molecules possessing, generally, no symmetry. The usual quantum-mechanical version of the Debye–Langevin susceptibility equation as given, for example, by Van Vleck,

$$K_\alpha = N_A \frac{\sum_i (E_i^I/kT - 2E_i^{II})e^{-E_i^0/kT}}{\sum_i e^{-E_i^0/kT}}, \tag{1.31}$$

where

$$\left. \begin{aligned} E_i^I &= \langle \psi_i^0 | \mu_\alpha | \psi_i^0 \rangle, \\ E_i^{II} &= \sum_{i \neq j} \frac{\langle \psi_i^0 | \mu_\alpha | \psi_i^0 \rangle \langle \psi_j^0 | \mu_\alpha | \psi_i^0 \rangle}{E_i^0 - E_j^0}, \end{aligned} \right\} \tag{1.32}$$

in which symbols have their usual significance within perturbation theory, provides an expression for the calculation of a susceptibility K_α parallel to one principal direction (α) of the magnetic tensor. In high-symmetry systems those directions are known but in the absence of symmetry-defining constraints they are not: we described some aspects of this problem in connection with crystal susceptibilities in §1.5. We therefore require an equivalent expression for the calculation of all six, independent components $K_{\alpha\beta}$ of the susceptibility tensor so as to avoid making unfounded assumptions about the orientation of the principal molecular susceptibilities in unsymmetrical molecules: this is described in chapter 7.

The second point we touch on briefly here concerns the overall philosophy and strategy of what we might call a 'modern ligand-field synthesis'. A package for the measurement and analysis of magnetic and spectral ligand-field properties should satisfy certain minimum conditions. These are far more exacting than could be reasonably required 20 years ago in view of the advances made with respect to both apparatus design and computer power. In many ways, it can be argued that the present synthesis has come about because the technological power necessary for its success has been made available. A corollary must be that no excuse remains for inadequate ligand-field analysis. The approximations entertained in the past within magnetochemical and ligand-field models, especially those concerned with unduly truncated basis sets or with invalid idealizations of molecular symmetry, have no place in today's environment. Used intelligently, the contemporary techniques really do allow the hopes of the pioneers in the chemical use of ligand-field theory increasingly to be fulfilled. The present system provides a means to calculate the transition energies and assignments of transition-metal, or lanthanide, complex electronic spectra; the principal molecular and crystal paramagnetic susceptibilities, together with their orientations and temperature variation; and the esr principal g values and their orientations; and all this may be done for complexes with virtually any coordination number and geometry, involving any type or combination of ligand, for metals with any d^n or f^n configuration. The system includes facilities for comparison of any of these properties with experimental data and for optimizing the system parameters for best fit. All molecules are treated on the same basis so that comparisons may be made directly. Most important, from a practical viewpoint, is the fact that explorations of parameter space can be made quickly and without the need for new theory or computer programs: when we ask how one or other ligand-field property changes as we alter one or more parameters, there must be no psychological barrier to performing the necessary computations deriving from a necessity for a 'fresh start'. Finally, behind the whole approach, theory is available which provides a sound basis for the technique and serves to clarify the significance of the parameters employed. In Part II we discuss paramagnetism, in Part III ligand-field theory and the detailed construction of the complete package, and in Part IV we look at how it works out in practice.

References

[1] Gerloch, M., *Prog. Inorg. Chem.*, **26**, 1 (1979).

[2] Ballhausen, C.J., *J. Chem. Ed.*, **56**, 215, 294, 357 (1979).

[3] Howard, J.B., *J. Chem. Phys.*, **3**, 813 (1935).

[4] Bethe, H.A., *Ann. Phys.*, **3**, 143 (1929).

[5] Pauling, L., *J. Amer. Chem. Soc.*, **53**, 367, 3225 (1931); **54**, 988 (1932).

[6] Van Vleck, J.H., *J. Chem. Phys.*, **3**, 803, 807 (1935).

[7] Mulliken, R.S., *Phys. Rev.*, **40**, 55 and **41**, 49, 751 (1932); **43** 279 (1933); *J. Chem. Phys.*, **1**, 492 (1933); **3**, 375, 506 (1935).

[8] Gerloch, M. & Slade, R.C., *Ligand Field Parameters*, Cambridge University Press, 1973.
[9] Dewar, M.J.S., *Bull. Soc. Chim. France*, **18**, 71 (1951).
[10] Nyholm. R.S., *Rep. 10th Solvay Council, Brussels* (1956).
[11] Figgis, B.N., *Trans. Farad. Soc.*, **57**, 198, 204 (1961).
[12] Figgis, B.N., Lewis, J., Mabbs, F.E. & Webb, G.A., *J. Chem. Soc. A*, p. 1411, (1966).
[13] Figgis, B.N., Lewis, J., Mabbs, F.E. & Webb, G.A., *J. Chem. Soc. A*, p. 442 (1967).
[14] Figgis, B.N., Gerloch, M., Lewis, J., Mabbs, F.E. & Webb, G.A., *J. Chem. Soc. A*, p. 2086 (1968).
[15] Kotani, M., *J. Phys. Soc. Japan*, **4**, 293 (1949).
[16] Stevens, K.W.H., *Proc. Roy. Soc. (Lond.)*, **A219**, 542 (1954).
[17] Orgel, L.E., *An Introduction to Transition Metal Chemistry*, Methuen, 1962.
[18] Owen, J. & Thornley, J.H.M., *Rep. Prog. Phys.*, **29**, 676 (1966).
[19] Gerloch, M. & Miller, J.R., *Prog. Inorg. Chem.*, **10**, 1 (1968).
[20] Figgis, B.N., Gerloch, M., Lewis, J. & Slade, R.C., *J. Chem. Soc. A*, p. 2028 (1968).
[21] Gerloch, M., *J. Chem. Soc. A*, p. 2023 (1968).
[22] Gerloch, M., Lewis, J., Phillips, G.G. & Quested, P.N., *J. Chem Soc. A*, p. 1941 (1970).
[23] Owsten, P.G. & Rowe, J.M., *J. Chem. Soc.*, p. 3411 (1963).
[24] Falvello, L. & Gerloch, M., *Acta Cryst.* **B35**, 2547 (1979).
[25] Bentley, R.B., Mabbs, F.E., Smail, W.R., Gerloch, M. & Lewis, J., *J. Chem. Soc. A*, p. 3003 (1970).
[26] Falvello, L. & Gerloch, M., *Inorg. Chem.*, **19**, 472 (1980).
[27] Gerloch, M. & Quested, P.N., *J. Chem. Soc. A*, p. 2308 (1971)
[28] Gerloch, M., McMeeking R.F. & White, A.M., *J. Chem. Soc. (Dalton)*, p. 2454 (1975).
[29] Ballhausen, C.J., *Introduction to Ligand Field Theory*, McGraw-Hill, New York, 1962.
[30] Van Vleck, J.H., *The Theory of Electric and Magnetic Susceptibilities*, Oxford University Press, London, 1932.
[31] Cotton, F.A. & Wilkinson, G., *Advanced Inorganic Chemistry*, 4th edn, Wiley–Interscience, New York, 1980.
[32] Purcell, K.F. & Kotz, J.C., *Inorganic Chemistry*, W.B. Saunders, Philadelphia, USA, 1977.
[33] Burdett, J.K., *Molecular Shapes*, Wiley–Interscience, New York, 1980.
[34] Jørgensen, C.K., Pappalardo, R. & Schmidtke, H.H., *J. Chem. Phys.*, **39**, 1422 (1963).
[35] Schäffer, C.E. & Jørgensen, C.K., *J. Inorg. Nucl. Chem.*, **8**, 143 (1958).
[36] Schäffer, C.E. & Jørgensen, C.K., *Mat. Fys. Medd.*, **34**, 13 (1965).
[37] Wolfsberg, M. & Helmholz, L., *J. Chem. Phys.*, **20**, 837 (1952).
[38] Ballhausen, C.J. & Dahl, J.P., *Adv. in Quant. Chem.*, **4**, 170 (1968).
[39] Ballhausen, C.J., *Int. J. Quant. Chem.*, **5**, 373 (1971).
[40] Ballhausen, C.J. & Dahl, J.P., *Theor. Chim. Acta.*, **34**, 169 (1974).
[41] Schäffer, C.E., *Structure and Bonding*, **5**, 68 (1968).
[42] Ferguson, J., *J. Chem. Phys.*, **40**, 3406 (1964).
[43] Chatt, J., *Nature*, **165**, 637 (1950).
[44] Orgel, L.E., *Int. Conf. Coord. Chem. London: Chem. Soc. Spec. Pub.*, **13**, 93 (1959).

PART II

—

MAGNETISM

The exploitation of paramagnetism in transition-metal chemistry is a subject intimately linked with ligand-field theory. On the other hand, a description of the theory and practice of magnetism itself can be given without reference to the particular object of its ultimate application and so it is appropriate to begin with this topic and hence minimize the degree of cross-reference which would otherwise be required. We address two basic aims. One is to describe and derive relationships which express the principal magnetic susceptibilities and g values of molecules and crystals in terms of the eigenvalues and eigenvectors describing a system prior to the application of a magnetic field: we leave to Part III the business of defining what those energies and wavefunctions might be. The other is to show how such magnetic properties can be measured. The practical side of the subject is concerned with the macroscopic, classical view of magnetism, our treatment progressing, therefore, through the nature of a magnetic field, its interaction with matter, the definitions of magnetic moment and susceptibility, to the representation of crystal and molecular susceptibilities by tensors and how they are measured. On the theoretical side we observe how magnetism only arises as a quantum phenomenon which is best understood, however, in relation to early classical models.

We observe that all electric and magnetic phenomena are ascribable to stationary or moving electric charges. Magnetic charges or monopoles have never been observed and are generally believed not to exist. As there are no 'magnetic sources', the path to a discussion of magnetic fields is simplest and logically most satisfying through a review of electric phenomena. Stationary electric charges give rise to the phenomena of electrostatics, epitomized by Coulomb's law with which we begin.

2

Electric and magnetic fields

━━

We review here some of the basic principles of electricity and magnetism, introducing the quantities grad, div and curl; Coulomb's, Gauss', Stokes' and Ampère's laws; and so to the time-independent forms of Maxwell's equations that define electrostatics and magnetostatics. In following this through, we see how the magnetic field emerges as a relativistic consequence of moving electric charges – that is, of an electric current. The presentation closely follows the delightful texts of Purcell[1] and of Feynman,[2] except with respect to the choice of units. The SI system is used throughout the present book, though occasional reference to CGS units is made in passing: a brief discussion of some of the problems to do with units and nomenclature throughout this and the following chapter is presented in §3.11.

2.1 Coulomb's law, the principle of superposition and electric fields

Electric charges occur in nature in two forms – positive and negative – and we observe that like charges repel one another while unlike charges attract. Coulomb's law states that the force experienced by each of two charges, which are stationary relative to each other, varies inversely as the square of the distance r_{12} between them and is directed along their line of centres: thus

$$\mathbf{F}_1 = k \frac{Q_1 Q_2}{r_{12}} \mathbf{e}_{12} = -\mathbf{F}_2, \tag{2.1}$$

where $\mathbf{F}_1$ is the force on charge Q_1 and $\mathbf{e}_{12}$ is the *unit* vector in the direction to Q_1 from Q_2.

In the CGS system, we take the proportionality constant $k = 1$ and dimensionless, so providing a definition of the electrostatic unit of charge: two charges, each of one esu, repel each other with a force of one dyne when they are one centimetre apart. In the SI system we define the coulomb as the unit of charge, so that two charges, each of one coulomb, repel each other with a force of one newton when they are one metre apart. The constant of proportionality in the SI system is given by

$$k = \frac{1}{4\pi\varepsilon_0} = \frac{\mu_0 c^2}{4\pi} \tag{2.2}$$

and is measured in units of newton-metre^2/coulomb^2 or, equivalently, in volt-metre/coulomb. One coulomb is approximately 2.998×10^9 esu or, when c is measured in metre/sec, exactly $10c$ times larger than the esu. The permittivity ε_0 and permeability μ_0 of a vacuum are related by $\varepsilon_0\mu_0 = 1/c^2$: μ_0 is *defined* as $4\pi \times 10^{-7}$ henry/metre *exactly* and so the less common expression for k on the right of (2.2) is logically the more consistent in the SI system, as a revised measurement of the velocity of light c would necessitate an update in ε_0 rather than in μ_0. Nevertheless, we shall write some equations with ε_0 and some with μ_0, following convention.[4]

Of central importance in electromagnetism is the *principle of super-position* which states that the force experienced by one charge due to another is unaffected by the presence of any other charges. Equivalently, we can say that the total force on one charge within a group of charges is given simply by the vectorial sum of independent coulomb contributions from each other charge interacting with the reference charge. Now, suppose we have an arrangement of charges $Q_1, Q_2, \ldots, Q_N$ fixed in space and we are interested in their combined effect upon some test charge Q_0 brought into their vicinity. If all charges are stationary, we use Coulomb's law and the superposition principle to find the force $\mathbf{F}_0$ on the test charge as

$$\mathbf{F}_0 = \sum_{j=1}^{N} k\frac{Q_0 Q_j}{r_{0j}^2} \mathbf{e}_{0j}, \tag{2.3}$$

where r_{0j} is the distance between Q_0 and Q_j measured parallel to the unit vector $\mathbf{e}_{0j}$ joining their centres. The force is proportional to Q_0, so we may divide by Q_0 to obtain a vector quantity which depends only upon the configuration of the original system of charges $Q_1, \ldots, Q_N$ and on the coordinates (xyz) of the test location. The vector function,

$$\mathbf{E}(xyz) = \sum_{j=1}^{N} k\frac{Q_j}{r_{0j}^2} \mathbf{e}_{0j}, \tag{2.4}$$

is called the *electric field* at the point (xyz) arising from the electric sources

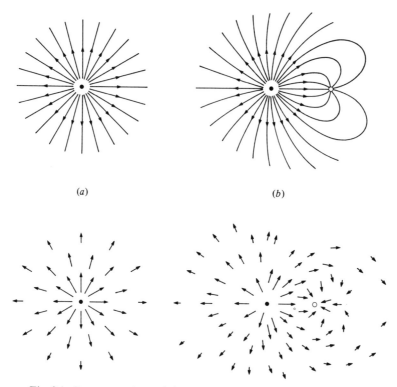

Fig. 2.1. Representations of electric fields (*a*) for a single charge of + 3 units, and (*b*) for a positive charge + 3 on the left with a negative charge − 1 on the right.

$Q_1, \ldots, Q_N$. The units of electric field strength in the CGS system are dyne/esu and in the SI system, newton/coulomb or, equivalently, volt/metre.

The concept of an electric field, as a mapping of the force experienced by a reference charge in the presence of source charges, derives directly from the superposition principle and its detailed form from Coulomb's law. Actually, the expression for **E** given in (2.4) must be understood with care, for it is only true if the source charges are rigidly fixed. Otherwise, the introduction of the test charge will generally induce shifts in the positions of the source charges and hence bring about a change in the electric field which it then experiences. Notice that a calculation of the force on the test charge in these circumstances would first require a computation of the new field of the source charges in the presence of the test charge: we shall refer back to this point when discussing the interaction of magnetic materials with magnetic fields later (§3.15). Meanwhile, it is quite proper to *define* the field of the source charges as in (2.4) *without* the presence of the test charge, equivalent to the source charges being fixed in the presence of the test charge. The electric field has thus been abstracted from the configuration of source charges and this abstraction provides a concept of greater generality and utility. In many applications, the precise details of the field sources need not be known as many equivalent charge configurations can give rise to the same field, at least within the volume of space under consideration; in which case the actual configuration is unassessable by probing with a test charge, and hence irrelevant.

It is frequently convenient to represent an electric field diagrammatically. Two methods of sketching vector fields are shown in figure 2.1 illustrating the electric fields of (*a*) a single positive charge and (*b*) positive and negative charges together. In the lower parts of the diagram, the field at any point is represented by an arrow whose orientation indicates the direction of the field at that point (say, at the tail-end of the arrow) and whose length describes the field strength. The alternative representation by field lines, or 'lines of force', shown in the upper part of the diagram, depicts field direction by the orientation of the lines again but field strength by the density of lines drawn. Both representations are necessarily only approximate. Around a single point charge, the field lines are necessarily radial, as required by Coulomb's law (2.1) and, *by convention*, are directed away from a positive charge and towards a negative one. The curved field lines shown in (*b*), do not violate this principle of course, arising as they do from the vector superposition of the fields from the two charges.

2.2 Flux and divergence

Coulomb's law together with the superposition principle are enough to characterize the phenomena of electrostatics completely. The abstraction of the electric field, however, as if describing something flowing through space, provides a concept of great generality and power. It is possible to establish some mathematical properties of fields in general, and of electric and magnetic fields in particular, from which can emerge a simpler expression of the laws of electrodynamics than the more obvious one based directly upon the forces acting on the original electric charges. While it is unnecessary to follow this path through here, it is useful for us to look, at least, at the properties of the *divergence* and *curl* of the fields, not only for their intrinsic interest but also for the simplicity they confer on later discussions.

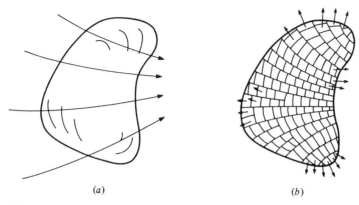

(a) *(b)*

Fig. 2.2. A closed surface immersed in a vector field (*a*); the scalar flux defined as the field component locally parallel to the outward surface normal (*b*).

Although the theorems we shall describe relate specifically to electric and magnetic fields, the properties with which they are concerned may be defined for any vector field. By way of introduction it is helpful to consider the case of a fluid flowing within or through a given volume of space: provided the analogy is not overused, we can think of electric and magnetic fields as vector maps of something flowing. In figure 2.2 we consider some arbitrary, closed surface (but one devoid of singularities) within a field and ask if anything is being lost from the inside: is there a quality of 'outflow'? Generally, we *define* the net outward flow or *flux* as the product of the average *outward* normal component of the vector field and the area of the surface. If we divide up the surface as in figure 2.2(*b*) into small parts, each of which is nearly flat, we may write the flux Φ, a *scalar* quantity, as

$$\Phi = \sum_j \mathbf{E}_j \cdot \mathbf{n}_j a_j, \tag{2.5}$$

where $\mathbf{n}_j$ is the unit vector normal to the jth part of the surface of area a_j, and $\mathbf{E}_j$ is the field at that location. If we consider the small pieces of the surface to decrease in size and become more numerous without limit, the sum in (2.5) tends to the surface integral,

$$\Phi = \int_{\text{surface}} \mathbf{E} \cdot \mathbf{n} \, da, \tag{2.6}$$

which is the definition of the flux just given.

Suppose the volume enclosed by the surface under consideration is divided into two parts, as in figure 2.3: the surface S_{ab}, which divides the

Magnetism

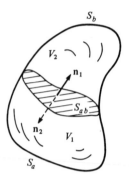

Fig. 2.3. The net flux through any internal surface is zero.

total surface S into two parts S_1 and S_2, need not be flat. The total volume V is thus divided into two volumes V_1 and V_2. Volume V_1 is bounded by the surface $S_a + S_{ab} = S_1$ while V_2 is bounded by $S_b + S_{ab} = S_2$. The flux through S_1 is given by

$$\Phi_1 = \int_{S_a} \mathbf{E} \cdot \mathbf{n} \, da + \int_{S_{ab}} \mathbf{E} \cdot \mathbf{n}_1 \, da, \qquad (2.7)$$

where $\mathbf{n}_1$ is the outward unit normal to S_{ab}. Similarly, the flux through S_2 is

$$\Phi_2 = \int_{S_b} \mathbf{E} \cdot \mathbf{n} \, da + \int_{S_{ab}} \mathbf{E} \cdot \mathbf{n}_2 \, da. \qquad (2.8)$$

As the definition of flux refers to the *outward* 'flow', $\mathbf{n}_1 = -\mathbf{n}_2$, so that

$$\int_{S_{ab}} \mathbf{E} \cdot \mathbf{n}_1 \, da = - \int_{S_{ab}} \mathbf{E} \cdot \mathbf{n}_2 \, da, \qquad (2.9)$$

and we have the result that the flux through the total surface S can be considered as the sum of the fluxes through the boundary surface of the two volumes into which the original volume was divided. If we so choose we can continue to subdivide the original volume into parts of ever-decreasing volume, knowing that the flux through the original surface can be expressed as a sum of the fluxes through the bounding surfaces of all the 'elemental' volumes.

Let us consider the case of such an elemental volume being the small box, or parallelopiped, shown in figure 2.4. Suppose the box, with origin (x, y, z) and with edge lengths $\Delta x, \Delta y, \Delta z$, to be oriented parallel

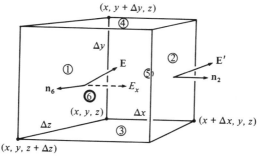

Fig. 2.4. The flux from an elemental parallelopiped.

to the reference cartesian frame. We compute the total flux through the surface of the box by adding the fluxes through each of the six faces. On face 1, the outward flux is the negative of the x component of **E** integrated over the area of the face; that is

$$\Phi_1 = - \int E_x \, dy \, dz, \tag{2.10}$$

but since we are considering a *small* box, we can approximate this integral by the value of E_x at the centre of the face multiplied by the area of the face $\Delta y \Delta z$,

$$\Phi_1 \approx - E_x^{(1)} \Delta y \Delta z. \tag{2.11}$$

Similarly, for the opposite face, 2,

$$\Phi_2 \approx E_x^{(2)} \Delta y \Delta z. \tag{2.12}$$

In general, of course, the fields at the two faces are slightly different. If Δx is small enough, we can write

$$E_x^{(2)} = E_x^{(1)} + \frac{\partial E_x}{\partial x} \Delta x + \dots, \tag{2.13}$$

where … refer to terms in a Taylor series involving $(\Delta x)^2$ and higher powers which will be negligible in the limit of small Δx. Therefore, on substitution into (2.12) and summing with (2.11), the fluxes through the opposite faces 1 and 2 are given by

$$\Phi_1 + \Phi_2 = \frac{\partial E_x}{\partial x} \Delta x \Delta y \Delta z. \tag{2.14}$$

(In the limit, it does not matter that the derivative was defined at (x, y, z) rather than at the midpoint of the faces.) Analogously, for the other pairs of opposite faces, we have

$$\Phi_3 + \Phi_4 = \frac{\partial E_y}{\partial y} \Delta x \Delta y \Delta z, \tag{2.15}$$

$$\Phi_5 + \Phi_6 = \frac{\partial E_z}{\partial z} \Delta x \Delta y \Delta z, \tag{2.16}$$

from which we calculate the total flux through the surface of the box as

$$\int_{\text{box}} \mathbf{E} \cdot \mathbf{n} da = \left(\frac{\partial E_x}{\partial x} + \frac{\partial E_y}{\partial y} + \frac{\partial E_z}{\partial z} \right) \Delta x \Delta y \Delta z. \tag{2.17}$$

Now the expression within the brackets can be written as a *scalar* product of two vectors:

$$\left(\frac{\partial}{\partial x}, \frac{\partial}{\partial y}, \frac{\partial}{\partial z} \right) \quad \text{and} \quad (E_x, E_y, E_z).$$

The former is a vector operator, $\mathbf{V}$, named del, and

$$\mathbf{V} \cdot \mathbf{E} = \frac{\partial E_x}{\partial x} + \frac{\partial E_y}{\partial y} + \frac{\partial E_z}{\partial z} \tag{2.18}$$

is called the *divergence* of $\mathbf{E}$, or div $\mathbf{E}$. Knowing that $\Delta x \Delta y \Delta z$ is the volume ΔV of the box, we rewrite (2.17) as,

$$\int_{\text{surface}} \mathbf{E} \cdot \mathbf{n} da = (\mathbf{V} \cdot \mathbf{E}) \Delta V, \tag{2.19}$$

and so observe that *the divergence of a vector at a point is the flux per unit volume of the vector out of an elemental volume, centred at that point*. For a general surface like that in figure 2.3, we obtain, by integration of (2.19),

$$\int_S \mathbf{E} \cdot \mathbf{n} da = \int_V \mathbf{V} \cdot \mathbf{E} dV, \tag{2.20}$$

a general relationship known as *Gauss' theorem*: V is any volume bounded by the surface S in any vector field $\mathbf{E}$.

Distinct from this is *Gauss' law* which is a field equation that depends explicitly upon the inverse square relationship of Coulomb's law. If we imagine a light source as light 'streaming out' from a centre, it is natural to observe that the intensity decreases as the square of the distance from the source. Thus, visualizing a cone whose apex is at the source and

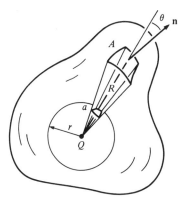

Fig. 2.5. A sphere enclosed within an arbitrary surface. The fluxes through both surfaces cut by the same cone subtended at the charge Q at the centre of the sphere are equal.

noting that all the light must pass through the surface opposite the apex, it is clear that the total amount of light flowing through that surface is independent of the radius of the cone because there is only one source and nothing is lost: the area of a normal end-surface of such a cone increases as the square of the radius and so the intensity, or amount of light per unit area, normal to the axis of the cone must decrease in the same proportion. Using such an example, the inverse square relationship appears obvious. That the same relationship should, and does, hold for electrostatic forces is somewhat less obvious when we recall that the electric field is an abstraction in that it does not actually refer to a flow of material or energy; so there is nothing to be conserved in the same sense as before. Nevertheless, Coulomb's law expresses the observation of an inverse square and radial behaviour of the forces between stationary electric charges: Gauss' law is thus derived from Coulomb's law.

We illustrate it first by reference to the simplest possible case and consider the flux of $\mathbf{E}$ through a spherical surface centred on a point charge Q. From Coulomb's law we have immediately that the magnitude of $\mathbf{E}$ at every point on the surface is kQ/r^2, where r is the radius of the sphere, and is directed radially. Using the definition (2.5), therefore, the total flux is

$$\Phi = \mathbf{E} \times \text{total area} = \frac{kQ}{r^2} \times 4\pi r^2 = 4\pi kQ. \tag{2.21}$$

We note that the flux is independent of the size of the sphere and proportional to the magnitude of charge Q, the source. Essentially, this is all there is to Gauss' law, except that (2.21) can be shown to apply to a surface of *any* shape, as follows.

Enclose a sphere with a closed surface of arbitrary shape, as in figure 2.5, and consider a small cone centred at the middle of the sphere where a charge Q is placed. Let the area of the inner, spherical, surface cut by the cone be a, and that of the outer arbitrary surface be A. For a small-angled cone, the area A is larger than a by two factors: by the ratio of the distances squared $(R/r)^2$ and, owing to the inclination of the piece of outer surface, by a factor $1/\cos\theta$, where θ is the angle between the outward normal of A and the radial direction of the small cone. The field at the outer surface has a magnitude correspondingly reduced by the factor $(r/R)^2$ but is still directed radially. Writing $\mathbf{E}_R$ for the field at the outer area A and $\mathbf{E}_r$ for that at the surface of the sphere, we have

$$\Phi_R = \mathbf{E}_R \cdot \mathbf{n}A = E_R A\cos\theta$$

$$= \left[E_r\left(\frac{r}{R}\right)^2 \right]\left[a\left(\frac{R}{r}\right)^2 \frac{1}{\cos\theta} \right]\cos\theta$$

$$= E_r a = \mathbf{E}_r \cdot \mathbf{n}a \equiv \Phi_r, \tag{2.22}$$

showing that the flux through the two end-surfaces of the cone are equal. Now each elemental area of the outer, arbitrary surface maps onto part of the spherical surface in this way, so that the total flux through the outer surface must be just $4\pi kQ$, from (2.21). Note that the proof does not depend upon the size of the sphere, so that we can always arrange that it really is enclosed by the surface in question rather than, perhaps, cut by it. Notice also that the total flux through the outer surface minus that through the sphere is zero: this observation hints at a further general result we now derive.

The surfaces we envisaged above were chosen to contain an electric source – the charge Q. Consider, instead, the arbitrary, closed surface in figure 2.6 which contains no charges, although it resides in the field of an external charge. We construct a representative cone, centred at the external charge Q, to cut the surface through areas 1 and 2 as shown. Using similar reasoning to that applied for the cone in figure 2.5, we know that the magnitudes of the flux through these two elemental surfaces are equal but, by the definition of flux being directed *outward* from a surface, of opposite sign. Accordingly, by summing over all such cones intersecting the given surface, we find that the total flux through a surface which does not contain any charges is zero. Thinking of the analogy of fluid flow, what flows in – flows out: the fluid is conserved. Analogously, we refer to

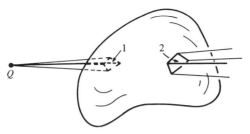

Fig. 2.6. The total flux through a surface not containing a charge is zero.

the electric field *outside* of source charges, as a *conservative field*.

In summary, Gauss' law states that the flux of the electric field **E** through any closed surface, that is, the integral $\int \mathbf{E} \cdot \mathbf{n} da$ over the surface, is proportional to the total charge enclosed by the surface:

$$\int \mathbf{E} \cdot \mathbf{n} da = 4\pi k \sum_i Q_i = \frac{1}{\varepsilon_0} \int \rho \, dv. \tag{2.23}$$

The summation sign introduced since (2.21) expresses the generality which results from a system of charges enclosed by a surface, a situation which may be treated directly using the superposition principle. In SI units (2.23) states that the total flux through a surface containing a charge Q is Q/ε_0 (using (2.2)). The last part of (2.23) provides an equivalent form of Gauss' law for electric sources considered in terms of a charge density ρ. Gauss' law can also be expressed in differential form. Consider an infinitesimal cubical surface. From (2.19) we have the flux of **E** out of the cube

$$\Phi = \mathbf{V} \cdot \mathbf{E} dV.$$

By the definition of ρ, the charge inside the volume dV is ρdV, so Gauss' law, in SI units, is

$$\mathbf{V} \cdot \mathbf{E} dV = \frac{\rho \, dV}{\varepsilon_0},$$

that is

$$\operatorname{div} \mathbf{E} \equiv \mathbf{V} \cdot \mathbf{E} = \frac{\rho}{\varepsilon_0} = \mu_0 c^2 \rho. \tag{2.24}$$

This particular form of Gauss' law is one of Maxwell's equations for electrostatics.

2.3 Circulation and curl

When introducing the concept of flux we referred to a quality of 'outflow' and in (2.5) defined flux in terms of the component of a vector field *normal* to a surface. We turn now to the other field components we call *tangential*. Particularizing once more to the case of a flowing fluid, we may enquire not only about the net outflow from a given column, but also about the quality of rotation or circulation. In much the same way that complex molecular motions may be decomposed into translations and rotations, so also may the properties of a vector field be separated into 'translation' or flux and 'rotation' or *circulation*.

Consider a fictitious, closed, yet permeable tube within the body of a swirling mass of liquid, as in figure 2.7(*a*). Suppose that we could instantaneously freeze all the liquid outside the tube while leaving it free to flow within it as in (*b*): the tube has now been made instantly impermeable. The liquid would flow around the tube in one direction or the other and we may define the circulation in the original body of fluid, or generally within a vector field of any kind, as the product of the average tangential component of the field and the distance around the chosen closed path. Notice that, in comparison with our introduction to flux, here we consider the tangential field component around a closed line as against the radial component out of a closed surface. Note too, that an extra convention will be required here in that we must pay heed to the sense of the circulation, analogously to our convention that flux is considered to be positive for flow outward from a surface. Symbolically, we write Γ, for the circulation around a closed loop C as

$$\Gamma = \oint_C \mathbf{E} \cdot d\mathbf{s}. \qquad (2.25)$$

As $d\mathbf{s}$ is an infinitesimal vector locally tangent to the curve C, the dot product with $\mathbf{E}$ ensures that the integral refers to the tangential component of $\mathbf{E}$ at each point. The symbol $\oint$ means an integral over the complete, *closed* curve C. The latter, incidentally, need not lie in a plane.

Proceeding as when discussing flux, we divide a closed loop into two parts, as shown in figure 2.8. If the circulation around the complete (outside) loop $C = C_a + C_b$ is Γ, we define Γ_1 and Γ_2 as the circulations around the separate loops $C_1 = C_a + C_{ab}$ and $C_2 = C_b + C_{ab}$. Now the contribution to Γ_1 and Γ_2 from the line integral between the intersection points 1 and 2 will be equal in magnitude but opposite in sign for, as

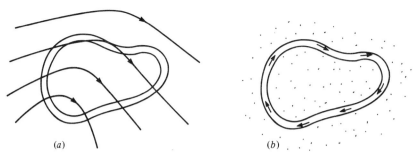

Fig. 2.7. Circulation is the tangential component of a vector field at any point along an arbitrary path.

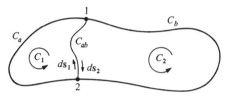

Fig. 2.8. The circulation around the outer loop equals the sum of the circulations around the inner loops.

mentioned above, we must work with a consistently defined sense of circulation. There results

$$\Gamma_1 = \int_{C_a} \mathbf{E} \cdot d\mathbf{s} + \int_{C_{ab}} \mathbf{E} \cdot d\mathbf{s}_1, \qquad (2.26)$$

$$\Gamma_2 = \int_{C_b} \mathbf{E} \cdot d\mathbf{s} + \int_{C_{ab}} \mathbf{E} \cdot d\mathbf{s}_2$$

$$= \int_{C_b} \mathbf{E} \cdot d\mathbf{s} - \int_{C_{ab}} \mathbf{E} \cdot d\mathbf{s}_1, \qquad (2.27)$$

so that,

$$\Gamma = \Gamma_1 + \Gamma_2 = \int_{C_a} \mathbf{E} \cdot d\mathbf{s} + \int_{C_b} \mathbf{E} \cdot d\mathbf{s} = \int_{C} \mathbf{E} \cdot d\mathbf{s}. \qquad (2.28)$$

In other words, the circulation around the outer loop is given simply by the sum of the circulations around the inner loops formed from it. We may continue to subdivide the loop in this way, knowing that the circulation around the whole loop is just the sum of the circulations

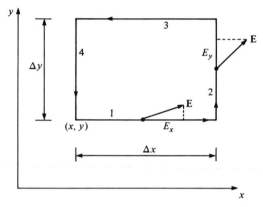

Fig. 2.9. Curl as the circulation around an elemental path.

around all the elemental loops. If the elemental loops are chosen small enough, they will bound small areas which are essentially flat (even though the original loop may not be) and we lose no generality by considering the case in which each area element is a smaller planar rectangle as shown in figure 2.9.

Let us orient the rectangle parallel to a chosen cartesian frame and suppose it to lie in the xy plane. We adopt the convention that, taking a right-handed set of cartesian axes – so that z is upwards from the plane of the page, positive circulation refers to a rotation from the positive x axis towards, or through, the positive y axis: this is shown by the small arrows around the rectangle. Also, as we shall consider an infinitesimal rectangle eventually, we suppose that the vector $\mathbf{E}$ varies little along any one side of the rectangle. The contribution to the circulation from side 1, given by the tangential component of $\mathbf{E}$ times the length of the path, is $E_x(1)\Delta x$. The contributions from sides 2, 3, and 4 are similarly, and respectively, $E_y(2)\Delta y$, $-E_x(3)\Delta x$ and $-E_y(4)\Delta y$, so that the total circulation is

$$\oint \mathbf{E} \cdot d\mathbf{s} = E_x(1)\Delta x - E_x(3)\Delta x + E_y(2)\Delta y - E_y(4)\Delta y. \qquad (2.29)$$

Consider the first two terms on the right. Bearing in mind our approximation about the near constancy of $\mathbf{E}$ in the area of the small rectangle, it might appear that $[E_x(1)-E_x(3)]\Delta x$ should vanish. Indeed, if $\mathbf{E}$ is *exactly* constant over the area, it does vanish. However, for the more interesting case where $\mathbf{E}$ varies, albeit slightly, we must take into account the rate of change of E_x across the rectangle and write

$$E_x(3) = E_x(1) + \frac{\partial E_x}{\partial y} \Delta y \tag{2.30}$$

as the first terms of a Taylor expansion. Hence,

$$[E_x(1) - E_x(3)] \Delta x = -\frac{\partial E_x}{\partial y} \Delta x \Delta y \tag{2.31}$$

and the derivative can be evaluated in the limit at (x, y). Similarly, for the remaining two terms in (2.29), we have

$$[E_y(2) - E_y(4)] \Delta y = +\frac{\partial E_y}{\partial x} \Delta x \Delta y, \tag{2.32}$$

and so the total circulation around the rectangle is

$$\oint \mathbf{E} \cdot d\mathbf{s} = \left(\frac{\partial E_y}{\partial x} - \frac{\partial E_x}{\partial y} \right) \Delta x \Delta y. \tag{2.33}$$

By definition, this is the circulation, in the prescribed sense, around z. Had we not arranged that the rectangle lay in the xy plane but was oriented arbitrarily with respect to a given cartesian frame, we could similarly find expressions for the circulations with respect to all three areas. The property of circulation is evidently a *vector* quantity and its components may be obtained from (2.33) by cyclic permutation of the axis labels, so maintaining the same sense of circulation with respect to a right-handed cartesian frame; thus

$$\left. \begin{array}{l} \left(\dfrac{\partial E_y}{\partial x} - \dfrac{\partial E_x}{\partial y} \right) \Delta x \Delta y \quad \text{for circulation around } z, \\[2ex] \left(\dfrac{\partial E_z}{\partial y} - \dfrac{\partial E_y}{\partial z} \right) \Delta y \Delta z \quad \text{for circulation around } x, \\[2ex] \left(\dfrac{\partial E_x}{\partial z} - \dfrac{\partial E_z}{\partial x} \right) \Delta z \Delta x \quad \text{for circulation around } y. \end{array} \right\} \tag{2.34}$$

The total circulation of $\mathbf{E}$ for the arbitrarily oriented rectangle may therefore be written, from (2.33), as

$$\oint \mathbf{E} \cdot d\mathbf{s} = \left[\left(\frac{\partial E_y}{\partial x} - \frac{\partial E_x}{\partial y} \right), \left(\frac{\partial E_z}{\partial y} - \frac{\partial E_y}{\partial z} \right), \left(\frac{\partial E_x}{\partial z} - \frac{\partial E_z}{\partial x} \right) \right] \cdot \mathbf{n} \Delta a \tag{2.35}$$

where $\mathbf{n}$ is the unit vector normal to the rectangle of area Δa. The form of the components of the vector in square brackets result from the vector product of $\mathbf{V}$ and $\mathbf{E}$: recalling that

$$\mathbf{V} = \left(\frac{\partial}{\partial x}, \frac{\partial}{\partial y}, \frac{\partial}{\partial z} \right),$$

we have

$$\left.\begin{aligned}
(\mathbf{V} \wedge \mathbf{E})_z &= \nabla_x E_y - \nabla_y E_x = \frac{\partial E_y}{\partial x} - \frac{\partial E_x}{\partial y}, \\
(\mathbf{V} \wedge \mathbf{E})_x &= \nabla_y E_z - \nabla_z E_y = \frac{\partial E_z}{\partial y} - \frac{\partial E_y}{\partial z}, \\
(\mathbf{V} \wedge \mathbf{E})_y &= \nabla_z E_x - \nabla_x E_z = \frac{\partial E_x}{\partial z} - \frac{\partial E_z}{\partial x}.
\end{aligned}\right\} \tag{2.36}$$

Therefore (2.35) may be rewritten in the more concise form,

$$\int \mathbf{E} \cdot d\mathbf{s} = (\mathbf{V} \wedge \mathbf{E}) \cdot \mathbf{n} \Delta a. \tag{2.37}$$

The quantity $\mathbf{V} \wedge \mathbf{E}$ is called the *curl* of $\mathbf{E}$ and sometimes written curl $\mathbf{E}$. The curl of a vector field at a point in space refers to the circulation around an infinitesimal area centred at that point, here with respect to the three orthogonal directions, and is a vector. The divergence of a vector field describes the flux out of the surface of an infinitesimal volume centred at the chosen place and is a scalar.

Analogous to Gauss' theorem for flux is *Stokes' theorem* for circulation. For a general closed loop C we have, by summing the circulations from all elemental loops inscribed on the surface S bounded by C,

$$\oint_C \mathbf{E} \cdot d\mathbf{s} = \int_S (\mathbf{V} \wedge \mathbf{E}) \cdot \mathbf{n} \, da. \tag{2.38}$$

This is a general mathematical expression relating a line integral to a surface integral, as Gauss' *theorem* relates a surface integral to a volume integral. Gauss' *law* arose from the inverse square dependence in Coulomb's law, and so gave a special relationship for the divergence of the electric field. Somewhat similarly, the other feature of Coulomb's law, that is, that the force between two charges is directed along their line of centres, is responsible for a special relationship for the curl of an electric field, as we now discuss.

2.4 The field gradient, potentials and curl-free fields

The abstraction of an electric field arose, in (2.4), from a consideration of the force experienced by a (unit) test charge in the presence of a configuration of fixed, source charges. An abstraction of a different kind emerges if we consider instead the work done in moving a test charge from one place to another. The work done *against* the electrical forces in carrying a charge along some path is the *negative* of the component of the electrical force in the direction of motion, integrated along that path. In moving a charge from point A to B, the work W done is

$$W = - \int_A^B \mathbf{F} \cdot d\mathbf{s}, \qquad (2.39)$$

where $\mathbf{F}$ is the electrical force on the charge at each point and $d\mathbf{s}$ is the differential vector displacement along the path. In deriving the expression (2.4) for the electric field from the Coulomb force, we divided out Q_0, the charge of the test charge. Analogously, we may consider here the work *per unit charge* and write.

$$W(\text{unit}) = - \int_A^B \mathbf{E} \cdot d\mathbf{s}. \qquad (2.40)$$

Now, for a *general* vector field it is possible for this quantity, $W(\text{unit})$, to depend upon the particular path traversed between points A and B. When this is so, we may extract energy from the field (or the reverse) by moving the unit charge from A to B by one path and then back to A by another. However, for *electrostatics*, the field ultimately derives from the forces between *fixed* charges and so, from the principle of the conservation of energy, no energy may be extracted from them. The electric field is a *conservative* field and the line integral in (2.40) has a value independent of the path taken for moving the test charge from A to B. This can be demonstrated more concretely, however, from the form of Coulomb's law as we now see.

Consider first the field arising from a single charge. We carry a different charge, of unit magnitude, from point A, distant r_1 from Q, to B distant r_2, by the simple path shown in figure (2.10a). Now the field is radial because Coulomb's law states that the force between two stationary charges is directed along their line of centres, so that no work is done in moving the test charge from A to A' along the path drawn as an arc of the circle

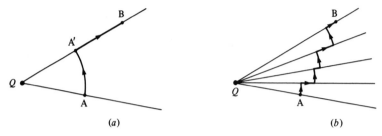

Fig. 2.10. The work done moving a charge from A to B in the field
of a charge Q is independent of the path taken.

centred at Q. Therefore the total work required to move the charge from
A to B is given by that required to traverse the radial displacement A' to
B. But any path from A to B, straight or curved, can be decomposed into
'workless arcs' and radial displacements (fig. 2.10b) so that the total work
in moving a test charge from A to B involves only the sum of all radial
displacements. So we see that the radial nature of an electric field from a
single charge implies that the field is conservative. The same is true of a
general electric field deriving from several source charges, as follows from
the application of the superposition principle: therefore

$$W(\text{unit}) = - \int_{\substack{A \\ \text{any path}}}^{B} \mathbf{E} \cdot d\mathbf{s}. \tag{2.41}$$
$$\scriptstyle A \to B$$

Since the work done depends only on the end-points, it can be represented
by the difference between two numbers (scalars). $W(\text{unit})$ is called the
electric *potential difference* between the points A and B. In the CGS system,
potential difference is measured in erg/esu, a unit known as the *statvolt*.
In the SI system, it is expressed in *volts*: one joule (10^7 erg) of work is
required to move a charge of one coulomb through a potential difference
of one volt. One volt is approximately $(1/299.8)$ statvolts.

Suppose we hold point A fixed at some reference location. The $W(\text{unit})$
becomes a function only of position B, with coordinates (x, y, z) say. We
may then write

$$\phi_{\text{B}}(x, y, z) = - \int_{A}^{B} \mathbf{E} \cdot d\mathbf{s} \tag{2.42}$$

as the *electric potential* at the point (x, y, z). The potential is a scalar
quantity which depends upon the coordinates (x, y, z) and so defines a

scalar field. Once the vector field **E** is given, the potential function ϕ is determined, except for an arbitrary constant arising from our choice of point A. The existence of the scalar potential field depends upon the conservative nature of the vector field – here, that the electric field is radial – so that (2.42) is an expression with some generality. For the electric field of a single charge, the relationship may be made more specific by using the other part of Coulomb's law, namely that the force between two charges varies as the inverse square of the distance between them. Then the work done in moving the charge in figure 2.10 from A′ to B is given, in the SI system, by

$$\int_{A'}^{B} \mathbf{E} \cdot d\mathbf{s} = \frac{Q}{4\pi\varepsilon_0} \int_{A'}^{B} \frac{dr}{r^2} = -\frac{Q}{4\pi\varepsilon_0}\left(\frac{1}{r_A} - \frac{1}{r_B}\right), \tag{2.43}$$

and the potential ϕ at a point (x, y, z) by

$$\phi(x, y, z) = \frac{Q}{4\pi\varepsilon_0}\frac{1}{r}. \tag{2.44}$$

Now consider two points (x, y, z) and $(x + \Delta x, y, z)$. The work per unit charge done in moving a charge between these two points is the potential difference between them:

$$\Delta W = \phi(x + \Delta x, y, z) - \phi(x, y, z) = \frac{\partial \phi}{\partial x}\Delta x, \tag{2.45}$$

while the work per unit charge done against the field for the same path is given from (2.41) as

$$\Delta W = -\int \mathbf{E} \cdot d\mathbf{s} = -E_x \Delta x, \tag{2.46}$$

and hence we have

$$E_x = -\frac{\partial \phi}{\partial x}. \tag{2.47}$$

For two points in a general relationship to one another, we find similarly that

$$E_y = -\frac{\partial \phi}{\partial y} \quad \text{and} \quad E_z = -\frac{\partial \phi}{\partial z}, \tag{2.48}$$

that is, that

$$\mathbf{E} = -\nabla \phi. \tag{2.49}$$

The electric field is given by the negative of the gradient of ϕ: when del operates on a scalar, $\nabla\phi$ is also called grad ϕ. Equation (2.49) is the differential form of (2.42).

We now return to the subject matter of the previous section and reconsider the curl of an electric field. Since $\mathbf{E} = -\,\text{grad}\,\phi$ from (2.49), we have

$$\text{curl}\,\mathbf{E} \equiv \nabla \wedge \mathbf{E} = -\,\nabla \wedge \nabla\phi. \tag{2.50}$$

By explicit substitution of the definition of ∇ as $(\partial/\partial x,\ \partial/\partial y,\ \partial/\partial z)$ into (2.50), it is simple to show that the curl of the gradient of a scalar function vanishes: here that *the electric field is curl-free*. Alternatively, the same result follows from (2.41). Thus curl $\mathbf{E}$ is defined as the line integral of $\mathbf{E}$ around a *closed* loop. Suppose we divide the loop into two parts and consider the line integral from point A to point B by path 1, and from B to A by path 2. From (2.41) we know that the line integral is independent of path so that the second integral is just the negative of the first and the integral around the complete, closed path must vanish.

In summary, *for electrostatics*, we have

$$\nabla \cdot \mathbf{E} = \rho/\varepsilon_0, \tag{2.24}$$

and

$$\nabla \wedge \mathbf{E} = 0. \tag{2.51}$$

Further, in charge-free regions of space, that is, away from electric sources, $\nabla \cdot \mathbf{E} = 0$, so that in regions of free space, electric fields are both curl-free and divergence-free. It is also interesting to note that the relation $\nabla \wedge \mathbf{E} = 0$ (everywhere) followed from the radial nature of Coulomb's law and, as such, is a symmetry property only. Because of this (2.51) expresses only part of the laws of electrostatics.

We have defined the mathematical entities of grad, div, and curl and illustrated their application to electric potentials, flux and circulation. The point has now been reached at which we may turn to the subject of magnetic fields and magnetostatics and so move closer to the main theme of this book.

2.5 Some basic phenomena of magnetostatics

Electrostatic phenomena are properties of stationary electric charges. The further and more varied phenomena of electrodynamics arise with moving charges but we need only consider a subset of these, namely the area of *magnetostatics* which describes the consequences of the special case of

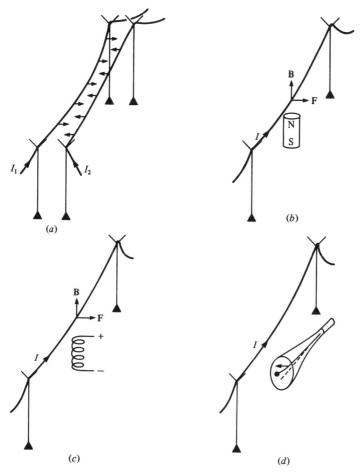

Fig. 2.11. A summary of experiments illustrating the relationship between electric currents and magnetic fields.

steady electric currents – that is, of many charges whose average velocity is constant.

Some basic magnetic phenomena are reviewed in figure 2.11. Firstly, recall how two parallel wires carrying electric currents in the same direction are attracted to each other:[††] with currents in opposite directions they are repelled. The force on one of the two wires, per unit length of wire, proves to be proportional to the product of the two currents and inversely proportional to the distance between them. As these forces are not observed

[††] This experiment is used to define μ_0: see reference [1], p. 450 and reference [4], p. 101.

when no current flows in the wires, they are clearly not electrostatic in origin: they are called *magnetic forces*. Like electrostatic forces, they involve action-at-a-distance and it is possible to define, empirically, a magnetic vector field **B** which describes the magnitude and direction of the force which may be sensed throughout space. In figure 2.11(*b*) is shown an experiment in which a current-carrying wire experiences a force which is simultaneously at right-angles to a magnetic field and to the electric current. The relationship between electric currents and magnetic fields is further suggested by figure 2.11(*c*) in which the same effect is produced if the bar magnet in figure 2.11(*b*) is replaced by a coil of current-carrying wire – a solenoid. Another manifestation of the situation in figure 2.11(*a*) is shown in figure 2.11(*d*) in which the electron beam in a cathode-ray tube is displaced by the current in the wire. It turns out that the force experienced by each charge (here electron) in the cathode ray involves a velocity-independent part proportional to the electric field within the tube and a strictly velocity-proportional part which is also related to the magnetic field created by the current in the wire. That magnetic fields are created by moving charges and currents is thus amply demonstrated by the experiments shown throughout figure 2.11.

The total force on a moving charge situated within both an electric and a magnetic field is called the *Lorentz force* and is given by

$$F = Q(E + v \wedge B),\qquad(2.52)$$

where **v** is the velocity of the charge Q. This equation serves to *define* the magnetic field **B**. It does not *explain* it, however, and the next few sections are intended to provide a qualitative 'explanation' of magnetic forces and fields in terms of the relativistic dynamics of moving electric charges. Rather too much space would be required to provide fully self-contained proofs of the various results we shall describe. The aim is more to sketch in the shape of a development which reveals magnetic phenomena to be relativistic consequences of electric currents.

2.6 Electric charge invariance

We must first examine how to measure the amount of electric charge on a moving particle. Only then can we enquire what effect the motion has upon the charge itself. A charge is measured by the effects it produces and in the case of a stationary charge Q, Coulomb's law states that a test charge experiences a force proportional to the charge Q, varying inversely with the charge separation squared and directed towards (or away from) that charge: in turn, the electric field of a stationary point charge is necessarily radial and spherically symmetric. Since we cannot be sure that any of these properties will hold for measurements made on a moving charge, it is essential to define charge in a way that does not depend on the position of the observer (test charge) with respect to the direction of the motion of the charge. The convention adopted is to define Q by averaging over all directions. Thus, imagine a large number of infinitesimal test charges distributed and fixed evenly over a sphere. At the instant a moving charge Q passes through the centre of the sphere, suppose we measure the radial component of force on each test charge and use the average of these forces to compute the magnitude of Q. This is the process of determining the surface integral of the electric field over that sphere at time t, and so suggests that Gauss' law, rather than Coulomb's law, offers a natural way to define the amount of charge on a moving particle. The technique is not mandatory and the laws of nature do not depend upon its use: this method does, however, yield the simplest algebra.

Accordingly, we define the amount of electric charge in a region by the surface integral of the electric field $\mathbf{E}$ over a surface enclosing that region. This surface S is fixed in some coordinate frame F and the field $\mathbf{E}$ is measured at any point (x, y, z) at time t in F, by the force on a test charge at rest in F, at that time and place. The surface integral is thus only defined for a particular time t. As S is stationary in F, there is no difficulty in measuring the field values simultaneously by observers deployed all over S. The amount of charge defined in this way is given by

$$Q = \frac{1}{4\pi} \int_{S(t)} \mathbf{E} \cdot \mathbf{n} da. \tag{2.53}$$

For a stationary charge, Gauss' law also states that the value of Q determined by (2.53) is independent of the shape of the surface S. It is *an experimental fact* that the same holds true for moving charges, which is also fortunate if (2.53) is to serve as a definition for Q. This property is a part of a more general fact of fundamental significance, namely that the surface integral in (2.53) depends only upon the number and variety of charged particles in S and not on how they are moving: *charge is invariant* with respect to a transformation from one inertial frame F to another, F';

$$\int_{S(t)} \mathbf{E} \cdot \mathbf{n} da = \int_{S'(t')} \mathbf{E}' \cdot \mathbf{n}' da. \tag{2.54}$$

The same is *not* true of mass, for example. The special theory of relativity tells us that the observed mass of a moving object is related to its rest-mass by the factor $1/\sqrt{(1 - v^2/c^2)}$, where v and c are the speeds of the body and of light respectively.[3] That charge *is* invariant can be demonstrated by a number of simple and elegant experiments: we briefly illustrate two.

Suppose we arrange to heat an electrically insulated metal bar which at the beginning of the experiment is electrically neutral, known by measuring the electric field nearby. As the temperature increases, the 'free' electrons move faster, while there is relatively little effect on the motions of the protons or core electrons. If charge were velocity dependent, the electric neutrality of the bar would be destroyed to a slight extent and a local electric field created: no such effect is observed. Lest it be doubted that such a hypothetical effect could be observed, let us assume that the charge might vary with velocity by some factor like $1/\sqrt{(1 - v^2/c^2)}$: then it should be noted that (i) electric forces are very strong and an imbalance of positive and negative charges to one part in 10^{10} should be fairly easily detected, and (ii) magnetic forces, which we shall show to arise from relativistic effects, are quite capable of turning electric-motor armatures and so on.

An even more convincing demonstration of charge invariance arises from a consideration of the exact electrical neutrality of atoms. It is known, from various elaborate experiments, that the protons and electrons in a hydrogen molecule carry the same charge within an experimental accuracy of about one part in 10^{20}. The same experimental information is available for helium atoms also. The relative motion of the protons in the hydrogen molecule, where they dawdle around each other only $0.7\,\text{Å}$ apart, is many orders of magnitude less than that in the helium atom where, being tightly bound in the nucleus, they possess kinetic energies of several million electron volts. Nevertheless, despite the extremely different velocities of the charges in these two species, both are electrically neutral within very high experimental accuracy.

So charge is invariant. This is not the same thing as being conserved: conservation implies constancy with respect to a transformation from one inertial frame to another. Energy is conserved but is not a relativistic invariant: charge, on the other hand, is both conserved and a relativistic invariant.

2.7 Electric fields measured in different reference frames

The invariance of charge under a relativistic or Lorentz transformation directly or indirectly determines the transformation properties of other

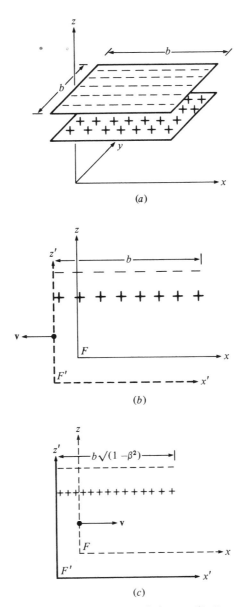

Fig. 2.12. (*a*) Two stationary sheets of charge. (*b*) Cross-section of sheets as seen in frame *F*, and (*c*) as seen in frame *F′*. *F* and *F′* are inertial frames moving with velocity **v** relative to one another. $\beta = v/c$

electric quantities. In particular, the electric field transforms in a most interesting way. Consider two stationary sheets of charge of uniform density $+\sigma$ and $-\sigma$ C m^{-2}. Let the sheets be square, of side b m, separated by a sufficiently small distance that the field between may be treated as uniform: conversely, for a given separation, let b be large enough to determine the same result. We define a cartesian reference frame F, such that the sheets lie parallel to the xy plane, as in figure 2.12(a). The magnitude of the electric field between the plates is then σ/ε_0. Now consider an inertial frame F' oriented parallel to F and moving parallel to x in a negative sense with speed v. To an observer in F', the charged plates are no longer square because their x' dimension is relativistically contracted, under the Lorentz transformation, by the factor $\sqrt{(1 - \beta^2)}$, where β is v/c. However, since the total charge is invariant to a Lorentz transformation, the charge *density* measured in F' is greater than that measured in F by the ratio $1/\sqrt{(1 - \beta^2)}$, as sketched in figure 2.12(c). From this it is a short step to show [1] that the electric field between the two plates as measured in F' is given by

$$E'_z = E_z/\sqrt{(1 - \beta^2)} = \gamma E_z,\tag{2.55}$$

where we use the abbreviation

$$\gamma = 1/\sqrt{(1 - \beta^2)} = 1/\sqrt{(1 - v^2/c^2)}.\tag{2.56}$$

Consider now a situation in which the two sheets are oriented parallel to the yz plane, while retaining the same definitions of F and F'. This time, the sheets do not appear contracted in F'. Their separation does decrease by the usual relativistic factor but, since the field between close, parallel plates is independent of the separation of the plates, the field observed in F' is the same as that in F and we write

$$E'_x = E_x.\tag{2.57}$$

Now if the concept of a field is to have any meaning, the detailed origin of a field should be irrelevant and we may take the equations (2.55) and (2.57) to describe the general behaviour of field components perpendicular and parallel to the direction of the relative motion of two frames, regardless of whether the field in any frame were produced by parallel plates, point charges or any other charge configuration. The relationships may be generalized, therefore, as

$$E'_\parallel = E_\parallel, \quad E'_\perp = \gamma E_\perp.\tag{2.58}$$

It is interesting to apply this result to the case of a moving point charge.

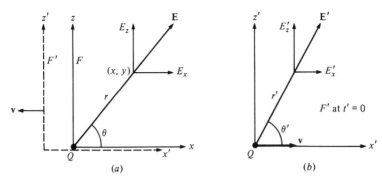

Fig. 2.13. The electric field of a point charge Q (*a*) in a frame F in which the charge is stationary, and (*b*) in a frame F' in which the charge moves with a constant velocity **v**.

In figure 2.13(*a*) is shown a point charge Q at rest at the origin of frame F. Coulomb's law tells us that the field is radially directed with components, at (x, z) related to the figure, given by

$$E_x = k\frac{Q}{r^2}\cos\theta = \frac{kQx}{(x^2 + z^2)^{3/2}}$$ (2.59)

and

$$E_z = k\frac{Q}{r^2}\sin\theta = \frac{kQz}{(x^2 + z^2)^{3/2}}.$$ (2.60)

Consider another frame F', moving in the negative x direction with speed v, and the components of **E** measured with respect to that frame: figure 2.13(*b*). The relationship between the space–time coordinates in these two frames, F and F', are given by the Lorentz transformation[3] as

$$x = \gamma(x' - \beta ct'), \quad y = y', \quad z = z', \quad t = \gamma\left(t' - \frac{\beta x'}{c}\right),$$ (2.61)

where β and γ are defined in (2.56). The time origin is set as zero when $x = 0$ and $x' = 0$ coincide. Now, from (2.58), $E_z' = \gamma E_z$ and $E_x' = E_x$ and so, using (2.60) and (2.61), we write E_z and E_x in terms of the coordinates in F'. At $t' = 0$, when Q passes the origin in F'.

$$E_x' = E_x = \frac{\gamma kQx'}{[(\gamma x')^2 + z'^2]^{3/2}}$$ (2.62)

and

$$E_z' = \gamma E_z = \frac{\gamma kQz'}{[(\gamma x')^2 + z'^2]^{3/2}}.$$ (2.63)

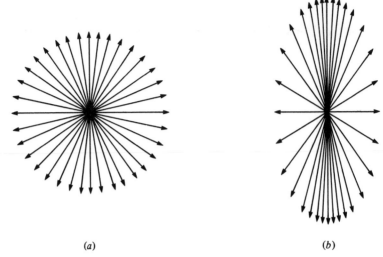

(a)   (b)

Fig. 2.14. The electric field of a stationary charge (*a*), and (*b*) of the same charge moving sideways with constant velocity; $v/c \approx 0.8$.

We observe immediately that $E'_z/E'_x = z'/x'$ which means that the vector **E**′ makes the same angle with respect to the x' axis as does the radius vector **r**′. Therefore **E**′ is directed radially along a line drawn from the instantaneous position of Q. (This may not be true if the charge is accelerating relative to F, but we confine our discussion to the case of constant velocities here.) Some straightforward algebra now gives the resultant of E'_x and E'_z; that is, the magnitude of the field **E**′ in F', as

$$E' = \frac{kQ}{r'^2}\left[\frac{1-\beta^2}{(1-\beta^2\sin^2\theta')^{3/2}}\right]. \tag{2.64}$$

This formula is like Coulomb's law but modified by the expression in square brackets – a function of angle θ and charge velocity **v**. The field magnitude is greatest for observation points at right-angles to the direction of charge motion and least directly ahead of, or behind, the moving charge. We may represent[††] this situation by the field lines drawn in figure 2.14. Note that, while the field strength along the line of travel is less than that of the same charge when stationary, that perpendicular to the direction of motion is greater: when integrated over a surface surrounding the charge, the result is the same whether the charge is moving or not as we know from the invariance of that property. The effect is quite small for

[††] The present graphical field representation is an (obvious) amalgam of these shown in figure 2.1.

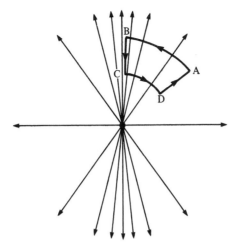

Fig. 2.15. The field of a moving charge is not curl-free.

normal velocities, of course, and the expression in square brackets in (2.64) tends to unity as v tends to zero.

So we observe that the electric field of a moving charge is still radially directed towards or away from that charge, but is not spherically symmetric. If we were to observe such a field, without having noticed the moving charge causing it, we would know that it had been so caused. This is because no combination of stationary charges can produce a field with these same characteristics. In particular, consider the curl of the field in figure 2.14(*b*). Suppose we calculate the line integral around the box ABCD in figure 2.15: for simplicity the box is constructed so that AB and CD trace out circular arcs around the charge and so contribute nothing to the line integral. Note, however, that the contributions of the radial parts of the path, DA and BC, do *not* cancel in this non-spherical field so that the field of a moving charge is *not* curl-free. This is not a situation that can arise from a stationary charge, nor by any configuration of stationary charges with superposition.

2.8 The force on a moving charge in an electric field

The force experienced by a stationary charge in the field of another charge that is moving at constant velocity is given from (2.64). We now enquire about the converse, that is, the force acting on a charge moving in the field of some other, stationary charges. We begin by looking at the system in a coordinate frame F' which moves along with the particle so that in this 'particle frame', the particle is at rest. We may now use our earlier results (2.58): the force on the stationary charge Q is just $\mathbf{E}'Q$ where $\mathbf{E}'$ is the electric field observed in the frame F' and given by (2.58). All we now require is to transform this force, observed in the particle frame F', back into the 'laboratory frame' F in which the original field source charges were at rest. We look at only the major steps in this calculation.

By force is meant a rate of change of momentum. In directions parallel and perpendicular to the direction of motion, the relationships between the components of momentum change with respect to time between a stationary frame F and a uniformly moving frame F' are given[1] by

$$\frac{dp_{\parallel}}{dt} = \frac{dp'_{\parallel}}{dt'} \quad \text{and} \quad \frac{dp_{\perp}}{dt} = \frac{1}{\gamma}\frac{dp'_{\perp}}{dt'}. \tag{2.65}$$

Now if the component of $\mathbf{E}$ parallel to the instantaneous direction of motion of the charged particle is $E_{\parallel}$, then, with respect to the frame F', moving with the particle at that instant, $E'_{\parallel} = E_{\parallel}$ from (2.58). So, in frame F', the force is

$$\frac{dp'_{\parallel}}{dt'} = QE'_{\parallel} = QE_{\parallel} \tag{2.66}$$

and, using the transformation laws (2.65), the force in frame F is

$$\frac{dp_{\parallel}}{dt} = QE_{\parallel}. \tag{2.67}$$

For the perpendicular component, $E'_{\perp} = \gamma E_{\perp}$ from (2.58) and hence

$$\frac{dp'_{\perp}}{dt'} = QE'_{\perp} = Q\gamma E_{\perp},$$

and in frame F, using (2.65) once more, we have

$$\frac{dp_{\perp}}{dt} = \frac{1}{\gamma}(Q\gamma E_{\perp}) = QE_{\perp}, \tag{2.68}$$

and γ *has dropped out*. Our conclusion, therefore, is that the force acting on a charged particle in motion through F is Q times the electric field $\mathbf{E}$ in that frame, *strictly independent of the velocity* of the particle. We have thus built up the first part of the Lorentz force equation (2.52). This result derives directly from the law of charge invariance and is, in a way, a statement of that law. The second part of the Lorentz force, and hence the notion of a magnetic field which is our principal concern here, emerges from the interactions between several moving charges, as we now discuss.

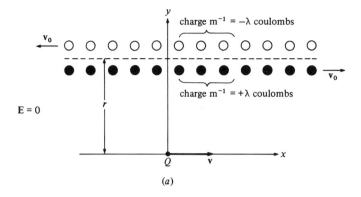

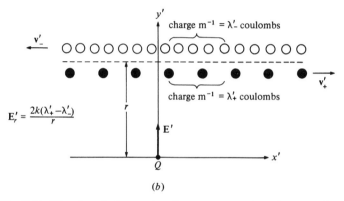

Fig. 2.16. The electric interaction between a moving charge and a current. The test charge Q is moving in (a) but at rest in (b). The line charge densities and associated fields are shown relative to the laboratory frame in (a), and to the particle frame in (b). Negative charges are shown as open circles, positive as solid.

2.9 The magnetic force

The magnetic interaction of electric currents, outlined in figure 2.11, arise from the conjunction of Coulomb's law, the postulates of special relativity and the invariance of charge. They emerge when we examine the *electric* interaction between a moving charge and other moving charges.

In the laboratory frame, as shown in figure 2.16(a), imagine an infinitely long linear array of positive charges moving to the right with speed v_0 and superimposed on it a procession of negative charges moving to the left at the same speed. We suppose these charges are sufficiently numerous and closely spaced that their discreteness can be ignored at the distances we shall consider. The positive and negative charges are shown separated in the diagram for reasons only of clarity. Now for this system there exists no frame of reference in which all charges are at rest. Let the density of positive and negative charges, as measured in the laboratory frame, be equal at λ C m^{-1}. The net density of charge on the line in the laboratory frame is therefore zero, from which it follows that the electric field E in that frame is also zero. We have here the equivalent of an uncharged wire carrying a steady electric current. In a metallic wire, only the negatively charged electrons would move, while the positive nuclei are held at rest in the lattice: the present, more symmetrical, situation is investigated to simplify the discussion slightly.

We now examine the force on a test charge Q near the 'charge line', or wire, moving with velocity **v** parallel to it. Figure 2.16(*b*) shows the same situation referred to the frame of the moving rest particle. The speeds of the positive and negative charges in the wire now appear unequal and so the two charge densities suffer different Lorentz contractions. As observed in the particle frame, the wire is charged! Consequently, there is an electric field between the moving test charge and the wire and a force of attraction or repulsion, depending upon the sense of the particle motion relative to that of the like charges in the wire. *In essence*, this is the second part of the Lorentz force in (2.52). There is one more remarkable fact about this force, which, though not difficult to prove,[1] is not derived here. It concerns the magnitude of the force between the test charge Q and the wire given by

$$F_y = \frac{4kQ\lambda vv_0}{rc^2};$$ (2.69)

but, since the current in the wire, in amperes, is just $2\lambda v_0$, we can write the force with magnitude,

$$F = \frac{2kQvI}{rc^2}.$$ (2.70)

This result is remarkable, not only for revealing the force of attraction (or repulsion) between a moving charge and a current-carrying wire, and that this force depends exactly linearly on the velocity of the test charge, but that it does *not* depend upon the velocity of the charge carried in the 'wire'. What matters is the *current I* in the wire and not what carries that current: the current could consist of a few fast-moving electrons as in a cathode-ray tube, or slow-moving electrons as in a length of metal wire,

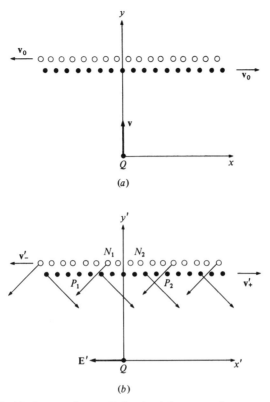

Fig. 2.17. (*a*) A test charge Q in the laboratory frame moves at right-angles to the current-carrying wire. (*b*) As seen in the particle frame, the wire moves towards the test charge which experiences a field $\mathbf{E}'$. P and N label positive (open circles) and negative charges in the wire.

or of positive and negative ions moving with unequal speeds as in a conducting electrolyte solution: in each case, the force developed on a nearby moving charge only depends upon the total current carried.

Now let us investigate the case of a charged test particle moving at right-angles to a current-carrying wire. Once again we shall simplify the discussion by envisaging the current as equal but opposite velocities of positive and negative charges in the wire, as shown in figure 2.17(*a*). The same situation referred to the frame of a charged particle moving at right-angles towards the wire is shown in figure 2.17(*b*). The picture looks symmetrical: however, consider pairs of positive and negative charges on the wire, symmetrically located with respect to the test charge Q. In figure 2.18 are sketched the electric fields of these pairs of positive (P_1

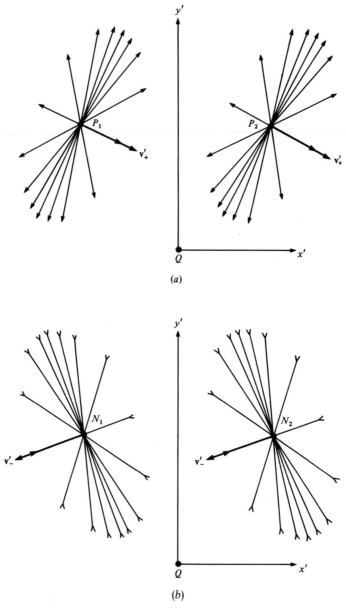

Fig. 2.18. Pairs of positive (*P*) and negative (*N*) charges from figure 2.17(*b*), symmetrically placed relative to the test charge *Q*. The diagram illustrates why the field **E**′ is directed parallel to the wire and to the left in figure 2.17(*b*).

and P_2) and negative (N_1 and N_2) charges. Similar diagrams could be constructed for all pairs of particles so related with respect to Q. The relativistic contractions of the fields of these moving charges destroy their spherical symmetry so that the field of P_2 is stronger at Q than that of P_1, as shown by the position and densities of the lines of force in figure 2.18. Similarly, the field of N_1 is stronger at Q than that of N_2. The *net* field in figure 2.18(*a*) is directed down and to the left: that in figure 2.18(*b*) is directed up and to the left. Clearly, the y' components cancel when the fields from all particles are summed, but the x' components do not. We are left with a field $\mathbf{E}'$ in figure 2.17(*b*) directed parallel to the wire and at *right-angles* to the motion of the test charge.

If we had to analyse every system of moving charges by transforming back and forth between various coordinate systems, the subject would grow tedious and confusing. Certainly these 'transverse' forces between moving charges that we have discussed are solely electrical in origin and arise as a relativistic effect. It is simpler, however, to describe a new quantity which properly *maps* these effects, and so avoid much unnecessary algebra. The 'new' entity, which was of course discovered thousands of years before the theory of relativity, is the magnetic field.

2.10 The magnetic field

We have seen in the last few sections how the electric interaction between moving charges can give rise to a velocity-proportional force component which is directed at right-angles to the direction of motion. The defining equation of magnetic field $\mathbf{B}$, the Lorentz force equation

$$\mathbf{F} = Q(\mathbf{E} + \mathbf{v} \wedge \mathbf{B}), \tag{2.52}$$

incorporates these features. Thus, parallel to the direction of motion, the force on a charge is just Q times the electric field through which it passes. The perpendicular component arises from the cross-product $\mathbf{v} \wedge \mathbf{B}$ in (2.52) and is proportional to the speed v. For (2.52) to be consistent with (2.70), the magnitude of the magnetic field must be defined as

$$B = \frac{2kI}{rc^2} = \frac{\mu_0 I}{2\pi r}. \tag{2.71}$$

From (2.52), the defining equation for $\mathbf{B}$, we see the unit of magnetic field in the SI system is the newton.second/coulomb.metre also called the volt.second/metre^2 or the newton/ampere.metre and, most briefly, the tesla (T). In the CGS system, the magnetic field unit is the dyne.second/esu.cm.

However, the practice in that system is to take a practical unit of *current* as the ampere (3×10^9 esu s^{-1}, *approximately*) and hence define a unit for magnetic field which always was 'half-way' towards that ultimately chosen in the SI system. The unit is called the gauss and 1T equals 10^4 gauss, *exactly*.

2.11 Properties of magnetic fields

The magnetic field associated with a current-carrying wire is represented in figure 2.19 by concentric circles (or for extended wire, by sheets). The spacing of the field lines, which form closed loops, is intended to convey the fact that the field strength decreases inversely with distance from the wire, as given in (2.71). According to the Lorentz force equation (2.52), a charged particle moving parallel to the wire in figure 2.19 experiences a force proportional to $\mathbf{v} \wedge \mathbf{B}$; that is, inversely proportional to its distance from the wire, proportional to its velocity and directed simultaneously at right-angles to both $\mathbf{B}$ and $\mathbf{v}$ (by the definition of a vector cross-product). The latter feature means that the force is directed radially towards the wire when the particle is moving parallel to I and radially away from it if $\mathbf{v}$ is antiparallel. Thus we see that the description of the field and field lines around the wire are consistent with the earlier discussion of the relativistic dynamics of charged particles. From now on, therefore, we shall use such a description of a magnetic field as if it had an existence separate from its electric and relativistic origins, in the knowledge that the Lorentz force equation provides a proper mapping for it. Like the electric field, the magnetic field is a device for describing how charged particles interact with each other; and like the electric field, it is a *vector* field.

It is immediately obvious that the magnetic field is *not* curl-free. The magnetic field lines in figure 2.19 form closed loops with the field directed tangentially at any point in a sense determined by the direction of the current in the wire using, say, a right-hand rule. The line integral around that loop does not vanish. However, consider the path ABCD lying in a plane perpendicular to the wire, as shown in figure 2.20(*a*). The path is constructed such that AB and CD are circular arcs centred on the current-carrying wire. The line integrals along AD and BC are therefore equal in magnitude and opposite in sign. But the same is true for the arcs AB and CD for, while CD is longer than AB by the ratio $r_2 : r_1$, the strength of the magnetic field at distance r_2 is less than that at r_1 by the same ratio. So the integral over the closed path ABCD vanishes and the

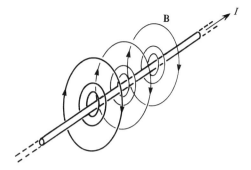

Fig. 2.19. The magnetic field associated with a current-carrying wire.

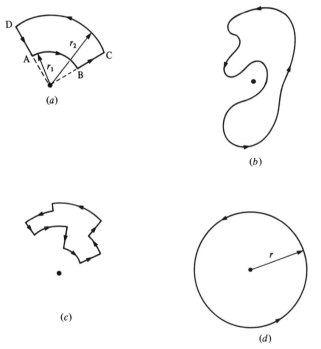

Fig. 2.20. The line integral of the magnetic field **B** over any closed path depends only upon the electric current enclosed. In each case the current flows down into, and normal to, the plane of the paper at the point indicated by the black dot.

loop encloses a curl-free region of the magnetic field. But *any* closed loop, as in figure 2.20(*b*), *not* enclosing the wire can be made up from such elemental loops as in figure 2.20(*c*). In all such cases the total line integral vanishes and the magnetic field is curl-free in those regions. Only paths

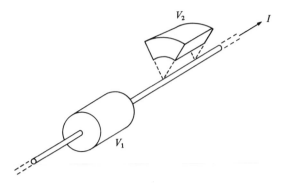

Fig. 2.21. The flux of **B** from either volume is zero.

which enclose a current have non-vanishing line integrals. Consider the
circular path in figure 2.20(d). The circumference is $2\pi r$, the field is $2kI/rc^2$
and everywhere parallel to the path, so that the total line integral around
the path is $2\pi r \times 2kI/rc^2$ or $4\pi kI/c^2$: in the SI system, $I/\varepsilon_0 c^2$. The same
result holds for any closed path enclosing the conductor because an
irregular one can be subdivided into an inner circle which encloses the
wire plus outer fragments of irregular shape which do not. Therefore, we
have the general result (in the SI system) that

$$\int \mathbf{B} \cdot d\mathbf{s} = \frac{I}{\varepsilon_0 c^2} = \mu_0 I. \tag{2.72}$$

Generally, using Stokes' theorem for an elemental surface through which
current is flowing, we have *Ampère's law*,

$$\text{curl } \mathbf{B} = \mu_0 \mathbf{J}, \tag{2.73}$$

in which **J** is the current *density* normal to the elemental surface. In
current-free regions outside the conductor, curl **B** $= 0$. While (2.73) provides
the relationship by which curl **B** is determined from the current density
J, the magnetic field **B** itself is only fully determined when we have also
established its divergence.

Consider, therefore, the small boxes shown in figure 2.21. Remembering
that the magnetic field lines form closed loops around the wire, the net
flux through either of these boxes is zero. The flux entering the box V_2
leaves it on the other rectangular side: for V_1 there is no flux out of the
surface at all. Since a volume of any shape and size may be constructed
from a combination of elemental volumes like these, we have immediately
the absolutely general relationship,

$$\text{div } \mathbf{B} = 0, \quad \text{everywhere} \tag{2.74}$$

This equation effectively states that there are no magnetic sources – no magnetic monopoles. In an electric field, lines flow from or towards a source of electric charge: magnetic field lines flow *around* the (electric) source. There is a fundamental asymmetry in the field equations of electromagnetism because of the observed lack of magnetic sources.

2.12 Induction

So far we have only considered the phenomena of electrostatics and magnetostatics. Induction is concerned with how electric fields and currents are caused by magnetic fields which vary with time. We shall confine our study of induction to a consideration of a skeletal derivation of Faraday's law of induction.

We know from the Lorentz force equation (2.52) that electric charges inside pieces of conducting wire which move in a magnetic field experience forces given by

$$\mathbf{F} = Q\mathbf{v} \wedge \mathbf{B}. \tag{2.75}$$

A piece of wire oriented normal to both the directions of its motion and of a surrounding magnetic field will experience a charge polarization, as shown in figure 2.22(*a*): equilibrium will be achieved when the electric field set up in the wire by this polarization is such that

$$Q\mathbf{E} = -\mathbf{F}. \tag{2.76}$$

Consider next the situation illustrated in figure 2.22(*b*) where a rectangular wire loop moves parallel to one of its edges in a uniform magnetic field. We observe a similar charge polarization arising from similar effects in the two edges of the loop that are perpendicular to the direction of motion. A much more interesting experiment, however, is summarized in figure 2.22(*c*), in which this same rectangular loop moves through an *inhomogeneous* magnetic field. At any given instant, the magnetic fields experienced by the leading and trailing edges of the wire loop are unequal and so an electric current will flow in the loop. Consider the line integral of the force $\mathbf{F}$ acting on each charge Q moving in this loop – an integral which receives contributions only from the edges normal to both $\mathbf{v}$ and $\mathbf{B}$:

$$\int \mathbf{F} \cdot d\mathbf{s} = Qvw(B_1 - B_2), \tag{2.77}$$

where w is the width of the loop and B_1 and B_2 are the magnitudes of the magnetic field, all as shown in figure 2.22(*c*). The line integral per unit

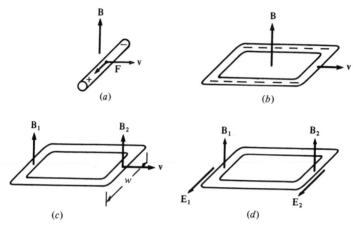

Fig. 2.22. A conducting rod (*a*) or loop (*b*) move at constant velocity through a uniform magnetic field. The movement of a loop at constant velocity in an inhomogeneous field is viewed in the laboratory frame in (*c*), and in the frame of the moving loop in (*d*).

charge defines, incidentally, the electromotive force ε or emf:

$$\varepsilon = \frac{1}{Q}\oint \mathbf{F}\cdot d\mathbf{s} \qquad (2.78)$$

so, in the present case,

$$\varepsilon = vw(B_1 - B_2). \qquad (2.79)$$

Now as the loop moves through the inhomogeneous field, the flux through the loop changes with time. The magnetic flux, of course, is given by the surface integral of $\mathbf{B}$ over the surface having the loop as its boundary. The loop moves a distance vdt in a time increment dt, as shown in figure 2.23: therefore, the leading edge of length w traces out an area $wvdt$ in a magnetic field essentially of strength B_2 while the trailing edge 'loses' a similar area in a field B_1, as illustrated. The change in flux Φ in this time interval is, therefore,

$$d\Phi = -vw(B_1 - B_2)dt \qquad (2.80)$$

and hence, by comparison with (2.79), we find

$$\varepsilon = -\frac{d\Phi}{dt} = -\frac{d}{dt}\int_S \mathbf{B}\cdot \mathbf{n}\, da, \qquad (2.81)$$

where, as usual, $\mathbf{n}$ is the unit vector normal to the elemental area da. The

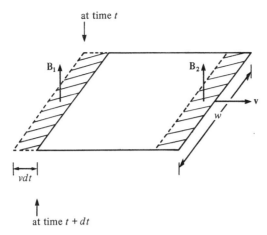

at time t

B_1　B_2

v

w

vdt

at time $t + dt$

Fig. 2.23. Illustrating the change in flux through the loop moving with velocity $\mathbf{v}$ through a non-uniform magnetic field.

foregoing derivation is both rudimentary and specialized: it is possible to show, however, that the relationship (2.81) – that the emf is equal to the rate of attenuation of magnetic flux – holds quite generally for a loop of any shape moving in any inhomogeneous magnetic field in any manner.

An alternative view of the experiment illustrated in figure 2.22(c) focusses on the electric fields induced in the loop as it moves through the magnetic field. In the trailing and leading edges of the loop, respectively, different electric fields are generated[††]:

$$\mathbf{E}_1 = \mathbf{v} \wedge \mathbf{B}_1; \quad \mathbf{E}_2 = \mathbf{v} \wedge \mathbf{B}_2. \tag{2.82}$$

The line integral of the electric field around the whole loop will vanish if $B_1 = B_2$ (corresponding, for example, to the situation pictured in figure 2.22(b)) but is given in the present more general case as

$$\oint_C \mathbf{E} \cdot d\mathbf{s} = vw(B_1 - B_2). \tag{2.83}$$

Comparing this with (2.79) and (2.81), we have

$$\oint_C \mathbf{E} \cdot d\mathbf{s} = -\frac{d}{dt} \int_S \mathbf{B} \cdot \mathbf{n} \, da, \tag{2.84}$$

[††] This description and equations are satisfactory insofar as the speed of the loop is small compared with that of light; $v \ll c$. Though not difficult to recast the treatment into a relativistically exact form, our present purposes do not require that we do so.

and, using Stokes' theorem (2.38), we rewrite the left-hand side of (2.84) to get

$$\int_s (\nabla \wedge E) \cdot n \, da = -\frac{d}{dt} \int_s B \cdot n \, da \tag{2.85}$$

and thence express (2.84) in its differential form,

$$\nabla \wedge E = -\frac{dB}{dt}. \tag{2.86}$$

However, **B** may generally vary with both position and time, so we should replace the total derivative in (2.84) with the partial derivative $\partial B/\partial t$, so that the final differential form of (2.84) becomes

$$\nabla \wedge E = -\frac{\partial B}{\partial t}. \tag{2.87}$$

Equations (2.84) and (2.87) make equivalent statements of the *law of induction*, also known as *Faraday's law*. We shall have need of these equations later.

2.13 Maxwell's equations

All electromagnetism is contained in Maxwell's famous equations:[1,2]

$$\left. \begin{array}{ll} \nabla \cdot E = \rho/\varepsilon_0, & \nabla \wedge E = -\dfrac{\partial B}{\partial t}, \\[3mm] c^2 \nabla \wedge B = \dfrac{\partial E}{\partial t} + \dfrac{J}{\varepsilon_0}, & \nabla \cdot B = 0. \end{array} \right\} \tag{2.88}$$

The first – Gauss' law – was derived in §2.2; the second – Faraday's law – in §2.12; the fourth was described in §2.11. We shall not use the third equation in this book and so merely quote the result: it is not entirely unfamiliar, however, as becomes clear when we recognize that the special equations of electrostatics and magnetostatics emerge from (2.88) in the cases where nothing depends upon time; that is, that charges are either at rest or move with a constant average velocity. Under these circumstances, all time derivatives vanish and the Maxwell equations become

$$\nabla \cdot E = \rho/\varepsilon_0, \qquad \nabla \wedge E = 0, \tag{2.89}$$

and

$$\nabla \wedge B = \mu_0 J, \qquad \nabla \cdot B = 0. \tag{2.90}$$

The first of these last two equations was derived in §2.11. It is also interesting to note how, under these same circumstances, the equations separate into two pairs; (2.89) referring to the subject of electrostatics and (2.90) to magnetostatics. Note too, that in charge-free and current-free regions of space, *both* electric and magnetic fields are *both* divergence-free and curl-free.

In this chapter we have been concerned with the origins and fundamental properties of electric and magnetic fields. In describing the relationships (2.89) and (2.90) we have become acquainted with Coulomb's law, the principle of superposition and Gauss' law. A qualitative examination of the relativistic dynamics of moving electric charges showed how the phenomena of magnetostatics arise as relativistic effects of moving charges. We have also seen how the algebraic complexities of transformations between inertial frames may be avoided by defining, through the Lorentz force equation (2.52), the magnetic field. In Maxwell's equations we observe the fundamental asymmetry of electromagnetism which derives from the fact of nature that only electric sources have been observed – never magnetic ones. Electric and magnetic fields are vector fields possessing the qualities of flux and circulation. In large regions it is appropriate to represent these by surface integrals and line integrals respectively: Gauss' and Stokes' theorems provide general mathematical relationships between such integrals and certain volume integrals. At points in these fields, or rather in elemental regions surrounding points, flux and circulation are described by the qualities called divergence and curl, so providing differential analogues of the surface and line integrals.

The scalar potential field of §2.5 was established as a different abstraction from Coulomb's law and the superposition principle, such that the electric vector field is given by the negative of the gradient of the scalar potential. An analogous *vector potential* field related to the magnetic field can be defined and it is with this that we begin the next chapter in which we turn at last to study the interaction of magnetic fields and magnetic materials.

References

[1] Purcell, E.M., *Electricity and Magnetism, Berkeley Physics Course*, vol. 2, McGraw-Hill, New York, 1963.
[2] Feynman, R.P., Leighton, R.B. & Sands, M., *The Feynman Lectures on Physics*, vol. 2, Addison-Wesley, 1964.
[3] *Idem, ibid*, vol. 1.
[4] Bleaney, B.I. & Bleaney, B., *Electricity and Magnetism*, 3rd edn, Oxford University Press, 1976.

3

The interaction of magnetic fields and matter

3.1 The magnetic vector potential

We begin this chapter by examining a field which is related to the magnetic field as the scalar potential of §2.5 is related to the electric field. Our gain in doing so is a greater simplicity in several matters which follow.

In electrostatics we may represent the electric field as the (negative of the) gradient of a scalar field ϕ: that is possible since the curl of **E** is always zero and because of the mathematical relationship[7] that $\nabla \wedge \nabla \phi = 0$. On the other hand, the curl of the magnetic field does not generally vanish so that **B** cannot be written as the gradient of a scalar. However, the divergence of **B** is always zero so the mathematical identity $\nabla \cdot \nabla \wedge \mathbf{A} = 0$ makes possible the expression of **B** as the curl of another vector field, and we write

$$\mathbf{B} = \nabla \wedge \mathbf{A}. \tag{3.1}$$

The field **A** is called the magnetic *vector potential.*

Recall that the scalar potential ϕ was not completely specified by the definition

$$\mathbf{E} = -\nabla \phi, \tag{3.2}$$

as the same field **E** arises by taking the (negative) gradient of another potential $\phi + C$, where C is some arbitrary constant. Choosing that constant is equivalent to selecting a point of origin (for example, at infinity) from which to measure the potential. There is an analogous arbitrariness with the definition (3.1) of the vector potential. As **B** is obtained from **A** by differentiation, the addition of a constant to **A** changes nothing physical. In fact there is even more latitude in defining **A** for we can add to it any field which is the gradient of some scalar field, as we now show. Thus, in addition to (3.1), let $\mathbf{B}$ also equal the curl of another field $\mathbf{A}'$: then

$$\mathbf{B} = \nabla \wedge \mathbf{A}' = \nabla \wedge \mathbf{A}, \qquad (3.3)$$

and so

$$\nabla \wedge \mathbf{A}' - \nabla \wedge \mathbf{A} = \nabla \wedge (\mathbf{A}' - \mathbf{A}) = 0. \qquad (3.4)$$

However, if the curl of a vector vanishes, that vector must be expressible as the gradient of some scalar field, say ψ, so $\mathbf{A}' - \mathbf{A} = \nabla \psi$. Therefore, if $\mathbf{A}$ satisfies (3.1), then for any scalar ψ at all, $\mathbf{A} + \nabla \psi$ will also satisfy (3.1). In this context, $\nabla \psi$ is for the vector potential $\mathbf{A}$ what the constant C was for the scalar potential ϕ.

For magnetostatics it is convenient and conventional to define $\mathbf{A}$ as having zero divergence. As $\mathbf{A}$ and $\mathbf{A}'$ have the same curl – (3.3) – and hence give rise to the same magnetic field $\mathbf{B}$, there is no restriction on the divergence of $\mathbf{A}$, so we take

$$\nabla \cdot \mathbf{A} = 0. \qquad (3.5)$$

The selection of a value for div $\mathbf{A}$ is called 'choosing a gauge' and the choice (3.5) is called the Coulomb gauge: the reason for this choice in magnetostatics will become clear in the next section. Even the choice of a gauge does not uniquely define $\mathbf{A}$ and to do so it would be necessary to stipulate its behaviour on some boundary. *We* need not bother with this, however, as, on obtaining $\mathbf{A}$ in a given situation, we soon, thereafter if not immediately, differentiate it to derive $\mathbf{B}$. Should all this seem unnecessarily circuitous, it must be pointed out that one of the main advantages of working with the vector potential is that many equations of magnetostatics can thereby be made to closely resemble those of electrostatics. This is clearly a boon for those familiar with various electrostatic problems: for those who have seen the contents of the last chapter for the first time, the advantages are obviously much less! Nevertheless, we shall persist, not least so that we may make contact with most other texts on this subject.[1,2,3]

We end this introductory section by recasting three of the equations discussed in the preceding chapter in terms of the vector potential: again, we shall require these results later. Firstly, the equation of Faraday's law (2.87) may be written

$$\nabla \wedge \mathbf{E} = -\frac{\partial \mathbf{B}}{\partial t} = -\frac{\partial (\nabla \wedge \mathbf{A})}{\partial t} \qquad (3.6)$$

using (3.1). Then, where an electric field arises from both stationary charges *and* from time-dependent magnetic fields, we may use (2.49) and (3.6) to express a general electric field in terms of both scalar and vector potentials:

$$\mathbf{E} = -\nabla\phi - \frac{\partial \mathbf{A}}{\partial t}. \tag{3.7}$$

Finally, we may express the Lorentz force in terms of these same potentials by substitution of (3.7) into (2.52):

$$\mathbf{F} = -Q\nabla\phi - Q\frac{\partial \mathbf{A}}{\partial t} + Q\mathbf{v} \wedge \nabla \wedge \mathbf{A}. \tag{3.8}$$

3.2 The Biot–Savart law

Since **B** is determined by currents, so also is **A**. To find **A** in terms of the currents, we begin with Ampère's law,

$$\nabla \wedge \mathbf{B} = \mu_0 \mathbf{J}, \tag{2.73}$$

that is,

$$\nabla \wedge (\nabla \wedge \mathbf{A}) = \mu_0 \mathbf{J}. \tag{3.9}$$

This equation is for magnetostatics what Gauss' law,

$$\nabla \cdot \nabla\phi = -\rho/\varepsilon_0, \tag{3.10}$$

is for electrostatics. Consider the x component of (3.9):

$$\frac{\partial}{\partial y}\left(\frac{\partial A_y}{\partial x} - \frac{\partial A_x}{\partial y}\right) - \frac{\partial}{\partial z}\left(\frac{\partial A_x}{\partial z} - \frac{\partial A_z}{\partial x}\right) = \mu_0 J_x. \tag{3.11}$$

On rearranging, we have

$$-\frac{\partial^2 A_x}{\partial y^2} - \frac{\partial^2 A_x}{\partial z^2} + \frac{\partial}{\partial x}\left(\frac{\partial A_y}{\partial y}\right) + \frac{\partial}{\partial x}\left(\frac{\partial A_z}{\partial z}\right) = \mu_0 J_x, \tag{3.12}$$

and by adding and subtracting $\partial^2 A_x/\partial x^2$ on the left, there results,

$$-\frac{\partial^2 A_x}{\partial x^2} - \frac{\partial^2 A_x}{\partial y^2} - \frac{\partial^2 A_x}{\partial z^2} + \frac{\partial}{\partial x}\left(\frac{\partial A_x}{\partial x} + \frac{\partial A_y}{\partial y} + \frac{\partial A_z}{\partial z}\right) = \mu_0 J_x. \tag{3.13}$$

Repeating the same process for the y and z components, and recombining, (2.90) becomes

$$-\nabla^2 \mathbf{A} + \nabla(\nabla \cdot \mathbf{A}) = \mu_0 \mathbf{J}, \tag{3.14}$$

where ∇^2, the operator called the Laplacian, is defined by

$$\text{div grad} \equiv \nabla^2 = \frac{\partial^2}{\partial x^2} + \frac{\partial^2}{\partial y^2} + \frac{\partial^2}{\partial z^2}. \tag{3.15}$$

In (3.5), however, we chose to *define* $\mathbf{V} \cdot \mathbf{A} = 0$, for the very reason of simplifying the present relationship, so that (3.14) becomes

$$\nabla^2 \mathbf{A} = -\mu_0 \mathbf{J}. \tag{3.16}$$

Each component of (3.16),

$$\nabla^2 A_x = -\mu_0 J_x, \quad \nabla^2 A_y = -\mu_0 J_y, \quad \nabla^2 A_z = -\mu_0 J_z, \tag{3.17}$$

is *mathematically identical* with the relationship

$$\nabla^2 \phi = -\rho/\varepsilon_0 \tag{3.18}$$

for electrostatics, known as Poisson's equation, which latter follows immediately from (3.2), (2.89) and (3.15); namely,

$$\mathbf{E} = -\operatorname{grad} \phi, \quad \operatorname{div} \mathbf{E} = \rho/\varepsilon_0 \text{ and div grad} \equiv \nabla^2.$$

The isomorphism between (3.17) and (3.18) lays bare the technique we may now employ as a result of introducing the vector potential: first solve for the scalar potential ϕ in terms of the charge density ρ and then write down each component of the vector potential $\mathbf{A}$, replacing ρ by J_x/c^2 etc.; finally, find the curl of $\mathbf{A}$ – that is, $\mathbf{B}$ – by differentiation of the components of $\mathbf{A}$.

We can now solve (3.9). From §2.5 and (2.44) we know that the potential at any point (x_1, y_1, z_1) in a charge distribution $\rho(x, y, z)$ is given by the volume integral

$$\phi(x_1, y_1, z_1) = \frac{1}{4\pi\varepsilon_0} \int \frac{\rho(x_2, y_2, z_2)}{r_{12}} dV_2 \tag{3.19a}$$

or, in a more brief notation,

$$\phi(1) = \frac{1}{4\pi\varepsilon_0} \int \frac{\rho(2)}{r_{12}} dV_2, \tag{3.19b}$$

and so, from the analogy between (3.18) and (3.17), we have

$$A_x(1) = \frac{\mu_0}{4\pi} \int \frac{J_x(2)}{r_{12}} dV_2, \text{ etc.} \tag{3.20}$$

or, in general

$$\mathbf{A}(1) = \frac{\mu_0}{4\pi} \int \frac{\mathbf{J}(2)}{r_{12}} dV_2. \tag{3.21}$$

If we now consider a current-carrying wire rather than a general current density, we can replace $\mathbf{J} \, dV_2$ by $I \, d\mathbf{l}$ because $dV_2 = a \, dl$, where a is the cross-sectional area of the wire and the vector $d\mathbf{l}$ is taken to point in the direction of the positive current. Hence, at distance r_{12} from the wire, carrying current I, the vector potential is given by

$$\mathbf{A} = \frac{\mu_0 I}{4\pi} \int \frac{d\mathbf{l}}{r_{12}}. \tag{3.22}$$

From the definition (3.1), we have

$$d\mathbf{B} = \mathbf{V} \wedge d\mathbf{A}, \tag{3.23}$$

where

$$d\mathbf{A} = \mu_0 I \, d\mathbf{l}/4\pi r_{12}, \tag{3.24}$$

from (3.22). As $d\mathbf{l}$ is a constant for a given elemental length of wire, we can rearrange (3.28) to get

$$d\mathbf{B} = \frac{\mu_0}{4\pi} \mathbf{V} \wedge \frac{I d\mathbf{l}}{r_{12}} = -\frac{\mu_0 I}{4\pi} d\mathbf{l} \wedge \mathbf{V} \left(\frac{1}{r_{12}} \right). \tag{3.25}$$

Since $\mathbf{V}(1/r) = -\mathbf{e}/r^2$, where $\mathbf{e}$ is the unit vector from point 1 to point 2, on substitution into (3.25) we find

$$d\mathbf{B} = \frac{\mu_0}{4\pi} \frac{I d\mathbf{l} \wedge \mathbf{e}}{r^2}, \tag{3.26}$$

which is known as the *Biot–Savart law*. A more obviously significant form of this law, relating to the magnetic field generated by a current in a closed circuit, is obtained by integrating over the complete circuit:

$$\mathbf{B}(1) = -\frac{\mu_0}{4\pi} \int \frac{I \mathbf{e} \wedge d\mathbf{l}_2}{r^2}, \tag{3.27}$$

where the minus sign appears again because we have reversed the order of the cross-product. The relative simplicity in using (3.22) rather than (3.27) is demonstrated in the following section where we examine the field of a current loop.

3.3 Current loops and magnetic dipoles

We consider a closed conducting loop of arbitrary, flat shape lying in the xy plane and encircling an origin, as in figure 3.1. A steady current I flows around the loop, maintained externally in a way we do not enquire into. We seek to determine the magnetic field $\mathbf{B}$ created by this current loop at *distant* points like P_1 in the figure. We may therefore take r_1, the

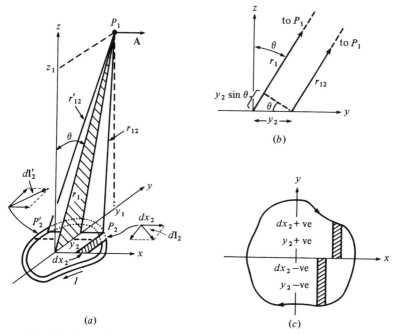

Fig. 3.1. Determination of the vector potential **A** at a distant point P_1 from a current loop.

distance from the origin to P_1, as much larger than any dimension in the loop. Without loss of generality, as we shall see, we may locate P_1 in the yz plane. We begin by calculating the vector potential **A** at P_1; that is, $\mathbf{A}(0, y_1, z_1)$.

Consider the variation of the denominator r_{12} in (3.22) as P_2 moves round the loop in figure 3.1. As P_1 is far away, the first-order variation in r_{12} depends only on the coordinate y_2 of the segment dl_2, and not on x_2: figure 3.1(b) emphasizes this point. So, neglecting quantities proportional to $(x_2/r_{12})^2$, we may take $r_{12} \approx r'_{12}$. Also, to first order in the ratio (loop size/distance to P_1), we have

$$r_{12} \approx r_1 - y_2 \sin\theta, \quad r_1 \gg y_2. \tag{3.28}$$

Now compare the two elements of the loop dl_2 and dl'_2 shown in figure 3.1(a). For these segments, the two dy_2s are equal and opposite and, from above, the r_{12}s are essentially equal. Therefore, to this order, their contributions to the line integral (3.22) will cancel. A similar argument may be made for all similar pairs of segments, approximately related by reflection in the yz plane. Hence at P_1, **A** will have no y component.

Obviously it has no z component either, as the current path itself has no z component. The remaining x component of the vector potential derives from the dx part of the path integral, thus

$$A(0, y_1, z_1) = e_x \frac{\mu_0 I}{4\pi} \int \frac{dx_2}{r_{12}}, \tag{3.29}$$

where e_x is the unit vector parallel to x. Now, to the same order of approximation, (3.28) can be rewritten

$$\frac{1}{r_2} \approx \frac{1}{r_1}\left(1 + \frac{y_2 \sin\theta}{r_1}\right), \tag{3.30}$$

and so, on substitution into (3.29), there results

$$A(0, y_1, z_1) = e_x \frac{\mu_0 I}{4\pi r_1} \int \left(1 + \frac{y_2 \sin\theta}{r_1}\right) dx_2, \tag{3.31}$$

in which integration r_1 and θ are constants. Now $\int dx_2$ around the loop vanishes, or course. The quantity $\int y_2 \, dx_2$ around the loop is just the area of the loop, regardless of its shape: figure 3.1(c) illustrates this point. Therefore

$$A(0, y_1, z_1) = e_x \frac{\mu_0 I \sin\theta}{4\pi r_1^2} \times (\text{area of loop}). \tag{3.32}$$

Note that the *shape* of the loop is irrelevant, which is why the choice of placing P_1 in the yz plane brings no essential restriction. Accordingly, (3.32) describes a general result: the vector potential of a current loop of any shape, at a distance r from the loop (where r is much greater than the size of the loop) is a vector whose direction is perpendicular to the plane containing r and the normal to the plane of the loop, and whose magnitude is

$$A = \frac{\mu_0 I a \sin\theta}{4\pi r^2}, \tag{3.33}$$

where a is the area of the loop and θ is the angle subtended by r and the normal to the loop. In this general expression, the x and y coordinates need not be mentioned and hence this vector potential is symmetric around the axis of the loop. We infer that B is cylindrically symmetric also. Of course, this result emerges simply because the details of the shape of the loop seen at distant points are not essentially discernable. All loops with the same product *current* × *area* produce the same *far* field.

Let us define a vector **m** to summarize this property of the current

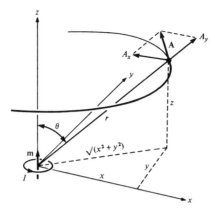

Fig. 3.2. With a small current loop placed in the xy plane – and hence the equivalent magnetic dipole **m** along z – the resultant vector potential **A** is everywhere directed tangentially to a circle drawn parallel to the xy plane and around the z axis.

loop: the vector has *magnitude* given by

$$m = Ia \qquad (3.34)$$

and, since a plane may be defined by its normal, its *direction* is taken perpendicular to the plane of the current loop with a *sense* given by the right-hand rule with respect to the current direction. We shall name the vector **m** shortly. Incorporating this definition into (3.33), we have

$$A = \frac{\mu_0}{4\pi} \frac{m \wedge e}{r^2}, \qquad (3.35)$$

where **e** is the unit vector in the direction *from* the loop to the point for which **A** is being calculated: the direction of **A** will always be that of the current in the *nearest* part of the loop.

We now seek to calculate the magnetic field **B** created by this current loop. In figure 3.2 is sketched the position reached so far. The small current loop centred at the origin is equivalenced by the vector **m**. The vector potential field is cylindrically symmetric with respect to **m**: compare how the magnetic field **B** in figure 2.19 was similarly symmetric with respect to a current-carrying straight wire. With respect to the coordinate frame shown in figure 3.2, consider the vector potential at some point (x, y, z). Remembering that $r^2 = x^2 + y^2 + z^2$ and $\sin \theta = \sqrt{(x^2 + y^2)}/r$, the magnitude of **A** at that point is

$$A = \frac{\mu_0 m \sin\theta}{4\pi r^2} = \frac{\mu_0 m \sqrt{(x^2 + y^2)}}{4\pi r^2}. \qquad (3.36)$$

Since **A** is tangent to a circle normal to z, its components are

$$\left.\begin{aligned}
A_x &= A\left(\frac{-y}{\sqrt{(x^2+y^2)}}\right) = -\frac{\mu_0 my}{4\pi r^3}, \\
A_y &= A\left(\frac{x}{\sqrt{(x^2+y^2)}}\right) = \frac{\mu_0 mx}{4\pi r^3},
\end{aligned}\right\} \qquad (3.37)$$

and

$$A_z = 0.$$

The components of the magnetic field **B** are then given by taking the components of curl **A** in the usual way, and so

$$\left.\begin{aligned}
B_x &= (\nabla \wedge \mathbf{A})_x = \frac{\partial A_z}{\partial y} - \frac{\partial A_y}{\partial z} = -\frac{\partial}{\partial z}\frac{k'mx}{(x^2+y^2+z^2)^{3/2}} = \frac{3k'mxz}{r^5}, \\
B_y &= (\nabla \wedge \mathbf{A})_y = \frac{\partial A_x}{\partial z} - \frac{\partial A_z}{\partial x} = \frac{\partial}{\partial z}\frac{-k'my}{(x^2+y^2+z^2)^{3/2}} = \frac{3k'myz}{r^5}, \\
B_z &= (\nabla \wedge \mathbf{A})_z = \frac{\partial A_y}{\partial x} - \frac{\partial A_x}{\partial y} \\
&= k'm\left[\frac{-2x^2+y^2+z^2}{(x^2+y^2+z^2)^{5/2}} + \frac{x^2-2y^2+z^2}{(x^2+y^2+z^2)^{5/2}}\right] \\
&= \frac{k'm(3z^2-r^2)}{r^5},
\end{aligned}\right\} \qquad (3.38)$$

where $k' = \mu_0/4\pi$. Recalling the cylindrical symmetry of **B**, we recast these equations in spherical polar coordinates and find

$$\left.\begin{aligned}
B_\perp &= 3\mu_0 m\sin\theta\cos\theta/4\pi r^3, \\
B_\parallel &= \mu_0 m(3\cos^2\theta - 1)/4\pi r^3,
\end{aligned}\right\} \qquad (3.39)$$

where the $\parallel$ and $\perp$ signs refer to field components parallel and perpendicular to the axis of the vector **m**. Figure 3.3 provides a sketch of the magnetic field of a current loop in three dimensions. The detailed form of this field in a plane containing the axis of **m** is shown in figure 3.4 where it is compared with the electric field which may be shown[1,2] to arise from a pair of stationary electric charges of opposite sign. The electric field is called a dipolar field, arising as it does from a pair of opposite charges. The magnetic field of the current loop has *identical form at distant points* in space and is also called a dipolar field by analogy. This field arose from the current loop represented by the product of current × area

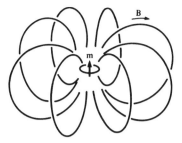

Fig. 3.3. Representing the three-dimensional magnetic field at large distances from a small current loop.

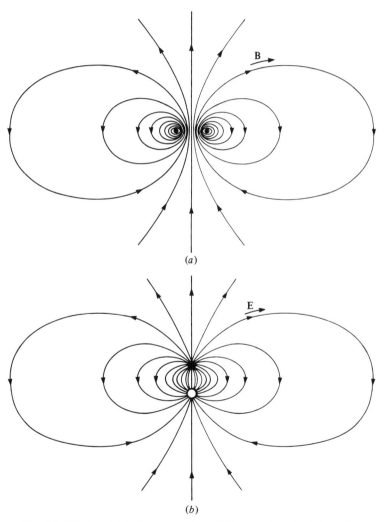

Fig. 3.4. The far field of a current loop (*a*) has an identical character to the far field of an electric dipole (*b*).

we labelled **m**: accordingly, the vector **m** is called the *magnetic dipole moment* or, more simply, the *magnetic moment* of the current loop, which latter is, of course, the real source of the magnetic field.

Close to the current loop, the magnetic field is quite different from the electric field close to the electric dipole. The electric field lines begin and end at the source charges; div **E** $\neq 0$: the magnetic field lines form continuous closed loops and div **B** $= 0$, everywhere.

3.4 A current loop in a uniform external magnetic field

We have just discussed the magnetic field produced by a current loop. As a totally distinct exercise we now investigate the effect upon such a current loop of another 'external' magnetic field. These two fields are quite separate from one another: there can be no question of the magnetic field produced by a current loop exerting forces upon that loop itself if only because of the thermodynamic impossibility of picking onself up by one's bootstraps! The magnetic field produced by one current loop *can* exert a force on *other* magnetic loops, but that is a quite separate matter to which we shall return in due course. Meanwhile we first examine the effect of an external homogeneous magnetic field on a current-carrying loop: throughout we assume that the current in the loop is *maintained* at the expense or otherwise of some source like a battery.

For illustration, consider the simple case of the rectangular loop shown in figure 3.5. Suppose the loop to be inclined with respect to the magnetic field **B** in the symmetrical way shown in the diagram. A current **I** flows in the loop or, equivalently, charges move along each arm of the circuit. By the Lorentz force law (2.52), these charges experience forces perpendicular to both **B** and **I** locally. The forces on each arm of the loop are indicated in the diagram; $\mathbf{F}_2$ and $\mathbf{F}_4$ are diametrically opposed and equal in magnitude, because the opposite edges of the rectangle are of equal length and hence carry equal numbers of charges, but moving in opposite directions. The forces $\mathbf{F}_1$ and $\mathbf{F}_3$ are also equal in magnitude to each other and antiparallel, but are clearly not diametrically opposed. Therefore, the (rigid) loop experiences a torque **G**, directed along $\mathbf{F}_2$ and of magnitude

$$G = F_1 b \sin \theta = IaBb \sin \theta = (Iab \sin \theta)B; \qquad (3.40)$$

and hence, by the definition (3.34) of the magnetic moment of a current loop,

$$\mathbf{G} = \mathbf{m} \wedge \mathbf{B}. \qquad (3.41)$$

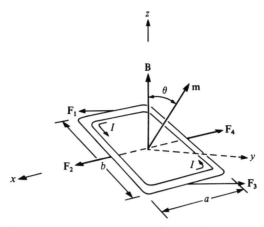

Fig. 3.5. The torque on a current loop in a uniform magnetic field: the loop is tilted such that **m** and **B** both lie in the *yz* plane.

We may draw two conclusions from this simple exercise: firstly that, in general, a current-carrying loop suffers a torque in an external magnetic field, even in a uniform field; and secondly, that the concept of the magnetic dipole moment continues to provide a useful abstraction of the qualities of a current loop.

3.5 A current loop in an inhomogeneous external magnetic field

It is obvious that a current loop will suffer a torque in any external magnetic field – homogeneous or not. However, suppose we allow the loop to orient itself so as to relieve that torque and then enquire whether any other, 'non-torque', forces remain to be studied. It should be clear that there are none in the case of a uniform magnetic field as all forces in figure 3.5, together with those implicit in the assumed rigidity of the loop, cancel exactly. The situation is different for an inhomogeneous field, however, as illustrated by the following description of a special, but typical, case.

Let us place a small current loop in the inhomogeneous magnetic field produced, say, at the end of a solenoid: the cylindrically symmetric situation is shown in figure 3.6. As usual, the field lines in the diagram do not include the contribution from the field produced by the current loop itself. The current loop produces no force on itself and provided the field it creates does not disturb the sources of the external field (which it will not if that field derives from currents in a rigidly constructed solenoid, for example), then we may ignore it in this discussion.

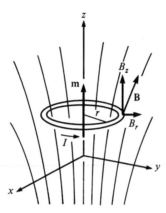

Fig. 3.6. A current loop in a plane perpendicular to the symmetry axis of an axial but otherwise inhomogeneous magnetic field.

The current loop experiences no torque because of the symmetrical experimental arrangement we have selected. It does, however, experience a force by interaction with the external field. It arises because the external field **B** has an outward, *radial* component B_r everywhere around the ring. Each element of the loop, $d\mathbf{l}$, must experience a downward (for the current and field senses shown in figure 3.6) force, $\mathbf{I} \wedge \mathbf{B}$, of magnitude $IB_r dl$ and, in the axially symmetric situation here, the total downward force will have the magnitude

$$F = 2\pi r I B_r, \tag{3.42}$$

where r is the radius of the current loop.

Now B_r is related to the gradient of B_z, where z is taken parallel to the axis of the solenoid. To determine this relationship, we consider the small cylindrical volume of radius r and height Δz in figure 3.7, and recall that div $\mathbf{B} = 0$ everywhere so that the net flux out of any volume is zero. The outward flux from the side of this cylinder is $2\pi r \Delta z B_r$ and the net outward flux from the circular end-faces is $\pi r^2 [-B_z(z) + B_z(z + \Delta z)]$, which to first order in the small distance Δz is just $\pi r^2 (\partial B_z/\partial z)\Delta z$. For vanishing total flux, therefore,

$$\pi r^2 \frac{\partial B_z}{\partial z} \Delta z + 2\pi r B_r \Delta z = 0 \tag{3.43}$$

and so

$$B_r = -\frac{r}{2}\frac{\partial B_z}{\partial z}. \tag{3.44}$$

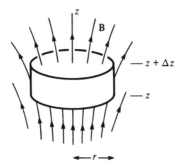

Fig. 3.7. The relationship between B_r and $\partial B_z/\partial z$.

After substitution into (3.42), we express the downward force on the current loop in terms of the external field gradient as

$$F = 2\pi r I \frac{r}{2} \frac{\partial B_z}{\partial z} = \pi r^2 I \frac{\partial B_z}{\partial z} \tag{3.45}$$

and, recognizing that πr^2 is the area of the loop, we again use the definition (3.34) and write

$$F = m \frac{\partial B_z}{\partial z}. \tag{3.46}$$

This special result is one example of a relationship which can be demonstrated for any small current loop in an inhomogeneous magnetic field of any profile. Thus, using the transformation[3] between surface and line integrals

$$\oint_C \mathbf{B} \wedge d\mathbf{s} = \int_S (\operatorname{div} \mathbf{B})\mathbf{n}\, da + \int_S (\operatorname{curl} \mathbf{B}) \wedge \mathbf{n}\, da - \int_S da(\mathbf{n} \cdot \operatorname{grad})\mathbf{B}, \tag{3.47}$$

we write

$$\mathbf{F} = I \oint_C d\mathbf{s} \wedge \mathbf{B} = -I \oint_C \mathbf{B} \wedge d\mathbf{s}$$

$$= -I \int_S (\operatorname{div} \mathbf{B})\mathbf{n}\, da - I \int_S (\operatorname{curl} \mathbf{B}) \wedge \mathbf{n}\, da + I \int_S da(\mathbf{n} \cdot \operatorname{grad})\mathbf{B} \tag{3.48}$$

and, since $\operatorname{div} \mathbf{B} = 0$ from (2.74) and $\operatorname{curl} \mathbf{B} = 0$ from (2.73) in the absence of external currents (we do not count the current loop here, which is to be equivalenced by m and will not exert a force on itself),

$$\mathbf{F} = I \int_S da\, \mathbf{n} \cdot \operatorname{grad} \mathbf{B} = \int d\mathbf{m} \cdot \operatorname{grad} \mathbf{B}. \qquad (3.49)$$

If $\operatorname{grad} \mathbf{B}$ is constant in the region occupied by the current loop, we get the generalized version of (3.46),

$$\mathbf{F} = \mathbf{m} \cdot \operatorname{grad} \mathbf{B}. \qquad (3.50)$$

There have been two purposes in this section. In the first place we have examined the result of placing a current loop in a magnetic field and discovered that, apart from any torque it may suffer as described in the preceding section, it experiences a force on its centre of gravity that is proportional to the field gradient. It follows immediately, of course, that such a force will vanish in a homogeneous magnetic field – so that only torques are relevant there. The relationship (3.50) will be useful in chapter 5 when we consider the measurement of magnetic moments. Secondly, we observe, once again, that a current-carrying loop may be represented by an equivalent magnetic dipole moment.

3.6 The phenomenological interaction of magnetic fields and matter

All materials experience a force in an inhomogeneous magnetic field. Diamagnets tend to move towards the weaker field regions – that is, they are repelled – while paramagnets and ferromagnets are attracted. The forces experienced by diamagnets and paramagnets are usually proportional to the field gradient and to the field strength: occasionally, at very low temperatures and/or high fields, the force experienced by paramagnets may be less than that linearly extrapolated from those at weaker fields. Ferromagnetic materials, with which we are not concerned in this book, show a completely different field dependence and their properties may not be explained in the way we shall describe for the much weaker diamagnetic and paramagnetic phenomena. While diamagnetism is very nearly independent of temperature, the forces on paramagnets vary considerably with temperature and, for many paramagnetic materials, the variations are linear with inverse absolute temperature, following the phenomenological summary of Curie's law.[4]

A very long time ago, Ampère suggested that the origins of magnetism lay in small current loops within the body of a material but it was only in this century that detailed – though false, as we shall see – classical support for the notion of amperian currents was forthcoming. On the other hand, ultimate and valid confirmation of the idea of molecular magnetic dipoles *has* emerged subsequently from quantum mechanics. Indeed, we shall see that all magnetic phenomena are quantum mechanical in origin, classical physics not only being unable to account for them but actually demonstrating their non-existence! Nevertheless, it is both interesting and instructive to follow through some of the attempts of classical physics to account for diamagnetism and paramagnetism, for the eventual success of the quantum-theoretical description relies not so much on a rejection of the classical foundations as on the establishment of supplementary conditions.[5] In the next two sections, therefore, we describe very briefly certain classical views of the microscopic origins of para- and diamagnetism, really only to provide the conventional background against which the development of the subject may be more familiar and apparent.

3.7 Magnetic moments and electronic angular momenta

From our chemists' background we are well aware of many of the quantum-mechanical features of atoms and molecules, including the concepts of quantized spin, and orbital, angular momenta. While keeping one eye on these matters, so to speak, we shall nevertheless suppose, *pro tem*, that an atom consists of electrons travelling around a heavy, and hence essentially stationary, nucleus. A time-average description from the classical viewpoint would represent the moving electrons as current loops within the atom; we consider one such current loop. Suppose as electron with speed v around a circular orbit of radius r to produce an effective loop current,

$$I = -\frac{ev}{2\pi r}, \tag{3.51}$$

where $-e$ is the electron charge. The far magnetic field is that of a magnetic dipole **m** of strength,

$$m = \pi r^2 I = -evr/2. \tag{3.52}$$

A central feature in the theory of magnetism is that we may relate the electronic orbital angular momentum **L** to the magnetic moment vector **m** because the magnitude of **L** about the normal to the orbit plane is just

$$L = m_e vr, \tag{3.53}$$

where m_e is the rest-mass of the electron (the velocities of electrons in atoms are quite small compared with the velocity of light and the actual mass of the moving electron is only about 1% greater than m_e). So, comparing (3.52) and (3.53), we find

$$\mathbf{m} = -\frac{e}{2m_e}\mathbf{L}, \tag{3.54}$$

a relationship which can be shown to hold for quite generally shaped orbits. A negative sign is associated with the negative charge of the electron and means that the orbital angular momentum of an electron is directed antiparallel to the associated magnetic moment. A positive sign appears for a positive charge; for example, when we consider the spin angular momentum of a nucleus. The relationship (3.54) derived originally, as here, from classical physics: remarkably, it also emerges from a quantum-mechanical treatment.

A similar relationship obtains for the quantum treatment of electron spin $\mathbf{S}$, for which there is no satisfactory classical analogy, but the ratio of $\mathbf{m}$ to $\mathbf{S}$ is twice as large,

$$\mathbf{m}_S = -\frac{e}{m_e}\mathbf{S}, \tag{3.55}$$

a feature that we discuss in §7.4. Again the crucial issue in the discussion so far is the intimate connection between magnetic moment and angular momentum: the present section is meant to provide a (albeit simplistic) classical rationale for what can validly only arise in the quantum-mechanical treatment. Looking ahead, our view of paramagnetic materials is one in which there is a net permanent magnetic dipole associated with the set of electrons within any atom or molecule (ultimately deriving from spin and/or orbital electronic angular momentum) and a tendency for this dipole to align with an external magnetic field. This tendency would be proportional to the magnetic field strength from our discussion in §3.4 and such is indeed the case experimentally. The randomizing effect of thermal collisions in an assemblage of such molecular magnets tends, however, to destroy this alignment so that the net paramagnetic effect is a sensitive function of temperature – again as found in practice. Apparently, we may base a detailed, statistical, thermodynamic treatment of the phenomenon of paramagnetism on these remarks, but the course we must follow is subtle and, if pursued consistently within the classical framework, leads to vanishing magnetism, as we shall see in §3.21. For the moment, we note that only (3.54) and (3.55) have validity within contemporary physics.

3.8 Diamagnetism and the Larmor precession

In most materials, the molecules and atoms possess no net, permanent magnetic moment. Again with one eye on the quantum-mechanical model of atoms and molecules, we know that most electrons are paired with respect to spin and, in that they mostly reside in closed subshells, with respect to orbital angular momentum too. The same is true for the majority of electrons in paramagnetic molecules. This does not mean, however, that electrons which are paired in this way fail to interact with an external magnetic field due to pairwise cancellations. The following classical argument (though not in so simplified a manner) was provided several years prior to the advent of even Bohr's theory of the atom.

Consider the atoms in a piece of material as containing revolving electrons whose associated orbital magnetic moments are distributed evenly through all orientations. Let us study those orbits which happen to lie in a chosen plane – xy, say – of which there will be essentially equal numbers with **m** parallel to z as antiparallel, and enquire what happens to one of these orbits when we introduce an external magnetic field in the z direction. We model this situation by first studying that shown in figure 3.8(a) in which an object of mass m_Q and electric charge Q is attached to a fixed point by a cord of fixed length r. We imagine the object to rotate with velocity v_0 around the fixed point, held in a circular path in the xy plane by the centripetal force $\mathbf{F}_0$ of magnitude

$$F_0 = m_Q v_0^2 / r. \tag{3.56}$$

Now we suppose a magnetic field **B** to be introduced gradually along the z axis, normal to the orbit of the object. While the field grows at the rate dB/dt, there will be an induced electric field **E** around the path of the rotating body, as shown in figure 3.8(b). The rate of change of flux Φ through the circular path is given by

$$\frac{d\Phi}{dt} = \pi r^2 \frac{dB}{dt}, \tag{3.57}$$

from which we have the line integral of the electric field as

$$\int \mathbf{E} \cdot d\mathbf{l} = \pi r^2 \frac{dB}{dt} = 2\pi r \mathbf{E}. \tag{3.58}$$

Hence,

$$E = \frac{r}{2} \frac{dB}{dt}. \tag{3.59}$$

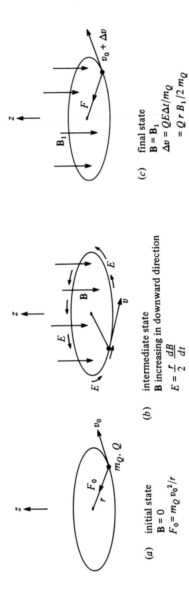

Fig. 3.8. The growth of the magnetic field induces an electric field which accelerates the orbiting charged particle.

The direction of this induced electric field is given by the right-hand rule and is such as to accelerate the body if Q is a positive charge. The acceleration along the path, dv/dt is determined by the force QE along the path:

$$m_Q \frac{dv}{dt} = QE = \frac{Qr}{2} \frac{dB}{dt}, \qquad (3.60)$$

so providing a relationship between the change in velocity and the change in magnetic field:

$$dv = \frac{Qr}{2m_Q} dB. \qquad (3.61)$$

As the radius r is fixed by the length of the cord, the factor $Qr/2m_Q$ is constant. Let Δv denote the net change in velocity in the whole process of increasing the magnetic field from zero to the final value B_1, say. Then,

$$\Delta v = \int_{v_0}^{v_0 + \Delta b} dv = \frac{Qr}{2m_Q} \int_0^{B_1} dB = \frac{QrB_1}{2m_Q}, \qquad (3.62)$$

from which we note that the time has dropped out and the final velocity is not dependent upon the rate at which the change is made.

However, the increased speed of the charged body in the final state implies an increase in the upward-directed magnetic moment **m**. A *negatively* charged body would have been *retarded* under similar circumstances which would, in turn, have *decreased* its *downward* moment. In *either* case, therefore, the application of the external field **B**$_1$ brings about a change of magnetic moment in a direction *opposing* the applied field; and this is just Lenz's law, of course. The change of magnitude in **m** is, using (2.52),

$$\Delta m = \frac{Qr}{2} \Delta v = \frac{Q^2 r^2}{4m_Q} B_1, \qquad (3.63)$$

and so the effect is linear in the field B_1 but varies as the square of the charge orbit radius. For charges of either sign, revolving in either direction, Lenz's law prevails and we may write

$$\Delta \mathbf{m} = \frac{-Q^2 r^2}{4m_Q} \mathbf{B}_1. \qquad (3.64)$$

Now the introduction of the magnetic field causes a change Δv in the particle velocity: this *apparently* implies a concomitant change in the centripetal force **F**. If so, this would imply that for the case of an arbitrary and untethered electron, for example, an altered centripetal force would in turn necessitate a changed radius and thence a fresh calculation of the effect just described; and so on. However, take the case when B_1 is small enough that $\Delta v \ll v_0$, which is certainly apposite for 'real' electron orbits (Bohr, perhaps) with practical magnetic fields. In the final state, the centripetal force acquires the magnitude

$$F_1 = \frac{m_Q(v_0 + \Delta v)^2}{r} \approx \frac{m_Q v_0^2}{r} + \frac{2m_Q v_0 \Delta v}{r}, \qquad (3.65)$$

neglecting terms in $(\Delta v)^2$. But now, of course, the external magnetic field itself exerts an inward force on the moving charge, via the Lorentz force equation (2.52), of magnitude $Q(v_0 + \Delta v)B_1$. Using (3.63) to express B_1 in terms of Δv, we find this extra inward force has the magnitude $Q(v_0 + \Delta v) \times 2m_Q \Delta v/Qr$ which, to first order in $\Delta v/v_0$, is $2m_Q v_0 \Delta v/r$. In other words, the tension in the cord remains unchanged at the value F_0.

The significance of this result is that the relationship (3.61) is valid for any kind of tethering force. It could be the simple Coulomb attraction of a nucleus for an electron or the net attraction to one electron in a many-electron atom, which will have a complicated dependence on radius: in all cases, (3.64) should describe the change in magnetic moment due to an external field. An alternative view of (3.64) is to note that each electron in this classical model continues to revolve at the same radius, but that its angular velocity changes from v_0/r by a small amount $\Delta \omega = \Delta v/r$, where

$$\Delta \omega = \frac{\Delta v}{r} = \frac{-eB_1}{2m_e}, \qquad (3.66)$$

an angular velocity depending only upon the strength of the applied field and the charge–mass ratio of the electron. The new system looks like the old one in the absence of the magnetic field, but viewed from a rotating frame of reference. The angular velocity (3.66) is called the Larmor angular velocity, or the *Larmor frequency*. The concept can be established[3,5] in the classical framework in much more general terms than of the circular orbit considered here.

If the foregoing is to provide a useful model for a real atom in an external magnetic field, one extension which must be included is the consideration of all possible orientations of the electron orbits and not just those perpendicular to the field, as in figure 3.8. Now the distance r in figure 3.8 is given by $\sqrt{(x^2 + y^2)}$ and the spherical average of $x^2 + y^2$

is $\frac{2}{3}$ of the average of the square of the true radial distance from the centre point of the atom, because $r^2 = x^2 + y^2 + z^2$. Accordingly, (3.64) is modified to

$$\Delta\mathbf{m} = \frac{-Q^2}{6m_Q} \overline{r^2} \mathbf{B}, \qquad (3.67)$$

although the same qualitative features of our discussion survive.

The interaction just described should occur for *all* electronic current loops[††] in atoms and molecules and hence be a universal property of electron-containing materials. The effect would also overlie any paramagnetism due to a permanent net magnetic moment in the molecules. It would survive any cancellations of individual contributions from electronic moments for, as emphasized already, the effect obeys Lenz's law, regardless of the sense of rotation or, equivalently, of the sense of each individual magnetic dipole moment. As the *induced* moment, $\Delta\mathbf{m}$, opposes the applied field, the macroscopic result is that the material is repelled by the magnetic field; and this is the phenomenon of diamagnetism. There is no question in the foregoing of thermal distribution over different orientations of magnetic dipoles. Therefore, diamagnetism is essentially temperature independent: such slight variation with temperature that is observed is attributable to small changes in effective electron-orbit radii in the atoms or molecules. Note that the induced moment in (3.67) is proportional to the external magnetic field strength: in turn the interaction of that induced dipole is proportional to the field strength again, if we are considering torques as in §3.4, or to the field gradient if an experiment involves an inhomogeneous field as in the Faraday method, for example. *In toto*, therefore, the torques on diamagnets are predicted (and observed) to be proportional to B^2 and simple forces of repulsion to be proportional to $B\partial B/\partial z$. To emphasize this point: one power of the external field strength induces the moment, and the other turns or repels that moment.

In this section we have briefly sketched the outline of what was once considered to be a considerable success for the understanding of magnetic materials within the concepts of classical physics. Later in this chapter we shall study the classical model a little more rigorously both in preparation for the modern quantum-mechanical view of the subject and also to see why a fully self-consistent classical approach eventually fails

---

[††] …and, of course, for nuclei: but the very small effective loop radii involved there lead to effects that are negligible *for us*.

despite this promising beginning. Meanwhile it is enough to know that the concept of molecular magnetic dipoles survives the classical–quantum transition and the classically incorrect statistical model is eventually replaced by a valid and amazingly similar quantized treatment. With this confidence, we leave the subject of the microscopic origins of magnetism and consider aspects of the macroscopic phenomena which can, of course, be treated perfectly well within the classical framework.

3.9 Magnetization

Material placed in a magnetic field becomes magnetized. We have set aside for the moment any discussion of the precise microscopic mechanisms by which this occurs. The result, however, is that a magnetic moment is induced in the material and that, in turn associated with this moment, is an induced magnetic field. In the present discussion we shall ignore the cause of the magnetization and concentrate exclusively upon the field produced by the magnetized sample which may thus be regarded as a small magnet in its own right.

We first study the case of a uniformly magnetized block of material; that is, one which contains an essentially equal number of oriented molecular magnetic dipoles per unit volume. We represent the *intensity of magnetization* in the sample by the vector **M** considered as the product of the number of oriented dipoles per unit volume and the magnetic moment of each molecular dipole. Let us define z as the direction parallel to the magnetization vector **M** and consider a thin slice of the material, cut normal to z and of thickness dz, as shown in figure 3.9(a). We further consider this slab as divided up into small elements or 'tiles'. A tile of area da, contains a total dipole moment of $M\,da\,dz$: **M** is also called the *dipole moment per unit volume*. The far magnetic field produced by this elemental tile – at points which are distant compared with the dimensions of the *tile* – is equivalent to that of any dipole with the same magnetic moment. An equivalent dipole would result, for example, from a current-carrying ribbon, as shown in figure 3.9(b), in which a current $I = M\,dz$ flows, for then the ribbon dipole moment, given by $I \times$ area, is just $M\,dz\,da$ as before. Suppose we substitute such a current ribbon for every tile in the slab. As the given slab is uniformly magnetized, these currents will cancel at each intertile boundary, as illustrated in figure 3.9(c). The magnetization of the whole slab is therefore equivalent to that of a single ribbon around the outside, as in figure 3.9(d), carrying the equivalent

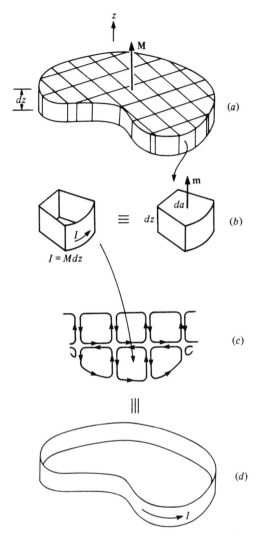

Fig. 3.9. Illustrating the steps by which a thin slab of magnetized material may be equivalenced by a current ribbon round its periphery, with respect to the external magnetic field it produces.

surface current $M\,dz$. The elemental tiles can be made very small, though not down to molecular size if we wish to maintain the uniformity of magnetization on average: therefore, the field at any *external* point, even very close indeed to the slab, is the same as that of the current ribbon.

A complete sample of material may be reconstructed by summing over elemental slabs so that the total magnetic moment is equivalenced by a wide ribbon carrying a current $M\,dz$ in each elemental ribbon width dz: altogether, therefore, the magnetization $\mathbf{M}$ is equivalenced by a surface current *density* $\mathcal{J}$, in a surface parallel to $\mathbf{M}$, so that $\mathcal{J} = M$. However, although we have assumed that the sample here is uniformly magnetized, this can never really be true if only because the sample has a boundary. Even if the magnetization is uniform (on the non-molecular scale we consider here) within the bulk of the body, its value falls to zero outside the surface. Therefore, in the region close to the boundary there must be large magnetization gradients. We might guess at this stage that the current circulating around the surface is related to these gradients, the word 'circulating' suggesting a dependence on the curl of $\mathbf{M}$. Indeed, this turns out to be the case, though it is simplest to show by studying the current density distribution for a sample in which the magnetization is considered to be generally non-uniform everywhere.

Qualitatively, it is obvious that the mutual cancellation of the elemental ribbon currents, described in figure 3.9(c), will not be exact in a non-uniformly magnetized sample, so that non-zero currents will appear in the body of the material. We may quantify this idea by considering a small rectangular block of material oriented parallel to some cartesian frame, as shown in figure 3.10(a). No special direction of the total magnetization with respect to these reference axes is presumed. The total magnetization in the block may be resolved parallel to the chosen axis frame; the z component, for example, being given by $M_z abc$, where abc is the volume of the block. As magnetic moment is defined as current $\times$ area, and the area of the surface perpendicular to z is ab, we can equivalence the z component of total magnetization in the block by a ribbon, around the block planes which are parallel to the z component of magnetic moment, in which a current

$$I = M_z c \tag{3.68}$$

flows perpendicular to z, as shown in figure 3.10(a). Now consider two small adjacent blocks as in figure 3.10(b). The magnetization in each of the two blocks is slightly different in general and on the surface between them will arise opposite but unequal currents as shown. Block 1 produces current I_1 flowing in the positive y direction while block 2 produces I_2 in the negative y direction so that the net current on this surface is given by

$$I = I_1 - I_2 = M_z c - (M_z + \Delta M_z)c = -\Delta M_z c. \tag{3.69}$$

Provided the blocks are taken small enough relative to the rate of change

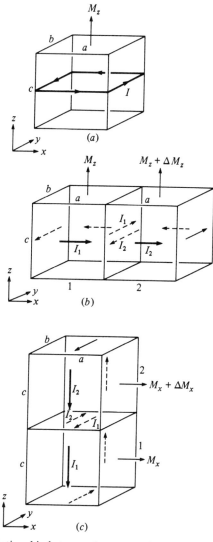

Fig. 3.10. The relationship between the current density and the curl of the magnetization.

of magnetization, we may write

$$\Delta M_z = \frac{\partial M_z}{\partial x} a,$$ (3.70)

and hence

$$I = -\frac{\partial M_z}{\partial x} ac.$$ (3.71)

Now I, flowing parallel to y on the interblock boundary surface, may be related to an average *volume* current *density* $\mathbf{J}$ when we recall that the division of the material into small blocks occurs at arbitrary boundaries. Thus I *is* the component of the volume current parallel to y and the corresponding current *density* J'_y is given by dividing I by the area of the face perpendicular to y:

$$J'_y = I/ac = -\frac{\partial M_z}{\partial x}. \tag{3.72}$$

But there is another contribution to J_y; namely, that arising from the variation of magnetization with respect to z, the other direction perpendicular to y. A similar calculation may be performed using two blocks adjacent in the z direction, as shown in figure 3.10(c), from which results

$$J''_y = +\frac{\partial M_x}{\partial z}, \tag{3.73}$$

the signs in (3.72) and (3.73) being established from a consistent observance of the handedness of the reference frame. Combining (3.72) and (3.73), therefore, we get

$$J_y = \frac{\partial M_x}{\partial z} - \frac{\partial M_z}{\partial x}. \tag{3.74}$$

Values for the current density components J_x and J_z may be derived analogously by constructing similar diagrams to those in figure 3.10; or by cyclic permutation of the axis labels, remembering that our coordinate frame was oriented arbitrarily. Altogether, therefore, the vector current density is given by

$$\mathbf{J} = \nabla \wedge \mathbf{M}. \tag{3.75}$$

This equation relates the average magnetic moment ($\mathbf{M}$) per unit volume to an average current density $\mathbf{J}$ in matter and we have reached the stage at which we can compare (3.75) with Ampère's law (2.73), namely

$$\text{curl } \mathbf{M} = \mathbf{J} \text{ and curl } \mathbf{B} = \mu_0 \mathbf{J}.$$

3.10 Free and bound currents: the field H

The current densities in (2.73) and (3.75) are *not* the same thing. The quantity $\mathbf{J}$ in (3.75) refers to the currents which produce the magnetization in the sample: they correspond to the small molecular, 'amperian' currents which are equivalent to the various aligned permanent, and induced, molecular magnetic dipoles. Equation (2.73) is more general, referring to the total current within the volume of the material. It is convenient to separate the total current which may be present in a system into two parts: $\mathbf{J}_{\text{free}}$, referring to free currents arising from, say, conduction electrons and/or holes in the medium, and directly measurable with an ammeter; and $\mathbf{J}_{\text{bound}}$ describing the 'amperian'-like currents, permanent or induced by a magnetic field. For example, if a medium is both conducting and magnetizable, the total current density will be the sum of the real current density $\mathbf{J}_{\text{free}}$ and the equivalent magnetization current density $\mathbf{J}_{\text{bound}}$, and both must be counted in Ampère's law. So the total current density is written

$$\mathbf{J} = \mathbf{J}_{\text{free}} + \mathbf{J}_{\text{bound}} = \mathbf{J}_{\text{free}} + \nabla \wedge \mathbf{M} \tag{3.76}$$

and, on substitution into (2.73), Ampère's law takes the form

$$\operatorname{curl} \mathbf{B} = \mu_0 (\mathbf{J}_{\text{free}} + \operatorname{curl} \mathbf{M}) \tag{3.77}$$

or

$$\operatorname{curl}(\mathbf{B} - \mu_0 \mathbf{M}) = \mu_0 \mathbf{J}_{\text{free}}. \tag{3.78}$$

We now introduce a new vector $\mathbf{H}$ such that

$$\mathbf{B} - \mu_0 \mathbf{M} = \mu_0 \mathbf{H} \tag{3.79a}$$

or

$$\mathbf{B} = \mu_0 (\mathbf{H} + \mathbf{M}) \tag{3.79b}$$

when (3.78) becomes

$$\operatorname{curl} \mathbf{H} = \mathbf{J}_{\text{free}}, \tag{3.80}$$

an expression of Ampère's law which holds *in vacuo* or in a magnetizable medium: in this sense, (3.80) is more general than (2.73). The general form (3.80) collapses to (2.73) *in vacuo*, of course, when

$$\mathbf{B} = \mu_0 \mathbf{H}, \quad \textit{in vacuo}. \tag{3.81}$$

The integral expression equivalent to (3.80) is

$$\int_C \mathbf{H} \cdot d\mathbf{s} = \int_S \operatorname{curl} \mathbf{H} \cdot \mathbf{n} \, da = \int_S \mathbf{J}_{\text{free}} \cdot \mathbf{n} \, da = I_{\text{free}}, \tag{3.82}$$

here I_{free} is the total free current (that is, measurable by an ammeter) enclosed by the path C.

Two issues come to the fore at this stage, at least so far as the magnetochemist is concerned as user. The first is a very noisy affair to do with units and names: the second is much more important from our point of view and concerns the relationship between macroscopic and microscopic fields. We shall try and dispose of the units question first.

3.11 Units, dimensions and nomenclature

It is clear from (3.81) and (3.79) that the *dimensions* of **B** and **H** differ in the SI scheme. From (3.82) we have the dimensions of **H** as ampere/metre while from (2.52) and (2.71), **B** has dimensions of newton/ampere-metre. In turn, we have the dimensions of μ_0 as newton/ampere^2. The SI *units* for these quantities, consistent with the equations above, are: for **B**, tesla (T); for μ_0, henry/metre (H m^{-1}); and for **H**, ampere/metre (A m^{-1}). Now consider the dimensions of the intensity of magnetization **M**: from (3.79) these are the same as those for **H** – ampere/metre. This is consistent with the dimensions of the magnetic moment **m** – ampere-metre^2, from its definition as current $\times$ area – recalling that **M** is also defined as the magnetic dipole moment per unit volume.

There remains the vexed problem of the naming of parts, which for us really concerns only the quantities **B** and **H**. In the present system they have different dimensions and hence units: Bleaney & Bleaney[3] call **B** the *magnetic flux* and **H**, the *magnetic field strength*. On the other hand, Purcell[2] and Feynman[1], using CGS and MKS systems respectively, prefer to maintain the name *magnetic field* for **B** while referring to **H** either simply as **H** or as *the field* **H**; and the present author follows suit, for several reasons. Firstly, the convention, followed in our discussion of electric fields, of naming the field component per unit area normal to a given surface as the *flux* is well established and so preempts the naming of **B** by that name. Further, since both **B** and **E**, and **H** for that matter, are all mathematical vector quantities which describe how the force on a charge, or the torque on a current loop, vary in space – quantities with local properties of divergence and curl – there seems no good reason not to refer to them as fields. No problem is caused by **B** and **H** having different dimensions and units here, for **E** has different dimensions again. So we may think of both **B** and **H** as magnetic fields, but drop the word 'magnetic' from **H** so as to maintain the distinction otherwise conveyed by the symbols. In response to a suggestion that this stance would lead to **M** also being described as a field, because it has the same units as **H**, we would merely observe that the converse does not hold: the different units of **E**, **B** and **H** do not deny their natures as fields of various kinds. As Purcell[2] points out, 'it is only the names that give trouble, not the symbols'. It is not just for historical reasons, however, that it is worthwhile to summarize briefly some of the units systems and nomenclatures which have been used over the years, so far as they refer to **B** and **H**.

In the CGS system, (3.79) is written

$$\mathbf{B} = \mathbf{H} + 4\pi\mathbf{M}, \quad \text{CGS} \tag{3.83}$$

so that **B** and **H** have the same dimensions. Historically, however, they have been given different units – **B** in gauss and **H** in oersted! The usual relationship adopted in the MKS system, equivalent to (3.79), is

$$\mathbf{B} = \mu_0\mathbf{H} + \mathbf{M}, \quad \text{MKS(i)}, \tag{3.84}$$

so that the dimensions of **B** and **H** are the same as those adopted in the SI scheme. However, the units of **M** and **B** are now identical, so that those preferring this scheme redefine the magnitude of the magnetic moment as

$$m = \mu_0 Ia, \quad \text{MKS(i)} \tag{3.85}$$

instead of (3.34) for internal consistency. Then again, Feynman[2], working within the MKS system, prefers a third alternative, namely

$$\mathbf{B} = \mathbf{H} + \mu_0\mathbf{M}, \quad \text{MKS(ii)} \tag{3.86}$$

because the units of **B** and **H** are thereby made the same. Note that the choice (3.86) in both MKS(ii) and SI systems would replace (3.80) by

$$\text{curl}\,\mathbf{H} = \mu_0\mathbf{J}_{\text{free}}, \quad \text{MKS(ii)}, \tag{3.87}$$

so giving Ampère's law the same shape in terms of either **B** or **H**.

To summarize: apart from our choice of the names for **B** and **H**, which can hardly increase the level of confusion that already exists in this field, we adhere to the dimensional and unit definitions given in the first paragraph of this section.

3.12 The macroscopic mean field inside a magnetized sample

We consider now the field *inside* a magnetized body. Recall that, in §3.9, we demonstrated how the external magnetic field of a magnetized body is identical with that of an equivalent current ribbon around the periphery of the body. We cannot expect the same to be true of the *internal* fields, simply because, while that enclosed by a current ribbon will vary smoothly, the field within a molecular medium must vary sharply on the atomic scale. One thing that can be said of this *microscopic* field, however, is that it is simply a magnetic field in a vacuum, deriving from the assemblage

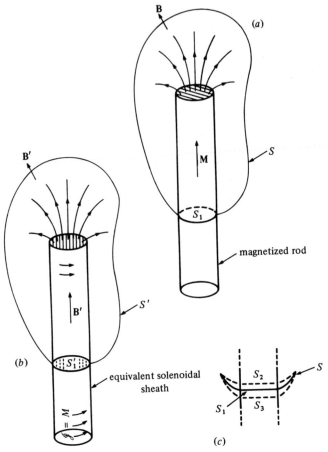

Fig. 3.11. The spatial average of the microscopic field **B** in the magnetized rod is equal to the macroscopic field **B** inside the equivalent current sheath.

of atomic and molecular magnetic dipoles, or current loops constituting the array we call matter: it must be described by the symbol **B**. Further, following our belief that magnetic charges (sources) do not exist, we expect that div **B** = 0 for this microscopic field. From this it follows that the *spatial average of the internal, microscopic magnetic field in a body is equal to the macroscopic field inside the equivalent current ribbon,* as we now show.

Consider a long rod of material magnetized uniformly and parallel to its length as in figure 3.11(*a*). The external field **B** produced by this magnet will be the same as that, **B'**, of the long cylinder of current shown in figure 3.11(*b*): if the intensity of magnetization in the rod is **M**, the surface current density in the equivalent current ribbon is given by $\mathcal{J} = \mathbf{M}$. Let us compare the total magnetic flux leaving identical surfaces S, S' which cut the magnetized rod, on the one hand, or the equivalent solenoidal current sheath, on the other. As all the magnetic fields under present consideration – internal, external, macroscopic and microscopic – satisfy $\operatorname{div} \mathbf{B} = 0$, Gauss' theorem requires that the surface integral over either closed surface must vanish. As the magnetic field *external* to the rod is identical to that external to the cylindrical current ribbon, the surface integral of the microscopic field over that internal portion S_1 of the surface dividing the rod must equal that of the macroscopic field over the equivalent surface S_1' cutting the inside of the current sheath. The surface integral of the microscopic internal field over a series of closely spaced surfaces $S_2, S_3, \ldots$, in the rod near to S_1, will be the same because the field outside a long solenoid (and hence rod) is negligibly small in the region shown in figure 3.11(c). As the mean surface integral over such a set of adjacent planes is equivalent to the volume average of the field $\mathbf{B}$ in that region, it follows overall that the spatial average of the microscopic field $\mathbf{B}$ in the magnetized rod is equal to the macroscopic field $\mathbf{B}'$ inside the equivalent current ribbon: formally, we write,

$$\bar{\mathbf{B}}_V = \frac{1}{V} \int_V \mathbf{B}\, dv. \tag{3.88}$$

This average field is a *macroscopic* quantity, relating to a volume of the material which is at least large enough to contain very many molecular magnetic dipoles – sufficient, for example, that in a uniformly magnetized sample, each elemental volume is equally magnetized and pervaded by an equal field. At the same time, the volume that $\mathbf{B}$ relates to can generally easily be small enough for the differential quantities div and curl to have meaning. The same is true for (3.79), (3.80) and all that goes with those definitions, so that, essentially by definition, the quantities $\mathbf{H}$ and $\mathbf{M}$ are also *macroscopic* entities, relating once more to volumes of material which are too small to be accessible by ordinary means, yet large enough to contain very large numbers of molecules. It is usual to omit the bar over $\bar{\mathbf{B}}$ when the context makes it clear that we imply that $\mathbf{B}$ inside matter is to be taken as the statistical average in (3.88).

3.13 Fictitious magnetic sources and the field H

It is to be emphasized immediately that we are **not** equating the field $\mathbf{H}$ with the spatial average $\mathbf{B}$ of the magnetic field in a sample. The distinction between $\mathbf{B}$ and $\mathbf{H}$ causes many problems and requires further discussion.

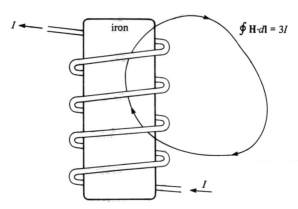

Fig. 3.12. A line integral of **H** is easily measured.

Firstly, we reiterate that the field **H** is an auxilary or derived vector field simply *defined* by the relation (3.79) and possessing the one certain property given by (3.80). One may discern two qualities of **H** which confer on it considerable utility. The first is a practical matter, relating to the integral form of (3.79); namely (3.82), which defines the line integral of **H** (*not* **B**) in terms of the *real*, free current enclosed by the path of that integral. That current is *measurable* using an ammeter so it is a simple task to determine the line integral of **H** around a chosen closed path. Consider, for example, an electromagnet consisting of many turns of conducting wire around an iron core. For a closed path which lies partly inside the iron core and partly outside and which encircles n segments of the solenoid, as in figure 3.12, the line integral of **H** around that path is given by nI amperes. By contrast, the equivalent line integral of **B** is much less readily determined. So (3.82) has utility. However, neither (3.82) nor its precursor (3.80) are to be misread as defining a vector field whose *source* is the free current density $\mathbf{J}_{free}$. Thus, while div **B** $= 0$ everywhere, div **H** has not been defined and is not necessarily zero. So a knowledge of the line integral of **H** from (3.82) does not imply a complete knowledge of **H** itself. To emphasize this point: considering a magnetized sample placed in a magnetic field in an experiment to measure the magnetic susceptibility, we would normally have the situation where there are no free currents in the sample at all and so curl **H** $= 0$. This does *not* mean, however, that **H** $= \mathbf{0}$: **H** will have magnitude and direction within a typical element of the sample, given by the defining relation (3.79).

The second useful aspect of the field **H** concerns a parallel between **H** and **E** that is not shown by **B** and **E**. We approach this by considering the boundary conditions imposed on **B** and **H** at the interface between

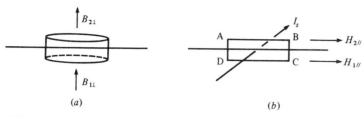

Fig. 3.13. (a) The normal components of the field **B** are continuous at a boundary; (b) the tangential components of **H** are continuous, but only in the absence of a surface current.

two different magnetic media (one of which may later be taken as a vacuum, of course). We apply Gauss' theorem to an elementary box which encloses a portion of the boundary between the different materials, as in figure 3.13(a). If the box is taken to approach the shape of a flat disc, with a height which is very small relative to its cross-section, the only contributions to the total flux arise from components of the magnetic field normal to the medium interface. Therefore

$$B_{1\perp} = B_{2\perp}, \tag{3.89}$$

so that the normal components of the magnetic field **B** at a boundary are continuous, a result which stems from the absence of magnetic sources. On the other hand, the boundary conditions for **H** are determined by applying Ampère's law (3.80) to an elemental rectangular circuit ABCDA whose sides AD, BC are very small compared with AB and CD, as shown in figure 3.13(b). In the presence of a surface current I_s per unit length of the medium interface, normal to the elemental circuit, (3.82) gives

$$H_{2\|} - H_{1\|} = I_s \tag{3.90}$$

for the components of $H_\|$ which are perpendicular to the surface current, while those components parallel to I_s are continuous. In vector form, we write

$$\mathbf{n} \wedge (\mathbf{H}_2 - \mathbf{H}_1) = \mathbf{I}_s, \tag{3.91}$$

where **n** is the unit normal to the media boundary, and corresponding to (3.89) we have

$$\mathbf{n} \cdot (\mathbf{B}_2 - \mathbf{B}_1) = 0. \tag{3.92}$$

In summary: the normal components of **B** are always continuous while the tangential components of **H** are continuous only in the absence of a surface current: these boundary conditions are similar to those for the interface between dielectrics in electrostatics[3]. It is **H** and **E** which behave similarly, however, not **B** and **E**. The analogy with electrostatics is enlarged in those cases (normal for our applications) when no real currents are present, for then curl **H** = 0 and *under these circumstances* we can write **H** as the gradient of a (magnetic) scalar potential ϕ_m,

$$\mathbf{H} = -\operatorname{grad}\phi_m, \tag{3.93}$$

analogous to (3.2). Now, by substitution into (3.79), we find

$$\operatorname{div}\mathbf{B} = -\mu_0\operatorname{div}\operatorname{grad}\phi_m + \mu_0\operatorname{div}\mathbf{M} = 0, \tag{3.94}$$

and so

$$\operatorname{div}\operatorname{grad}\phi_m \equiv \nabla^2\phi_m = \operatorname{div}\mathbf{M}. \tag{3.95}$$

Compare this relationship with Poisson's equation for electrostatics,

$$\nabla^2\phi = -\rho/\varepsilon. \tag{3.18}$$

which, using Gauss's law (2.24), may be written

$$\nabla^2\phi = -\operatorname{div}\mathbf{E}. \tag{3.96}$$

Apart from the sign difference between (3.95) and (3.96), which arises from the *conventions* adopted in defining electric and magnetic potentials (ϕ and ϕ_m) and ultimately in the choice of a positive sign for **M** in paramagnetic materials, the equations are isomorphous. As already discussed in §§3.1 and 3.2, electrostatic problems are usually rather easier to solve than magnetostatic ones, so that this isomorphism has considerable practical convenience.

The comparison between (3.95) and (3.96) also suggests the use of a simple fiction for those who like to associate a physical model with their mathematical equations. It is to note that the electric–magnetic analogy between **H** and **E** would be almost complete if we were to imagine magnetic monopoles as sources for **M** in (3.95) as electric charges are the sources for **E** in (3.96). Actually, in some older texts, magnetic sources were accepted as real and **H** was introduced as the primary magnetic field. In turn, **B** was defined as in (3.83) and called the *magnetic induction*. Some contemporary authors still use this name, even while treating **B** as the *primary* field, because the name 'magnetic field' was preempted by **H**. This all adds to the confusion which has attached to these quantities and, in the present book, we shall not use the term magnetic induction in this way at all.

As an example of regarding the field **H** as arising from magnetic pole sources, we review the boundary conditions for **B** and **H** discussed above. There are no true magnetic sources and so the component of magnetic

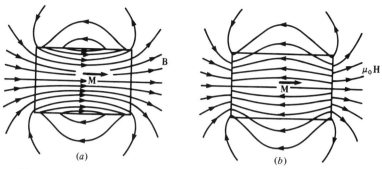

Fig. 3.14. A magnetized cylinder *in vacuo*, showing (*a*) the field lines of **B**, and (*b*) of μ_0**H**. The external fields are identical. The internal field in (*b*) may be viewed as arising from fictitious magnetic poles assembled at the ends of the cylinder: field-direction arrows therefore consistently point towards north poles and away from south. The boundary conditions of figure 3.13 are manifest throughout this diagram.

field **B** perpendicular to a medium boundary is continuous, as in (3.89). However, if we regard a bar magnet, say, as containing magnetic north poles at one end and south poles at the other, as was once believed, the field **H** so established will have all the same geometric properties as the field **E** defined in an electret (the electrostatic equivalent of a magnet): in particular, **H** will be discontinuous normal to the ends of the magnet just as **E** would be discontinuous at a surface layer of electric charge. In figure 3.14 are shown field lines of (*a*) **B** and (*b*) μ_0**H** both inside and outside a (given) uniformly magnetized cylinder. Note straightway that the external field lines appear identical[††] (external meaning *in vacuo* here). So they should, for this is a manifestation of the way the *far* fields of electric and magnetic dipoles look alike, as shown in figure 3.4, and the 'sources' of **H** here resemble an electric dipole. Incidentally, 'far' in the present context only means relative to molecular dimensions, for the total field anywhere arises, of course, from a superposition of myriads of molecular dipoles: so the external fields assume their form within, perhaps, 20 Å of the surface of the magnet. The internal fields are both *macroscopic* quantities, once more representing averages of microscopic fields as in (3.88) for **B**; although one cannot meaningfully talk of a microscopic field **H** anyway. Now note the manifestation of the boundary conditions of (3.91) and (3.92) in figure 3.14. In truth, there are no accumulations of magnetic charge at the ends of the cylinder and the magnetic field **B** is equivalenced by a similarly-sized solenoid: there are no real or equivalent currents at the

[††] Only when **H** is scaled by μ_0 in this way, because of (3.81).

ends of the magnet and the **B** field lines are continuous there. By contrast, field lines of **H** associated with fictitious magnetic poles are necessarily discontinuous where these are accumulated at the ends of the magnet, but continuous around the cylindrical surface where there are presumed to be no equivalent magnetic charges.

3.14 Definitions of magnetic susceptibility and permeability

Let us look ahead for a moment and identify our interest in magnetostatistics as magnetochemist users. It is surely twofold. We shall be concerned to measure the magnetic susceptibility of a substance, a process which ultimately devolves on the determination of the magnetic moment of a body by measuring, for example, the force or torque on that macroscopic magnetic dipole in an applied field. Having accomplished that, it will be our business to provide an interpretation of the results via a model, one aspect of which will involve reference to the local magnetic field experienced by a representative molecule: we must therefore address the question of defining such a local field. Progress on both these matters relies on *empirical* observation. First, in the present section, we consider the definition of magnetic susceptibility.

Excluding ferromagnets and considering, in particular, diamagnets and paramagnets, it is *observed* that under most circumstances, magnetization is proportional to field strength. In less common circumstances – for example, those involving samples at very low temperatures in very strong magnetic fields – saturation effects occur which progressively destroy this linear relationship but, unless specifically stated, we shall not be concerned with those cases. The more usual proportionality *could* be expressed as

$$\mathbf{M} = \chi' \mathbf{B} \qquad (3.97)$$

but is normally presented as

$$\mathbf{M} = \chi \mathbf{H}, \qquad (3.98)$$

where χ is called the *magnetic susceptibility*. Note that throughout the present discussion we shall assume that we are dealing with isotropic media only: the case of generally anisotropic materials is dealt with at length in the following and later chapters. We emphasize that (3.98) is an *empirical* relationship. However, **B** and **H** are related by *definition*, (3.79), and so we have

$$\mathbf{B} = \mu_0(\mathbf{H} + \mathbf{M}) = \mu_0 \mathbf{H}(1 + \chi)$$
$$= \mu_r \mu_0 \mathbf{H} = \mu \mathbf{H}, \qquad (3.99)$$

which expressions serve to define μ as the *magnetic permeability* of the medium and μ_r,

$$\mu_r = \mu/\mu_0 = 1 + \chi, \tag{3.100}$$

as the *relative magnetic permeability*. More interesting than these definitions, however, is the linear relationship between **B** and **H** that emerges in isotropic media obeying (3.98). Thus, comparing (3.97) and (3.99), we have

$$\chi' = \chi/\mu, \tag{3.101}$$

an expression which serves to confirm that either definition of susceptibility – (3.97) or (3.98) – is acceptable in principle: we reiterate, however, that (3.98) is that actually chosen.

The susceptibility defined by (3.98) is more properly called a *volume susceptibility*, related as it is to the intensity of magnetization **M**, describing the magnetic dipole moment per unit volume. As **M** and **H** have the same dimensions, from (3.79), volume susceptibility in the SI system is a dimensionless ratio. The same is not true within the MKS(i) scheme, defined by (3.84), of course. It *is* the case, however, in the CGS system but we note from (3.83) that the volume susceptibility (per cubic metre) in the SI system is a number larger by a factor of 4π than the corresponding number (per cubic centimetre) in the CGS emu scheme.

Mass susceptibility is defined as

$$\chi_m = \chi/d, \tag{3.102}$$

where d is the density of the material. The unit of χ_m in the SI system is therefore metre^3/kilogram. More useful in magnetochemistry is the *molar susceptibility*, defined by

$$\chi_M = \chi_m W_M = \chi W_M/d, \tag{3.103}$$

where W_M is the molecular weight of the material. The unit of χ_M in the SI system is metre^3/mole and molar susceptibilities in the SI system are $4\pi \times 10^{-6}$ times values given in CGS emu (cm^3/mole).[††]

3.15 A magnetizable sphere in a uniform field

We shall shortly look at the second way in which empirical observation makes conventional magnetochemical measurements sensible and useful.

[††] To check this relationship, compute the volume susceptibility for 1 cm^3 in the SI system, and then for N_A molecules (where N_A, Avogadro's number, is the same in either system).

A path is conveniently prepared by considering the magnetic analogue of a system that is well documented in standard texts[1,2,3] on electrostatics; namely, the fields associated with a polarizable sphere.

We place a sphere of a homogeneous magnetizable material in a uniform magnetic field. One characteristic of the latter, of course, is that its sources are presumed to be unaffected by the presence of the sample. We label this *applied* magnetic field $\mathbf{B}_0$. The spherical body becomes magnetized and so generates an *induced* field $\mathbf{B}'$. The *total* field, $\mathbf{B}$, in the vicinity of the sphere is no longer necessarily uniform and is given as the sum of the applied and induced fields,

$$\mathbf{B} = \mathbf{B}_0 + \mathbf{B}'. \tag{3.104}$$

The field $\mathbf{B}'$ depends upon the magnetization $\mathbf{M}$ of the material of the sample, which in turn depends upon the value of the total field $\mathbf{B}$ inside the sphere: and we do not yet know what that total field is. Note, therefore, the cyclic or self-consistent nature of the calculation that is necessary. Two tactics are useful here. The first is to simplify the process by making it virtually identical to that followed for the electrostatic analogue (even though *we* have not studied that) by setting up the problem again but in terms of the field $\mathbf{H}$ rather than $\mathbf{B}$, because of the very close analogy of $\mathbf{H}$ with $\mathbf{E}$, as discussed in §3.13. In so doing, we illustrate how all problems involving $\mathbf{B}$ can be expressed in terms of the associated $\mathbf{H}$ field when desired: there is no question of one basis being correct and the other incorrect. Analogous to (3.104), we now have

$$\mathbf{H} = \mathbf{H}_0 + \mathbf{H}', \tag{3.105}$$

although, to be pedantic, one would write (3.105) multiplied throughout by μ_0 in order to preserve the identity of $\mathbf{B}_0$ and $\mu_0\mathbf{H}_0$ outside the sample, which we presume to be *in vacuo*. Note, however, that (3.81) does not hold inside the sample so that neither $\mathbf{H}$ nor $\mathbf{H}'$ are related to $\mathbf{B}$ and $\mathbf{B}'$ by this simple proportionality. In other words, (3.105) represents a physically different problem to (3.104), though its solution is all we require. However, while we make this point, let us also note that it is perfectly proper to mean by 'the applied field', *either* $\mathbf{B}_0$ or $\mathbf{H}_0$, as convenient.

The second tactic we employ in the solution of (3.105) is to guess the answer and thence proceed by verification. Let us assume that the sphere becomes *uniformly* magnetized, parallel to the direction of $\mathbf{H}_0$. For the moment, then, we confine our attention to the magnetized sphere and so may temporarily disregard the origin of that magnetization: what is called the 'induced field' in the complete problem is to be regarded simply as

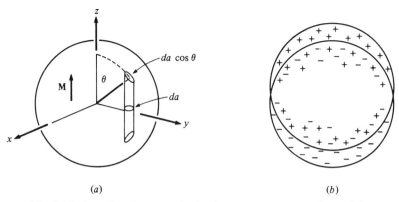

(a) (b)

Fig. 3.15. A uniformly magnetized sphere: two representations of the equivalent surface pole distribution varying as $1/\cos\theta$. + and − signs signify north and south magnetic monopoles.

the 'field H'' of the magnetized sphere, for our present purposes. As we are working with H' rather than B', we can refer to magnetic poles or charges as a working fiction. So, if we divide up the magnetized sphere into rods parallel to the magnetization vector M in the body as shown in figure 3.15(a), we can equivalence each of these rods by magnetic charge, of magnitude ($M \times$ rod cross-section), placed at the ends of the rod. Therefore the induced field H', *produced by the magnetized body*, is that of a surface magnetic charge distribution with density $M\cos\theta$ over the surface of the sphere. The factor $\cos\theta$ arises from the fact that the area at the ends of the elemental rod varies as $1/\cos\theta$ as the elemental surface moves from the axis of magnetization. An alternative representation of this surface magnetic pole distribution is shown in figure 3.15(b) which, by emphasizing how the z components of the elemental surfaces vary with $1/\cos\theta$, demonstrates how the whole magnetization configuration may be *equivalenced* by a superposition of two spheres of magnetic charge, one of north poles and the other of south, but displaced relative to each other parallel to the direction of magnetization. Now the external field of a spherical electric charge distribution is the same as if the charge were accumulated at its centre; and the same applies for magnetic charge distributions which are similarly subject to an inverse square law of interaction. Therefore, the uniformly magnetized sphere produces an *external* field, H'_{out}, which is the same as that of two (closely spaced) magnetic charges of opposite polarity; in short, of a magnetic dipole. The magnitude of this equivalent magnetic dipole $\mathscr{M}$ is just equal to the volume of this uniformly magnetized body times its intensity of magnetization,

$$\mathcal{M} = \frac{4\pi a^3}{3} M, \tag{3.106}$$

and its sense will be parallel to **M** (and to the applied field **H**$_0$ as we define M positive for a paramagnet). Hence, that part of the induced field external to the sphere, **H**$'_{out}$, is that of a central magnetic dipole $\mathcal{M}$: and this is so to within molecular dimensions of the surface of the sphere because, as usual, we know the body surface to be built from a myriad of small, current loops of those dimensions.

The internal (induced) field, **H**$'_{in}$, is different because the uniformly magnetized sphere is not *really* a magnetic dipole: it only *appears* so from the *outside*. We can determine the internal field from a consideration of the magnetic potential ϕ_m. The potential on the boundary of the sphere may be seen as giving rise to the field **H**$'$ both outside and inside the sphere. The external field is that of a dipole and so the surface potential must be a dipolar potential of the form **m**· **r**$/4\pi r^3$ which in the present case takes the form, for points on the boundary of the sphere, using (3.106),

$$\phi_m = \frac{\mathcal{M}\cos\theta}{4\pi a^2} = \frac{M}{3}a\cos\theta = \frac{M}{3}z. \tag{3.107}$$

Thus we see that the magnetic potential on the surface of a uniformly magnetized sphere depends only on the coordinate parallel to the direction of magnetization. Equivalently, the potential difference between any two points on the surface of the sphere is a function of their difference in that coordinate. Using (3.93), we find that the magnitude of the internal field of the uniformly magnetized sphere is given by

$$H'_{in} = -\frac{\partial \phi_m}{\partial z} = -M/3 \tag{3.108}$$

or, since the direction of magnetization determines the direction of **H**$'_{in}$,

$$\mathbf{H}'_{in} = -\mathbf{M}/3. \tag{3.109}$$

The configurations of **H**$'$ both inside and outside the uniformly magnetized sphere are shown in figure 3.16. Again, we emphasize that this field is a property of this body, regardless of the origin of its magnetization.

Now we return to the original problem in (3.105). *If* the applied field **H**$_0$ induces a magnetization in the sphere which is uniform, then **H**$'$ of (3.105) is given within the sample by (3.109) and so,

$$\mathbf{H} = \mathbf{H}_0 - \mathbf{M}/3. \tag{3.110}$$

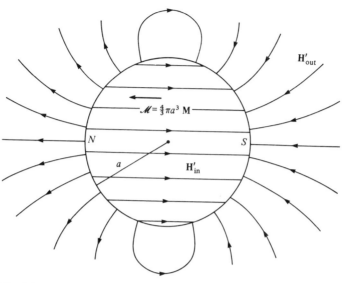

Fig. 3.16. The internal and external field **H′** produced by a uniformly magnetized sphere.

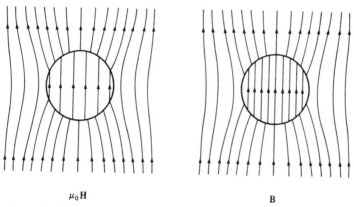

Fig. 3.17. The total internal and external fields associated with a homogeneous spherical paramagnet in a uniform applied field. The field **B** is augmented while the field **H** is diminished.

But $\mathbf{M} = \chi\mathbf{H}$ from (3.98) where **H** is the *total* field causing the magnetization **M**: hence, using the definition (3.100) we find

$$\mathbf{H} = \left(\frac{3}{2 + \mu_r}\right)\mathbf{H}_0. \tag{3.111}$$

So we have the result that the total field inside the sphere is simply proportional to the applied field. However, the applied field was given as uniform and therefore so also is the total field. In turn, the magnetization of the sample is uniform and so we observe that the assumption providing (3.110) is consistent with the result (3.111).

Note that, since $\chi > 0$ for a paramagnet and hence $\mu_r > 1$, the total field **H** inside a spherical isotropic paramagnet is less than the applied field $\mathbf{H}_0$: conversely, (3.111) gives $\mathbf{H} > \mathbf{H}_0$ for a diamagnet. The opposite situation pertains for the magnetic fields **B** and $\mathbf{B}_0$, however, as is readily seen from (3.108) and the definition (3.79). These different situations are shown in figure 3.17, the external field **H** having been established by superposition of the applied field $\mathbf{H}_0$, together with $\mathbf{H}'_{out}$, the dipolar field of $\mathcal{M}$ in (3.106). Note that, like the situation in figure 3.14, the fields we compare here are **B** and $\mu_0\mathbf{H}$ and again, outside the spherical body (presumed to be a vacuum), these two fields are the same.

3.16 The shape of a magnetizable sample

The self-consistent nature of the analysis we have just studied indicates a considerable technical difficulty which must be solved in principle if any practical measurement of susceptibility is to be completed. It is generally the case that few bulk shapes, *even* of magnetically isotropic materials, are uniformly magnetized *even* in a uniform applied field. It occurs, in fact,[3] only in ellipsoidally shaped samples, of which the sphere, flat disc and needle are limiting cases (the latter two, for extreme oblate and prolate ellipsoids respectively). Note, in passing, that the situation shown in figure 3.14 cannot occur for an induced magnetization, though it can be closely approximated in a good permanent magnet.[2]

While we leave to chapter 5 the business of the experimental measurement of susceptibility, we may note here that most such measurements derive from a determination of the magnetic moment $\mathcal{M}$ of the whole sample as a discrete object, followed by a division by the volume or mass of the sample on the assumption that the magnetization is uniform. If such is not the case, we shall merely obtain some sort of average which then may be compared with results from theoretical models only in a qualitative and generally unsatisfactory way. Alternatively, we must solve the problem (3.105) explicitly for the particular shape of the sample to hand: yet again, we might consider grinding all samples into a spherical shape. In fact, almost no heed is paid to this problem at all for, in practice, a second *empirical* feature of magnetic susceptibility allows an important simplification to be made. It is that, *unlike* electric susceptibilities, magnetic

Fig. 3.18. **B** and μ_0**H** are very nearly equal in magnitude and direction.

susceptibilities are very small ($\chi \ll 1$) so that the difference between the applied and total fields can be neglected in most circumstances. We sketch this in figure 3.18 to emphasize that the empirical idealization that

$$\mathbf{B} \approx \mathbf{B}_0; \quad \mathbf{H} \approx \mathbf{H}_0 \qquad (3.112)$$

does not imply any neglect of **B**′ or **H**′, of course: the situation is akin to the approximation $\sin \theta \approx \theta$ for small θ. The approximation in (3.112) is to be understood to refer to both magnitude and direction of the relevant vectors. Hence the total field in a sample placed in a uniform applied field is also uniform in this approximation, and the same is therefore true of the induced magnetization. Clearly, (3.112) represents a tremendous simplification, one that is not available for ferromagnets, of course. Although the approximation is satisfactory for paramagnets and dia-magnets with a high degree of accuracy, some experiments can be performed which reveal the approximation even qualitatively. For example, an asymmetrically shaped crystal of a magnetically isotropic material can suffer a torque in a homogeneous field: the effect is small but makes the point. These matters are discussed a little further in chapter 5.

3.17 The local field

We have a similar problem when we come to enquire about the field experienced by an individual molecule in a sample. This, of course, is the field which must enter the quantum-mechanical models we describe later. Strictly it must be a microscopic field because, on the scale of the reference dipole, molecular sources give rise to the rapidly modulatory, free-space field. So, as discussed in §3.12, we are concerned with a field **B**. Moreover, the magnetic dipole moment induced in a molecule of a diamagnet on the one hand, or the degree of alignment of the permanent magnetic dipole of a paramagnet on the other, are determined by a magnetic field which arises from all sources *other* than that of the reference molecule. The reference is subject, therefore, to a microscopic field given by

$$\mathbf{B}_{\text{rest}} = \mathbf{B}_{\text{mic}} - \mathbf{B}_{\text{ref}} \qquad (3.113)$$

where $\mathbf{B}_{\text{mic}}$ is the total microscopic field of (3.88), $\mathbf{B}_{\text{ref}}$ is that of the reference dipole itself, and $\mathbf{B}_{rest}$ includes both the microscopic fields of all other induced or aligned dipoles *and* the applied field $\mathbf{B}_0$. Once again, therefore, we have a potentially difficult problem in determining the precise field acting on a molecular dipole: here the result depends, not only upon the volume and shape of the crystal as before, but also upon the detailed crystal and molecular structure. However, it is again quite sufficiently satisfactory for our purposes to use the empirical fact that χ is very small and so recognize that $\mathbf{B}_0 \gg \mathbf{B}'$. The difference between $\mathbf{B}_{rest}$ and $\mathbf{B}_{mic}$ is therefore, trivial and, given that the sources of $\mathbf{B}'$ are ultimately just the molecular dipoles with which we are concerned, it is clearly reasonable also to take $\mathbf{B}_{rest} \approx \mathbf{B}_{mic}$. In summary therefore: we shall write equations concerned directly with molecular dipoles in terms of a microscopic $\mathbf{B}$ field. The connection with experimental results is then made by using the approximate equalities

$$\mathbf{B}_{rest} \approx \mathbf{B}_{mic} \approx \bar{\mathbf{B}}_{mic} = \mathbf{B} \approx \mathbf{B}_0 = \mu_0 \mathbf{H}_0 \approx \mu_0 \mathbf{H}, \qquad (3.114)$$

in which only the first two quantities are true microscopic magnetic fields. The practical use of (3.114) means that one should not take too pedantic a stand on 'molecule-scaled' equations being written in terms of $\mathbf{B}$ rather than $\mathbf{H}$ (provided μ_0 is included in the SI scheme) for, although only $\mathbf{B}$ can refer to a microscopic field, no significant differences in numerical results of susceptibility calculations will occur if the macroscopic field $\mathbf{H}$ is used instead.

3.18 Generalized coordinates; the Lagrangian and Hamiltonian functions

We leave the subject of the magnetic fields to which microscopic and macroscopic objects are subject and consider the connection between statistical descriptions of molecular properties and the experimental quantity we call susceptibility. In preparation for this and, indeed, for many other matters, it is useful at this point to summarize, very briefly, some principles of classical mechanics expressed within so-called 'generalized coordinates'.

In classical mechanics, Newton's law of motion determines the motion of a particle, for example, according to the relationships

$$m\frac{d^2x}{dt^2} = f_x, \quad m\frac{d^2y}{dt^2} = f_y, \quad m\frac{d^2z}{dt^2} = f_z. \qquad (3.115)$$

It is not always convenient, however, to work with these cartesian coordinates and it is well known that, for complicated problems involving many particles, some of which may be connected rigidly or non-rigidly, the simplicity of Newton's law can easily be obscured. Using generalized coordinates, it is possible to construct a small number of powerful expressions of so general a nature that they are immediately applicable in any coordinate scheme. We do not take the space to develop these equations here, however: instead, we quote and review them, meanwhile referring the reader to the literature.[6, 7] As *we* shall really only use these expressions in a small number of closely related special circumstances, it may seem perverse to introduce the generalized methods at all. We do so, however, partly in the belief that the technique will be familiar to many anyway, and partly so as to make a fuller contact with the standard literature of magnetism and, ultimately, of quantum mechanics.

In general, the various particles in a classical system may not be totally free: the coordinates of some may be fixed relative to this because they form parts of a rigid body; or two or more particles may be constrained to move on some fixed surface; and so on. In three-dimensional space, a free particle has three degrees of freedom. A system of n particles will generally possess k degrees of freedom ($k < 3n$) and so be subject to $3n - k$ independent constraints of one kind or another. Hence, only k independent coordinates are required to describe the motion of the system. Generally, the $3n$ cartesian coordinates of such a system, or the $3n$ coordinates of any other type, may be written in terms of an unspecified set of so-called *generalized coordinates* $\{q\}$,

$$x_i = \phi_i(q_1, \ldots, q_{3n}), \tag{3.116}$$

where i runs over all coordinates (x, y and z, in the cartesian system, of all particles) and there will generally be $3n - k$ equations of constraint between these qs. Newton's equation now takes the form

$$m\frac{d^2 x_i}{dt^2} \equiv m\ddot{x}_i = f_i. \tag{3.117}$$

We now introduce the kinetic energy T of the system,

$$T = \sum_i \tfrac{1}{2} m\dot{x}_i^2, \tag{3.118}$$

the potential energy V and finally the combination L of these,

$$L = T - V, \tag{3.119}$$

called the *Lagrangian function*. One of the major expressions of Newton's law (3.117), in terms of generalized coordinates q, is given by the *Lagrangian equations* of motion, the set of k equations,

$$\frac{d}{dt}\frac{\partial L}{\partial \dot{q}_i} - \frac{\partial L}{\partial q_i} = 0, \quad i = 1, \ldots, k. \tag{3.120}$$

These equations express Newton's laws in a powerful and generalized way. Hamilton derived an alternative set of generalized equations in terms not just of generalized coordinates but also of generalized momenta. In this way the k second-order Lagrangian equations are transformed into $2k$ first-order differential equations, as we now review.

Generalized momenta p_i may be introduced as variables of the system which are *canonically conjugate* with respect to the generalized coordinates q_i, by which is meant that

$$p_i = \frac{\partial L}{\partial \dot{q}_i}, \quad \dot{p}_i = \frac{\partial L}{\partial q_i}. \tag{3.121}$$

For example, the generalized momentum of a free particle moving in the cartesian x direction is given from (3.121) by

$$p_x = \frac{\partial L}{\partial \dot{x}} \rightarrow \frac{\partial T}{\partial \dot{x}} = \frac{\partial}{\partial \dot{x}}(\tfrac{1}{2}m\dot{x}^2) = m\dot{x}, \tag{3.122}$$

as usual. The *Hamiltonian function*, $\mathscr{H}$, is defined by

$$\mathscr{H} = \sum_{i=1}^{k} p_i \dot{q}_i - L \tag{3.123}$$

and the magnitude of $\mathscr{H}$ is given by $T + V$, that is, the total energy of the system, though $(T + V)$ does *not define* $\mathscr{H}$ and must be used with care. From (3.123) and (3.121) we find

$$d\mathscr{H} = \sum_{i}^{k} (\dot{q}_i dp_i - \dot{p}_i dq_i), \tag{3.124}$$

and so, by the definition of a partial derivative, there result *Hamilton's equations of motion*:

$$\frac{\partial \mathscr{H}}{\partial p_i} = \dot{q}_i, \quad \frac{\partial \mathscr{H}}{\partial q_i} = -\dot{p}_i, \quad i = 1, \ldots, k. \tag{3.125}$$

Either (3.120) or (3.125) may be used to express Newtonian mechanics in a generalized way, and they do so in terms of energies rather than forces, a feature which is especially valuable in making the transition from Newtonian to quantum mechanics. We apply these equations now to the case of electromagnetism.

3.19 The classical dynamics of molecular electrons in external fields

In (3.8) we expressed the Lorentz force equation (2.52) in terms of the electric and magnetic potentials to which a charge Q is subject. If that particle is a molecular electron we may usefully discriminate between those potentials that are externally applied and those that arise from within the molecule itself. We may safely neglect molecular magnetic forces along with any relativistic corrections and so retain only internal electric potentials (this idea fits well with the observed, relatively very small value of magnetic susceptibilities discussed earlier). If we express the internal electric potential for the ith electron[†] by

$$\phi_i^{\text{int}} = \frac{1}{4\pi\varepsilon_0} \sum_{j \neq i} \frac{Q_j}{r_{ij}}, \tag{3.126}$$

using (2.44), the potential energy V may be written as

$$V = -\frac{1}{4\pi\varepsilon_0} \sum_{i<j} \frac{eQ_j}{r_{ij}} = -\frac{1}{8\pi\varepsilon_0} \sum_j e\phi_j^{\text{int}}, \tag{3.127}$$

where e is the modulus of the electronic charge and Q_j is the charge on the jth charged particle. The Lorentz force equation we seek, therefore, is for the ith electron and, using (3.8),

$$\mathbf{F}_i = -\operatorname{grad}_i V + e\operatorname{grad}\phi_i + e\frac{\partial \mathbf{A}_i}{\partial t} - e\mathbf{v} \wedge \operatorname{curl} \mathbf{A}_i, \tag{3.128}$$

where the first term refers to the internal electric potential energy (the suffix implying differentiation with respect to the coordinates of the ith electron in the molecule) and the rest to the externally applied electric and magnetic potentials. Equating force to mass times acceleration, the x component of $\mathbf{F}_i$ in cartesian coordinates, for example, is

$$m_e \ddot{x}_i = -\frac{\partial V}{\partial x_i} + e\frac{\partial \phi_i}{\partial x_i} + e\frac{\partial A_{x_i}}{\partial t}$$

$$- e\left[\dot{y}_i\left(\frac{\partial A_{y_i}}{\partial x_i} - \frac{\partial A_{x_i}}{\partial y_i}\right) - \dot{z}_i\left(\frac{\partial A_{x_i}}{\partial z_i} - \frac{\partial A_{z_i}}{\partial x_i}\right)\right], \tag{3.129}$$

where m_e is the mass of an electron.

[†] In that the nuclei in molecules are so massive compared with electrons, we neglect their contribution to the molecular magnetic moment and susceptibility in the present treatment. We need retain mention of them only in (3.126) and (3.127) in defining the internal electric potential to which the electrons are subject, which is, as we see below, virtually irrelevant to *our* purposes anyway.

These differential equations of motion, together with those for the y and z components, form a set equivalent to the Lagrangian equations (3.120) if the Lagrangian function is taken as

$$L = \tfrac{1}{2}\sum_i m_e v_i^2 - V + \sum_i e\phi_i - \sum_i e\mathbf{v}_i \cdot \mathbf{A}_i. \tag{3.130}$$

This may be verified by specializing the k ($= 3n$, where n is the number of electrons in the molecule) generalized coordinates q of (3.120) to the cartesian set $x_1, \ldots, x_n,\ y_1, \ldots, y_n,\ z_1, \ldots, z_n$ and by using the relationships

$$\frac{\partial}{\partial x_i}(\mathbf{v}_i \cdot \mathbf{A}_i) = \dot{x}_i \frac{\partial A_{x_i}}{\partial x_i} + \dot{y}_i \frac{\partial A_{y_i}}{\partial x_i} + \dot{z}_i \frac{\partial A_{z_i}}{\partial x_i} \tag{3.131}$$

and

$$\frac{d}{dt}\frac{\partial}{\partial \dot{x}_i}(\mathbf{v}_i \cdot \mathbf{A}_i) = \frac{dA_{x_i}}{dt} = \frac{\partial A_{x_i}}{\partial t} + \dot{x}_i \frac{\partial A_{x_i}}{\partial x_i} + \dot{y}_i \frac{\partial A_{x_i}}{\partial y_i} + \dot{z}_i \frac{\partial A_{x_i}}{\partial z_i}. \tag{3.132}$$

These expressions are obtained by recalling (i) the nature of the partial derivatives which are to operate here within the coordinate set $\{q_1, \ldots, q_k;\ \dot{q}_1, \ldots, \dot{q}_k\}$ specialized to the cartesian system, and (ii) that the last part of (3.132) arises out of the fact that in the total differentiation with respect to t, the vector potential $\mathbf{A}$ must be considered to involve time implicitly through the positional coordinates x, y, z as well as explicitly through t. Direct substitution of (3.131) and (3.132) into (3.120) yields (3.129).

We now construct the form of the *generalized* momentum p_i for the ith electron in the atom from its definition (3.121) and the Lagrangian (3.130): by differentiation we immediately find that

$$p_{x_i} = m_e \dot{x}_i - eA_{x_i}. \tag{3.133}$$

Note that this generalized momentum, the canonical variable to x_i, is *not* the same as the ordinary momentum $m_e v_i$, except in the absence of a magnetic field or of a charge on the particle (electron).

Now, using the definition of the Hamiltonian function (3.123), together with the Lagrangian (3.130) and the generalized momenta (3.133), we obtain the Hamiltonian function of the n electrons within a molecule exposed to electric and magnetic fields as

$$\mathscr{H} = \sum_i^n \frac{1}{2m_e}[(p_{x_i} + eA_{x_i})^2 + (p_{y_i} + eA_{y_i})^2 + (p_{z_i} + eA_{z_i})^2] + V - \sum_i^n e\phi_i. \tag{3.134}$$

This expression is appropriate for the most general case of variable and inhomogeneous applied electric and magnetic fields, while our interests lie in the situation described by a time-independent, uniform magnetic field only.[††] A convenient, though non-unique, way of expressing this is to take the potentials

$$\phi = 0, \quad A_x = -\tfrac{1}{2}By, \quad A_y = +\tfrac{1}{2}Bx, \quad A_z = 0. \tag{3.135}$$

These clearly correspond to a vanishing applied electric field and by the definition $\mathbf{B} = \text{curl } \mathbf{A}$, to a uniform magnetic field of magnitude B, oriented parallel to z. On substitution of (3.135) into (3.134), we obtain

$$\mathscr{H} = \sum_i^n \left[\frac{1}{2m_e}(p_{x_i}^2 + p_{y_i}^2 + p_{z_i}^2) + \frac{eB}{2m_e}(x_i p_{y_i} - y_i p_{x_i}) + \frac{e^2 B^2}{8m_e}(x_i^2 + y_i^2) \right] + V \tag{3.136}$$

as the Hamiltonian function for n molecular electrons in a uniform external magnetic field along z. However, the magnitude of $\mathscr{H}$ is just the total energy of the system, of course, and can therefore be expressed as the sum of kinetic plus scalar potential energy:

$$T + V = \tfrac{1}{2} \sum_i^n m_e v_i^2 + V, \tag{3,137}$$

which does not involve B. Thus there seems to be a conflict between the expressions (3.136) and (3.137). Two points can be made here. Firstly, we note that *magnetic forces do no work* as follows immediately from the Lorentz force equation (2.52) in that $(\mathbf{v} \wedge \mathbf{B})$ is a vector directed normal to $\mathbf{F}$. Therefore, no difference in the energy arises from the extra terms in (3.136) involving $\mathbf{B}$. If it be objected that the Zeeman effect demonstrates energy changes in a magnetic field, it is to be noted that, rather like the discussion in §3.8, the *achievement* of a given magnetic field *must* proceed via a period in which the field changes. During that period, Faraday's law (2.87) requires that there arise a concomitant electric field; which field does precisely that amount of work involved in the observed Zeeman effect. The second point counters any objection one may have about the inclusion of terms in $\mathscr{H}$ of (3.136) which do no work. It is that the canonical momenta required to preserve the form of the Hamiltonian equations are defined by (3.121) as $\partial L / \partial \dot{q}_i$ and *not* $\partial T / \partial \dot{q}_i$; and in a magnetic field, this means p_i is given by (3.133) rather than by $m_e \dot{x}_i$.

We now consider the change in the Hamiltonian as a function of $\mathbf{B}$ by

---

[††] This is true even in the Faraday experiment involving applied fields with non-zero gradients. This is because, as usual, the molecules under discussion here fall within an 'elemental' volume (containing many, many molecules, nevertheless) in which the applied field may be considered uniform.

differentiating (3.136) with respect to the applied field:

$$\frac{\partial \mathcal{H}}{\partial B} = \sum_i^n \left[\frac{e}{2m_e}(x_i p_{y_i} - y_i p_{x_i}) - \frac{e^2 B}{4m_e}(x_i^2 + y_i^2) \right], \qquad (3.138a)$$

$$= \sum_i^n \frac{e}{2}(x_i \dot{y}_i - y_i \dot{x}_i) = -m_z, \qquad (3.138b)$$

using (3.133), (3.135); and (3.54), written as

$$\mathbf{m} = -\frac{e}{2m_e}\mathbf{L} = -\frac{e}{2}\sum_i^n (\mathbf{r}_i \wedge \mathbf{v}_i). \qquad (3.139)$$

Thus, the magnitude of the total molecular magnetic moment in the direction of an applied magnetic field in terms of the classical theory of bound electrons is just

$$\frac{\partial \mathcal{H}}{\partial B} = -m. \qquad (3.138c)$$

Note again the distinction between the 'true' momenta $m_e \dot{x}$ and the canonical momenta p, as described in (3.133). In (3.138a) the canonical momenta appear explicitly and we observe the distinction between para- and diamagnetic contributions, whereas this separation is lost in (3.138b) which serves to define the total molecular magnetic moment – permanent plus induced.

3.20 Elements of Langevin's susceptibility formula

We return at last, therefore, to the question left at the end of §3.8 concerned with the relationship between induced or permanent molecular dipole moments and the macroscopic property of susceptibility. As advertised there, we shortly look at how a consistent application of classical statistical thermodynamics predicts identically zero magnetism. Let us begin by summarizing what we need to know about the Boltzmann distribution function.

If $q_1, \ldots, q_f, p_1, \ldots, p_f$ are a set of canonical generalized coordinates and momenta relating to the various particles in a molecule, the probability P that a molecule be in the configuration $(q_1, q_1 + dq_1), \ldots; \ldots, (p_f, p_f + dp_f)$, where $(q_i, q_i + dq_i)$ means that q_i lies between q_i and $q_i + dq_i$ etc., is given by

$$P = Ce^{-\mathcal{H}/kT}dq_1, \ldots, dq_f dp_1, \ldots, dp_f, \qquad (3.140)$$

where $\mathcal{H}$ is the Hamiltonian function, k the Boltzmann constant and T, the temperature. The requirement that the total probability within the distribution be unity determines C as

$$1/C = \int \cdots \int e^{-\mathscr{H}/kT} dq_1, \ldots, dq_f dp_1, \ldots, dp_f. \tag{3.141}$$

From (3.140), it follows that the average value of *any* function f of the qs and ps is

$$\bar{f} = C \int \cdots \int f e^{-\mathscr{H}/kT} dq_1, \ldots, dq_f dp_1, \ldots, dp_f. \tag{3.142}$$

We can write down immediately a *formal* expression for the molar magnetic susceptibility; namely

$$\chi_M = \frac{\mathbf{M}}{\mathbf{H}} \approx \mu_0 \frac{\mathbf{M}}{\mathbf{B}} = -\frac{\mu_0 N_A C}{B} \int \cdots \int \frac{\partial \mathscr{H}}{\partial B} e^{-\mathscr{H}/kT} dq_1, \ldots, dp_f \tag{3.143a}$$

or

$$\chi_M = \frac{\mu_0 N_A kT}{B} \frac{\partial \log Z}{\partial B}, \tag{3.143b}$$

where

$$Z = \int \cdots \int e^{-\mathscr{H}/kT} dq_1, \ldots, dp_f \tag{3.143c}$$

is called the 'partition function'. Here we have used (3.138b) with (3.142) to give (3.143a). Strictly, the field $\mathbf{B}$ of the integral in (3.143a) is that given as $\mathbf{B}_{\text{rest}}$ in (3.113) but for all the reasons discussed in §3.17 and not least because of the averages involved in the present statistical treatment, all the magnetic fields in the expression may be satisfactorily approximated by the applied macroscopic fields $\mathbf{B}_0$ or $\mu_0 \mathbf{H}_0$. Note that an important consequence of the present formulation in terms of the Hamiltonian function and of the canonical set of generalized coordinates and momenta is that the results are independent of the particular variables used, provided that such sets of variables are related by a so-called 'contact transformation', which is one having a unit functional determinant[7] and hence

$$dq_1, \ldots, dp_f = dQ_1, \ldots, dP_f \tag{3.144}$$

for the old and new variables.

We now exemplify the formal expression (3.143) by a brief and cursory study of some aspects of Langevin's susceptibility formula. We suppose paramagnetic materials to comprise an assemblage of molecules possessing

Magnetism

identical permanent magnetic dipole moments $\mathbf{m}_p$. A given molecular dipole inclined at an angle θ with respect to a reference direction contributes a magnetic moment to the system, with respect to that direction, of $m_p \cos \theta$. In one mole of N_A such magnetic dipoles, the total magnetic moment, or magnetization per mole, is $N_A m_p \overline{\cos \theta}$, where the bar signifies an average over all dipole orientations. As we shall assume the usual condition of uniform magnetization, this average is supposed to be the same as an average defined over the many molecules in a small volume element[††]. Hence the susceptibility is given by

$$\chi_M = \frac{\mathbf{M}}{\mathbf{H}} \approx \mu_0 \frac{\mathbf{M}}{\mathbf{B}} = \frac{\mu_0 N_A}{B} m_p \overline{\cos \theta} \qquad (3.145)$$

and we are left with the problem of determining the mean value of $\cos \theta$ over the molecular assemblage. In the absence of an applied field, of course, all values of θ are equally probable (for the reference direction is not *physically* defined), and so $\overline{\cos \theta}$ vanishes. In the presence of a magnetic field, however, the mean is non-zero. We now treat this using the Boltzmann distribution (3.143) but in a very loose way for we shall ignore the contribution of kinetic energy to the Hamiltonian function. This gross neglect is not essential and the same result is obtained when the kinetic energy is included, but the proof is lengthier.[5] The purpose of the present sketchy treatment is to provide first, the essence of the Langevin formula and second, a background against which the ultimate failure of classical physics to explain magnetism can be discussed. A more careful description of the equivalent formula within quantum theory, which we use in practice, is given in chapter 7. So, we shall replace the Hamiltonian function in (3.143) by the potential energy, $-m_p B \cos \theta$, of the reference dipole in the applied field. The probability that a dipole axis is oriented within the solid angle $d\Omega = \sin \theta d\theta d\phi$ is proportional to $e^{m_p B \cos \theta / kT} d\Omega$ and so (3.145) takes the form

$$\chi_M^p = \frac{\mu_0 N_A m_p \iint \cos \theta \, e^{m_p B \cos \theta / kT} d\Omega}{B \iint e^{m_p B \cos \theta / kT} d\Omega}, \qquad (3.146a)$$

$$= \frac{\mu_0 N_A m_p \iint (\cos \theta + m_p B \cos^2 \theta / kT + \ldots) d\Omega}{B \iint (1 + \ldots) d\Omega}, \qquad (3.146b)$$

where we have approximated the exponential terms in (3.146a) by the usual expression, $e^x \approx 1 + x; \ x \ll 1$. Now the average value of $\cos \theta$ over

---

[††] See previous footnote.

a sphere is zero while that of $\cos^2 \theta$ is $1/3$. Therefore, on substitution into (3.146b) we find

$$\chi_M^p = \mu_0 N_A m_p^2 / 3kT, \tag{3.147}$$

which is Langevin's formula for paramagnetism and generates Curie's law ($\chi^p \propto 1/T$).

Suppose, however, that the molecules in an assemblage possess no permanent magnetic dipole moments, so that there will be no terms in (3.136) that are linear in B. The magnetic moment of a molecule in a field now arises only by induction and (3.138b) is replaced by

$$\frac{\partial \mathcal{H}}{\partial B} = + \sum_i^n \frac{e^2 B}{4m_e}(x_i^2 + y_i^2), \tag{3.148}$$

and since the average of $x^2 + y^2$ is $\frac{2}{3}r^2$, there results

$$\chi_M^d = -\frac{\mu_0 N_A e^2}{6m_e} \sum_i^n \overline{r_i^2}. \tag{3.149}$$

This represents a diamagnetic contribution to Langevin's formula, the expression for the magnetic moment (3.148) being comparable with (3.67). The diamagnetic susceptibility is independent of temperature, provided that $\overline{r^2}$ does not change with temperature, and is, of course, a property of all species possessing charge, and so is universal. We have ignored some technical points in the derivations of both χ^p and χ^d, and for these the reader is referred elsewhere,[5] but these matters pale into insignificance in comparison with the central problem with the whole discussion as first described independently by Bohr[8] and van Leeuwen[9] and, at last, to be described here.

3.21 Vanishing classical magnetism

As promised, we now demonstrate that if we follow *classical* mechanics far enough, all magnetic effects cancel identically. It seems that the first proof of this was given by Bohr[8] in his dissertation in 1911, but a comprehensive summary of the matter and an alternative proof were presented by van Leeuwen[9] in her dissertation in 1919; and Van Vleck played no small part in broadcasting her results and, indeed, in independently rediscovering[5] the proof first given by Bohr.

Van Leeuwen's proof is along the following lines. The conventional Langevin theory assumes from the beginning that each molecule possesses a definite, permanent magnetic dipole moment as a characteristic of the molecular composition. As magnetic moment is proportional to angular momentum, the assumption implies a supposition that the electronic angular momentum of each molecule has one specific value. Of course, we believe that within the Bohr model of the atom or within the modern quantum picture, such quantization of angular momentum – such stability and permanence of molecular angular momentum – actually pertains, so that the path to the rescue of Langevin's formula, or something very like it, is clear: and we take it in chapter 7. But from the classical viewpoint, unadulterated with *ad hoc* quantization, quantities like the electronic angular momentum, used in the paramagnetic part of Langevin's theory, or the electronic radius, used in the diamagnetic part, *must* take *continuous* ranges of values. Further, the distribution of these molecular properties within the appropriate ranges must be determined by the Boltzmann probability factor $e^{-\mathscr{H}/kT}$. In effect, the survival of the classical Langevin theory requires the supplementation of classical statistics by an auxiliary condition which denies the infinite number of possible electronic motions allowed by a consistent and correct application of classical theory. Consider what happens when we remove this supplementary condition.

Recall, from (3.54), that the total magnetic moment of a molecule is related to its electronic angular momentum which, in cartesian coordinates, may be exemplified by

$$m_z = -\frac{e}{2} \sum_i (x_i \dot{y}_i - y_i \dot{x}_i). \tag{3.150}$$

In terms of the generalized coordinates of §3.18, we would write the molecular magnetic moment in the z direction as

$$m_z = \sum_j^f a_j \dot{q}_j, \tag{3.151}$$

where the coefficients a_j are generally functions of the generalized coordinates, as in (3.150). Now, using (3.142), the classically correct expression for the molar magnetization $\mathscr{M}_z$ in the z direction is

$$\mathscr{M}_z = N_A C \int \dots \int \sum_j a_j \dot{q}_j e^{-\mathscr{H}/kT} dq_1, \dots, dq_f dp_1, \dots, dp_f. \tag{3.152}$$

Consider a typical term j in the *summation*. From Hamilton's equations (3.125), we have $\dot{q}_j = \partial \mathscr{H}/\partial p_j$ and so we can immediately perform the integration over p_j to get $-kT \partial(a_j e^{-\mathscr{H}/kT})/\partial p_j$ for this particular term. So the contribution to (3.152) becomes

$$- N_A CkT \int \ldots \int [a_j e^{-\mathcal{H}/kT}]_{p_j=a}^{p_j=b} dq_1, \ldots, dq_f dp_1, \ldots, dp_{j-1} dp_{j+1}, \ldots, dp_f.$$

(3.153)

The key step in van Leeuwen's proof now states that, in a truly classical system, the energy assumes infinite values as the momentum takes its extreme values a or b. For example, in cartesian coordinates, the limiting values of p_j are $\pm \infty$ and the energy is clearly infinite here. Hence the contribution $[a_j e^{-\mathcal{H}/kT}]_a^b$ vanishes identically regardless of the values of all remaining variables q, p. The same is true of all terms in (3.152) and so the magnetization is exactly zero. Notice that no explicit mention of an applied magnetic field has been made, so that the proof stands irrespective of the presence or magnitude of any such field: in fact the average value of any quantity of the general form (3.151) vanishes.

It is also worth reviewing an alternative demonstration of vanishing classical magnetism in order to clarify the nature of what is, in effect, a catastrophe somewhat analogous to the more celebrated 'ultraviolet catastrophe'. In fact, this second proof, following those of Van Vleck[5] and Bohr,[8] is rather more general than van Leeuwen's. This time we work within the cartesian system when the magnetization, equivalent to (3.152), is written

$$\mathcal{M}_z = \frac{- N_A \int \ldots \int \sum_i \frac{2}{2m_e}(x_i p^0_{y_i} - y_i p^0_{x_i}) e^{-\mathcal{H}/kT} dx\,dy\,dz\,dp_x\,dp_y\,dp_z}{\int \ldots \int e^{-\mathcal{H}/kT} dx\,dy\,dz\,dp_x\,dp_y\,dp_z},$$

(3.154)

where dx is short for $dx_1\,dx_2, \ldots$, etc. and the same for dp_x, etc.: also $p^0_{x_i}$ means $m_e \dot{x}_i$ and *not* the canonical generalized momentum of (3.133), of course; and hence, by (3.138a), both dia- and paramagnetism are included. When expressed in terms of the canonical variables x, y, z, p_x, p_y, p_z, the Hamiltonian function $\mathcal{H}$ involves the magnetic field **B** as a parameter, as in (3.136). By contrast, if written in terms of the variable set $x, y, z,$ p^0_x, p^0_y, p^0_z, the total energy (3.137) does not involve **B** explicitly and, indeed, that expression is not a true Hamiltonian function as defined by (3.123), for the variables $x, \ldots, p^0_x, \ldots$ do not satisfy Hamilton's equations. We shall write the total energy in (3.137) as $\mathcal{H}^*$ to distinguish it from $\mathcal{H}$ of (3.136): thus,

$$\mathcal{H}^* = \mathcal{H}^*(x, y, z, p^0_x, p^0_y, p^0_z) \text{ but } \mathcal{H} = \mathcal{H}(x, y, z, p_x, p_y, p_z, B). \quad (3.155)$$

We emphasize that the momenta inside the bracket of (3.154) must be the ordinary momenta p^0 by the definition (3.54) and (3.139), while $\mathscr{H}$ in the exponent can be either of those in (3.155) because the classical statistical treatment merely requires this $\mathscr{H}$ to be the total energy: it is not a requirement of the thermodynamic theory that *that* $\mathscr{H}$ satisfy Hamilton's equations. However, as $\mathscr{H}^*$ is independent of **B**, it is *convenient* to change the integration variables of (3.154) from the $p_x p_y p_z$ to the $p_x^0 p_y^0 p_z^0$. Now, because these two sets of momenta are related by the general expression

$$p_{x_i}^0 = p_{x_i} + f(x, y, z), \tag{3.156}$$

as exemplified by (3.133), the functional determinant[7] $\partial(p_x, p_y, p_z)/\partial(p_x^0, p_y^0, p_z^0)$ for the transformation is unity. Hence,

$$dp_x \, dp_y \, dp_z = dp_x^0 \, dp_y^0 \, dp_z^0, \tag{3.157}$$

a result which can be obtained without the explicit use of the Jacobian, by straightforward partial differentiation, recalling that coordinates and momenta comprise a set of *independent* variables. Note that this transformation could not have been introduced immediately into (3.154) as the fundamental equation (3.142) of statistical mechanics refers to the so-called 'phase space' of the *canonical* variables, $x, \ldots, p_x$. This remark does not invalidate the use of either $\mathscr{H}$ in (3.155), however, for, as discussed above, the distribution function depends upon the energy value rather than on the Hamiltonian function.

So we have reached the point at which (3.154) may now be written in terms of $\mathscr{H}^*$ and the p^0:

$$\mathscr{M}_z = -\frac{N_A \int \ldots \int \sum_i \frac{e}{2m_e}(x_i p_{y_i}^0 - y_i p_{x_i}^0)e^{-\mathscr{H}^*/kJ}dx \, dy \, dz \, dp_x^0 dp_y^0 dp_z^0}{\int \ldots \int e^{-\mathscr{H}^*/kT}dx \, dy \, dz \, dp_x^0 dp_y^0 dp_z^0} \tag{3.158}$$

Now, as the p^0 are independent of **B**, the limits of integration for the p^0 are also independent of **B**. Further, by (3.155), $\partial\mathscr{H}^*/\partial B = 0$ and so, altogether, the right-hand side of (3.158) is independent of **B**. Therefore the magnetic moment of the assemblage of molecules is the same in the absence or presence of an applied magnetic field. Since non-ferromagnetic bodies carry no residual magnetic moment in the absence of a magnetic field, it follows from the classical theory that they should display zero moments even in the presence of such a field.

Van Vleck's proof may seem less general than van Leeuwen's in that it demonstrates identical magnetization in the presence and absence of an applied magnetic field – then supplemented by the empirical observation of zero moment outside a field – while van Leeuwen's proof yields identically *zero* magnetizations directly. However, its greater generality stems from the method of proof. Thus it follows from the immediately preceding discussion that the probability of the system being in the element $dp_x^0 dp_y^0 dp_z^0 dx dy dz$ of ordinary space (where, as usual, we mean that $p_{x_i}^0$ falls between $p_{x_i}^0$ and $p_{x_i}^0 + dp_{x_i}^0$, etc.) is just $Ce^{-\mathscr{H}*/kT} dp_x^0 dp_y^0 dp_z^0 dx dy dz$. This corresponds, as we have seen, to the distribution of velocities and coordinates in the general case; that is, even in the presence of a field. However, exactly the same probability results for $Ce^{-\mathscr{H}/kT} dp_x dp_y dp_z dx dy dz$ if $B \to 0$. This conclusion is not restricted to the case of bound electrons. For example, the classical distribution of translation velocities of free particles is similarly unaffected by a magnetic field. In general, therefore, we find that the classical statistical mean of *any* functions of velocity and coordinate variables, which do not explicitly involve B are unchanged by the application of a magnetic field. In this sense, van Leeuwen's proof is to be seen as a special case of the present result corresponding to the function being that given by (3.151). Note that in both proofs it has been demonstrated that the para- and diamagnetic parts of the classical susceptibility *cancel* identically, this following from the use of the p^0 in (3.154), for example, because of the relationship in (3.138a).

In summary: the probability that a system be in a given state of motion is proportional to $e^{-E/kT}$, where E is the energy of that motion. Classically, that energy is given by the kinetic energy plus the ordinary potential energy: there is no contribution from a magnetic field, because magnetic fields do no work. Hence the probability of a particular state of motion is the same in the presence of a magnetic field as in its absence and so, provided thermal equilibrium has been established, a magnetic field has no effect. The latter condition is necessary, for the process of applying the field necessarily involves a non-zero $\partial B/\partial t$ and hence the generation of an electric field which does do work. Once thermal equilibrium has been allowed to reestablish itself, however, the classical distribution of motion in a system becomes identical with that existing before the magnetic field was applied.

So we see that in order to avoid this null result, which certainly does not accord with experiment, we must invoke a supplementary condition that disallows the application of classical statistics to all coordinates and momenta, and confines the ranges of some to allowed values instead. Such is one aspect of quantum theory, of course, and we shall see, in chapter 7, how this works out in detail. While recognizing, therefore, that magnetism is a purely quantum phenomenon at the molecular level, we return now to the macroscopic scale and look at more technical and experimental aspects of susceptibility.

References

[1] Feynman, R.P., Leighton, R.B. & Sands, M., *The Feynman Lectures on Physics*, vol. 2, Addison-Wesley, 1964.

[2] Purcell, E.M., *Electricity and Magnetism, Berkeley Physics Course*, vol. 2, McGraw-Hill, New York, 1963.

[3] Bleaney, B.I. & Bleaney, B., *Electricity and Magnetism*, 3rd edn, Oxford University Press, 1976.

[4] Curie, P., *Ann. de Chim. et Phys.*, **5**, 289 (1895).

[5] Van Vleck, J.H., *The Theory of Electric and Magnetic Susceptibilities*, Oxford University Press, 1932.

[6] Eyring, H., Walter, J. & Kimball, G.E., *Quantum Chemistry*, John Wiley, New York, 1944.

[7] Margenau, H. & Murphy, G.M., *The Mathematics of Physics and Chemistry*, Van Nostrand, Princeton, 1956.

[8] Bohr, N., Dissertation, University of Copenhagen, 1911.

[9] Van Leeuwen, J.H., Dissertation, University of Leiden, 1919.

4

The susceptibility tensor

We remarked in the last chapter on the experimental observation that the intensity of magnetization in a dia- or paramagnetic material is usually proportional to the strength of the applied field, which fact led to the definition

$$M = \chi H \tag{3.98}$$

of the magnetic susceptibility χ. We also observed that the very small magnitudes of χ measured for such media allow several approximations to be made; in particular, that H in (3.98) may be taken as the applied field or as that inside the sample, as convenient. In chapter 5 we discuss an example of the inapplicability of this approximation and in §4.4, we shall consider circumstances in which (3.98) might be inappropriate. Meanwhile, however, we confine our attention to the regime, most common by far in magnetochemistry, in which the susceptibility simply expresses the slope of the linear relationship between M and H.

There is a further property of susceptibility to which we only barely alluded in chapter 3. It is that no restriction is imposed upon the relative *orientations* of the magnetization and field vectors in (3.98). In so-called isotropic materials M and H are always coparallel (*pace* the usual assumptions following on the very small magnitude of χ), while this only occurs in special directions with anisotropic systems. The quantity χ is not a scalar multiplier but a property which relates a vector of action (H) and a vector of effect (M) which need not point in the same direction: the susceptibility property may be represented by a *tensor*.

4.1 The tensor property

When two vectors are related by some property there is no universal reason for these vectors to be coparallel, although they may be. The vectors related in (3.98) are completely described by their components in ordinary three-dimensional space, and so that expression summarizes the three separate equations

$$
\left.\begin{aligned}
M_1 &= \chi_{11}H_1 + \chi_{12}H_2 + \chi_{13}H_3, \\
M_2 &= \chi_{21}H_1 + \chi_{22}H_2 + \chi_{23}H_3, \\
M_3 &= \chi_{31}H_1 + \chi_{32}H_2 + \chi_{33}H_3,
\end{aligned}\right\}
\tag{4.1a}
$$

in which *each* component of **M** is linearly related by the coefficients χ_{ij} to *each* component of **H**. When written in matrix form,

$$
\begin{pmatrix} M_1 \\ M_2 \\ M_3 \end{pmatrix} =
\begin{bmatrix}
\chi_{11} & \chi_{12} & \chi_{13} \\
\chi_{21} & \chi_{22} & \chi_{23} \\
\chi_{31} & \chi_{32} & \chi_{33}
\end{bmatrix}
\begin{pmatrix} H_1 \\ H_2 \\ H_3 \end{pmatrix},
\tag{4.1b}
$$

the defining equation (3.98) reveals that the susceptibility property χ must generally be represented by an array of nine numbers (scalars). So χ itself is not a scalar but rather an example of what is called a tensor property. More exactly, we *represent* the property of susceptibility by a mathematical entity called a tensor of second rank (referring to the number of suffixes in (4.1)).

In a standard form of abbreviation, (4.1) is written as

$$
M_i = \sum_j^3 \chi_{ij}H_j
\tag{4.2}
$$

but an even more condensed form, due to Einstein, is extremely convenient and in common use. In the so-called '*summation convention*' the summation sign, together with its arguments and limits, is omitted: a repeated suffix in the same term is taken to imply summation over all possible values for that suffix (here, = 1, 2, 3 corresponding to the three dimensions of ordinary space). Accordingly, (4.2) is written

$$
M_i = \chi_{ij}H_j,
\tag{4.3}
$$

in which j, being the summation index, is dummy and can be replaced at any time with another index (other than i).

4.2 Transformation of the susceptibility tensor

We later require the seminal relationships (4.1a) referred to different reference frames which are related to one another by rotation. It is both conventional and convenient to describe crystal susceptibilities, with which we shall be greatly concerned, within cartesian frames of reference: and we refer then to cartesian tensors. In other circumstances, the rectilinear scheme is less convenient and, in chapter 8, we introduce the concept of the spherical tensor. For the present, consider two orthogonal cartesian frames: the 'old' frame x_1, x_2, x_3 and the 'new' frame x'_1, x'_2, x'_3. They may be related conveniently by the cosines of the angles between any pair of axes; that is, by the matrix of *direction cosines*,

$$
\begin{array}{c}
\phantom{\text{'new'}} \quad \text{'old'} \\
\phantom{\text{'new'}} \quad
\begin{array}{ccc}
x_1 & x_2 & x_3
\end{array} \\
\text{'new'} \quad
\begin{array}{c}
x'_1 \\
x'_2 \\
x'_3
\end{array}
\begin{pmatrix}
a_{11} & a_{12} & a_{13} \\
a_{21} & a_{22} & a_{23} \\
a_{31} & a_{32} & a_{33}
\end{pmatrix} \equiv \mathbf{a}.
\end{array}
\tag{4.4}
$$

Only three of these nine numbers are independent, once the handedness of the frames has been agreed, the others being related by the unitary property of the array, meaning that vectors formed by any row (or column) of the matrix are normalized to unity and orthogonal upon the vector formed by any other row (or column):

$$
\sum_j^3 a_{ij}^2 = 1 = \sum_i^3 a_{ij}^2
\tag{4.5a}
$$

and

$$
\sum_i^3 a_{ij}a_{ik} = 0 = \sum_j^3 a_{ij}a_{kj},
\tag{4.5b}
$$

or, together in summation convention

$$
a_{ki}a_{kj} = a_{ik}a_{jk} = \delta_{ij},
\tag{4.6}
$$

where δ_{ij} is the usual Kronecker function.

Now consider the field vector $\mathbf{H}$ as having components H_1, H_2, H_3 in the 'old' frame, and H'_1, H'_2, H'_3 in the 'new'. By resolving H_1, H_2, H_3 along x'_1, and then along x'_2, etc., we obtain

$$
\left.
\begin{aligned}
H'_1 &= a_{11}H_1 + a_{12}H_2 + a_{13}H_3, \\
H'_2 &= a_{21}H_1 + a_{22}H_2 + a_{23}H_3, \\
H'_3 &= a_{31}H_1 + a_{32}H_2 + a_{33}H_3,
\end{aligned}
\right\}
\tag{4.7a}
$$

or, in summation convention,

$$H'_i = a_{ij}H_j, \tag{4.7b}$$

expressing the field in the 'new' frame in terms of that in the 'old' one: (4.7) thus provides the transformation rule for a vector **H**. For the reverse transformation, we find

$$H_i = a_{ji}H'_j, \tag{4.8a}$$

or, in matrix algebra,

$$\mathbf{H} = \mathbf{a}^\dagger\mathbf{H}', \tag{4.8b}$$

where the dagger indicates the transpose of the matrix **a** in (4.4).

We seek the transformation rule for (4.3). Corresponding to (4.3) in the 'old' frame, we have

$$M'_i = \chi'_{ij}H'_j, \tag{4.9}$$

in the 'new'. Now, by analogy with (4.7b),

$$M'_k = a_{ik}M_k \tag{4.10}$$

and from (4.3),

$$M_k = \chi_{kl}H_l. \tag{4.3}$$

From (4.8a), we have

$$H_l = a_{jl}H'_l, \tag{4.8a}$$

and so, by combining these last three equations,

$$M'_i = a_{ik}M_k = a_{ik}\chi_{kl}H_l = a_{ik}\chi_{kl}a_{jl}H'_j \tag{4.11}$$

when, by comparison with (4.9), we obtain the rule which expresses the susceptibility tensor in the 'new' frame in terms of that in the 'old':

$$\chi'_{ij} = a_{ik}\chi_{kl}a_{jl}. \tag{4.12a}$$

In matrix notation we write this tensor transformation rule as

$$\chi' = \mathbf{a}\chi\mathbf{a}^\dagger, \tag{4.12b}$$

also known as a *similarity transformation*. The physical quantity (susceptibility) does not change on rotation of the reference frame, of course, but the nine coefficients (χ_{ij}) making up the second-rank tensor, that represents it, do.

4.3 Summary of transformation laws

Tensors represent collections of coefficients in simultaneous equations and are defined by their transformation properties. The rank of a tensor is given by the number of suffixes to describe it and on this basis we may relate scalars, vectors and tensors. For example:

	Transformation law	Representing, for example,
zero-rank tensors (scalars)	$t' = t$	temperature
first-rank tensors (vectors)	$H'_i = a_{ij}H_j$	vectors like force, momentum: vector fields like $\mathbf{E}$ and $\mathbf{A}$
second-rank tensors	$\chi'_{ij} = a_{ik}\chi_{kl}a_{jl}$	susceptibility, electrical conductivity

Other physical properties, like piezoelectricity, elasticity, etc., may be represented by third-, fourth-, and higher-rank tensors, whose transformation laws build up in this table in a similar manner: the excellent book by Nye[1] describes these in considerable detail.

Not only may physical properties be represented by tensors of one rank or another, for these entities are completely defined in the present context by their transformation properties. For example, a complete set of coordinates transforms like a vector, or first-rank tensor: the 'new' coordinates of (4.4) are related to the 'old' ones by the equation

$$x'_i = a_{ij}x_j, \tag{4.13}$$

which is mathematically isomorphous with (4.7b). By repeated application of this same transformation, it may be shown that a set of binary, coordinate products transforms like products of vectors; that is, as a tensor of second rank, so that

$$x'_i x'_j = a_{ik}x_k x_l a_{jl}, \tag{4.14}$$

and so on. Note, however, that not all arrays of coefficients (matrices) transform according to one or other of these rules and so they need not be tensors: for example, the matrix (4.4) of direction cosines does *not* comprise a tensor quantity. We emphasize this distinction typographically by presenting matrices in round brackets but tensors in square brackets: the distinction is only required by us for second-rank tensors.

4.4 Tensor rank and the physical relationship between M and H

The coefficients in the equations (4.1) relating induced magnetic moment and applied field transform under a coordinate rotation as the components of a second-rank tensor. Somewhat less exactly we refer to the susceptibility as a second-rank tensor property. It is *second*-rank because of the empirically observed *linear* relationship between **M** and **H**. In less common circumstances (3.98) and (4.1) do not provide an adequate summary of the dependence of the magnetization on the field. For example, a paramagnetic sample at very low temperatures placed in a very strong magnetic field may be subject to the phenomenon called *saturation*, when the observed magnetization is less than that extrapolated from the linear relationship (3.48) determined at lower field strengths. A different instance of the inapplicability of (3.98) is provided by ferromagnets which support a residual magnetization even in the absence of an applied field. Phenomenologically, these more complicated behaviours of magnetization may be expressed in terms of a power series in **H**:

$$M_i = M_{0_i} + \chi^{(1)}_{ij} H_j + \chi^{(2)}_{ijk} H_j H_k + \chi^{(3)}_{ijkl} H_j H_k H_l + \ldots, \tag{4.15}$$

where M_0 represents a permanent magnetic moment and the various $\chi^{(n)}$ ($n \geqslant 2$) describe 'hypersusceptibility' tensors of third and higher rank. However, paramagnets and diamagnets possess no permanent moment. Actually, all terms in even powers of the field vanish identically for such materials because of time-inversion symmetry. **M**, **B** and **H**, are all axial vectors and odd under time reversal while para- and diamagnetic susceptibilities are even. A ferromagnet, on the other hand, is odd under this transformation, although, if only because of the universality of the diamagnetic overlay, it must give rise to terms of both parities in practice. So, while all terms in (4.15) may be present in a ferromagnetic system, only those in odd powers of **H** survive in the materials with which *we* are concerned. We discuss axial vectors in §4.15 and time-inversion symmetry in chapter 7. For the moment all we require is the more restricted form of (4.15);

$$M_i = \chi^{(1)}_{ij} H_j + \chi^{(3)}_{ijkl} H_j H_k H_l + \ldots. \tag{4.16}$$

The summation convention has been used in (4.16), of course, but it is also possible to perform some of the summations explicitly and write

$$M_i = \chi^{(1)}_{ij} H_j + \chi'_{ij}(\mathbf{H}) H_j + \ldots. \tag{4.17}$$

in which $\chi'_{ij}(\mathbf{H})$ is a second-rank tensor. It is not independent of the field, however. Nevertheless, it is formally possible to expand the magnetization in this way and write

$$M_i = \chi^{(\mathbf{H})}_{ij} H_j \tag{4.18}$$

analogous to (4.3), except that we now refer to a 'field-dependent susceptibility',

$$\chi^{(\mathbf{H})} = \chi^{(1)} + \chi'(\mathbf{H}) + \ldots. \tag{4.19}$$

In magnetochemistry, however, it is doubtful if this formal manoeuvre has any value. When, under most conditions of interest to us, only the linear term in (4.16) has any significant magnitude, the concept of a susceptibility is useful for, by expressing the slope of a linear relationship between **M** and **H**, the system may be completely characterized at one value of the field strength. When this is no longer true, we might just as well work with the quantities actually observed, namely, the explicit magnetization values as functions of field strength. It is certainly simplest to make contact between theory and experiment along these lines.

4.5 Principal axes

While introducing the concept of a tensor property in §4.1 we remarked that the vectors of 'action' and 'effect' need not be coparallel. That does not preclude their ever being so, however. Being coparallel is just one special case of the general tensorial relationship between the two vectors. We therefore seek conditions under which **M** and **H** may be coparallel, that is, under which their components are proportional, $M_i \propto H_i$, or, in summation convention, when

$$\chi_{ij}H_j = \lambda H_i, \tag{4.20}$$

where λ is a constant. This expression represents three simultaneous linear equations in the variables H_j,

$$\left. \begin{array}{l} (\chi_{11} - \lambda)H_1 + \chi_{12}H_2 + \chi_{13}H_3 = 0, \\ \chi_{21}H_2 + (\chi_{22} - \lambda)H_2 + \chi_{23}H_3 = 0, \\ \chi_{31}H_3 + \chi_{32}H_2 + (\chi_{33} - \lambda)H_3 = 0, \end{array} \right\} \tag{4.21}$$

and the condition that these equations be linearly independent, other than the trivial case $H_i = 0$, is that the determinant of the coefficients of the variables vanish. There emerges the well-known secular equation

$$|\chi_{ij} - \lambda\delta_{ij}| = 0, \tag{4.22}$$

which, in the present case of three-dimensional space, is a cubic equation with three roots. Associated with each root or principal susceptibility value is a vector which, by (4.20), defines a direction, parallel to which an applied field will induce a coparallel magnetization in the sample. In order that the roots correspond to measurable quantities – that the eigenvalues of (4.22) be generally real, rather than complex or pure imaginary – the χ tensor must be symmetric. A proof[1,2] of this experimentally verifiable fact is based on thermodynamic arguments and in essence is as follows.

Susceptibility and magnetic polarization involve a thermodynamically reversible process. Therefore the net energy change over the following cycle must be zero: (i) apply a magnetic field parallel to x, (ii) apply a field in the y direction, (iii) remove the field along x, and (iv) remove the field along y. The net work done in this cycle can only vanish generally if $\chi_{xy} = \chi_{yx}$: similar arguments can be applied to any other directions. In chapter 7 we show how the same symmetry property emerges from the quantum-mechanical model we describe. The symmetry is that, within the whole array of nine coefficients, $\chi_{ji} = \chi_{ij}$. The relationship is not quite as obvious as noting that a measurement of susceptibility does not depend on the polarity of the applied field; rather that the susceptibility component found by applying **H** parallel to x_1 and measuring the component of the induced moment **M** parallel to x_2 will be the same as that found by applying **H** parallel to x_2 and measuring the component of **M** along x_1. The centrosymmetric property of the susceptibility tensor is consistent with the quite different condition, that in $\mathbf{M} = \chi\mathbf{H}$, both **M** and **H** are odd with respect to spatial inversion.[††]

In summary: given the extra fact (experimental and thermodynamic) that the susceptibility tensor is symmetric, it inevitably emerges from a linear relationship like (4.3) that there exist three mutually perpendicular directions in space parallel to which an applied field induces a real coparallel moment. If these directions had been taken initially to be the axes of the reference frame, the susceptibility tensor would take the form

$$\begin{bmatrix} \chi_1 & 0 & 0 \\ 0 & \chi_2 & 0 \\ 0 & 0 & \chi_3 \end{bmatrix} \tag{4.23}$$

and we refer to the *principal magnetic axes* and to the *principal suscepti-bilities* χ_1, χ_2, χ_3. In that reference frame, when the tensor has the diagonal form (4.23), (4.3) takes the particularly simple form of (4.20) we sought:

$$\left.\begin{aligned} M_1 &= \chi_1 H_1, \\ M_2 &= \chi_2 H_2, \\ M_3 &= \chi_3 H_3. \end{aligned}\right\} \tag{4.24}$$

In general, the process of putting a tensor into diagonal form ('dia-gonalization') is equivalent to solving the cubic equation (4.22) for the principal susceptibilities and their orientations relative to the given reference frame.

[††] but see §4.15.

4.6 Crystal symmetry and special forms of the susceptibility tensor

In magnetochemistry we are concerned with materials in the form of single crystals. The physical properties of a single crystal are consequences of its structure and we therefore expect that the symmetry of the directional variation of any physical property be consistent with the symmetry of the crystal structure. Different experimental techniques, referring to different physical properties, will not necessarily reveal all, or indeed the same set of, symmetry elements in the structure but those defined by any one technique must be compatible with those defined by another. A fundamental axiom of crystal physics is that *the symmetry elements of any physical property of a crystal must include the symmetry elements of the point group of the crystal.* This is called Neumann's principle and it restricts the form of the susceptibility tensor in various crystal systems in ways we now discuss.

The susceptibility tensor property is centrosymmetric and possesses three mutually perpendicular characteristic, or principal, axes. For our present purposes we shall classify crystals according to their seven crystal systems and require that the principal magnetic axes coincide with appropriate symmetry axes of the respective point groups. The results are summarized in table 4.1 and the entries in the last column were made as follows.

Triclinic system. The susceptibility tensor is centrosymmetric and so no distinction is made between the two space groups belonging to this system. In neither case are any directions in space made special by symmetry and therefore there are no restrictions whatever upon the orientation of the principal magnetic axes nor upon the relative magnitudes of the principal susceptibilities. The susceptibility tensor is therefore defined by the maximum number – six – of independent coefficients χ_{ij}.

Monoclinic system. Monoclinic crystals are characterized by possessing only one direction made special by symmetry. The unique symmetry axis is conventionally labelled b. The 14 monoclinic space groups arise out of the various distinguishable combinations of symmetry elements which can give rise to a single crystal axis of symmetry. These comprise two-fold rotation or screw axes parallel to b and/or mirror or glide planes perpendicular to b: there may or may not be a centre of inversion. For purposes of comparison with the susceptibility tensor we are not concerned with translations nor with the absence or not of an inversion

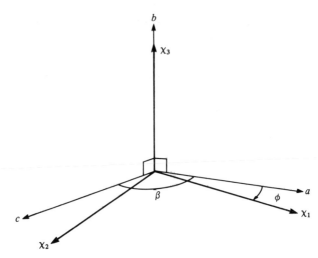

Fig. 4.1. Conventions used to define principal susceptibilities for monoclinic crystals.

centre. We may therefore add the centre, and group all monoclinic space groups together under the single Laue group of the monoclinic system, $2/m$.

One principal magnetic axis must be oriented parallel to the unique crystal axis. We follow the long-accepted, but undoubtedly confusing, convention of chosing χ_3 parallel to b in the monoclinic system. No other directions are special in monoclinic crystals and so there are no restrictions upon the orientations of χ_1 and χ_2 in the crystal ac plane. When referred to the standard crystallographic frame, therefore, the monoclinic crystal susceptibility tensor takes the general form shown in the last column of table 4.1 and there are now only four independent tensor coefficients. The situation is illustrated in figure 4.1. By convention,[3] $\chi_1 < \chi_2$ and ϕ is the angle subtended by χ_1 and the crystal a axis, measured from the positive direction of a towards or through the positive direction of c. An equivalent set of four independent quantities to the χ_{11}, χ_{22}, χ_3 and χ_{12} in the table are the principal crystal susceptibilities χ_1, χ_2, χ_3 and the angle ϕ.

Orthorhombic system. Orthorhombic crystals possess three mutually perpendicular symmetry axes. The same is true of the susceptibility tensor and, by Neumann's principle, the two sets of axes must coincide. There are no special relationships between the three symmetry axes so no relationships between the magnitudes of the principal crystal susceptibilities are imposed. By convention, the three independent tensor coefficients are taken as the principal crystal susceptibilities; $\chi_1 = \chi_a$, $\chi_2 = \chi_b$, and $\chi_3 = \chi_c$.

Uniaxial system. Trigonal, tetragonal and hexagonal crystals are defined by a unique direction associated with three-, four-, or six-fold rotational symmetry. By convention, this direction is labelled c or $\parallel$. One principal susceptibility must lie parallel to c and this is labelled χ_3; that is, $\chi_3 = \chi_c = \chi_\parallel$. The susceptibility in the ab, or $\perp$, plane is isotropic which means that a magnetic field applied in any direction in that plane will induce a magnetic moment (of magnitude $\chi_\perp H$) parallel to the applied field. This comes about as follows.

First, consider the tetragonal case. Consider any one set of two directions in the ab plane which are mutually perpendicular and equivalent. By Neumann's principle, two principal susceptibilities must lie parallel to these but, since the crystal directions are equivalent by symmetry, the principal susceptibilities must be of equal magnitude. The crystal tensor in this frame therefore takes the form shown in table 4.1: in particular, two eigenvalues of (4.20) are degenerate. But we know from a simple theorem[4] that any arbitrary linear combination of degenerate eigenfunctions is also an eigenfunction with the same eigenvalue. A two-fold degeneracy is defined by two orthogonal functions, one of which may be chosen within the given space in an entirely arbitrary way. Accordingly, we may select any direction in the ab plane for χ_1 and, by orthogonality, construct χ_2 (of equal magnitude) perpendicular to it. Since the eigenfunctions in (4.20) correspond to directions in which the applied magnetic field and induced magnetic moment are coparallel, it follows that application of the magnetic field in *any* direction perpendicular to the c axis of a tetragonal crystal will induce a magnetic moment (of magnitude $\chi_1 H = \chi_2 H = \chi_\perp H$) parallel to the applied field.

The cases of trigonal or hexagonal crystals are only a little more complicated. In these systems, the symmetry axes in the crystal ab plane are not mutually perpendicular but they are equivalent. Only if the susceptibility tensor in the ab plane is isotropic can it be guaranteed that, regardless of which direction in that plane is chosen to be parallel to χ_1, the three- or six-fold symmetry is maintained, as required by Neumann's principle.

Cubic system. The characteristic three-fold axes of the cubic system render the three cartesian axes equivalent. If we choose to orient the principal susceptibilities along these cartesian axes, their magnitudes are necessarily equal. Accordingly, we have a three-fold degeneracy in three dimensions

Table 4.1. *Relationships between the crystal system and the susceptibility tensor*

Optical classification	System	Characteristic symmetry	Number of independent tensor coefficients	Tensor referred to crystal axes
biaxial	triclinic	$P1$ or $P\bar{1}$	6	$\begin{bmatrix} \chi_{11} & \chi_{21} & \chi_{31} \\ \chi_{21} & \chi_{22} & \chi_{32} \\ \chi_{31} & \chi_{32} & \chi_{33} \end{bmatrix}$
biaxial	monoclinic	1 2-fold axis	4	$\begin{bmatrix} \chi_{11} & 0 & \chi_{21} \\ 0 & \chi_{3} & 0 \\ \chi_{21} & 0 & \chi_{22} \end{bmatrix}$
biaxial	orthorhombic	3 mutually orthogonal 2-fold axes	3	$\begin{bmatrix} \chi_{1} & 0 & 0 \\ 0 & \chi_{2} & 0 \\ 0 & 0 & \chi_{3} \end{bmatrix}$
uniaxial	trigonal / tetragonal / hexagonal	1 3-fold axis / 1 4-fold axis / 1 6-fold axis	2	$\begin{bmatrix} \chi_{1} & 0 & 0 \\ 0 & \chi_{1} & 0 \\ 0 & 0 & \chi_{3} \end{bmatrix}$
anaxial (isotropic)	cubic	4 3-fold axes	1	$\begin{bmatrix} \chi & 0 & 0 \\ 0 & \chi & 0 \\ 0 & 0 & \chi \end{bmatrix}$

and the preceding arguments apply once again. Therefore, application of a magnetic field along any direction whatever induces a magnetic moment, of magnitude χH, parallel to that field. The susceptibility tensor is therefore totally isotropic (and thus scalar), as shown in the table.

4.7 Précis

A magnetic moment is induced in a diamagnetic or paramagnetic crystal when it is placed in a magnetic field. Under most experimental conditions, the induced moment is very small and has a magnitude which varies linearly with the strength of the applied field. The direction of magnetization does not generally coincide with the direction of the magnetic field. The property relating applied field and induced moment, called magnetic susceptibility, involves nine components which transform as the components of a second-rank tensor. Thermodynamic arguments as well as experimental observation show that the susceptibility tensor is symmetric. The existence of principal axes of the tensor; that is, of three, mutually orthogonal special directions in the sample parallel to which an applied field induces a coparallel moment, follows essentially from the linear relationship between field and moment and from the properties of free space. The orientations and relative magnitudes of the principal susceptibilities of crystals are determined to a greater or lesser degree by the symmetry of the crystal system.

This constitutes the basic information we require to proceed in magnetochemistry. However, it is common in the literature to see mention of the 'susceptibility ellipsoid' and otherwise to maintain a mental picture of the tensor property in terms of some geometrical figure or construction. Such geometric *representations* are not without interest but they are not essential to an understanding of susceptibility. The reader should note that all of the required material has been described so far without any reference whatever to such figures or mention of ellipsoids. However, the following sections are presented partly in the spirit of broadening our acquaintance with tensor properties and partly in explanation of jargon commonly found in the literature.

4.8 The representation quadric

We consider the equation

$$S_{ij}x_ix_j = 1, \tag{4.25}$$

in which the S_{ij} are coefficients. Written out explicitly, and with the further proposition that $S_{ij} = S_{ji}$, (4.25) becomes

$$S_{11}x_1^2 + S_{22}x_2^2 + S_{33}x_3^2 + 2S_{23}x_2x_3 + 2S_{31}x_3x_1 + 2S_{12}x_1x_2 = 1, \tag{4.26}$$

which is the general equation of a second-degree surface, or quadric, referred to its centre as origin. In general, it may be an ellipsoid or hyperboloid. Now (4.26) transforms as a second-rank tensor, whether $S_{ij} = S_{ji}$ or not. This follows directly from the fact that the product of two coordinates (here, x_i) so transforms, as in (4.14), and (4.25) expresses the fact that S_{ij} times such products yields the scalar 1. More explicitly, we have

$$x_i = a_{ki}x_k' \quad \text{and} \quad x_j = a_{lj}x_l', \tag{4.27}$$

so that (4.25) becomes

$$S_{ij}a_{ki}a_{lj}x_k'x_l' = 1 = S_{kl}'x_k'x_l', \tag{4.28}$$

whence

$$S_{kl}' = a_{ki}S_{ij}a_{lj}, \tag{4.29}$$

which is the transformation rule for second-rank tensors. Hence we see

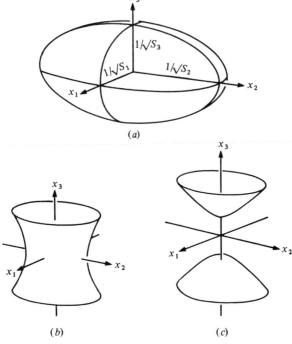

Fig. 4.2. Physically realistic representation quadrics: (*a*) an ellipsoid
for all semi-axis lengths of the same sign; (*b*) an hyperboloid of one
sheet when two coefficients are positive; (*c*) an hyperboloid of two
sheets if only one principal susceptibility is positive.

that the coefficients of the quadric (4.25) transform like the components
of a second-rank tensor: the surface (4.26) may therefore be used to map
the property of susceptibility.

Such a figure possesses three mutually perpendicular principal symmetry
axes and, when referred to these axes, the quadric (4.26) takes the form

$$S_1 x_1^2 + S_2 x_2^2 + S_3 x_3^2 = 1, \qquad (4.30)$$

which may be compared with the standard equation of a conicoid with
semi-axis lengths a, b, c:

$$\frac{x^2}{a^2} + \frac{y^2}{b^2} + \frac{z^2}{c^2} = 1. \qquad (4.31)$$

Hence, the semi-axis lengths of the representation quadric (4.30) are $1/\sqrt{S_1}$,
$1/\sqrt{S_2}$, $1/\sqrt{S_3}$. If S_1, S_2, S_3 are all positive, the surface is an ellipsoid: if
one principal coefficient is negative, it is a hyperboloid of one sheet: if

two coefficients are negative, it is a hyperboloid of two sheets: if all S_i are negative, there results an imaginary ellipsoid. The first three cases are illustrated in figure 4.2. It is important to recognize that the representation quadric of a susceptibility tensor *can be any* of these figures. The imaginary nature of any or all of the semi-axis lengths is an artefact of the representation and does not imply that the quantities they represent are unmeasurable. The representation quadric takes the form of a hyperboloid, for instance, if the crystal is diamagnetic in one principal direction and paramagnetic in the others. Lest this mapping seem little more than quaint, we turn now to consider two very useful geometric properties of the quadric.

4.9 The magnitude of the susceptibility and the length of the radius vector

In general, the induced magnetic moment is not parallel to the applied field. It is sometimes convenient to resolve the magnetic moment into components parallel and perpendicular to the applied field and to define the susceptibility in the direction of the applied field as the coparallel component of the induced moment divided by the magnitude of the field; that is, $M_\parallel/H$.

Let l_i be the direction cosines of $\mathbf{H}$ referred to general axes, so that $H_i = Hl_i$. The component of $\mathbf{M}$ parallel to $\mathbf{H}$ is $(\mathbf{M} . \mathbf{H})/H$, or in suffix notation $(M_iH_i)/H$. Therefore, the susceptibility in the direction l_i is

$$\chi = \frac{M_iH_i}{H^2} = \frac{\chi_{ij}H_jH_i}{H^2} \tag{4.32}$$

or

$$\chi = \chi_{ij}l_il_j. \tag{4.33}$$

In order to construct a geometrical interpretation of χ, we consider a general point P on the representation quadric centred at origin O,

$$\chi_{ij}x_ix_j = 1, \quad \chi_{ij} = \chi_{ji}. \tag{4.34}$$

If the direction cosines of the radius vector OP are l_i, we have

$$x_i = rl_i, \tag{4.35}$$

where $r = \text{OP}$. On substitution in (4.34), there results

$$r^2\chi_{ij}l_il_j = 1 \tag{4.36}$$

and from (4.33),

$$\chi = 1/r^2 \text{ or } r = 1/\sqrt{\chi}. \tag{4.37}$$

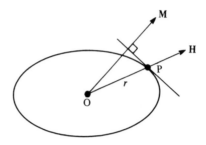

Fig. 4.3. The representation quadric provides a ready means for establishing the relative orientations of applied field and induced moment.

Thus the length of any radius vector of the representation quadric is equal to the reciprocal of the square root of the magnitude χ of the susceptibility in that direction. For the special case when OP is parallel to each of the principal axes of the representation quadric, the magnitudes of the susceptibility (the principal susceptibilities) are $1/\sqrt{\chi_1}, 1/\sqrt{\chi_2}, 1/\sqrt{\chi_3}$, as shown in figure 4.2.

4.10 The radius-normal property

The form of the representation quadric referred to axes parallel to the principal axes is

$$\chi_1 x_1^2 + \chi_2 x_2^2 + \chi_3 x_3^2 = 1. \tag{4.38}$$

Let **H** be oriented with direction cosines l_1, l_2, l_3 so that

$$\mathbf{H} = [l_1 H, l_2 H, l_3 H], \tag{4.39}$$

and hence

$$\mathbf{M} = [\chi_1 l_1 H, \chi_2 l_2 H, \chi_3 l_3 H]. \tag{4.40}$$

Therefore the direction cosines of **M** are proportional to $\chi_1 l_1, \chi_2 l_2, \chi_3 l_3$. Let P be the point on the quadric (4.38), centred at O, such that OP is parallel to **H** and has magnitude r: P has coordinates (rl_1, rl_2, rl_3). The tangent plane to the quadric at P is given by fixing one of the variables in the second-order expression (4.38) as P, giving

$$\chi_1 rl_1 x_1 + \chi_2 rl_2 x_2 + \chi_3 rl_3 x_3 = 1 \tag{4.41}$$

and therefore the normal to the quadric at O has direction cosines proportional to $l_1 \chi_1, l_2 \chi_2, l_3 \chi_3$, giving the important result that the normal at P is parallel to **M**.

The importance and utility of the representation quadric, therefore, is that it provides a ready means for establishing the relative *orientations* of the applied field and induced moment by geometrical construction. As illustrated in figure 4.3 for the case[tt] of the quadric being an ellipsoid, after choosing the field direction, we construct a tangent plane (or simply a tangent in the principal plane shown in the figure) at the point of intersection with the representation quadric. The induced moment **M** is then found as parallel to the normal to this tangent plane.

4.11 The magnitude ellipsoid

The representation quadric provides a graphical means of finding the orientation of the induced moment but not its *magnitude*. Another, quite *different*, surface may be constructed for that purpose. Let us suppose the applied field **H** is normalized to unity and then determine the magnitude of the induced moment **M** as the direction of **H** varies throughout angular space. We write

$$\mathbf{H} = [H_1, H_2, H_3] \tag{4.42}$$

with respect to the principal axes of the susceptibility tensor, so that

$$\mathbf{M} = [M_1, M_2, M_3] = [\chi_1 H_1, \chi_2 H_2, \chi_3 H_3]. \tag{4.43}$$

Therefore, since $H_1^2 + H_2^2 + H_3^2 = 1$, by construction, we find that

$$\frac{M_1^2}{\chi_1^2} + \frac{M_2^2}{\chi_2^2} + \frac{M_3^2}{\chi_3^2} = 1, \tag{4.44}$$

that is, that the extremity of the vector **M** lies on the surface

$$\frac{x_1^2}{\chi_1} + \frac{x_2^2}{\chi_2} + \frac{x_3^2}{\chi_3} = 1. \tag{4.45}$$

Note that the surface (4.45) is an ellipsoid – *never* a hyperboloid – regardless of the relative signs of the principal susceptibilities χ_1, χ_2, χ_3. It is called the *magnitude ellipsoid*.

In figure 4.4 are shown the magnitude ellipsoid and the representation quadric (the latter for the case of an ellipsoid) which between them provide for the geometrical construction of both the magnitude and direction of the induced magnetic moment in a sample as the applied field rotates in space. Note that, while the semi-axis lengths of the representation quadric

[tt] Reference [1] shows analogous constructions for hyperboloids.

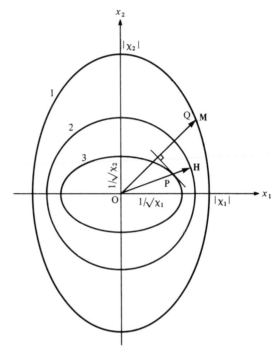

Fig. 4.4. A principal section through a magnitude ellipsoid (1), a unit sphere (2), and an ellipsoidal representation quadric for susceptibility (3).

(3) $(1/\sqrt{\chi_i})$ may be real or imaginary, giving rise to hyperboloids on occasion, those of the magnitude ellipsoid are necessarily real and positive. As **H** rotates on the sphere (2), **M** rotates and changes in magnitude: the direction of **M** is given by the radius-normal property of the representation quadric, and its magnitude by the length OQ on the magnitude ellipsoid (1). It is important to recognize that neither surface gives information about both the magnitude and direction of the induced moment.

4.12 An heuristic example

All materials show a diamagnetic contribution to susceptibility. In paramagnets the diamagnetic effect is usually relatively small and is treated as a 'correction' to the gross experimental susceptibility. Diamagnetic and paramagneitc effects are conveniently regarded as distinct effects and the total experimental susceptibility of a paramagnetic material is usually decomposed into a sum of diamagnetic and paramagnetic tensors. Neither

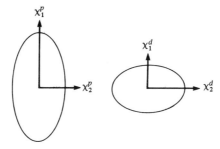

Fig. 4.5. Magnitude ellipses for the para- and diamagnetic contributions to some crystal susceptibility for the case of parallel orientation and $\chi_1^p > \chi_1^d$ but $\chi_2^p < \chi_2^d$.

tensor needs, in general, to be isotropic, nor to have coparallel principal axes. Let χ^p and χ^d represent the para- and diamagnetic susceptibilities respectively, referred to the *same* frame. The rule for tensor addition is that we add the tensors element-for-element, provided they are referred to a common frame. This follows directly from the form of the seminal equations (4.1). It also follows that the sum of two tensors is itself a tensor. Thus for a second-rank tensor **A**, for example, defined by the transformation

$$A'_{ij} = a_{ik}A_{kl}a_{jl},\tag{4.46}$$

and a second such **B**, referred to the same frame and defined by

$$B'_{ij} = a_{ik}B_{kl}a_{jl},\tag{4.47}$$

we have

$$A'_{ij} + B'_{ij} = a_{ik}(\mathbf{A} + \mathbf{B})_{kl}a_{jl} = (\mathbf{A}' + \mathbf{B}')_{ij},\tag{4.48}$$

which defines $(\mathbf{A} + \mathbf{B})$ as a second-rank tensor. Note that neither the sum nor the constituent tensors *need* share common principal axes. However, for the point to be illustrated here, it is enough to consider the special case of diamagnetic and paramagnetic tensors whose principal axes *are* parallel.

Suppose the magnitude ellipsoids for the two properties are proportional to those shown in figure 4.5, in which we consider principal sections (ellipses) for illustration. The diagrams describe a situation in which the net susceptibility parallel to x_1 is paramagnetic while that parallel to x_2 is diamagnetic. Although uncommon, the circumstances are by no means unreal. Geometrical representations of tensor properties can be misleading: as emphasized in the preceding section, the magnitude ellipsoid is a

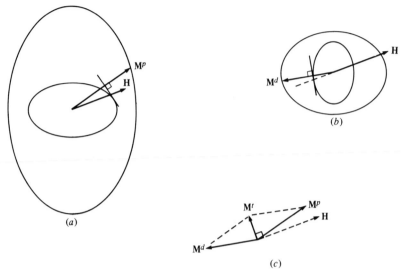

Fig. 4.6. Magnitude ellipses and representation quadrics for (a) the paramagnetic contribution and (b) the diamagnetic part, showing magnitudes and orientations of para- and diamagnetic contributions to the magnetic moment for an applied field oriented at right-angles to the total, resultant magnetic moment in (c).

construction whose sole purpose is to map the magnitude of an induced moment. We may certainly not 'add' the ellipses in figure 4.5 to produce some 'clover-leaf' figure. The total susceptibility is still a tensor property, as demonstrated above and because it relates two vectors. It is irrelevant that each of these vectors can (arbitrarily) be decomposed into two or more separate parts: the total tensor can still be represented by a magnitude ellipsoid. Of course, in the present example, application of a field **H** parallel to x_1 induces a moment $(\chi_1^p + \chi_1^d)H$ which is positive (paramagnetic), while application of **H** along x_2 induces the negative (diamagnetic) gross moment $(\chi_2^p + \chi_2^d)H$. The representation quadric of the total susceptibility in this case is, therefore, a hyperboloid (of one or two sheets, depending upon the properties perpendicular to the plane of figure 4.5), but the magnitude ellipsoid is still an ellipsoid: recall that the semi-axis lengths of a magnitude ellipsoid are equal to the *moduli* of the principal susceptibilities.

Given the situation of gross diamagnetism in one principal direction and gross paramagnetism in another, it is natural to enquire about the property along some direction in between. Certainly, there is some orientation of an applied field in the x_1x_2 plane for which the coparallel component of the induced moment vanishes. However, except in the case of identical para- and diamagnetic ellipsoids giving rise to a totally zero resultant susceptibility, this vanishing coparallel component of magnetic moment is associated with a non-vanishing transverse component: and this is of such a magnitude that $|\mathbf{M}|$ lies between $|M_1^p + M_1^d|$ and $|M_2^p + M_2^d|$, as required by the magnitude ellipsoid of the total tensor. The situation is summarized in figure 4.6. On the left (*a*) are shown projections of the magnitude ellipsoid and representation quadric for the paramagnetic contribution, and on the right (*b*), those for the diamagnetic part. For the same magnitude and direction of the applied field in each case are drawn the directions and magnitudes of the induced para- and diamagnetic moments. The resultant moment $\mathbf{M}^t$ is shown in (*c*). The situation depicted in these diagrams corresponds to the case of the vanishing parallel moment, so that $\mathbf{M}^t$ is oriented perpendicular to the applied field. Finally, we can enquire whether $\mathbf{M}^t$ in (*c*) might acquire zero magnitude. This will happen if $\mathbf{M}^p$ and $\mathbf{M}^d$ are antiparallel vectors of equal magnitude, but, from the rules by which these representations are constructed, we can see that this can only occur if the dia- and paramagnetic tensors are identical, except for sign. In that case, the gross susceptibility is a null property and, though not impossible, is obviously too unlikely an accident to occur in practice.

4.13 Molecular susceptibilities

The magnetochemist measures the magnetic properties of a bulk sample like a crystal, ultimately in order to probe the nature of the chemical bonding in the molecules from which the substance is built. It is natural, therefore, to try to define and determine magnetic properties of molecules analogous to those of crystals. If it can be assumed that there are no interactions between the molecules in a crystal, other than their 'thermal contact', one might envisage independent but identical behaviour for each molecule and so consider the crystal property as a sum of the corresponding molecular properties. It is somewhat along these lines that the concept of a *molecular susceptibility* has been introduced in the literature[3,5,6] although, as we now discuss, there are fundamental difficulties with this idea from the outset.

The clause, 'other than their thermal contact', identifies the problem, for magnetic susceptibility is a statistical concept. Suppose we were to write

$$m_i = \kappa_{ij} H_j \tag{4.49}$$

in an attempt to construct a molecular analogue for the physics described by (3.98). We ignore the fact that an individual molecule will experience a microscopic field, and make the usual approximation (3.114). So κ is supposed to be a molecular susceptibility tensor property relating the applied field **H** and the induced magnetization **m**. Of course, since we are considering a single molecule, the magnetization is to be read as an induced magnetic moment. This idea may raise no serious questions in the case of a diamagnet but does so for a paramagnet, whether we persist with an (incorrect) classical view or embrace a quantum-mechanical picture. From the standpoint of the classical approach we have described so far, we note that the induced molecular magnetic moment of (4.49) would not be the same thing as the permanent molecular dipole of (3.147), for example. We would have to imagine an 'average molecule' as one whose permanent magnetic dipole were oriented in some 'average' way with respect to the applied field, where this average would, of course, be determined by a Boltzmann distribution function. Within a quantum-mechanical description, the value of m observed experimentally would be statistical twice over: once because of the thermal distribution and once by virtue of the statistical interpretation of quantum expectation values via the superposition principle.

However, we need not wriggle to extract ourselves from these difficulties, for, despite the well-established but unfortunate name, the quantity normally referred to as a 'molecular susceptibility' is actually no such thing! The situation is clarified when we ask why we seek to establish a molecular property equivalent to the crystal susceptibility χ. It is, surely, that we wish to associate the magnitudes and orientations of the principal crystal susceptibilities with structural features in the molecules comprising the crystal. Then imagine the most simple case of all in which a crystal is built up from identical molecules in identical orientation. An association of the crystal properties directly with the (mean) molecular frame follows immediately and is of significance provided that the individual molecular entities can be meaningfully identified. If the molecules in a given crystal act cooperatively with respect to an applied field, there follows a reduced *significance* of any directions in a particular molecule so that the exercise of defining a molecular property loses interest correspondingly. In this book we are concerned exclusively with so-called *magnetically dilute* systems where such cooperative magnetic interaction is either absent completely or else too small to be detected by the experiments we perform: this last point is discussed further in chapter 5. Now, in systems without magnetic concentration, the magnetic behaviour of a molecule may quite properly be factorized into one part, intrinsic to the electronic character of the molecule, and another related to its overall orientation relative to some (arbitrary) given frame of reference. Since *we* are ultimately concerned with the electronic properties of a molecule, this second factor is without interest in magnetically dilute systems. It is therefore a sensible, useful and legitimate process to divide up the total crystal susceptibility, not into molecular entities, but into parts, each of which *refers* to all molecules with the same orientation; and, for purposes of standardization, these properties are each normalized to the same macroscopic unit, most usually the mole. This is how the concept of a molecular susceptibility is always used; never as in (4.49).

Accordingly, we define a molecular susceptibility tensor, which we label **K**, to distinguish it from the crystal property χ, by the relationship

$$M_i = K_{ij}H_j, \tag{4.50}$$

it being understood that the 'molecular' property describes the statistical property of one mole, say, of identically aligned molecules. The 'molecular' tensor is centrosymmetric and also symmetric with respect to interchange of suffixes. It is again characterized by three principal axes which we take to bear a definite but unknown relationship to the molecular geometry. Let the orientation of these principal molecular susceptibilities with respect to the (orthogonalized) crystal axes be

$$\begin{matrix} & & \text{molecular magnetic axes} \\ & & x \quad\quad y \quad\quad z \\ \text{orthogonal} & a \\ \text{crystal} & b \\ \text{axes} & c \end{matrix} \begin{pmatrix} \alpha_1 & \alpha_2 & \alpha_3 \\ \beta_1 & \beta_2 & \beta_3 \\ \gamma_1 & \gamma_2 & \gamma_3 \end{pmatrix} \equiv \mathbf{c}. \tag{4.51}$$

With respect to the molecular magnetic frame, the susceptibility tensor takes the form

$$\begin{pmatrix} K_1 & 0 & 0 \\ 0 & K_2 & 0 \\ 0 & 0 & K_3 \end{pmatrix} \equiv \mathbf{K}^M, \tag{4.52}$$

but with respect to the orthogonalized crystal frame, we have

$$\mathbf{K}^C = \mathbf{c}\mathbf{K}^M\mathbf{c}^\dagger, \tag{4.53}$$

where the superscripts refer to crystal and molecular frames, **c** is the matrix of direction cosines (4.51), and we have used the transformation law (4.12b). In general, not all molecules in a given crystal lattice have the same orientation and, in recognizing this, we may rederive the information in table 4.1 as follows.

Triclinic system. In the space group $P1$ all molecules in the lattice have identical orientations. Although half the molecules in $P\bar{1}$ are related to the other half by centres of inversion, the orientations of the molecular susceptibility tensors for each set of molecules are identical because of the centrosymmetric nature of the tensor. There follows the identity,

$$\mathbf{K} \equiv \boldsymbol{\chi}. \tag{4.54}$$

The orientation of $\boldsymbol{\chi}$ in the crystal lattice is determined entirely by those factors determining the orientation of $\mathbf{K}$ and hence is not fixed in any way by the crystal symmetry (or, rather, lack of it): the generality of the crystal $\boldsymbol{\chi}$ tensor in table 4.1 follows.

Monoclinic system. In all crystal systems, spacegroup translations are of no concern to us for we have already assumed a regime of magnetic dilution and the so-called molecular property is defined as representative of all molecules in common orientation. Similarly, as noted above, the principal magnetic axes of centrosymmetrically related groups of molecules are parallel, so removing the need to pay heed to centres of inversion. Accordingly, in the monoclinic system we only consider molecules related by the actual or – from the $2/m$ Laue symmetry, implicit – diad parallel to the unique b axis. Therefore, if one-half (A) of the molecules in the lattice are associated with principal magnetic axes oriented as in (4.51), the other half (B) are characterized by a 'molecular' susceptibility tensor with principal directions given by

molecular magnetic axes for set B

$$
\begin{array}{ll}
\text{orthogonal} & a \\
\text{crystal} & b \\
\text{axes} & c'
\end{array}
\begin{array}{ccc}
x & y & z \\
\end{array}
\left(
\begin{array}{ccc}
\alpha_1 & \alpha_2 & \alpha_3 \\
-\beta_1 & -\beta_2 & -\beta_3 \\
\gamma_1 & \gamma_2 & \gamma_3
\end{array}
\right). \tag{4.55}
$$

The total crystal tensor is formed by addition of the molecular tensors for the sets A and B. Such tensor addition only has meaning if the tensors are referred to the same frame, as the following discussion shows. Suppose we apply a magnetic field along some arbitrary direction in the crystal. The induced moment in molecules (A), oriented as in (4.51) can be directed along some (generally different) directions determined by the tensor relationship (4.53). This direction will generally not be the same as that of the induced moment of the set B of molecules oriented as in (4.55). The total moment for all molecules in the lattice is given by the vector sum of the moments from the two differently oriented sets of molecules and there results the relationship between the applied field and the total induced moment that we have previously written as

$$\mathbf{M} = \chi\mathbf{H} \tag{3.98}$$

in the crystal frame. Just as vectors are summed by adding their components in a common frame, so also must we operate with second-rank tensors.

In their respective frames, both sets of molecules have the same diagonal susceptibility tensors (4.52). Using (4.53), we transform each into a common crystal frame: for reasons of convenience[††] we choose an orthogonal system and in the monoclinic system this might be a, b, c'. For the A set of molecules, we find

$\mathbf{K}^C(\mathrm{A}) =$

$$
\begin{bmatrix}
K_1\alpha_1^2 + K_2\alpha_2^2 + K_3\alpha_3^2 & K_1\alpha_1\beta_1 + K_2\alpha_2\beta_2 + K_3\alpha_3\beta_3 & K_1\alpha_1\gamma_1 + K_2\alpha_2\gamma_2 + K_3\alpha_3\gamma_3 \\
K_1\beta_1\alpha_1 + K_2\beta_2\alpha_2 + K_3\beta_3\alpha_3 & K_1\beta_1^2 + K_2\beta_2^2 + K_3\beta_3^2 & K_1\beta_1\gamma_1 + K_2\beta_2\gamma_2 + K_3\beta_3\gamma_3 \\
K_1\gamma_1\alpha_1 + K_2\gamma_2\alpha_2 + K_3\gamma_3\alpha_3 & K_1\gamma_1\beta_1 + K_2\gamma_2\beta_2 + K_3\gamma_3\beta_3 & K_1\gamma_1^2 + K_2\gamma_2^2 + K_3\gamma_3^2
\end{bmatrix},
$$

$$\tag{4.56}$$

while the tensor for the B set, $\mathbf{K}^C(\mathrm{B})$, in the crystal frame is numerically the same but with sign changes given by

$$
\begin{bmatrix}
+ & - & + \\
- & + & - \\
+ & - & +
\end{bmatrix}, \tag{4.57}
$$

corresponding to the change in the sign of β_i between (4.51) and (4.55). Recalling that we have $\frac{1}{2}$ mole each of A- and B-type molecules in a lattice containing one mole, there follows

$$\chi = \tfrac{1}{2}[\mathbf{K}^C(\mathrm{A}) + \mathbf{K}^C(\mathrm{B})], \tag{4.58}$$

that is,

$\chi =$

$$
\begin{bmatrix}
K_1\alpha_1^2 + K_2\alpha_2^2 + K_3\alpha_3^2 & 0 & K_1\alpha_1\gamma_1 + K_2\alpha_2\gamma_2 + K_3\alpha_3\gamma_3 \\
0 & K_1\beta_1^2 + K_2\beta_2^2 + K_3\beta_3^2 & 0 \\
K_1\alpha_1\gamma_1 + K_2\alpha_2\gamma_2 + K_3\alpha_3\gamma_3 & 0 & K_1\gamma_1^2 + K_2\gamma_2^2 + K_3\gamma_3^2
\end{bmatrix},
$$

$$\tag{4.59}$$

reproducing the form of the monoclinic crystal tensor deduced from Neumann's principle in table 4.1.

[††] See also chapter 6 and Appendix A.

Orthorhombic system. Similar procedures and arguments pertain here, except that in the orthorhombic system molecular susceptibility tensors are related by *mmm* symmetry, so that one-quarter of the molecules are associated with principal magnetic axes oriented as in (4.51), one-quarter as in (4.55), and one-quarter each as in (4.51) but with the signs of either α_i or γ_i reversed, corresponding to the presence of explicit or implicit diads parallel to the crystal *a* and *c* axes respectively. When transformed into the orthorhombic crystal frame, the tensors for each set of molecules are numerically equal but occur in pairs such that, when summed to give the total crystal tensor, all non-diagonal terms vanish identically to give,

$$
\chi = \begin{bmatrix}
K_1\alpha_1^2 + K_2\alpha_2^2 + K_3\alpha_3^2 & 0 & 0 \\
0 & K_1\beta_1^2 + K_2\beta_2^2 + K_3\beta_3^2 & 0 \\
0 & 0 & K_1\gamma_1^2 + K_2\gamma_2^2 + K_3\gamma_3^2
\end{bmatrix},
$$

$$(4.60)$$

corresponding to the entry in table 4.1.

Uniaxial systems. In tetragonal crystals, molecular tensor orientations can be taken as those given for the orthorhombic system, together with a condition which relates molecular sets in pairs via the four-fold rotation axis parallel to *c*. Thus, for each molecule oriented as in (4.51) there is another whose direction cosines are given by

molecular magnetic axes

$$
\begin{array}{cc}
 & \begin{array}{ccc} x & y & z \end{array} \\
\begin{array}{ll}
\text{orthogonal} & a \\
\text{crystal} & b \\
\text{axes} & c
\end{array} &
\begin{pmatrix}
\beta_1 & \beta_2 & \beta_3 \\
\alpha_1 & \alpha_2 & \alpha_3 \\
\gamma_1 & \gamma_2 & \gamma_3
\end{pmatrix}
\end{array}, \qquad (4.61)
$$

corresponding to the crystallographic equivalence of the *a* and *b* axes. After summation, the crystal susceptibility tensor takes the form

$$
\chi =
$$

$$
\begin{bmatrix}
\frac{1}{2}K_1(\alpha_1^2 + \beta_1^2) + \frac{1}{2}K_2(\alpha_2^2 + \beta_2^2) & 0 & 0 \\
\quad + \frac{1}{2}K_3(\alpha_3^2 + \beta_3^2) & & \\
0 & \frac{1}{2}K_1(\beta_1^2 + \alpha_1^2) + \frac{1}{2}K_2(\beta_2^2 + \alpha_2^2) & 0 \\
 & \quad + \frac{1}{2}K_3(\beta_3^2 + \alpha_3^2) & \\
0 & 0 & K_1\gamma_1^2 + K_2\gamma_2^2 + K_3\gamma_3^2
\end{bmatrix}
$$

$$(4.62)$$

reproducing the form given in table 4.1. Exactly the same form arises for trigonal and hexagonal systems. Recall that the direction cosines for a molecule related to the reference molecule by an n-fold rotation axis in the crystal parallel to c are given by those in (4.51) multiplied by the rotation matrix $\mathbf{R}^{(n)}$,

$$\mathbf{R}^{(n)} = \begin{pmatrix} \cos\left(\dfrac{2\pi}{n}\right) & \sin\left(\dfrac{2\pi}{n}\right) & 0 \\ -\sin\left(\dfrac{2\pi}{n}\right) & \cos\left(\dfrac{2\pi}{n}\right) & 0 \\ 0 & 0 & 1 \end{pmatrix} \tag{4.63}$$

For example, the direction cosines for the three rotationally related molecules in a trigonal system with respect to the crystal a axis, say, are α_i, $-\frac{1}{2}\alpha_i + \frac{\sqrt{3}}{2}\beta_i$, $-\frac{1}{2}\alpha_i - \frac{\sqrt{3}}{2}\beta_i$ so that the first entry in the summed tensor χ, from (4.56), is

$$\chi_{aa} = \frac{1}{3}\sum_i K_i[\alpha_i^2 + (-\frac{1}{2}\alpha_i + \frac{\sqrt{3}}{2}\beta_i)^2 + (-\frac{1}{2}\alpha_i - \frac{\sqrt{3}}{2}\beta_i)^2] = \frac{1}{2}\sum_i K_i(\alpha_i^2 + \beta_i^2),$$

$$\tag{4.64}$$

and so on. A similar calculation yields the same result (4.62) for the hexagonal system also.

Cubic system. For each molecule with the reference direction cosines (4.51), there is one with direction cosines which are cyclically permuted over the equivalent crystal axes a, b, c. Accordingly, the crystal tensor is diagonal with each diagonal element given by

$$\chi_{ii} = \frac{1}{3}[K_1(\alpha_1^2 + \beta_1^2 + \gamma_1^2) + K_2(\alpha_2^2 + \beta_2^2 + \gamma_2^2) + K_3(\alpha_3^2 + \beta_3^2 + \gamma_3^2)], \tag{4.65}$$

where the summation convention has *not* been used.

More than one molecule in the asymmetric unit. In some crystal structures there is more than one molecule in the asymmetric unit. From the point of view of our describing 'molecular' susceptibilities, we must assign different orientation matrices like (4.51) to each set of molecules within the set of asymmetric units. Hence all the preceding expressions for crystal susceptibilities are modified by replacing each term by a sum over the crystallographically inequivalent molecules. For example, the χ_{aa} term for the orthorhombic system, given by

$$\chi_{aa} = \sum_i^3 K_i\alpha_i^2, \tag{4.66}$$

is to be replaced by

$$\chi_{aa} = \sum_{j}^{\text{indep mol sets}} \sum_{i}^{3} K_i^{(j)} \alpha_i^{(j)2}, \qquad (4.67)$$

where, again, the summation convention has *not* been used.

4.14 The calculation of molecular susceptibilities from crystal measurements

Ultimately, our purpose in measuring the susceptibilities of single crystals is to deduce, by way of a quantum-mechanical model, something of the nature of the bonding in the molecules of which the crystal is composed. Having defined molecular susceptibility with this goal in mind in such a way that the principal values and orientations of the **K** tensor relate to a unique set of structural information – that is, to parallel-oriented molecules, it is natural to enquire whether it is possible to derive the molecular magnetic susceptibility tensor directly from the measurements made on crystals. In general, however, the answer is 'no', simply because, in most cases, the crystal symmetry reduces the number of observables below the six necessary to define a second-rank tensor. Consider the case of the orthorhombic system, for example, where experiment yields values of χ_a, χ_b and χ_c only. Even if there is only one molecule in the asymmetric unit, these three pieces of experimental data are insufficient to make feasible a calculation of three principal molecular susceptibilities *plus* their orientations.

In some special circumstances, however it *is* posssible to deduce some or all of the characteristics of a 'molecular' tensor from the observations in the crystal frame, although even in these cases we still rely on the assumption of magnetic dilution if the crystal property is to correlate with the structural features of individual molecules. In principle, the most straightforward case concerns triclinic crystals. Provided that there is no more than one molecule in the asymmetric unit (less than one would correspond to a centrosymmetric molecule in $P\bar{1}$ and this occasions no problems), the orientations of the principal molecular and crystal susceptibilities are identical: indeed χ and **K** are the same by the identity (4.54). Therefore an experimental determination of the magnitudes and orientations of all three principal crystal susceptibilities in this case yields the 'molecular' quantities by identity. Unfortunately, the practical measurement of the susceptibilities of triclinic crystals is not simple and it is usually preferable to work with crystals of higher symmetry, if possible; however, techniques are available and are discussed in chapter 6.

For crystal systems with higher symmetry than triclinic, the molecular magnetic properties can be derived from the crystal measurements only if the orientations of the principal molecular susceptibilities are known from another source. That source might be an esr experiment, but only if it is accepted that the principal susceptibility axes coincide with the directions of the principal *g* values, and that need not be true. Otherwise, we must rely on symmetry. If the molecules lie in crystallographically special positions, some *rigorous* symmetry will attach to them which may define the directions of some or all of the principal molecular susceptibilities. Where such information *is* available, it is possible to 'work backwards', so to speak, and calculate the *magnitudes* of the principal molecular susceptibilities from the crystal measurements. Even this is not always a straightforward matter, however, as the non-linear nature of the equations relating crystal and molecular susceptibilities can yield ambiguities of sign, or worse. Information on the algebra of these 'back-calculations' is available in various reviews[3,5,6] but is not reproduced here – quite deliberately. Thus, while it is true that some systems possess sufficient molecular symmetry to permit the direct calculation of some or all of the molecular susceptibilities, most do not. Experience has taught us, as discussed briefly in the introductory review of chapter 1, that *approximate*, or non-rigorous, molecular symmetry is an utterly unreliable guide to the selection of principal molecular magnetic axes: occasionally it works, but we cannot depend on it. In any case, as also discussed in our introduction, magnetochemistry is generally most informative when applied to molecules of little or no symmetry. Altogether, therefore, this author's preference is to eschew all calculations of molecular susceptibilities from the raw experimental data. Furthermore, there seems little point in the exercise for we may always compare theory and experiment in the crystal frame. Ultimately, we make further assumptions in constructing a quantum-mechanical (ligand-field) model to interpret the measurements: the models we use refer to *molecules* and *with no further* approximations (other than the additivity we call 'magnetic dilution') we may transform and sum the calculated 'molecular' tensors and so make comparisons between theory and experiment in the crystal frame. Our interests and goals lie in the parameters of the quantum-mechanical model and their determination may be achieved by fitting in either frame: we should therefore choose the one that involves less assumptions.

4.15 Polar and axial vectors

We describe susceptibility as a second-rank tensor property because it is natural and convenient to represent the physical property by that mathematical entity. As discussed in §§4.2 and 4.3, tensors are defined, not in relation to their applications but by the manner of their behaviour under coordinate transformation: and the same, of course, goes for vectors. However, we have attached the name 'vector' to a number of quantities in the last three chapters which do not quite transform according to the rule given in §4.3. Although this is not a serious matter for most purposes, it is occasionally very important to distinguish so-called *polar* and *axial* vectors.

At a most basic level a discrimination can be made between these two types of quantities in terms of the number of *ad hoc* conventions that are required to complete their definition. Only one such is used to describe a polar vector. For example, *by convention*, we designated the lines of force representing an electric field in §2.1 as radiating *outwards* from a *positive* charge: so the electric field maps the variation of a polar vector thoroughout space. On the other hand, two conventions are used in the definition of a magnetic field. A magnetic field is produced by the movement of charge in a manner summarized for example, by the Biot–Savart law (3.27). But that expression involves a *vector product*: physically, this is exemplified by the direction of the magnetic field in a solenoid or current loop, being normal to the axis of that coil or loop. So **B** results from the vector product of two polar vectors. In addition to the conventions used to define their polarity, we now require a further one to decide in which of the two possible, antiparallel directions, perpendicular to both polar vectors, **B** points. The magnetic field **B** as the vector product of two polar vectors is, then, an example of an axial vector. For analogous reasons and because equality signs only relate species of precisely the same kind, a review of the material in chapters 2 and 3 shows that **B**, **H** and **M** are all axial vectors, while **E** and **A** are polar vectors. The magnetic vector potential **A** is polar, for example, because the axial vector **B** is expressed as its curl: the very name 'curl' makes this point, of course.

Wooster[7] describes transformations of polar and axial vectors in some detail. Let us briefly review these properties. In figure 4.7 are shown pairs of (*a*) polar and (*b*) axial vectors related by rotation by an arbitrary angle ϕ about some (arbitrary) reference axis. The transformation rule for the polar vectors is just that given in (4.7) and the same is true of the axial vectors in figure 4.7(*b*) which have been represented, not by an arrow but

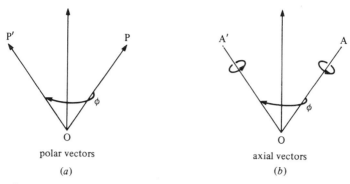

Fig. 4.7. Pairs of polar and axial vectors related by rotation about an arbitrary axis.

by a line associated with a sense of rotation: note that the sense of rotation is unchanged by the transformation just described. By contrast, however, consider the behaviour of polar and axial vectors on reflection in a plane, as in figure 4.8. Here we observe different behaviour of the two kinds of vectors in that the reversed sense of rotation of OA' on reflection effectively determines a reversed direction related to OA: had we placed arrow heads on these vectors to represent the sense of rotation, OA would point from O to A while OA' would point from A to O. Finally, consider the action of a centre of symmetry upon the two types of vector. It is obvious that the polarity of a polar vector is reversed. There is no change, however, for an axial vector, as may be seen be recalling that an inversion centre can be generated by the combined operations of reflection in a mirror plane, together with a two-fold rotation about the normal to that plane: this may be effected by taking figure 4.7(b) with $\phi = \pi$, in conjunction with figure 4.8(b). All this may be summarized by two rules. The transformation (4.7)

$$v'_i = a_{ij}v_j \qquad (4.7)$$

holds for *rotations* of *all* vectors: if a transformation involves inversion, the same rule continues to hold for polar vectors but axial vectors transform according to

$$w'_i = -a_{ij}w_j. \qquad (4.68)$$

Another way to express the transformation law for axial vectors is to write

$$w'_i = \pm a_{ij}w_j, \qquad (4.69)$$

taking the positive sign for pure rotations and the negative for inversions

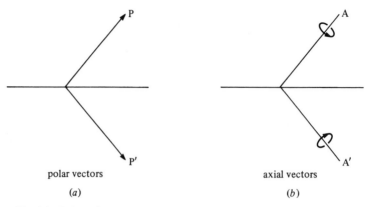

polar vectors   axial vectors

(*a*)   (*b*)

Fig. 4.8. Pairs of polar and axial vectors related by reflection in an arbitrary plane.

(in mirrors or a centre). Either way, the behaviour of axial vectors does not conform to the strict definition of a vector as first given in (4.7).

Should the idea of using the same name – vector – for two different entities be considered somewhat unsatisfactory, we can look again at the meaning of a vector product. Up till now, we agreed as a matter of definition to name the result of performing the operation known as a vector product a vector, though we have seen that this object does not transform like an 'ordinary' vector. So lets forget the business about names and recall the vector product of two polar vectors **r** and **s** expressed in the standard form:

$$m_{ij} = r_i s_j - r_j s_i. \qquad (4.70)$$

But we know from (4.14) that products like $r_i s_j$ transform as a second-rank tensor:

$$r_i' s_j' = a_{ik}(r_k s_l)a_{jl}. \qquad (4.71)$$

Under the same transformation, the reverse product transforms as:

$$\begin{aligned} r_j' s_i' &= a_{jk} r_k s_l a_{ij} \\ &= a_{jl} r_l s_k a_{ik} \\ &= a_{ik}(r_l s_k)a_{jl}, \end{aligned} \qquad (4.72)$$

in which we have used the fact that a dummy index may be renamed at will. Hence both $r_i s_j$ and $r_j s_i$ transform identically and so m_{ij} of (4.70) transforms as a second-rank tensor. However, we also note that

$$m_{ji} = r_j s_i - r_i s_j = - m_{ij}, \qquad (4.73)$$

so that the tensor is *antisymmetric*. Written in full, m_{ij} takes the form

$$[m_{ij}] = \begin{bmatrix} 0 & (r_1 s_2 - r_2 s_1) & (r_1 s_3 - r_3 s_1) \\ -(r_1 s_2 - r_2 s_1) & 0 & (r_2 s_3 - r_3 s_2) \\ -(r_1 s_3 - r_3 s_1) & -(r_2 s_3 - r_3 s_2) & 0 \end{bmatrix} \quad (4.74a)$$

$$= \begin{bmatrix} 0 & p_3 & -p_2 \\ -p_3 & 0 & p_1 \\ p_2 & -p_1 & 0 \end{bmatrix}, \quad (4.74b)$$

where

$$p_1 = r_2 s_3 - r_3 s_2, \quad p_2 = r_3 s_1 - r_1 s_3, \quad p_3 = r_1 s_2 - r_2 s_1 \quad (4.75)$$

have been chosen in such a way as to form a cyclic set.

In summary: the entity resulting from the operation of a vector product of two ordinary, polar vectors is an antisymmetric second-rank tensor. In conventional vector analysis, the three independent components of this tensor – the p_i of (4.75) – are taken to define an (axial) vector $\mathbf{p} = \mathbf{r} \wedge \mathbf{s}$. Nye[1] points out that it is only in three dimensions that the number of independent components of such an antisymmetric tensor 'coincidentally' equals the rank. We note further, that it is not too surprising to find that a vector formed from the off-diagonal elements of a skew-symmetric tensor does not transform exactly like an ordinary, polar vector.

References

[1] Nye, J.F., *Physical Properties of Crystals*, Oxford University Press, 1957.
[2] Feynman, R.P., Leighton, R.B. & Sands, M., *The Feynman Lectures on Physics*, vol. 2, Addison–Wesley, 1967.
[3] Krishnan, K.S. & Lonsdale, K., *Proc. Roy. Soc.*, **156A**, 597 (1936).
[4] Eyring, H., Walter, J. & Kimball, G.E., *Quantum Chemistry*, John Wiley, New York, 1944.
[5] Mitra, S., *Transition Metal Chemistry*, **7**, 183 (1972).
[6] Mitra, S., *Prog. Inorg. Chem.*, **22**, 309 (1977).
[7] Wooster, W.A., *Tensors and Group Theory for the Physical Properties of Crystals*, Oxford University Press, 1973.

5

Experimental arrangements

—

Having introduced the basic physical quantities of magnetism, we turn now to the business of their measurement. From the beginning we stress that our purposes in the subject defined by this book involve the measurement of the susceptibility tensors of dia- and paramagnetic compounds which might be expected to form the objects of ligand-field analyses. The practical aspects of magnetochemistry may be seen to fall under four main subheadings – the principles of experimental methods, instrumentation, measurement procedures, and sample preparation and handling – which are all interrelated to a degree. The subject of apparatus design is enormous, both historically and with respect to contemporary construction and so no attempt is made in this chapter to be representative, let alone comprehensive, in our discussions of how magnetic properties are measured. We look at the main classical methods in principle and examples of equipment that works well, focussing attention on the type of data which can and cannot be obtained and upon its accuracy, in such a way that procedures for the determination of susceptibility can be properly designed. We shall consider measurements for single crystals in particular, but also for powdered samples. *Procedures* for the measurement of susceptibilities of single crystals are described in the next chapter, once it has been established what is experimentally possible. In introducing all these matters, I take the opportunity to make a few subjective, and perhaps contentious, observations.

The apparatus employed in this area are broadly of two kinds; those involving the measurement of mechanical forces or couples, and those employing induction techniques. Over the years, many instruments using the same basic principles have been derived. Some have been discarded as general mechanical, electronic or cryogenic progress has favoured their competitors: others, though ingenious, were never satisfactory in the first place. Some were designed for measuring the susceptibilities of liquids and solutions: others, though excellent for studies of ferromagnets, were too insensitive for much weaker paramagnets. The last few years have seen the introduction of several commercial magnetometer systems, some capable of the detection and measurement of extremely feeble magnetization, which have the general property of being very expensive, and the occasional feature of not being designed to do the job claimed or even of not doing the job for which they were designed. This is not to say that all apparatus on offer in the market place are a waste of money: I merely wish to make the point that, for the prices now charged, manufacturers, on the one hand, ought to provide equipment which is robust, simple to operate, up to specification and utterly reliable, while, on the other, the user and buyer ought to think carefully first about what quantities he wishes to measure and not imagine that racks of electronic equipment, superconducting magnets and undoubted cryogenic sophistication will necessarily have been designed for the scientific task in view. The commercial norms of the consumer society are increasingly evident in the design of laboratory instrumentation – so *caveat emptor*.

Only a few instruments are described in this chapter and only one or two in any detail. The writer is not without prejudice, of course, but his preferences will be identified and justified by reference to two central objectives. Firstly, in relation to our subject matter as outlined in the Introduction, we are primarily concerned with the measurement of the magnetic susceptibilities of paramagnetic transition-metal and lanthanide complexes, in the form of powder or perhaps concentrated solutions, but especially as single crystals whenever these are available. Secondly, we seek an understanding of chemical interactions and bonding as revealed by magnetochemistry and ligand-field analysis: whatever the intellectual, or other, fascination to be had from electronic and mechanical gadgets, for us the magnetometer is merely a tool which we should be able to use with confidence and a minimum of special manipulative and cryogenic skill.

We begin with a discussion of two techniques involving the measurement of the forces acting upon a magnetic sample in an inhomogeneous magnetic field. The very well-known Gouy method is briefly reviewed and then the principles of the Faraday method are described. The application of the Faraday technique to the measurement of the susceptibilities of single crystals is discussed in some detail and a comparison is made between those instruments employing transverse magnetic fields on the one hand, and longitudinal ones, on the other. A magnetic crystal experiences only a couple in a uniform magnetic field, the measurement of which by a torsion balance yields the crystal anisotropy. The Krishnan 'critical-torque' method and its variants accordingly form the subject matter of the next section of this chapter. After that, a brief description of induction methods is provided, including some remarks on the most sensitive magnetometer yet devised – that employing a superconducting quantum-interference device (SQUID) at the heart of the detection system. The chapter concludes with some remarks on sample selection and handling.

5.1 The force on a sample in an inhomogeneous field

We saw in §3.5 that the force $\mathbf{F}$ on a current loop in a non-uniform magnetic field $\mathbf{B}$ is given by

$$\mathbf{F} = \mathbf{m} \cdot \operatorname{grad} \mathbf{B}, \tag{3.50}$$

where $\mathbf{m}$ is the magnetic moment of the loop. Using (3.114), an equivalent expression for a crystalline sample of volume v and magnetization $\mathbf{M}$ placed in a field $\mathbf{H}$ is

$$\mathbf{F} = \mu_0 \int_v \mathbf{M} \cdot \operatorname{grad} \mathbf{H} dv, \tag{5.1}$$

so that the force in the ith direction on an elemental volume dv is written

$$dF_i = dv \, \mu_0 \chi_{jk} H_k \frac{\partial H_j}{\partial x_i}, \tag{5.2a}$$

$$= \tfrac{1}{2} dv \, \mu_0 \chi_{jk} \frac{\partial}{\partial x_i} (H_j H_k), \tag{5.2b}$$

in terms of the susceptibility tensor defined by (4.3). This equation provides the principle for both the Gouy and Faraday methods for measuring susceptibilities.[††]

5.2 The Gouy method

In this well-known technique, the sample is investigated in the form of a long, thin cylindrical rod suspended between the poles of a magnet, as

---

[††] We are using volume susceptibility at this point – see §3.14.

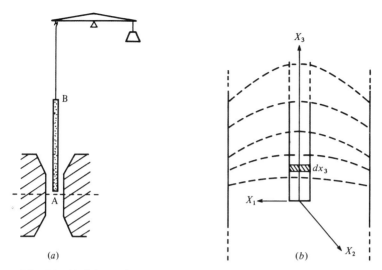

Fig. 5.1. (*a*) Schematic set-up for Gouy's method, and (*b*) reference frame and field gradient.

shown schematically in figure 5.1(*a*). In magnetochemical studies, the sample is almost invariably taken in the form of a finely ground powder, packed as uniformly as possible into a glass or perspex tube. However, the rod can equally well (and, from the standpoint of obtaining even-packing density – better) be machined from, or grown as, a large single crystal of the material – for example, for studies of the paramagnetism of metals and alloys – and so appropriate equations will be established for this more general situation first, while the simpler one for an isotropic powdered sample then follows as a special case. The lowest point A of the sample rod is taken to lie in a strong region of the field and this is most conveniently, and usually, taken to be very near where the field is maximal – say H_m. The highest point B of the sample is taken to lie in a region where the minimal value of the field is virtually zero. By taking the sample in the form of a thin cylinder and by suspending it at or very near the origin of the coordinate frame shown in figure 5.1(*b*), we may take $H_2 = H_3 = 0$ everywhere along the sample axis X_3. Therefore, we require only terms with $k = 1$ in (5.2a) and we note further that the experimental arrangement just described ensures that the only relevant non-zero first derivatives of the field are $\partial H_1/\partial x_3 = \partial H_3/\partial x_1$. The components of force experienced by each incremental volume dv of the sample are then given by

$$dF_1 = dv\,\mu_0\chi_{31}H_1\frac{\partial H_3}{\partial x_1}, \; dF_2 = 0, \; dF_3 = dv\,\mu_0\chi_{11}H_1\frac{\partial H_1}{\partial x_3}. \qquad (5.3)$$

Taking a as the cross-sectional area of the sample rod we have $dv = a\,dx_3$ and so the components of the total force acting on the rod are

$$F_1 = \mu_0 \int_A^B a\chi_{31} H_1 \left(\frac{\partial H_1}{\partial x_3}\right) dx_3, \quad F_2 = 0, \quad F_3 = \mu_0 \int_A^B a\chi_{11} H_1 \left(\frac{\partial H_1}{\partial x_3}\right) dx_3, \quad (5.4)$$

and hence

$$F_1 = \mu_0 a\chi_{31} \int_A^B H_1 dH_1 = -\tfrac{1}{2}\mu_0 a\chi_{31} H_m^2 \qquad (5.5)$$

and

$$F_3 = -\tfrac{1}{2}\mu_0 a\chi_{11} H_m^2. \qquad (5.6)$$

The negative sign implies that a paramagnet will suffer a force F_3 downward. The usual experimental arrangement is to suspend the sample from a sensitive balance working in a null-deflection mode so as to maintain the integration constants between measurement and calibration runs. Of course, as the balance serves only to measure vertical forces, the component (5.5) is of no interest in the Gouy experiment provided it is sufficiently small that the suspension is not significantly displaced from the vertical (against a restoring force arising from the sample weight) or, indeed, that the sample fouls the surrounding apparatus (cryostat or draught shield). When the sample is in the form of a powder whose crystallites are randomly oriented, (5.6) is modified to read

$$F_3 = -\tfrac{1}{2}\mu_0 a\bar{\chi} H_m^2, \qquad (5.7)$$

where $\bar{\chi}$ is the trace of the susceptibility tensor, called 'the mean or powder susceptibility' and is given in general by $\bar{\chi} = \tfrac{1}{3}\chi_{ii}$, or in particular by

$$\bar{\chi} = \tfrac{1}{3}(\chi_1 + \chi_2 + \chi_3). \qquad (5.8)$$

It is instructive to comment on the origins of the two components of force, (5.5) and (5.6), arising from a field applied along a single (X_1) direction. In general, the field H_1 induces magnetization in the sample parallel to all three principal magnetic axes. There is, for example, one moment along X_1 proportional to $\chi_{11} H_1$ and the force F_3 arises by the interaction of the field component H_3 with this moment. Although $H_3 = 0$ along the X_3 axis, the field gradient $\partial H_3/\partial x_1$ there is not generally zero. for H_3 is changing sign. H_3 therefore exerts a different force on the various parts of the moment $M_1 = \chi_{11} H_1$ and there results the force F_3 proportional to $\chi_{11}H_1(\partial H_3/\partial x_1) = \chi_{11}H_1(\partial H_1/\partial x_3)$. Similarly, the force F_1 describes the action of H_1 on the moment proportional to $\chi_{31}H_1$ induced along X_3.

For reasons made clear in the Introduction and elsewhere in this book, we are concerned more with the magnetism of single crystals of inorganic complexes than with the averaged information to be obtained from experiments on powders. We do not, therefore, elaborate on the more practical aspects of the Gouy method which are very well documented elsewhere.[1 – 5]

5.3 The Faraday method

In this method also, susceptibility is most usually determined from the vertical force (weight change) on a sample subjected to an inhomogeneous field. The most immediate difference from the Gouy method concerns the size of the sample, being of the order of 1 g in the Gouy experiment, while only about 0.1–10 mg of a transition-metal paramagnetic complex is required in contemporary Faraday apparatus. Two different field geometries may be employed with this technique, which we shall refer to as 'transverse' and 'longitudinal' according, respectively, to whether the field is perpendicular or parallel to the sample suspension. The longitudinal arrangement is usually effected with a superconducting solenoid, together with gradient coils, while the transverse field is commonly provided by a simple electromagnet with specially machined pole caps to produce an appropriate field gradient. It should be obvious why the latter setup has the longer pedigree and is still the more frequently employed Faraday technique. We discuss this arrangement, shown schematically in figure 5.2, first.

The principle of this method is virtually the same as that of the Gouy method, except that the sample is small and lies totally within a region of large, though highly inhomogeneous, field. With a large cross-sectional area of the magnet cores (say, $10 \, cm^2$ or more) and a sample volume of some $50 \, mm^3$ or less, we may once more assume that $H_2 = H_3 = 0$ and that $\partial H_2/\partial x_3 = \partial H_3/\partial x_3 = 0$ so that the vertical force on the sample is

$$F_3 = \mu_0 \int_v \chi_{11}H_1 \frac{\partial H_1}{\partial x_3} dv \qquad (5.9)$$

by analogy with (5.3). Further, the special profile of the pole tips indicated in figure 5.2(b) is such as to make the product $H_1(\partial H_1/\partial x_3)$ a constant – say C – over the sample-containing region (and beyond; typically over a volume in excess of $500 \, mm^3$): this region is often referred to as a 'region of

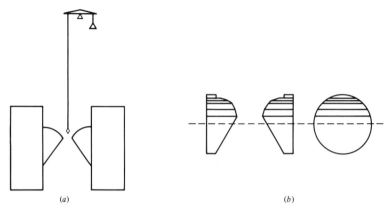

Fig. 5.2. The Faraday method (*a*) sample position and (*b*) a pole-piece design providing a region of constant force: the side and front views show a series of carefully machined flats, details being given in references [5, 7, 8].

constant force'. With this arrangement, we have

$$F_3 = \mu_0 \chi_{11} v C. \tag{5.10}$$

The constant C is normally calibrated using a sample of known susceptibility rather than determined explicitly via measurements of field strength and gradient. The experiment may be performed on a powdered sample, contained in a 'bucket' or some other appropriate vessel, in which case (5.10) is replaced by

$$F_3 = \mu_0 \bar{\chi} v C, \tag{5.11}$$

but we are more concerned with measurements on single crystals.

The form of (5.10) is especially suitable for single-crystal measurements of susceptibility, for we note that the measured force F_3 in the vertical direction provides a value for the magnetic susceptibility in the horizontal field direction. If we arrange that the magnet is free to rotate about the sample suspension axis X_3, we may investigate the magnetization of the sample in all directions perpendicular to that along which the force is measured without needing to reorient the sample. For example, if an orthorhombic crystal is suspended with the c axis vertical, appropriate orientations of the magnet in the horizontal plane will yield values for both χ_a and χ_b. We leave to chapter 6 the business of deciding which crystal orientations are appropriate in all cases. However, we note straightaway that, with the arrangement typical for many commercial magnetometers employing superconducting solenoids to provide the magnetic field and gradient, the force on a crystal is proportional to H_3 and $\partial H_3/\partial x_3$ so that the experiment yields the magnitude of the susceptibility *parallel* to the direction of force measurement. Therefore, three separate sample orientations are required to determine the complete susceptibility tensor of an orthorhombic crystal, say, rather than two as needed using the transverse field. The comparison is even more unfavourable to the longitudinal field experiment when measurements on lower-symmetry systems are undertaken, as detailed in the next chapter. This author has found the transverse field version of the Faraday experiment to be most successful and convenient and the following section describes a number of instrumentation details of his apparatus. Before looking at these, however, it is appropriate to review some cautionary remarks about the method, made by Wolf[6] some years ago.

Wolf reminds us that it is difficult to measure the field homogeneity directly so that we prefer to compare the force on a specimen with that exerted on a standard sample in the same field and thence to *assume* that the ratio of forces is equal to the ratio of the susceptibilities. He points out that this may not be valid and illustrates his argument as follows. Consider an anisotropic crystal having principal susceptibilities χ_a, χ_b, χ_c suspended in a transverse field Faraday magnetometer with its c axis vertical and parallel to the z axis of a laboratory-fixed frame. The laboratory xy plane and the crystal ab plane are coincident: let θ be the angle between the a and x axes in this plane. Then the force F_z on the sample is given by

$$F_z = \tfrac{1}{2}\int_v \frac{\partial}{\partial z}[H_x^2(\chi_a\cos^2\theta + \chi_b\sin^2\theta) + H_z^2\chi_c]\,dv, \qquad (5.12)$$

which, as the magnet is rotated around z, takes on two stationary values equal to

$$F_{1z} = \frac{v}{2}\left[\chi_b\left\langle\frac{\partial H_x^2}{\partial z}\right\rangle + \chi_c\left\langle\frac{\partial H_z^2}{\partial z}\right\rangle\right], \qquad (5.13)$$

and

$$F_{2z} = \frac{v}{2}\left[\chi_b\left\langle\frac{\partial H_x^2}{\partial z}\right\rangle + \chi_c\left\langle\frac{\partial H_z^2}{\partial z}\right\rangle\right], \qquad (5.14)$$

where $\langle\ldots\rangle$ denote average values (of force) over the volume of the crystal. For an isotropic specimen, these equations reduce to

$$\bar{F}_z = \frac{v}{2}\bar{\chi}\left[\left\langle\frac{\partial H_x^2}{\partial z}\right\rangle + \left\langle\frac{\partial H_z^2}{\partial z}\right\rangle\right]. \qquad (5.15)$$

If we write $\langle \partial H_z^2 / \partial z \rangle = A \langle \partial H_x^2 / \partial z \rangle$,

$$\frac{F_{1z}}{\bar{F}_z} = \frac{\chi_a + A\chi_c}{\bar{\chi}(1 + A)} \quad \text{and} \quad \frac{F_{2z}}{F_z} = \frac{\chi_b + A\chi_c}{\bar{\chi}(1 + A)}, \tag{5.16}$$

and so, when $\chi_a, \chi_b \neq \chi_c$ and $A \neq 0$,

$$\frac{F_{1z}}{F_z} \neq \frac{\chi_a}{\bar{\chi}} \quad \text{and} \quad \frac{F_{2z}}{F_z} \neq \frac{\chi_b}{\bar{\chi}}, \tag{5.17}$$

contrary to what is normally assumed in the calibration process. Wolf points out that it is difficult to estimate the magnitude or even sign of A. Of course, turning points exist at which the gradient $\partial H_z^2 / \partial z$ vanishes and, near these, no significant error would result, *for a sufficiently small crystal*, in comparing an anisotropic specimen with an isotropic standard. With larger crystals the contributions to A need not average out to zero and the effect of $\langle \partial H_z^2 / \partial z \rangle$ (and maybe even gradients of field components in the y direction with respect to z) may well be marked if $\chi_c \gg \chi_a$. Wolf quotes an experiment with cerium magnesium nitrate crystals, for which $\chi_b = \chi_c > 50\chi_a$ at low temperatures. In his conclusion, Wolf asserts his preference for other methods of measurement, like induction bridges or the vibrating-sample magnetometer, which employ uniform magnetic fields.

If has been worth reviewing Wolf's arguments so fully, if only because they identify the nature of the potential difficulty with the Faraday technique so clearly. However, we note that the results quoted by Wolf to illustrate his thesis were obtained using a Sucksmith ring balance and a design of pole piece which predated the marked advance made by Garber et al.,[7] and Heyding et al.,[8] and which is illustrated in figure 5.2(b). Not only has this latter design greatly reduced the unwanted term A in (5.16) *et seq.*, but contemporary Faraday equipment employing very sensitive electromicrobalances permits measurements to be performed on crystals of very small volume indeed. Finally, the ultimate empirical test shows that negligible errors arise from the neglect of this term in the instrument described below: this has been demonstrated by series of repeated determinations of the principal susceptibilities of fairly anisotropic crystals (10 : 1) by measurements on crystals of different shape and size, mounted along different axes, as appropriate. The instrument is believed accurate to about 1.5%.

5.4 A transverse field Faraday balance

The instrument design[9] we now review was guided by two main requirements: to measure the magnetic susceptibilities of small samples as powders and, especially, as single crystals; and to cover a large temperature range but with a minimal need of cryogenic skills so that the machine may be used as routinely as possible. No claim is made that this is the most sensitive or accurate magnetometer available: it is, however, robust, cheap and easy to use. These qualities may commend themselves to those more interested in the results of susceptibility measurements. It is worth recalling, for example that the model – ligand-field theory – within which such results are to be used and understood, is unlikely to be valid beyond the 5% level at best, as discussed in chapter 11. While this is no defence for careless experimental work, it does provide a guide for sensible compromise. The present, simple apparatus will usually furnish results accurate to about 1.5%; very much more care, effort and delicacy are required to decrease this figure by one order of magnitude. Sensitivity is a different question, to which we return in due course: suffice it to say that the instrument under consideration may be considered to be 'sensitive enough'.

The apparatus is shown in some detail, though schematically, in figure 5.3. On the left are sketched two orthogonal elevations of the system showing how the magnet and cryostat can rotate about the axis of sample suspension. Sample cooling is achieved with the consumption of electrical power but no cryogens, using a commercial closed-cycle, helium-gas heat pump. The 'Cryocooler' (Cryogenic Technology Inc.) is attached directly to the sample chamber between the magnet poles and held in a simple cradle that rotates with the magnet. The cryostat itself, shown in more detail in figure 5.3(*b*), was supplied on special order by The Oxford Instrument Co. The sample chamber comprises a stainless steel tube of about 8 mm diameter (id) having a copper tail and a sapphire window at the bottom for inspection of the sample via a simple mirror system. Cooling is achieved through the thick copper strap connecting the Cryocooler and sample tube; heating by means of a non-inductively wound nichrome coil around the copper tail-piece. A gold–iron (0.03% Fe)/chromel thermocouple, soldered with Wood's metal to the copper flange at the base of the sample tube, serves for temperature measurement and control. An external liquid-nitrogen reference junction completes the thermocouple. Temperature control is effected using a 'Precision Temperature Controller'

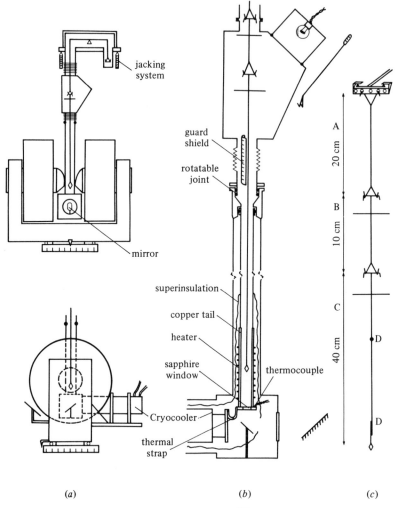

jacking
system

guard
shield

rotatable
joint

mirror

superinsulation

copper tail

heater

sapphire
window

thermocouple

Cryocooler

thermal
strap

A

20 cm

B

10 cm

C

40 cm

D

D

(a)   (b)   (c)

Fig. 5.3. A single-crystal Faraday balance[9]. (a) Elevations showing
rotatable magnet, cryostat and cooling system, supported beneath the
fixed microbalance. (b) Section of cryostat. (c) Torsionally rigid, but
demountable, light quartz suspension.

(Oxford Instruments) which incorporates both integrating and differentiating circuits.

Samples are weighed with an electromicrobalance (Sartorius, model
4433) accurate to 0.1 μg, supported on a commercial jacking system (Oxford
Instruments) with an additional facility for two-dimensional lateral
adjustment. Samples can thus be centred in the sample tube and moved

vertically in a reproducible fashion. The cryostat, cooler and magnet rotate together with respect to the microbalance and jacking system which are fixed. This is made possible by a simple rotating joint with O rings as shown in figure 5.3(*b*). The diagram also shows a device used for loading and removing samples from the cryostat (Oxford Instruments). The sample suspension (*vide infra*) is flexible enough to allow insertion at an angle through a removable light-bulb housing: this obviates procedures involving the swinging of the microbalance to and fro during sample changing. A loose-fitting metal sleeve inside the connecting bellows between cryostat and loading chamber prevents samples being caught up in the bellow folds while being raised and lowered in the cryostat.

The water-cooled electromagnet has special-profile Faraday pole caps, 10 cm in diameter with an adjustable gap about (3 cm here). The design of these caps is similar to those[8] in figure 5.2(*b*) in principle but varies in detail. The magnet, pole caps, magnet rotating base with protractor, and magnet power supply are all from Bruker Spectrospin.

Figure 5.3(*c*) shows the suspension system which was designed to permit easy loading of samples, to prevent rotation of the sample so that susceptibilities may be measured as functions of crystal orientation merely by rotation of the magnet subassembly, and to be lightweight. The latter condition is imposed by the design of the electromicrobalance which, in its most sensitive mode, takes a maximum of 3.0 g: the suspension shown in the diagram weighs rather less than 1 g. Only the lowest section (C) is removed when changing samples, the cross-piece serving both as a 'lifting handle' and as a guard against the whole section falling into the cryostat in case of accident. The vertical components of this section are made of quartz, by virtue of its low coefficient of expansion, poor thermal conductivity, and lack of paramagnetism. Cross-pieces and hooks are of stainless-steel wire. Two joints (D) are made with 'Durofix' (an ester-solvent-based glue) for ease of replacement – sample change or breakage – and the rest with ('Araldite') epoxy resin. The complete suspension is in three parts merely for ease in dismantling the superstructure in servicing operations. The overall length of about 70 cm ensures that this microbalance (enclosed in its own thick stainless-steel case) is completely unaffected by the magnetic field. No sample rotation is observed with this suspension as the magnet is rotated, even with very anisotropic crystals: also, balance readings are reproducible when a given magnet orientation is approached from either direction. The suspension, though light, is flexible (necessary for sample loading) and, with reasonable care, robust. Samples in the form of single crystals are attached to a short quartz extension as shown in the diagram and as will be discussed in §5.11. Powdered specimens are enclosed within a plastic container, in turn attached to the end of the main quartz rod by a short piece of quartz.

5.4.1 Operation and calibration

Field profile. Lateral adjustment of the crystal within the sample tube up to 3 mm either way makes no significant change in force. Vertical adjustment by ± 5 mm from the position of maximum force introduces a force change of about 2%. The jacking system, however, allows reproducible vertical setting positions to be achieved easily within 0.1 mm. Using samples with maximum dimensions of about 5 mm, the maximum difference in force between the extremities and middle of such crystals is less than 0.5%. As a matter of routine, the position of maximum force – virtually a 'plateau' with respect to vertical location – is established at the beginning of each run, including calibration runs.

Helium environment. Commercial-grade helium gas has sufficient impurities to cause noticeable condensation on the sample at low temperatures, as revealed by both the weight and appearance of the sample. This may be avoided by first passing the helium gas slowly, via a glass frit, through a column of molecular sieves (grade 5A) immersed in liquid nitrogen. The column consists of a series of U tubes of about 2 cm diameter and overall length 1 m. The sample chamber and electrobalance are evacuated and refilled, four times, with the helium so purified. The sieves are regenerated periodically by baking to 10^{-3} Torr at about 240° C for several hours. Difficulties in making the sample chamber, loading box, and balance sufficiently vacuum-tight to prevent ingress of air and subsequent condensation on cold samples under conditions of low helium-gas pressures (as is common practice) were overcome by running the experiment at about 3 psi overpressure. As so little volume of material actually becomes cold, an overpressure is found to be maintained throughout the temperature range. No problems with condensation occur under these conditions, no off-field weight drift with time, nor any time dependence of the force.

Temperature. Cooldown time with this system is about $1\frac{1}{4}$ hours compared with $\frac{3}{4}$ hour for the Cryocooler without load. A base temperature of some 20 K is achievable, compared to about 9 K without load. Temperature stabilization is better than ± 0.2 K. Temperature increases below 50 K are virtually instantaneous, but require 5–10 min near room temperature: these observations derive from monitoring both the precision temperature controller and microbalance readings. Experience indicates that the easier and quicker experimental routine is to take readings on warming from base temperature. A typical run of, say, 20 temperatures for two magnet orientations (field off-on-off) requires about six hours. Overall, the procedures are reliable, repeatable and routine.

Calibration. Samples are suspended in helium gas, an excellent heat conductor, within the copper tail-section of the sample tube. However, a large temperature gradient exists from the base of the tube, which is coupled directly to the cryocooler at 9 K, to the top of the tube at room temperature. Furthermore, although most of this gradient will occur along the stainless-steel top-section of the sample tube, it is hard to believe but that some thermal variation will exist within the region of the copper tail-piece itself, notwithstanding its excellent thermal conductivity. The thermocouple is mounted on the outside of the copper flange at the base of the sample tube and so cannot be relied upon to measure sample temperatures directly. Two objections can be raised to the idea of placing a supplementary thermocouple (or other thermometer which doesn't affect, and isn't affected by, the magnetic field) in the sample chamber just below the sample: firstly, that the obvious route for the electrical leads is (avoiding extra vacuum seals through the walls of the chamber) to be held, mechanically or with glue, to the inside walls of the sample chamber, thereby offering a possible obstruction when raising and lowering new samples; and secondly, that in the event of an accidental dropping of a crystal into the cryostat, its removal past such a thermocouple could be awkward. Anyway, in the interests of simplicity, as ever, together with a preference for characterizing the *system* in operation, temperature calibration is performed in *situ* against the known[10] temperature dependence of the susceptibility of manganese(II) Tutton salt, $Mn(SO)_4 \cdot (NH_4)_2 (SO_4) \cdot 6H_2O$. Crystals of this salt do not effloresce in dry helium gas (unlike those of ferric alum), are virtually magnetically isotropic, and exhibit Curie–Weiss behaviour for their susceptibility (with a Weiss constant, $\theta \approx 0.0$ K) down to about 1 K. The calibration procedure is best repeated for several crystals, of masses ranging from 1 to 5 mg. It is found that the reading of the precision temperature controller (that is, of the thermocouple) differs from the true temperature (of the sample) by up to $10°$, at the higher end of the experimental temperature range.

This calibration routine simultaneously furnishes an absolute calibration for the complete instrument, relating force and susceptibility for a given field strength. The process must be carried out for various magnet excitation currents, so providing the final feature of effective field calibration for the apparatus. True field calibration, probably required only for work at temperatures below those attainable with the Cryocooler system, can be effected by repeating these processes using a sample whose susceptibility varies with field (see §4.4) in a known way.

Diamagnetic corrections for sample holders. This involves a simple, direct and obvious procedure if the sample is in the form of a powder contained within a plastic 'bucket' or capsule: namely, the experiment is repeated with a full and empty container. As this is most conveniently done at room temperature, the value for the container plus (usually paramagnetic) sample is best obtained by careful interpolation of results from an extended series over a wide temperature range. Otherwise, the process is obvious. In the case of single crystals, some way of allowing for the diamagnetism of the glue, which attached the crystal to the suspension, must be devised: we have found the following technique satisfactory. As will be described in §5.11, the weighed crystal is glued with 'Durofix' onto a short length (about 2–3 cm) of thin quartz fibre (say, 0.3 mm diameter) in the desired orientation. This short length of fibre is similarly glued to the main, rather thicker, suspension by forming an overlap (see figure 5.3(c)) so as to maintain the proper sample orientation. Using the microbalance jacking system, the sample height is optimized and that vertical position is subsequently maintained throughout the following stages. The susceptibility measurements are carried out over the complete temperature range of interest and the sample is then left to warm to room temperature. The removable section (C) of the suspension is then lifted out of the cryostat and, by softening the glue with solvent (ester), the crystal is removed from the short, bottom piece of quartz: the latter is left precisely in place. All old glue at the bottom end of the suspension is dissolved away and finally replaced with a small 'blob' of glue guessed to be similar in amount to that originally used to attach the crystal. The diamagnetism of the whole suspension in this condition is taken as the required correction to the original raw data. The element of guesswork may seem somewhat careless, but a series of similar experiments with varying amounts of glue – ranging from none to an excess – show that the deviation of these results about the guessed mean is wholly acceptable within the accuracy of this experiment. By far the largest contribution to the overall diamagnetic correction of the sample support comes from the quartz extension piece; and that is one reason why a rather thinner piece is used here. One final cautionary note: the total force on the suspension without sample is not a maximum at the vertical setting chosen for the complete system but this is irrelevant, of course, and should not prompt any readjustment of the suspension height: it does mean, however, that diamagnetic corrections should be carried out in the way described at each field strength used to collect the raw data.

5.4.2 Comments

There follow, in note form, a few remarks about details of this apparatus and indeed about the philosophy of the instrument as a whole.

(i) The choice of a heat pump for sample cooling was made for the author's apparatus mainly because of the extreme simplicity of operation. After flushing the sample chamber with helium gas as described above, an electric switch is thrown and the room can then be left. Less than $1\frac{1}{2}$ hours later, the sample has achieved base temperature and the measurements may begin. There are no problems with transferring liquid cryogens, or of requiring them to be ordered in advance, or of finding a sensible use for them in the event of sample or apparatus failure. Before construction of the equipment in the first instance, we had some concern that the vibration of the heat pump itself would couple to the microbalance. These components are in contact through (i) the helium gas in the sample chamber, (ii) the pair of O rings in the rotatable joint, (iii) an O ring which separates the inner sample chamber from the outer vacuum jacket, and (iv) the metal bellows beneath the balance and between the loading box and the cryostat. No problems due to vibration were experienced at all.

Perhaps the greatest objection to the use of the Cryocooler device is its near-vanishing power at the lowest temperatures so that even the small thermal load of the sample chamber (well insulated by vacuum plus aluminized mylar 'superinsulation') is sufficient to raise the base temperature from about 8 K to about 20 K. If it is desirable to measure susceptibilities down to 4 K or less, the heat pump may be replaced by a 'Continuous-Flow Cryostat', as supplied commercially by Oxford Instruments, for example. That company are pleased to modify their standard equipment to meet the needs of the individual customer. Probably the simplest arrangement is to use a cold finger, cooled by the Joule–Thomson effect of the continuous-flow device, kept in thermal contact with the sample chamber via helium exchange gas. In this way the cryostat need not rotate with the magnet and so this obviates the need for a rotating-joint assembly. The cooling power of liquid helium is large, so that very low temperatures are easily obtained and, by pumping on the exhaust, sample temperatures down to about 2 K are fairly easily achievable. Further, the continuous-flow devices may be operated using either liquid helium or liquid nitrogen as cryogens, which offer, therefore, greater flexibility if cheaper runs over lesser temperature ranges are desired. Finally, it is worth remarking that it may not necessarily be desirable to measure susceptibilities down to 10 K or less. If the object of the study is to determine the ligand-field characteristics of a paramagnet, the temperature range provided by the convenient heat pump arrangement may well suffice. At the same time the onset of the intermolecular exchange – coupling effects–so common at the lowest temperatures–might not be revealed. Such effects are frequently exaggerated manifestations of very low energy processes, having rather little bearing upon theories of bonding, and in any case are extremely difficult to interpret in unambiguous, significant, physical detail. So, unless the investigator's interest lies with the physics of such second-order effects, an ignorance of these overdramatic cooperative artefacts might help prevent sight of the main chemical aims of a ligand-field study being lost.

(ii) So many designs of Faraday balance, both commercial and of 'private' construction, have in the past, and even today, been plagued by a seemingly trivial inconvenience. Thus it has been usual to separate magnet and balance, presumably to prevent any effect on the latter by stray fields, by a large distance, so causing samples to be hung on the ends of very long suspensions – 1.5 m or more. There are a number of operational objections to this: (i) long suspensions, whether fibres or chains, are much heavier than short ones and may reduce the sensitivity of the microbalance through overloading; (ii) they are more difficult to handle, especially to introduce and withdraw from a narrow sample tube; (iii) they aggravate the difficulty in centering a sample in a narrow cryostat; (iv) they are more subject to breakage, and (v) as demonstrated by the apparatus described above and in figure 5.3, long suspensions are simply not necessary. The steel enclosure of a Satorius microbalance easily eliminates the influence of stray fields some 70 cm distant; alternatively, if using a glass-enclosed electro-balance like those by Cahn, for example, a thick steel plate placed between magnet and balance will suffice equally well.

(iii) Though not essential, we have found a small window at the bottom of the sample chamber to be very convenient. It is then so easy to check that the sample is intact, reasonably centred and not fouling the sides of the sample chamber. Sapphire windows soldered to copper or other flanges with Wood's metal are standard 'off-the-shelf' components from various cryogenic companies today.

(iv) It is useful to place a small sample of, say, uranyl nitrate or of some other β emitter at the bottom of the sample chamber to help dissipate static electricity which tends to build up or be maintained on the sample and suspension in the very dry atmosphere. Incidentally, this author has found no need to coat suspensions with graphite for this purpose.

(v) Sometimes powdered samples do not orient randomly. An example of this has been demonstrated[111] for gillespite, a mineral containing planar-coordinated iron (II) as is discussed in chapter 12, where the rather flakey crystallites tend to align parallel and even to reorient in a magnetic field so yielding unreproducible results. All this was prevented,[111] simply by grinding the crystals in a small quantity of petroleum jelly (Vaseline). The susceptibility of this sticky mass, which freezes solid at low temperatures, of course, was measured in a Faraday 'bucket' and the sample subsequently weighed by dissolving away the petroleum jelly. Of course, this option may not be available for all substances, especially if they are soluble in the same solvents as the jelly: but, clearly, many variations on this theme may be tried.

(vi) The manufacture of the suspensions: the problem here is to arrange that the various 'double-hook' linkages shown in figure 5.3(c) hang snugly (rather than balanced on only one hook) while simultaneously causing the sample to be placed essentially central within the sample chamber and pole gap; and to do this whilst accepting that individual lengths of drawn quartz rod may not be perfectly straight. Although all reasonable efforts are made to construct the various parts of the suspension accurately, the final assembly includes one step which permits the system to correct any remaining errors itself. Referring to figure 5.4, the joint X between vertical and horizontal wires in the T part of the double-hook arrangement is made in the form of a freely swinging hook. The whole suspension is loaded with a typical crystal as shown, and surrounded by a draught shield. A drop of liquid epoxy resin is placed over and around this hook and the joint allowed to set hard over a period of hours. The 'front' of the suspension so constructed is marked with a fleck of bright paint so that the correct one of the two ways of hanging the suspension is readily identified. Once again, the philosophy of constructing and calibrating the various parts of the instrument *in situ* reduces the need for total accuracy in their manufacture.

(vii) Overall, the instrument we have reviewed is not sophisticated, if that is taken to mean high technology. However, it works well and reliably:

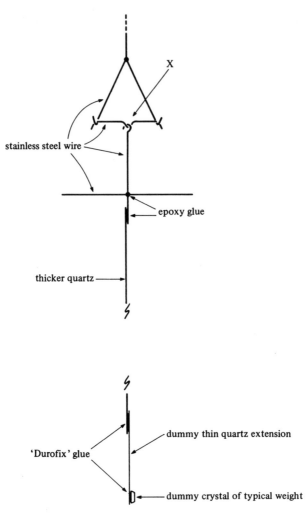

Fig. 5.4. Construction of the bottom section of the sample suspension for the Faraday balance shown in figure 5.3. Slow-hardening epoxy glue is introduced at X.

it is not especially delicate, though accurate enough. Its sensitivity can be illustrated by noting that results accurate to about 1.5% can be achieved from a crystal of $CuSO_4.5H_2O$ weighing 2 mg at room temperature or from a typical d^7 Co(II) complex crystal weighing as little as 0.4 mg. Normally, crystals of 1–10 mg are used, however. Having once oriented a single-crystal sample (§5.11) a single, working day is sufficient to hang the crystal from the balance, flush four times with the purified helium gas, characterize the 'plateau' of the field profile, cool to base temperature, make susceptibility measurements at about 20 temperatures up to room temperature for two different magnet orientations, and finally determine the diamagnetism corrections. The weighing process is boring, of course, and some people prefer to automate this function using relays, timers and sensors. It is worth pointing out that this tends to increase the risk of instrument failure for trivial, but annoying, reasons and that a project of ligand-field analysis, at any rate, is unlikely to require the study of so many crystals that the effort in building an automatic machine would be certain to be sufficiently rewarded. Nevertheless, for those preferring to build an automatic instrument, it is worth drawing attention to a version of the Faraday method in which the force on the sample is arranged to be horizontal rather than vertical. The advantage here is that buoyancy corrections, which vary markedly with temperature and other ambient conditions, are totally irrelevant, so that the usual sequence – weigh in zero field/weigh with field on/weigh with field off – is unnecessary. However, as stated in the introduction to this chapter, no attempt is made even to be representative in the choice of apparatus described here and so we now leave the Faraday technique and move on to the measurement of magnetic anisotropies.

5.5 The couple on an anisotropic crystal in a magnetic field

Unless otherwise stated in this section, we consider couples exerted on specimens in *uniform* magnetic fields. We know from §3.4 that an elemental magnetic dipole $\mathbf{m}$ – which may be permanent or induced by the applied field – is subject to a couple $\mathbf{G}$ in a magnetic field $\mathbf{B}$ given by

$$\mathbf{G} = \mathbf{m} \wedge \mathbf{B}. \tag{3.41}$$

The corresponding expression for the magnetic moment $\mathbf{M}$ of a crystal in a field $\mathbf{H}$ is

$$\mathbf{G} = \mu_0 \mathbf{M} \wedge \mathbf{H}, \tag{5.18}$$

where, again, $\mathbf{H}$ is the field produced by sources external to the sample; and this relationship is valid for a magnetic moment $\mathbf{M}$ which is permanent or induced by the external field. In the case of a permanently magnetized sample (like a compass needle) there exists exactly one orientation of the dipole $\mathbf{M}$ in the field $\mathbf{H}$ – that is, when $\mathbf{M}$ and $\mathbf{H}$ are parallel – for which the cross-product (5.18) vanishes identically. Otherwise a permanent dipole of non-zero magnitude in a non-vanishing field suffers a couple and will tend to rotate so as to achieve the unique orientation for which **M** is parallel to **H**.

The case for para- and diamagnets is rather different, depending upon the isotropy, or otherwise, of the *induced* magnetism. Excepting a generally small effect due to shape anisotropy which is discussed in §5.9, the situation is as follows. In an isotropic material – powdered sample or cubic crystal, say – the induced moment **M** is invariably parallel to the applied field **H** and so, as discussed above, the sample suffers no torque in the field. On the other hand, in the more general *anisotropic* case, **M** and **H** are not usually coparallel and so **G** in (5.18) does not vanish. Writing (5.18) in suffix notation, we have

$$G_{ij} = \mu_0(-M_iH_j + M_jH_i), \qquad (5.19)$$

so that the torque is an antisymmetric tensor whose three independent non-zero components form an axial vector, as described in §4.15. Now,

$$M_i = v\chi_{ij}H_j \qquad (5.20)$$

as usual, where v is the volume of the sample whose volume susceptibility is χ, and on substitution into (5.19) there results

$$G_{ij} = v\mu_0(-\chi_{ik}H_kH_j + \chi_{jk}H_kH_i). \qquad (5.21)$$

If we work in a laboratory frame oriented parallel to the principal susceptibility directions, the components of **G** simplify to

$$v\mu_0 \begin{bmatrix} 0 & -(\chi_1-\chi_2)H_1H_2 & (\chi_3-\chi_1)H_3H_1 \\ (\chi_1-\chi_2)H_1H_2 & 0 & -(\chi_2-\chi_3)H_2H_3 \\ -(\chi_3-\chi_1)H_3H_1 & (\chi_2-\chi_3)H_2H_3 & 0 \end{bmatrix} \qquad (5.22)$$

and we observe how the couple depends only upon the differences between the principal crystal susceptibilities; that is, on the crystal magnetic *anisotropy*. Various methods for measuring such couples, and hence appropriate anisotropies, are based upon the idea of suspending a crystal in a magnetic field on a fine torsion fibre. Mitra[12] has reviewed these in some detail so we confine ourselves here to a brief summary of the most common variants, so as to provide a background against which our summarizing remarks in §5.12 may be made.

5.6 The critical-torque method

In 1935, Krishnan & Banerjee introduced[13] their technique for anisotropy determination. A crystal is suspended in a uniform magnetic field on a fine fibre, commonly of quartz, from a calibrated torsion head. The sample is, of course, shielded from draughts, most usually within the confines of an appropriate cryostat. The experiment begins with the torsion head turned to such an angle that the crystal suffers no torque on introducing a magnetic field. For the present discussion we assume that this position is found exactly: as described shortly, the condition need only be achieved approximately in practice. With the crystal in this so-called 'setting position', the magnet is energized and the torsion head slowly rotated (clockwise, say) from the equilibrium position. While the torsion head rotates from the starting point by an angle α, the crystal rotates by an angle δ which is much smaller because the suspension fibre (unlike that used in the Faraday apparatus) is torsionally non-rigid and the crystal itself is subject to a restoring torque arising out of the interaction of its magnetic anisotropy and the applied field via (5.21). Using the same coordinate frame shown in figure 5.1(b), the magnitude of the couple about The (vertical) X_3 axis is then given by

$$
\begin{aligned}
G_{(\alpha-\delta)} &= v\mu_0(\chi_{max} - \chi_{min})H_1 H_2 \\
&= v\mu_0(\chi_{max} - \chi_{min})H^2 \sin\delta\cos\delta \\
&= \tfrac{1}{2}v\mu_0(\chi_{max} - \chi_{min})H^2 \sin 2\delta,
\end{aligned} \tag{5.23}
$$

where χ_{max} and χ_{min} are the maximum and minimum susceptibilities in the crystal plane perpendicular to the torsion fibre.[††] An approximate treatment now takes the maximum value of the restoring couple as occurring when $\sin 2\delta$ is maximal; that is, for $\delta = 45°$. In that case we conclude that, as the torsion head is smoothly rotated, the crystal rotates up to 45° from the original equilibrium 'setting' position but thereafter the restoring couple (5.23) decreases from the maximum value

$$
G^{max}_{(\alpha-45°)} = \tfrac{1}{2}v\mu_0\Delta\chi H^2, \tag{5.24}
$$

where $\Delta\chi$ is the in-plane anisotropy $(\chi_{max} - \chi_{min})$, as the mechanical torque continues to increase. The crystal equilibrium is therefore lost and the sample suddenly spins round or 'flips': the technique is often referred to as the 'flip-angle method'. It is good experimental practice to repeat the

---

[††] Again, (5.23) is written in terms of volume susceptibilities: the equivalent relation in terms of molar susceptibilities is

$$
G_{(\alpha-\delta)} = \frac{1}{2}\frac{w}{W_M}\mu_0(\chi_{max} - \chi_{min})H^2 \sin 2\delta,
$$

where w is the mass of the crystal comprising molecules of molecular weight W_M.

measurement by rotating the torsion head, first clockwise from the setting position till the crystal flips, and then – after removing the field, resetting and reintroducing the field – rotating the torsion head anticlockwise. Indeed, further statistical improvement may be obtained by repeating both measurements for a magnetic field of reversed polarity also. We note in passing that, as indicated above, the origin position for the torsion head need not correspond exactly with the crystal setting position. Provided it is approximately correct (that is, within one quadrant), the mean of measurements performed clockwise and anticlockwise is normally equally satisfactory: this obviates the need to establish the setting position exactly – a process which can be rather tedious. Using very fine quartz torsion fibres, values for α can be several hundred degrees and reproducible to within $1-5°$, so the method can be extremely *sensitive*. Its accuracy, on the other hand, is much less good, as will be discussed.

Actually, the maximum restoring torque is not given exactly by the treatment above. The correct analysis is a little more subtle and involves a correction to the critical angle although this is frequently, but not always, trivially small. First, let us introduce C, the torsional constant of the fibre: by this is meant that a total twist θ in the fibre between torsion head and crystal involves a torque of $C\theta$. Then, if the crystal has rotated by δ while the torsion head has been turned by α, the equilibrium between mechanical torque in the fibre and magnetic restoring couple is written, following (5.23), as

$$C(\alpha - \delta) = \tfrac{1}{2}v\mu_0\Delta\chi H^2 \sin 2\delta. \tag{5.25a}$$

Rearranging, we have

$$\alpha = \frac{\Delta\chi v\mu_0 H^2}{2C}\sin 2\delta + \delta. \tag{5.25b}$$

Now, as the torsion head is rotated (as α increases), the point is reached at which an incrementally tiny change in crystal orientation δ is caused by an infinitesimally small increase in α; and at the limit of 'flip', $d\alpha/d\delta = 0$. The true condition for 'flip', therefore, is that

$$\frac{d\alpha}{d\delta} = 0 = \frac{\Delta\chi v\mu_0 H^2}{C}\cos 2\delta_c + 1 \tag{5.26a}$$

and hence

$$\cos 2\delta_c = -\frac{C}{\Delta\chi v\mu_0 H^2}, \tag{5.26b}$$

where δ_c is the value of δ at the critical point of 'flip'. All quantities on

Table 5.1 *Angular corrections σ as function of torsion angle α (degrees)*

$\alpha°$	819	413	169	87	54
$\sigma°$	1	2	5	10	20

the right-hand side of (5.26b) are intrinsically positive, so that $\cos 2\delta_c$ is negative and hence $\delta_c > 45°$. Let us define a 'correction' σ to the earlier, elementary treatment by

$$\delta_c = 45 + \sigma. \tag{5.27}$$

Then from (5.26b) and the fact that $\cos(2 \times 45) = 0$,

$$\sin 2\sigma = \frac{C}{\Delta\chi v\mu_0 H^2} = -\cos 2\delta_c. \tag{5.28}$$

Similarly, and using (5.25a),

$$C(\alpha - \delta_c) = \tfrac{1}{2}v\mu_0\Delta\chi H^2 \cos 2\sigma \tag{5.29a}$$

and hence

$$\cos 2\sigma = \frac{2C(\alpha - \delta_c)}{\Delta\chi v\mu_0 H^2}. \tag{5.29b}$$

Dividing (5.29b) by (5.28), we finally obtain

$$\cot 2\sigma = 2(\alpha - \delta_c) = 2(\alpha - 45° - \sigma). \tag{5.30}$$

Hence, if we use (5.24) in the form

$$\Delta\chi = \frac{2}{v\mu_0}\left(\frac{C}{H^2}\right)(\alpha' - 45°), \tag{5.31}$$

the reading of the torsion head α' is to be corrected by subtracting an angle σ whose value depends upon the value to be corrected: table 5.1 lists a few such values, derived from (5.30). It is clear that for torsion-head rotation greater than, say, 250°, the correction is negligible. On the other hand, when such twists are small – because of intrinsically small anisotropies, low crystal mass or stiff fibres – the correction can become large. Added to this is the fact that the nature of the 'flip' under such circumstances is more in the manner of a 'lazy turn' whose critical point cannot be estimated with much accuracy. For all these reasons, therefore, the critical-torque method begins to fail rather rapidly with decreasing gross effect. The null-deflection method, reviewed briefly now, does not suffer these drawbacks.

5.7 The null-deflection method

Stout & Griffel[14] described a variation of this anisotropy–torsion fibre technique which avoids the correction problem. It probably requires a little more manipulative care than the critical-torque method but is intrinsically more accurate. No approximation is involved in (5.23) and the maximum restoring couple does occur when $\delta = 45°$: the 'correction' term discussed above arises out of the recognition that the crystal does not flip exactly under that condition. In the 'null-deflection' method, the crystal is maintained at equilibrium at all times. The setting position is first determined – this time with care. Then, with the field off, the crystal is rotated by $45°$ so that χ_{max} lies at $45°$ to what will be the field direction. The position of the crystal is now recorded accurately through an appropriate optical arrangement. The magnetic field is applied and the crystal tends to rotate so as to align χ_{max} with $\mathbf{H}$. The torsion head is then rotated so as to return the crystal to its original orientation. If this rotation is by an angle α_c, we have from the relationship for maximum restoring torque, (5.24),

$$\Delta\chi_g = \frac{2\alpha_c}{v\mu_0}\left(\frac{C}{H^2}\right),\tag{5.32}$$

for the volume anisotropy in the plane perpendicular to the suspension, or

$$\Delta\chi_M = \frac{2W_M\alpha_c}{w}\left(\frac{C}{H^2}\right),\tag{5.33}$$

for the molar anisotropy of a crystal of mass w, comprising molecules of molecular weight W_M.

5.8 Crystal mounting and orientation

Earlier versions of both of the methods described above used metal strips or wires as torsion fibres; for example, of phosphor–bronze or beryllium–copper. Such suspensions are generally far too stiff to be of much use with small crystals of only moderate anisotropy and so fine quartz fibres, which may be drawn from rod with difficulty, or purchased commercially with ease, are now the norm. On occasion this author has used quartz fibres so fine that they float in the breeze made just by moving his hands near them: admittedly, these represent an extreme situation. However, it remains the case that these torsion fibres are very delicate and this immediately imposes a limit on the size and mass of any putative crystal support or goniometer. We have experience with two systems.

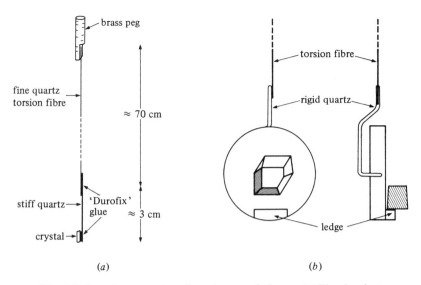

brass peg

fine quartz
torsion fibre

≈ 70 cm

stiff quartz

'Durofix'
glue

≈ 3 cm

crystal

torsion fibre

rigid quartz

ledge

(a)   (b)

Fig. 5.5. Sample suspensions for anisotropy balances. (a) The simplest arrangement, only requiring reasonably accurate alignment of a crystal axis parallel to the stiff quartz pole. (b) A single-axis goniometer head.

The first is extremely simple and illustrated in figure 5.5(a). The crystal is not stuck directly onto the torsion fibre because of the extreme probability of breaking this when changing samples and also because the lack of lateral strength in the fibre would allow a crystal to reorient itself under its own weight. The (relatively) heavy, torsionally rigid, quartz, mounting 'pole' avoids both of these problems. An ester-soluble glue ('Durofix' on the UK market) is ideal for these joints, provided plenty of time is allowed for setting before measurements begin (the glue tends to 'neck' as it dries out and may pull the crystal out of alignment – so care is required here). The simple mounting in figure 5.5(a) is good from the point of view of there being a minimum of extraneous material in the field region, having a low moment of inertia – an important consideration in the critical-torque mode when one's patience can be strained waiting for the system to settle after a rather violent 'flip'. It is adequate if a crystal is to be aligned along a principal magnetic axis (say, parallel to b in the monoclinic system) and the crystal possesses obvious developments (edges or faces) parallel or perpendicular to the chosen direction. In the time it takes for the glue to set – or become stiff really – it is easily possible to 'tease' the crystal into a desired orientation to within about 1–2° as judged by eye. As principal susceptibilities, and hence principal anisotropies, occur at 'stationary positions' in the tensor property, such errors are almost always tolerable. For other alignments, or when crystal faces and edges do not line up so conveniently with selected axes, the mounting jig shown in figure 5.5(b) may be more suitable.

This single-axis 'goniometer head' or 'wheel' was introduced[15] by Gerloch & Quested in a slightly more elaborate form. Ideally, it would be desirable to mount a crystal on a miniature goniometer head but any reasonably sophisticated device, even one made from a light plastic, would be too heavy to suspend from the delicate torsion fibres we use. The device, shown from the front and side, in figure 5.5(b) represents a reasonable compromise. The nylon wheel fits sufficiently tightly onto the thick (≈ 1 mm) quartz 'axle' that stable orientations may be achieved with finger-tip adjustment. The suspension axle is constructed so as to ensure verticality of the wheel when suspended with an average-sized crystal (≈ 5 mg) and also so that the centre of mass of the crystal should lie approximately vertically below the suspension point of the torsion fibre. The total weight is about 100 mg. The original device included a pointer and $10°$ marker engraved on the edge of the wheel. Subsequent models dispensed with those refinements for it was found simpler to adjust the wheel, suspend the whole at the bench top, steadied by a fragment of sponge underneath, and to measure the rotational alignment of the crystal about the horizontal axis by observation through a small telescope with a calibrated cross-wire. This obviates the earlier need for rather exact alignment of the wheel by hand – obviously a delicate affair – when all that is required is an accurate knowledge of the (arbitrary) crystal orientation. The crystal is glued either to the vertical face of the wheel or onto the small ledge, depending upon the crystal habit: both possibilities are illustrated in the figure. The anisotropy of the wheel itself (as opposed to its mean diamagnetism which is of no concern, of course) is reasonably small but large enough to render measurements on very small crystals of nearly isotropic compounds unreliable, even with prior calibration of the device. Otherwise, this crystal-mounting platform has been quite successfully used on several occasions. Figure 5.6 shows the results of its employment[15] to determine the ϕ angle in a monoclinic crystal (see figure 4.1) of cobalt ammonium Tutton salt. In principle this task might be accomplished by suspending a monoclinic crystal along the unique b axis using the simple arrangement of figure 5.5(a). By definition, $\chi_2 > \chi_1$ and so the setting position would have χ_2 parallel to the magnetic field and appropriate optical observations of the inclination of a well-developed crystal face with respect to the magnet axis would furnish a measure of ϕ. In practice, such experiments are difficult to perform well, even without

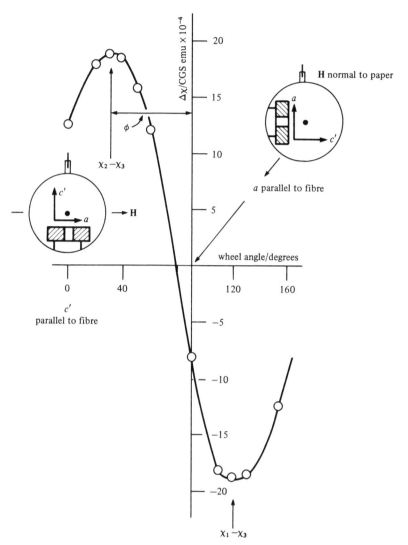

Fig. 5.6. For cobalt ammonium Tutton salt: variation of anisotropy with wheel angle, keeping b horizontal. (temp. = 291 K). Inserts show orientation of crystal and wheel with respect to the suspension fibre and magnetic field for wheel angles 0 and 90°, corresponding to c' and a parallel to the fibre respectively. ϕ is readily determined. The approximate numerical equality of $\Delta\chi_1$ and $\Delta\chi_2$ is coincidental.

the optical and other obstruction of a cryostat. It is one thing to be able to reproduce a crystal position, as in the null-deflection experiment; it is quite a different matter to measure in-plane angles using small, frequently imperfectly faceted crystals suspended on extremely fine fibres. A very approximate estimate of ϕ can be obtained in this way, certainly; and it is common practice to make such observations and so provide an order-of-magnitude check on a ϕ value determined in another way. One such is illustrated in figure 5.6, as is now described.

Advantage is taken of the uniqueness of the symmetry axis by mounting the crystal with b perpendicular to the wheel axis. Whatever the orientation of the wheel upon its own axis, b remains in the horizontal plane and is directed either exactly parallel or exactly perpendicular to the magnetic field vector. anisotropy measurements are made, either by the critical-torque or null-deflection methods, for a series of wheel angles. The results define a cosine-squared function, for the experiment is simply defining the principal ac elliptical section of the crystal-susceptibility, magnitude ellipsoid. As the orientation of the crystal on the wheel can be measured accurately, it is a simple matter to define the origin, so to speak, of the curve in figure 5.6 and hence the desired angle ϕ. With experience, it is found quicker and easier to make several measurements near the turning points and few elsewhere, and to fit these results by least squares to a cosine-squared function. Further examples of the technique are described in reference[16].

5.9 Calibration, sensitivity and accuracy

The torsion methods can be made absolute if those quantities intrinsic to the apparatus rather than to the specimen can be determined – the torsion constant C and the field strength(s) H. Torsion constants can be measured using the oscillation periods of accurately machined spheres or cylinders of known moments of inertia: such experiments are best done *in vacuo* and in the presence of β emitters and other precautions against static charge. Magnetic field strengths can be measured using search coils or other commercial gaussmeters. It should be clear, however, that both kinds of experiment require considerable care and that it is hardly desirable to go to such trouble for every fragile torsion fibre to be used, and perhaps soon broken.

Instead, the usual procedure is to calibrate the ratio C/H^2 (for several fields) for the actual experimental setup by reference to the known magnetic anisotropies of a few suitable calibrants. The process is obvious, of course, merely employing (5.32) or (5.33) in reverse. Mitra[12] has discussed some criteria for choosing calibrants. These include the desirability of having large, well-formed single crystals, free from gross imperfections and twinning; that the crystal habit be such as to permit simple and accurate alignment on the support; and that the calibrant be stable under the usual experimental conditions. He adds two further characteristics: that the calibration crystals always grow in the same habit and that they possess a 'medium' value of the anisotropy along one particular axis. The present author does not consider these latter conditions necessary. In the first place, crystal orientation should, in his view, always be rechecked using X-ray diffraction methods rather than reliance being placed on the external form. In the second, it is found that the linearity of the torsional properties of fine quartz fibres extends over several complete revolutions in a half-metre length, so that the calibrant and specimen anisotropies need not be very similar. Gregson & Mitra[17] have claimed that copper sulphate pentahydrate crystals probably provide the best calibrant for these torsion experiments, but this author does not agree. They certainly form large, well-developed crystals. Their composition with respect to water content (and hence to the concentration of magnetic centres) is known to be variable, however. Much more serious, though, is the fact that the crystals are triclinic so that the direction of the principal crystal susceptibilities – the 'turning points' of the susceptibility ellipsoid – are not established by symmetry. Therefore, especially accurate crystal orientation is required: the point is taken up in general terms in the next chapter.

We have preferred to calibrate our anisotropy balances using the values of either anthracene or triphenylbenzene. The diamagnetic anisotropy of monoclinic anthracene crystals have been determined by absolute methods by various workers[18−21] and fibres so calibrated agree within $1\frac{1}{2}\%$ of calibrations with triphenylbenzene.[22] The anisotropies are fairly large for diamagnetic systems, though obviously modest compared with those for many paramagnets. A great advantage of using diamagnets, however, is the virtually total lack of temperature dependence of the calibrating quantity so that the calibration procedure may be carried out at an unrecorded, unstabilized, ambient temperature. Calibrations with paramagnetic crystals should be performed over a temperature range on either side of that used for the standard value so that proper interpolation routines may be employed. A good paramagnetic calibrant is furnished[23] by Cs_3CoCl_5. The crystals are tetragonal and form large, square blocks with easily identified habit (the high symmetry virtually makes X-ray checking redundant here). Any crystal mounting perpendicular to the c axis provides for a measurement of $\Delta\chi_c = \chi_{\parallel} - \chi_{\perp}$: indeed, generally, this substance satisfies Mitra's criteria[12] rather better than $CuSO_4.5H_2O$.

The torsion methods for the measurement of magnetic anisotropy outlined here can be extremely *sensitive*. With care, very fine and quite long quartz fibres may be employed, so providing a large torsional constant C. Again with care, large crystals – up to 100 mg – can be used. Altogether, it is not impossible to detect and measure molar anisotropies as small as, say, 5×10^{-6} CGS emu. In short, these anisotropy methods can easily be made far more sensitive than any force method (Gouy or Faraday) or than the more conventional induction methods (vibrating-sample magnetometer, say) which would be required to provide a measure of the mean susceptibility $\bar{\chi}$ or of a principal crystal susceptibility, one of which is an essential complement to anisotropy measurements if a determination of the complete susceptibility tensor is desired (see the following chapter). On the other hand, the torsion methods are rather difficult to perform *accurately*. Errors arise from two main sources, associated with crystal orientation, on the one hand, and crystal shape, on the other.

The problems with crystal orientation have already been indicated and stem, essentially, from the fact that any crystal-mounting platform must be relatively insubstantial. Even if the same techniques as used for the Faraday method, namely transferring a crystal from the X-ray goniometer arcs to a short length of quartz and thence to the 'pole' of the arrangement in figure 5.5(a), the impossibility of countering any lateral forces means that the crystal can never be oriented better than $1-2°$. As stated earlier, this is frequently satisfactory for an overall accuracy of 2% in $\Delta\chi$, but no better. A further difficulty, which must have some bearing on the ultimate accuracy, is the fact that, even using a vertically scored mounting peg to attach the torsion fibre (again, using 'Durofix') to the torsion head, as shown in figure 5.5(a), it is very difficult to arrange that the fibre hangs perfectly coparallel with the rotation axis of the head. There is a marked tendency for the sample and support – 'wheel' or 'pole' – to precess as the torsion head is turned: this is a curious motion, for an apparently complete precession of the crystal's centre of mass can take place (over a circle of, say, 3 mm diameter) as the torsion head is rotated by one or more revolutions, even though the crystal itself rotates by less than 45°, of course. Anyway, claims for greater accuracy in the literature tend to be for experimental arrangements involving much thicker torsion fibres, longer mounting 'poles' (often 10 cm or more) and for measurements on much larger crystals: in other words, greater accuracy may be had but only at the expense of a lot of sensitivity.

The other major source of inaccuracy with these torsion methods tends to be a problem only if the true anisotropy is very small; it arises out of the difference between the applied and total magnetic fields and gives rise to a quantity known as *shape anisotropy*. Even a crystal of cubic symmetry, which is necessarily of isotropic material, can suffer a torque in a uniform applied magnetic field. We have discussed the difference between applied and total fields in §3.16. Now the derivation of (5.22) above was approximate in that the field in (5.19) is the applied field $\mathbf{H}_0$ while that in (5.20) is the total field $\mathbf{H}$. Therefore, when the small field $\mathbf{H}_C$ is taken into account (5.21) ceases to be exact. Even considering an isotropic material: if the total field $\mathbf{H}$ is not aligned parallel to a line of symmetry – here referring to the symmetry of the crystal shape – a crystal will suffer a torque in a uniform field. This is because $\mathbf{M}$, being parallel in such materials to $\mathbf{H}$, will otherwise not be exactly parallel to the applied field $\mathbf{H}_0$. Clearly, the couple experienced by a crystal of isotropic material will be a function of both shape and orientation. The magnitude of the effect may be estimated[24] as follows. The field $\mathbf{H}_C$ has magnitude of order M so that the angle between $\mathbf{M}$ and $\mathbf{H}_0$ will be about M/H, which equals χ_g: hence the couple has magnitude of order $vH^2\mu_0\chi_g^2$. Compare the order of magnitude of the torque on an anisotropic crystal, given by (5.21) as $vH^2\mu_0\chi_g$, if anisotropies are taken to be roughly of the same order of magnitude as the principal susceptibilities themselves. Altogether, therefore, the contribution to the anisotropy measured by a torque method from crystal-shape effects is of the order of χ_g – say one part in 10^5 – and so normally quite negligible. The sorts of circumstances where the effect can be important, however, are exemplified by high-spin d^5 complexes, whether the crystals are cubic or not. The intrinsic anisotropy of molecules having ^6S ground terms, energetically well separated from excited spin quarters and doublets, is very small. At the same time the mean susceptibilities of high-spin d^5 species are amongst the highest in the transition-metal block, so shape anisotropies, being roughly proportional to $\bar{\chi}$, are frequently important enough to require consideration here. Mitra[12] reviews various methods which have been used to help correct for such shape effects, including some quite detailed work by Majumdar[25] who studied the effect on crystals which were ground to various shapes and sizes.

If the applied field is at all non-uniform, samples may suffer an additional torque as a function of shape and size. The effect occurs in first order, unlike the second-order, shape–anisotropy effect just described. It arises for essentially the same reasons that the *force* on a magnetic sample arises in a non-uniform field. The only point we wish to make here, therefore, is that magnetic anisotropy experiments must be performed in uniform magnetic fields if calibrations are to be sound and the whole process is to be more than qualitative. It is quite possible that the curious sample precession sometimes observed in anisotropy–torque measurements arises out of very small non-uniformities in the magnetic field near the centre of a two-inch gap, say, between four-inch diameter pole caps. Empirically the problem does not seem to reduce the accuracy below the 1–2% level, but no doubt it has prevented its improvement beyond that. The same possibility of slight field inhomogeneity quite possibly renders the (relatively) bulky 'wheel', described above, useless for barely anisotropic specimens, due to its own innate tendency to align in the field: but, to emphasize; the effect is only very slight and, most often, insignificant.

5.10 Induction methods

Distinct from force methods of susceptibility measurement in which the measured quantity is a torque on, or change in weight in, the sample, are those techniques which exploit an altered induction within an electric circuit looped or coiled around the body being studied. Here we shall be confined to a few selected and generalized remarks on these methods, the reader being referred to the review literature[1,3,4,5,12] for more detail. Firstly, we recall from elementary electrical theory and, indeed, from the material of §2.12, that the phenomenon of induction is concerned with *changing* magnetic fields and electric currents. Perhaps the best-known example of the modification of the behaviour of an alternating current circuit by a body of matter which is electrically insulated from the circuit is the transformer, whose efficacy is greatly improved by the presence of iron cores within the solenoidal windings. Exploitation of the induction principle in susceptibility measurements began, naturally enough, with studies of ferromagnetic materials. However, the effects with paramagnets, at all but the lowest temperatures, were generally rather too small to provide the basis of a widely applicable magnetometer. By and large, these techniques were confined, for paramagnets, to systems with intrinsically larger susceptibilities and at temperatures well below 50 K. Apart from rather detailed refinements in apparatus design, concerned with coil geometry to improve the basic signal, and with amplifier design to improve its detection, the general situation changed relatively little until Foner[26] introduced his vibrating-sample magnetometer.

5.10.1 Vibrating-sample magnetometers

Foner enormously improved the sensitivity of the then current induction instruments by the simple device of vibrating the sample within a detection coil and so replacing the difficult measurement of a weak DC signal by a much easier amplification of an AC one. In his original apparatus the sample was glued to the end of a light, stiff tube (a drinking straw!) which was rigidly attached to the cone of a downward-pointing, large loudspeaker energized at about 60 hertz. The sample proper lay within the pole gap of a standard laboratory magnet, as also did the signal detection coil. Since then, this obviously delicate instrument has been commercialized and a totally professional appearance has been given to the original idea – at a price, of course. The mechanical part of the apparatus offered by Princeton Applied Research Corp. (PAR) is outlined in figure 5.7. The sample, in the form of an encapsulated powder or single crystal, is held in a diamagnetic plastic holder at the end of a long stainless-steel tube, about 1 m long, constrained to move vertically and oscillated sinusoidally by a large transducer driving coil at the top end. Also well away from the laboratory magnet, used to induce a dipole in the sample, a capacitor plate, attached to the drive rod, oscillates with respect to a pair of fixed plates. The signal produced in the detector coils is proportional to the moment of the sample plus holder and the amplitude and frequency of the vibration. At the same time a voltage is applied to the vibrating capacitor plate, so producing a signal on the fixed capacitor plates at a frequency identical with that from the sample. The two signals – from sample detector coil, and capacitor plates – are generally out of phase due to the geometries of the actual physical setups: the phase shift is cancelled by appropriate electronic equipment. All that remains–simple in principle, but obviously requiring quite sophisticated electronic circuitry – is to vary the voltage on the moving capacitor plate until the signal amplitude from the fixed plates equals that generated by the sample detector coils. The voltage readout from the moving plate, which may be displayed digitally, is directly proportional to the magnetic moment of the sample. Sensitivities of the same order as those for an average Faraday balance are available with these vibrating-sample magnetometers: accuracy and reproducibility are similarly comparable. In earlier models, these degrees of accuracy and sensitivity were only attainable by replacing the conventional laboratory

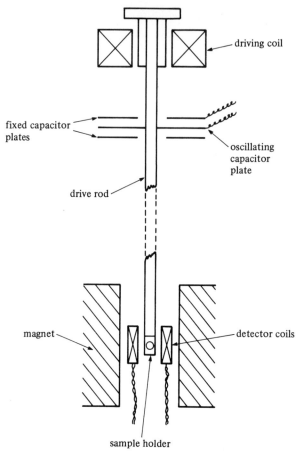

Fig. 5.7. Schematic illustration of the electromechanical parts of a vibrating-sample magnetometer.

electromagnet with a high-field superconducting solenoid. This carried with it the penalties of rather complex cryogenic hardware and techniques, high running costs and, depending upon the particular instrument supplied, a degree of unreliability; and all at very high capital cost. Since then, matters have improved on several fronts, including a redesign of detector coils that confers upon the setup with a conventional electro-magnet acceptable sensitivity and accuracy. Nevertheless, one is still very much in the hands of professional and commercial expertise which, as always, is a two-edged sword. On the other hand, there are many potential and actual advantages for this kind of instrument.

Firstly, the vibrating-sample magnetometer employs a homogeneous magnetic field so that no problems arise of the kind associated with non-uniform fields. This is probably more an advantage in principle than in practice, as discussed earlier. Secondly, when the instrument works, manipulations are simple and the magnetization of the sample may be read from a digital voltmeter simply by turning a switch. Most real advantages of the instrument, however, are associated with the allowable, and actual, bulk and rigidity of the sample holder; and, ultimately, with the fact that the technique does not involve the measurement of mechanical forces or couples. The diameter of the plastic sample holder is about 5 mm, so that various forms of sample mounting and rotation can be tried, the only remaining restriction being the size of the diamagnetic correction required. The sample need not be held in a still atmosphere, as with the force methods, so that sample cooling and temperature stabilization can be achieved using the best thermal bridge of all: namely, by immersing the sample and holder in a flowing stream of helium gas, taken directly from liquid boil-off and, as appropriate, subsequently heated. Temperatures lower than 4.2 K can be achieved quite simply by pumping on the tube containing the sample and cooling gas.

5.10.2 The superconducting, quantum-interference device (SQUID)

The SQUID is a device employing the Josephson effect[27] that can be used as part of an arrangement capable of measuring minute magnetic fields. The theory of these matters falls outside the scope of this book: useful references are given both as introductory[27,28] and as in the application of SQUIDs to magnetometry.[29] The broad outlines of the exploitation of a SQUID in current commercial magnetometer systems is as follows. Placed around the magnetized sample is a loop or coil of superconducting wire whose circuit is completed, using the same superconducting wire, with a second loop or coil a few centimeters away. This second loop enters the SQUID, consisting of a small cylinder of niobium metal with two connected holes of diameter typically about 3 mm and bridged with a niobium screw as the 'weak link' of the Josephson junction. The arrangement, not at all shown to scale in figure 5.8, is all held at around 4 K; that is, with both loops and SQUID in their superconducting states. A signal is induced in the loop surrounding the sample. This can be effected in several ways. The simplest is just to move the sample through the coil: no special gearing or other machinery is required for this – a gentle push on the sample rod is all that is required. Another method is to change the inducing magnetic field **H** surrounding the sample but this

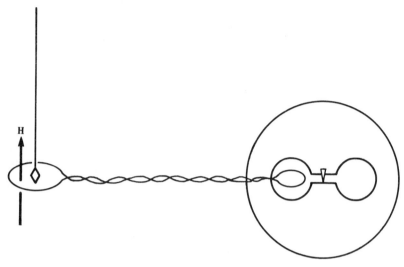

Fig. 5.8. The use of a SQUID in magnetometry.

is most inconvenient, as will be discussed. Yet another is to change the sample temperature (assuming its susceptibility to be temperature dependent): note that the experimental arrangements are such that the sample, together with its surrounding helium atmosphere, are contained within a sample tube which is thermally insulated from the main bath of liquid helium, which maintains the superconducting state of the main solenoid producing **H** as well as of the items shown in the figure. Finally, it is even possible to irradiate the sample with light, should this cause a change in the bulk susceptibility. However the signal in the sample coil is produced, it is held or 'trapped' in the double-loop circuit which, being super-conducting and, hence, of zero resistance, causes no damping or loss of the original signal. Therefore, there appears within the loop inside the SQUID a field which is directly related to that field produced by the sample. The SQUID device, together with an inductively coupled, ratio-frequency coil circuit, external to the cryostat, is capable of detecting, in principle anyway, changes equivalent to one quantum of magnetic flux. The idea is simple enough, given the Josephson effect, and the SQUID device itself is cheap – perhaps one or two hundred pounds. The technical problems arise out of the very advantage of the SQUID. It is so sensitive that changes in the earth's magnetic field or perturbations caused by the movement of a small piece of iron – like a bunch of keys, for example – in the laboratory will produce a totally unsatisfactory background signal in the magnetometer. So the practice is to surround the sample, superconducting solenoid and detector coil by a large cylinder of superconducting material (usually niobium), also maintained within the liquid-helium reservoir. The cylinder is first caused to make the transition to the normal state by switching on a small heater, the solenoid is energized to produce the desired field, and the niobium shield is then allowed to revert to the superconducting state. The upshot is that the last field present inside the shield – that produced by the solenoid, together with any small contributions from the earth's field, etc. – is trapped at that value, constant to about one part in 10^{14}. The susceptibility measurements can then be performed free of external interference. This explains, incidentally, why changing the solenoid field is *not* a suitable way of inducing a change in the moment of the sample. All in all, magnetometers employing SQUID technology (occasionally and awfully referred to as 'susceptometers') are capable of the detection and measurement of susceptibilities as small as 10^{-12} emu gm^{-1}, which is some 10^3 or 10^4 times more sensitive than the average Faraday balance. At the moment, these instruments are limited to the measurement of susceptibilities parallel only to the solenoid axis but there appears to be no fundamental reason why suitable solenoid and detector-coil geometries could not be developed to enlarge the capability of these instruments. They are, of course, very sophisticated, again dependent upon the manufacturer for repair and service; and, as ever, extremely expensive.

It is worth commenting briefly on the advantages of SQUID magnetometers. They are many and widely varied but, apart from one type of use in the biological area, we shall consider only those aspects strictly relevant to ligand-field studies. The 'biological application' is for the detection and measurement of the paramagnetism of certain metal-containing proteins and the like. A popular example is provided by haemoglobin. As the paramagnetism of the iron atoms in the paramagnetic form of this much-studied molecule is virtually swamped by the diamagnetism of the 'uninteresting' globin, it has always been very difficult to determine by conventional methods how this paramagnetism changes in response to various influences, like oxidation and reduction. The strength of the SQUID magnetometer, of course, lies in its being a 'differential' detector of magnetization, so that it is ideally suited to this kind of study[30,31]. The subject we address in this book, however, is ligand-field analysis and we therefore ask what advantages the increased sensitivity of the SQUID magnetometer can bring to conventional magnetochemistry. It may be that experiments involving very weak inducing fields may be important in studies of 'easy directions' of magnetization in exchange-coupled

materials. Similarly, investigations of doped systems as part of a programme to determine properties of molecular aggregation in antiferro-
magnetic species, for example, might well benefit from the high sensitivity
of the SQUID instruments. But for measurements of the susceptibilities
of pure, paramagnetic, transition-metal and lanthanide complexes, the
only apparent benefit to be had from this complicated and expensive
apparatus is the possibility of working with very much smaller crystals
than hitherto. The instruments would, without doubt, be capable of
providing for the measurement of the susceptibilities of crystals roughly
of the size normally used in X-ray structure analysis – say 0.2 mm cube.
As many interesting compounds defy chemists' attempts to grow crystals
much larger than this, here is a potential avenue of practical advance in
magnetochemistry. Unfortunately, the determinant factor in the process
of susceptibility measurement now ceases to be the detection of magnetization, but the most elementary of tasks – the weighing of the crystal!
Crystals of 'X-ray size' might typically weigh $1\,\mu g$ at most, so that
experiments requiring a typical accuracy of about 1% imply a capacity
to discriminate weights between 10^{-8} and $10^{-9}\,g$. The task is not quite
impossible but certainly difficult enough to deter this author.

In concluding these sections on induction methods of susceptibility
measurement, it should be acknowledged that great advances in the
appropriate technologies of electronics and cryogenics have been made
in recent years: even the more conventional mutual-induction bridge
systems have been made competitive with the best force methods available
today. As advertised at the outset, this author is not without prejudice.
This has arisen partly out of the contention that 'sufficient unto the day
is the evil thereof' in that the Faraday balance described in §5.4 allows
all necessary measurements to be made relatively easily, quickly and
cheaply; but also from unfortunate but expensive experiences with a
superconducting version of the vibrating-sample magnetometer being
supplied with a totally inadequate cryostat system. Conversations with
colleagues over the years suggest that the experience is not totally unique,
in kind anyway, so that the self-reliance that is possible with the simpler
machine seems to be a desirable attribute.

5.11 Crystal selection and orientation

For the sake of completeness it is worth making a few remarks that are
broadly held in common for all measurement techniques described here.
The author begs the indulgence of those who find their substance obvious.

Most of the experimental arrangements we have reviewed require single-crystal samples to weigh in the region 0.2–10 mg depending upon the intrinsic susceptibility of the material. Single crystals may then be typically 1–5 mm on edge and so are very large indeed by the standards of X-ray crystallographers. Even the best-formed crystals are imperfect, of course, if only because of a mosaic structure, and it is common to find crystals, of the size *we* must contemplate, possessing all manner of faults; cracks, striations, holes, etc. Most of these imperfections do not seem to be important for susceptibility measurements and considerable confidence about the quality of the sample can be gained by microscopic examination in polarized light. When extinction is expected, the great bulk of the crystal should extinguish cleanly under crossed polarizers. A few surface, crystallite growths are easily detected in this way and are usually of no consequence, representing but a fraction of 1% by weight as a rule. It is similarly unimportant if the crystal habit differs actually or, more likely, apparently, from that previously reported, for as we shall discuss, X-ray goniometry will be used to check crystal identity and alignment. Nor is it important if the crystal is broken, in the sense that a mechanical fracture has occurred so leaving the specimen short of some (or even all) of its natural faces and edges. What *does* matter is that the crystal be *single*, rather than twinned. Many forms of twinning are only approximate and simply involve several, easily identifiable crystals as having grown together: these are discarded. Rather more subtle twinning can sometimes be revealed by the polarizing microscope[32] but occasionally neither optical microscopy nor some X-ray diffraction techniques can detect the phenomenon: fortunately, such difficult cases are rather rare in the field of transition-metal complexes.

It is not difficult to *orient* a single crystal on a quartz pole, such as those described earlier for the Faraday or anisotropy balances. The simplest of all methods, having *always* checked the orientation by X-ray diffraction – rotational and Weissenberg photography are probably the most useful here – is simply to glue the crystal to the piece of quartz, oriented by eye. The eye is surprisingly sensitive to angular displacements and, provided the crystal has a suitably defined straight edge along which to make the alignment, the process can frequently be done to within 1–2°. A rather better arrangement, however, is to use a jig, rather like that shown in figure 5.9, for transferring a crystal, aligned parallel to the X-ray goniometer arcs, to a short length of quartz which may be introduced gradually via the threaded drive. Glue ('Durofix' diluted with a little solvent is often convenient) is introduced into the small gap between quartz fibre and crystal: after setting for a few hours, the crystal is parted from the

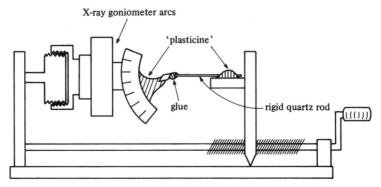

Fig. 5.9. Transferring an oriented crystal from goniometer arcs to a rigid quartz fibre.

arcs. This latter process can be helped by using 'plasticine' lightly scored with a finger nail: alternatively, the crystal can be glued to the arcs and dissolved off afterwards. Either way, the crystal is weighed before this whole process, of course, so none must be lost during the transfer. Subsequent glueing of the short length of quartz to the main piece of the Faraday suspension, say, as in figure 5.3(c), by overlapping length on length, can be done quite accurately enough by eye. The whole process can be demonstrated, by X-ray diffraction, to be accurate to $1-1\frac{1}{2}°$ at worst.

On the subject of alignment of these crystals by X-rays: the crystals are usually so large as to be opaque to the normal X-ray available in the crystallographer's laboratory. Common practice, therefore, is to arrange that the X-ray beam just catches the very tip of some appropriate part of the crystal. The ensuing photographs are always of poor quality because of absorption and scattering effects, but are quite sufficiently good for the purposes of crystal alignment. Of course, the information being collected only refers to the proportionately small region of the crystal being exposed. This points again to the desirability of prior examination of the crystal, especially under polarized light, so as to establish that the region chosen for X-ray exposure is likely to be representative.

5.12 Well-measurable quantities

The following chapter is concerned with *procedures* to be used in the determination of crystal susceptibility tensors: in short, in deciding along which directions crystals of various classes need be aligned in order to provide a sufficiently complete set of measurements to define the magnetic property. Sensible approaches can only be constructed with reference to the type of information which available apparatus makes possible to obtain. There exists an extended literature[12] on procedures for the measurement of triclinic crystals, for example, most of which were never tried out in practice, and some of which would probably fail through lack of recognition of the limited accuracy with which certain quantities may be measured. In this section, therefore, we summarize the quantities which can be determined well and those which cannot. In some respects, this has been the main purpose of placing this chapter between the description of the susceptibility tensor in the preceding chapter, and of its measurement in the next.

Using an electromicrobalance, crystals may be weighed very accurately; to better than 1 part in 1000, if necessary, for crystal weights of the order of 1 mg. The same balance provides for accurate measurement of vertical forces in the Faraday experiment. Crystals may be aligned parallel to the suspension fibres of Faraday or torque balances with sufficient accuracy. The positions of maximum and minimum susceptibility in the plane perpendicular to the suspension may be determined relative to the magnet protractor with great ease in the Faraday experiment; and with rather greater difficulty, using the 'wheel' device, for example, with an anisotropy balance. In the case of the Faraday experiment, this means that rotation of the magnet to the appropriate positions, once determined in a separate experiment, allows for the measurement of χ_{max} and χ_{min} in the plane perpendicular to the suspension. This does *not* mean, however, that the absolute orientations of these susceptibilities with respect to defined crystal axes are so readily determined. That requires some optical arrangement by which the orientation of a crystal face, usually whilst the crystal is inside the cryostat, can be determined relative to the fixed, laboratory frame. This is not an easy task at best, and is quite inaccurate for crystals of the size commonly used today and employing the experimental arrangements we have described. We prefer not to attempt such measurements and, as described in the following chapter, procedures can be devised for the complete definition of the required susceptibility tensor for which they are quite unnecessary. Broadly, the same remarks apply to the orientation of in-place susceptibilities in anisotropy experiments. They are less compelling, perhaps, for induction methods if these allow greater rigidity of the crystal-mounting platform.

Apart from these few remarks about what is and is not easy, it remains only to emphasize a preference for those experimental arrangements which provide for the measurement of several pieces of data from a single-crystal orientation on the sample support. Most instruments employing superconducting solenoids, like the longitudinal Faraday balance, for example, are restricted to the measurement of susceptibility parallel to the solenoid axis. Not only is it more tedious to have to make a separate mounting for each direction of interest, but that increases the possibility of inaccuracy: alternatively, the same number of crystal mountings using a transverse Faraday setup will yield a surfeit of data which may be used to improve the statistical accuracy of the tensor determination. Examples of this are given in the next chapter in which we now exploit the best features of these various experimental arrangements.

References

[1] Bates, L.F., *Modern Magnetism*, Cambridge University Press, 1951.
[2] Figgis, B.N. & Lewis, J. *In: Modern Coordination Chemistry*, J. Lewis & R.G. Wilkins (eds.), Interscience, New York, 1960.
[3] Figgis, B.N. & Lewis, J. *In: Technique of Inorganic Chemistry*, H.B. Jonassen & A. Weissberger (eds.), vol. 4, Interscience, New York, 1965.
[4] Mulay, L.N. *In: Physical Methods of Chemistry*, A. Weissberger & B.W. Rossiter (eds.), Wiley, New York, 1972.
[5] Mulay, L.N., *Magnetic Susceptibility*, Interscience, New York, 1963.
[6] Wolf, W.P., *J. Appl. Phys.*, **28**, 780 (1957).
[7] Garber, M., Henry, W.G. & Hoeve, H.G., *Can. J. Phys.*, **38**, 1595 (1960).
[8] Heyding, R.D., Taylor, J.B. & Hair, M.L., *Rev. Sci. Instr,*, **32**, 161 (1961).
[9] Cruse, D.A. & Gerloch, M., *J. Chem. Soc. Dalton Trans.*, p. 152 (1977).
[10] König, E., *In: Tables of Magnetic Susceptibilities*, Landolt-Börnstein, K.H. Hellwege & A.M. Hellwege (eds.), Springer-Verlag, Berlin, 1965.
[11] Mackey, D.J., McMeeking, R.F. & Hitchman, M.A., *J. Soc. Dalton Trans.*, p. 299 (1979).
[12] Mitra, S., *Transition Metal Chemistry*, **7**, 183 (1972).
[13] Krishnan, K.S. & Banerjee, S., *Phil. Trans. Roy. Soc.*, **A234**, 265 (1935).
[14] Stout, J.W. & Griffel, M., *J. Chem. Phys.*, **18**, 1449 (1950).
[15] Gerloch, M. & Quested, P.N., *J. Chem. Soc. A.*, p. 2307 (1971).
[16] Gerloch, M., McMeeking, R.F. & White, A.M., *J. Chem. Soc. Dalton Trans.*, p. 2452 (1975).
[17] Gregson, A.K. & Mitra, S., *J. Chem. Phys.*, **49**, 3696 (1968).
[18] Lonsdale, K. & Krishnan, K.S., *Proc. Roy. Soc.*, **A156**, 597 (1936).
[19] Lumbroso-Bader, N., *Ann. Chim. (France)*, **1**, 687 (1956).
[20] Gordon, D.A., *J. Phys. Chem.*, **64**, 273 (1960).
[21] Leela, M.R., PhD Thesis, University of London, 1958.
[22] Krishnan, K.S. & Banerjee, S., *Phil. Trans.*, **A234**, 265 (1935).
[23] Figgis, B.N., Gerloch, M. & Mason, R., *Proc. Roy. Soc.*, **A279**, 210 (1964).
[24] Nye, J.F., *Physical Properties of Crystals*, Oxford University Press 1957.
[25] Majumdar, M., *Ind. J. Phys.*, **36**, 111 (1962).
[26] Foner, S., *Rev. Sci. Instr.*, **30**, 548 (1959).
[27] Josephson, B.D., *Phys. Lett.*, **1**, 251 (1962).
[28] Bleaney, B.I. & Bleaney, B., *Electricity and Magnetism*, p. 423, 3rd edn, Oxford University Press, 1976.

[29] Swithenby, S.J., *Contemp. Phys.*, **15**, 249 (1974).

[30] Messana, C., Cerdonio, M., Shenkin, P., Noble, R.W., Fermi, G., Perutz, R.N. & Perutz, M.F., *Biochem.*, **17**, 3652 (1978).

[31] Perutz, M.F., *Ann. Rev. Biochem.*, **48**, 327 (1979).

[32] Hartshorne, N.H. & Stuart, A., *Crystals and the Polarizing Microscope*, Edward Arnold, London, 3rd edn, 1960.

6

The measurement of crystal susceptibilities

━━

The measurement of the average magnetic susceptibilities of solutions or powdered samples comes a rather poor second-best to the determination of the complete susceptibility tensor of a single crystal. Over and above the practical considerations, discussed in the last chapter, of apparatus and the selection and handling of suitable samples, the single-crystal experiment involves the choice of appropriate orientations for the sample with respect to which magnetometer measurements yield complete susceptibility tensors for each of the various crystal systems. In some cases, the choice of such alignments is determined entirely by considerations of crystal symmetry but, in others, the possibilities and limitations of the apparatus may also be important determinants of the procedure: indeed, that is the reason for interposing the last chapter between the introduction of the susceptibility tensor and the present discussion of measurement strategy. We begin with the simplest systems.

6.1 Uniaxial crystals

Trigonal, tetragonal and hexagonal crystals have susceptibility tensors characterized by magnitude ellipsoids which are figures of revolution. By convention, the unique susceptibility $\chi_{\parallel}$ lies parallel to the crystal c axis: that in the ab plane is labelled $\chi_{\perp}$. Apart from the completely isotropic cubic system, crystals in these systems yield the least informative susceptibility tensors which are, however, the simplest to measure. Using a transverse field Faraday balance, both characteristic susceptibilities may be determined from a single alignment or mounting of the crystal.

Orientation of a uniaxial crystal with *any* axis in the *ab* plane parallel to the suspension leaves both $\chi_\parallel$ and $\chi_\perp$ lying in the plane of the magnet protractor. With the magnetic field aligned parallel to *c*, the Faraday balance records forces proportional to $\chi_\parallel$; rotation of the magnet by 90° yields values for $\chi_\perp$. Using a longitudinal field Faraday balance, typical of those commercial devices employing a superconducting solenoid, two separate crystal orientations are required for uniaxial crystals – parallel and perpendicular to the unique crystal axis *c*.

The crystal anisotropy is completely determined by $\chi_\parallel - \chi_\perp$ so that measurements employing a Krishnan anisotropy balance or similar torsion device require that the sample be mounted along any direction perpendicular to *c*. As for the transverse Faraday setup, both $\chi_\parallel$ and $\chi_\perp$ now lie in the plane perpendicular to the torsion fibre and measured couples are thus proportional to $\Delta\chi = \chi_\parallel - \chi_\perp$. Anisotropy measurements alone can never provide a full description of the susceptibility tensor, of course. That requires at least one more piece of experimental information of a different kind: here, the value of either $\chi_\parallel$ or $\chi_\perp$ obtained from a Faraday balance, for example, or – more generally perhaps – the value of the mean susceptibility $\bar{\chi}$ obtained possibly from a Faraday or a Gouy balance. Since

$$\bar{\chi} = \tfrac{1}{3}(\chi_\parallel + 2\chi_\perp) \tag{6.1}$$

and

$$\Delta\chi = \chi_\parallel - \chi_\perp, \tag{6.2}$$

we have

$$\chi_\parallel = \bar{\chi} + \tfrac{2}{3}\Delta\chi \tag{6.3}$$

and

$$\chi_\perp = \bar{\chi} - \tfrac{1}{3}\Delta\chi. \tag{6.4}$$

6.2 Orthorhombic crystals

The susceptibility tensors of orthorhombic crystals are characterized by three, generally different, principal susceptibilities oriented strictly parallel to the crystal symmetry axes *a*, *b*, *c*. The procedures we employ to measure these magnetic properties are closely similar to those described for uniaxial crystals except that measurements are now required with respect to at least two crystal axes.

Using a transverse field Faraday magnetometer, orientation of an orthorhombic crystal parallel to any one symmetry axis provides for the determination of susceptibilities in both principal directions in the perpendicular (horizontal) plane. For example, values for χ_b and χ_c are obtained from a crystal mounted parallel to *a*. Accordingly, two separate experiments in which the crystal is oriented parallel to two different symmetry axes, yield all three principal crystal susceptibilities and, at the same time, provide a check on one of them. Of course, it is good discipline to make measurements for crystals mounted parallel to each of the symmetry axes and so improve the statistical accuracy of the results, but this is not essential. Consider again, however, the advantages of a transverse field Faraday balance as compared with a longitudinal field setup. Three separate crystal orientations are essential in the latter experiment to determine χ_a, χ_b and χ_c. Three similar experiments using the transverse field arrangement, though strictly unnecessary, yield the same principal susceptibilities plus one check for each.

It is convenient to define the crystal anisotropies cyclically:

$$\left.\begin{aligned}\Delta\chi_a &= \chi_b - \chi_c, \\ \Delta\chi_b &= \chi_c - \chi_a, \\ \Delta\chi_c &= \chi_a - \chi_b,\end{aligned}\right\} \tag{6.5}$$

where $\Delta\chi_a$ represents the anisotropy measured for an orthorhombic crystal oriented with *a* parallel to the torsion fibre, etc. It follows immediately that a determination of all three principal crystal susceptibilities, χ_a, χ_b, χ_c, cannot be based on anisotropy measurements alone, even when these are performed with respect to each of the symmetry axes, because equations (6.5) are not linearly independent: the sum of any two principal anisotropies equals the third. Of course, this relationship provides a useful check on the internal consistency of separate anisotropy measurements but it is *only* for this reason that it is worthwhile performing anisotropy measurements for all three principal crystal orientations. The complete susceptibility tensor can be derived, once again, only by recourse to separate measurements of a different type. A determination of only one of the principal susceptibilities appearing on the right-hand side of (6.5) suffices to establish the complete susceptibility tensor. Alternatively, any two principal anisotropies plus the average susceptibility, given by

$$\bar{\chi} = \tfrac{1}{3}(\chi_a + \chi_b + \chi_c), \tag{6.6}$$

may be used to determine the principal susceptibilities; for example, by inverting

$$\begin{pmatrix} 3\bar{\chi} \\ \Delta\chi_a \\ \Delta\chi_b \end{pmatrix} = \begin{pmatrix} 1 & 1 & 1 \\ 0 & 1 & -1 \\ -1 & 0 & 1 \end{pmatrix} \begin{pmatrix} \chi_a \\ \chi_b \\ \chi_c \end{pmatrix}. \tag{6.7}$$

The simplicity of the determinations of the susceptibilities of ortho-rhombic and uniaxial systems derives from the symmetry requirement (Neumann's principle) that all off-diagonal terms in the crystal tensor referred to the crystal symmetry axes necessarily vanish. Now, it is not essential to perform measurements on these crystals aligned parallel to symmetry axes. Measurements with respect to other, arbitrary but known, axes (but many more such, as we shall see in §6.3) would serve to determine the principal crystal susceptibilities and, at the same time, confirm that these off-diagonal terms do vanish. One might suppose that such a generalized strategy would be desirable, not least because the resulting values, being derived from many more – hopefully self-consistent – measurements, should have greater statistical reliability. It is a fact, however, that considerable tedium attaches to the determination of susceptibilities in general planes and, particularly importantly, their accuracy is likely to be far less than those determined in symmetry planes. This latter point follows from the form of the susceptibility tensor being an entity which varies as cosine squared (direction cosines appear to the second power in the tensor transformation law (4.12)). A cosine-squared function varies rather slowly for angles near zero but rapidly elsewhere. Accordingly, a small orientation error – say $1-2°$ – in the alignment of a crystal parallel to a principal magnetic axis will not generally give rise to serious errors in the ensuing estimates of principal susceptibility values. The situation might be very different for measurements based on general orientations, particularly if the crystal is highly anisotropic. There are therefore good reasons for measuring the susceptibilities of highly symmetrical crystals with respect to their symmetry axes, quite apart from general considerations of speed or indolence.

We turn now to consider the measurement of the susceptibilities of rather less symmetrical crystals than uniaxial and orthorhombic. It will be apparent later that the more obvious progression through monoclinic to the totally unsymmetrical triclinic systems is less suitable than the path we now take.

6.3 Triclinic crystals

Symmetry does not define the orientation of the crystal magnetic suscepti-bility ellipsoid in triclinic systems, so that the principal susceptibilities of these crystals are more difficult to measure than those of any other crystal system. A compensating advantage is that, once measured, the suscepti-bility tensor of a triclinic crystal will usually give the corresponding 'molecular' property directly by identity, as in (4.54): only when there is more than one magnetic centre in the asymmetric unit need the crystal and 'molecular' properties not coincide. Various methods for measuring the principal susceptibilities of triclinic crystals have been suggested, as reviewed by Mitra,[1,2] and by Hanton,[3] but most have received little experimental verification. Until very recently, virtually the only reported data on paramagnetic systems appears to be that of Krishnan & Mookerji[4-6] who determined the principal susceptibilities of $CuSO_4.5H_2O$ in the temperature range 80–300 K by measuring the magnetic anisotropy and direction of maximum susceptibility in eight different planes using a torsion balance. The direction of maximum susceptibility is particularly difficult to measure, however, even when the crystal possesses a well-developed morphology, and the method of Krishnan & Mookerji is tedious and rather inaccurate. As indicated in the previous section, measurements of susceptibilities in non-symmetry planes are particularly subject to error and an awareness of the experimental limitations of magnetometers is crucial to the development of a practical technique for the determination of the susceptibilities of triclinic crystals. Generally speaking, the many methods which have been proposed in the past suffer from the requirement to measure the *absolute* orientation of a maximum susceptibility in a plane, or of some other feature which is not generally, or has not, until recently, been measurable with sufficient accuracy. The question of accuracy is in no way pedantic for, explicitly or implicitly, the mathematical process of deriving the complete crystal susceptibility tensor from the experimental measurements requires the solution of simultaneous non-linear equations. It may be relatively straightforward to construct the appropriate equations: it is an entirely different matter to solve these equations with real, experimental data suffering the inevitable errors of measurement.

Accordingly, our discussion is restricted to the description of a technique, only recently developed,[7] and which, by taking advantage of the characteristics of the transverse field Faraday method, has been totally successful in five recently completed studies.[7-9] The technique relies *only* on those quantities which can be determined with reasonable accuracy; namely, the alignment of a chosen crystal axis parallel to a torsionally rigid suspension of quartz fibre, the weight of the crystal, and changes in weight in a magnetic field. Analogous procedures are available for longitudinal field magnetometers but the more satisfactory method employs the transverse field and is described first.

6.3.1 Transverse magnetic field

The first stage of this method for measuring the susceptibilities of triclinic crystals is based on that proposed by Ghosh & Bagchi[10] in which the mean susceptibility is measured in a crystal plane whose normal has direction cosines $\{\xi, \eta, \zeta\}$ with respect to a given cartesian reference frame. Let the susceptibility in the reference frame be $[\chi_{ij}]$. With respect to this 'old' frame, we define a 'new' one by the matrix of direction cosines,

$$
\begin{array}{c}
\text{'old' cartesian frame}
\end{array}
$$

$$
\begin{array}{cc}
 & \begin{array}{ccc} x & y & z \end{array} \\
\begin{array}{l} \text{'new'} \\ \text{cartesian} \\ \text{frame} \end{array}
\begin{array}{c} x' \\ y' \\ z' \end{array}
&
\begin{pmatrix}
a_{11} & a_{12} & a_{13} \\
a_{21} & a_{22} & a_{23} \\
\xi & \eta & \zeta
\end{pmatrix}
\end{array} , \qquad (6.8)
$$

where we have chosen the normal ξ, η, ζ to be the 'new' z' axis. The susceptibility tensor $[\chi'_{ij}]$, with respect to the 'new' frame, is given by

$$
[\chi'_{ij}] =
\begin{bmatrix}
\chi'_{11} & \chi'_{12} & \chi'_{13} \\
\chi'_{21} & \chi'_{22} & \chi'_{23} \\
\chi'_{31} & \chi'_{32} & \chi'_{33}
\end{bmatrix} , \qquad (6.9)
$$

and in the 'new' $x'y'$ plane (whose normal z' has direction cosines ξ, η, ζ), the average susceptibility is given by half the (invariant) trace $(\chi'_{11} + \chi'_{22})$; that is, by

$$
\bar{\chi}_{\perp(\xi\eta\zeta)} = \tfrac{1}{2}(\chi'_{max} + \chi'_{min}) \qquad (6.10)
$$

if χ'_{max} and χ'_{min} refer to the maximum and minimum values taken by the susceptibilities in the $x'y'$ plane.

In parentheses, let us elaborate this point. Suppose that, by good fortune, the 'new' cartesian frame had been chosen such that the principal susceptibilities in the $x'y'$ plane happened to lie parallel to the x' and y' axes (note that there is no requirement that these principal susceptibilities correspond to those of the whole tensor: we are considering only a *general* elliptical section of the magnitude ellipsoid here). Under these circumstances the top left-hand part of (6.9) is diagonal with $\chi'_{12} = \chi'_{21} = 0$ and $\{\chi'_{11}, \chi'_{22}\} = \{\chi'_{max}, \chi'_{min}\}$. Here, (6.10) follows immediately, of course. However, any general orientation of the 'new' frame in which $\chi'_{12} \neq 0$, which corresponds to a rotation of the above special frame about the suspension axis, is related to the special one by a similarity transformation (4.12) under which the trace $(\chi'_{11} + \chi'_{22})$ is invariant. In practical terms this means that the *sum* of *any* two *orthogonal* susceptibilities perpendicular to the suspension axis is independent of the absolute orientation of the apparatus magnet.

Using the tensor transformation law,

$$\chi'_{ij} = \chi_{kl} a_{ik} a_{jl},\tag{4.12}$$

we have

$$
\begin{aligned}
\chi'_{11} + \chi'_{22} &= \chi_{kl}(a_{1k}a_{1l} + a_{2k}a_{2l})\\
&= \chi_{11}(a_{11}^2 + a_{21}^2) + \chi_{22}(a_{12}^2 + a_{22}^2) + \chi_{33}(a_{13}^2 + a_{23}^2)\\
&\quad + \chi_{12}(a_{11}a_{12} + a_{21}a_{22}) + \chi_{13}(a_{11}a_{13} + a_{21}a_{23})\\
&\quad + \chi_{23}(a_{12}a_{13} + a_{22}a_{23}) + \chi_{21}(a_{12}a_{11} + a_{22}a_{21})\\
&\quad + \chi_{31}(a_{13}a_{11} + a_{23}a_{21}) + \chi_{32}(a_{13}a_{12} + a_{23}a_{22}).
\end{aligned}\tag{6.11}
$$

From the orthonormality of (6.8) and using (4.6), we have

$$(a_{11}^2 + a_{21}^2) = (1 - \xi^2),\ \text{etc.,}\tag{6.12a}$$

and

$$(a_{11}a_{12} + a_{21}a_{22}) = -\xi\eta,\ \text{etc.,}\tag{6.12b}$$

so that, remembering that the susceptibility tensor is symmetrical, there results

$$
\begin{aligned}
2\bar{\chi}_{\perp(\xi,\eta,\zeta)} &= \chi_{11}(1 - \xi^2) + \chi_{22}(1 - \eta^2) + \chi_{33}(1 - \zeta^2)\\
&\quad - 2(\chi_{12}\xi\eta + \chi_{13}\xi\zeta + \chi_{23}\eta\zeta).
\end{aligned}\tag{6.13}
$$

Since it is practicable to orient a crystal on the suspension fibre of the balance reasonably accurately parallel to some chosen direction ξ, η, ζ and to measure the average susceptibility in the plane perpendicular to that direction (which is, as was remarked above, simply the average of measurements in two, arbitrary but perpendicular, directions of the magnetic field in the horizontal plane), (6.13) provides the basis for a direct determination of the susceptibilities of a triclinic crystal. Measurement of the average susceptibilities in six different planes furnishes six simultaneous linear equations like (6.13), which may be solved for the six independent components of the susceptibility tensor. Of course it is necessary to ensure that the six equations are linearly independent: Ghose[11] has described circumstances under which this may not obtain. In practice, the problem is best solved by measuring susceptibilities in additional planes and then solving m equations in the six unknowns ($m > 6$) by the method of least squares.

Although Ghosh & Bagchi[10] proposed this method over 20 years ago, it is only recently that the procedure has been checked experimentally, presumably because, until the advent of a sufficiently sensitive, transverse field, single-crystal magnetometer, a measurement of $(\chi_{\max} + \chi_{\min})$ for any crystal plane has required two separate measurements. While one, for the anisotropy $(\chi_{\max} - \chi_{\min})$, is straightforward with a torsion balance, the other, for $\chi_{\max}$, using a Faraday or Curie balance, was formerly quite difficult. However, the single-crystal Faraday balance recently described by Cruse & Gerloch[12] permits individual measurement of $\chi_{\max}$ and $\chi_{\min}$ in the plane perpendicular to any chosen suspension direction, simply by rotation of the magnet. Further, if only the average susceptibility in a given plane is required, it is even unnecessary to determine the setting angle of the magnet corresponding to $\chi_{\max}$, as discussed above. However, we prefer to determine $\chi_{\max}$ and $\chi_{\min}$ individually, for reasons discussed shortly.

6.3.2 Checks, simplifications and limitations

If the direction of one of the principal crystal magnetic axes can be predetermined, perhaps from molecular symmetry considerations, much experimental effort can be saved. Thus, by choosing the 'old' cartesian frame such that the z axis is parallel to the presumed principal direction, we may set $\chi_{13} = \chi_{23} = 0$ and so find that data from only four crystal planes are then sufficient, in principle, to define the complete susceptibility tensor. However, as discussed in §4.14, we should not rely on assumptions concerning special molecular magnetic axes based merely on *approximate* molecular symmetry. Nevertheless, the simplification can be useful for monitoring progress during the rather lengthy process of complete data collection.

The consistency of the calculated tensor elements with the raw data, and the magnitude of the experimental errors associated with the determination of crystal mass and alignment, can be estimated in two main ways. First, individual values of $\chi_{\max}$ and $\chi_{\min}$ for each plane may be calculated from the tensor elements χ_{ij} determined at any stage and compared with the experimentally observed values (provided, of course, that values for $\chi_{\max}$ and $\chi_{\min}$ have been measured, rather than just their sum). In this connection, it is convenient to define a discrepancy index

$$R = \sum |\chi_{\mathrm{obs}} - \chi_{\mathrm{calc}}| / \sum \chi_{\mathrm{obs}}, \tag{6.14}$$

where χ_{obs} is an observed value for either $\chi_{\max}$ or $\chi_{\min}$, and χ_{calc} is the same quantity calculated from the appropriate tensor derived from the set of raw data referring to their sums. Secondly although the absolute orientation of χ_{max} in a given plane is difficult to measure, as discussed in the last chapter – and is not attempted – the variation of that orientation with temperature is easily determined by observing the orientation change of χ_{max} with respect to the laboratory-fixed frame of the magnet protractor. To be clear about this point: note how the absolute orientation of χ_{max} requires an accurate measurement of the *sample* orientation, while the measurement of the variation in χ_{max} does not. Finally, any such variation may be compared with that calculated from the derived $[\chi_{ij}]$ throughout the temperature range.

It is essential to perform these checks, for the procedure described above may not always succeed in practice. In systems exhibiting small anisotropies, for example, the differences between values of the sums ($\chi_{max} + \chi_{min}$) for different planes may be small and even comparable with experimental errors. In such cases the tensor $[\chi_{ij}]$ derived from one subset of the complete data (say from n planes, where $6 \leqslant n \leqslant m$ and m is the total number of separate crystal alignments parallel to the suspension fibre) may be in sharp disagreement with that from another subset. Collecting data from an ever-increasing number of planes will not necessarily improve the situation if experimental errors are comparable with actual anisotropies. There is little that can be done if the crystal is very nearly isotropic. However, a far more disappointing circumstance arises if no result can be achieved simply because half the experimental information has been thrown away: and this happens in the method so far described, for only the sums of χ_{max} and χ_{min} are used rather than their individual values. It is this consideration which led to the development[7] of the second stage of the new technique that we now describe.

6.3.3 The completed procedure

An obvious and considerable advantage of the transverse field system of magnetometer is that, simply by rotation of the magnet about the vertical axis, the susceptibility in any direction perpendicular to the sample suspension may be measured directly. In the first stage of this technique we only required the sum of χ_{max} and χ_{min}, obtainable from any two measurements at right-angles in the plane of the magnet protractor. It is better, however, to choose these directions parallel to the principal axes of the susceptibility ellipse in that plane. This may be achieved either by trial-and-error rotation of the magnet until χ_{max} is found, probably with the aid of graphical interpolation; or by plotting the relationship between susceptibility and magnet orientation with subsequent fitting of this curve to a cosine-squared function. Having once measured values for χ_{max} and χ_{min} separately in any given plane, we may proceed as follows.

We begin by noting that the susceptibility in the 'new' frame now takes the form

$$[\chi_{ij}] = \begin{bmatrix} \chi_{max} & 0 & \chi'_{13} \\ 0 & \chi_{min} & \chi'_{23} \\ \chi'_{13} & \chi'_{23} & \chi'_{33} \end{bmatrix}, \qquad (6.15)$$

rather than (6.9). Again, using the tensor transformation rule (4.12), we find, in comparison with (6.11),

$$\left.\begin{array}{c} \chi_{max} \\ \\ \\ \chi_{min} \end{array}\right\} = \left\{\begin{array}{l} a_{11}^2\chi_{11} + a_{12}^2\chi_{22} + a_{13}^2\chi_{33} + 2a_{11}a_{12}\chi_{12} \\ + 2a_{11}a_{13}\chi_{13} + 2a_{12}a_{13}\chi_{23}, \\ \\ a_{21}^2\chi_{11} + a_{22}^2\chi_{22} + a_{23}^2\chi_{33} + 2a_{21}a_{22}\chi_{12} \\ + 2a_{21}a_{23}\chi_{13} + 2a_{22}a_{23}\chi_{23}, \end{array}\right. \qquad (6.16)$$

where χ_{max} is taken as the greater of these two expressions, and χ_{min} as the lesser. Now the $\{a_{ij}\}$ are different for each crystal plane studied and, furthermore, are incompletely known because the absolute orientation of χ_{max} is unknown. Thus (6.16) relates measured quantities χ_{max} and χ_{min}, the unknown quantities $\{\chi_{ij}\}$ and, say, one unknown angle relating the 'old' and 'new' coordinate frames. Simultaneous solution of sets of equations like (6.16) is difficult because of their non-linear nature: indeed, this problem lay behind the adoption of (6.13) by Ghosh & Bagchi,[10] so reducing the problem to one of simultaneous *linear* equations only. However, a new approach[7] to this problem retains the handling of only linear equations while simultaneously utilizing the separate pieces of experimental information in χ_{max} and χ_{min}.

The process is an iterative one. Those who attach great importance to the elegance of explicit algebra, particularly with respect to analytical expressions in closed form, may find little appeal in this method. It has the compensating advantage, however, of being workable in principle and, more importantly and virtually uniquely, of having been proved several times in practice. As with all iterative procedures, we begin by assuming that we are given a reasonable first estimate for the $\{a_{ij}\}$ in (6.16). Given these $\{a_{ij}\}$, (6.16) reduces to a pair of linear equations in the $\{\chi_{ij}\}$: altogether there result $2n$ simultaneous linear equations, derived from experiments in n planes, which may be solved by the method of least squares to give an estimate of the $\{\chi_{ij}\}$. These are then used to provide a better estimate of the $\{a_{ij}\}$ and the cycle is repeated until self-consistency is achieved, when the $\{\chi_{ij}\}$ yield the $\{a_{ij}\}$ from which they were determined. The problems of solving simultaneous non-linear equations are thus replaced by a linear least-squares process. All that is required is a sufficiently accurate starting estimate of the $\{a_{ij}\}$. An obvious source, of course, are those $\{a_{ij}\}$ which may be derived from the $\{\chi_{ij}\}$ obtained from the values of $(\chi_{max} + \chi_{min})$, as described above. Experience so far suggests that the second stage – the iterative procedure – begun with estimates of the $\{a_{ij}\}$ obtained in this way, effectively converges after three or four cycles: the calculations take very little computer time, however, so that it is worth while performing ten cycles as a matter of routine.

In summary: the determination of the complete susceptibility tensor of a triclinic crystal by the procedure described here involves the measurement of χ_{max} and χ_{min} in a minimum of six different crystal planes. The absolute orientations of these maximum and minimum susceptibilities are *not* required. The process divides into two parts. In the first – essentially that of Ghosh & Bagchi[10] – estimates for the $\{\chi_{ij}\}$ are derived directly from the average susceptibilities in the chosen crystal planes. In the second, these estimates are improved by way of an iterative procedure in which the individual values of χ_{max} and χ_{min} in each plane, rather than just their sum, determine the result. The ultimate success of the approach obviously depends upon the magnitude of the magnetic anisotropy actually present in a given crystal. Success in the first part of the process rests upon the differences between the values of $(\chi_{max} + \chi_{min})$ obtained from the various planes. Even if the χ_{max} and χ_{min} values individually differ sufficiently for the second, iterative, process to converge with sufficient agreement between observed and calculated values for these quantities, the average susceptibilities in each plane may not differ enough to provide a suitable beginning for that second stage. In difficult cases, recourse to hints from features of molecular geometry, which may suggest the orientation of one principal molecular susceptibility, might help to provide that starting point. If so, the second process, utilizing all experimental data, should produce an optimized result. Further, once suitably initiated, the iterative procedure is self-contained so that the final result need not be prejudiced by the initial 'working hypothesis' of some presumed approximate molecular geometry. In this sense, we can retract the word 'minimum' from the summarizing statement that opened the present paragraph. The least-squares process, strictly, only requires six independent pieces of experimental data – perhaps values of χ_{max} and χ_{min} from three different planes – although the greater amount of data from six or more crystal orientations may obviously improve the statistical accuracy of an iterative procedure. However, if the starting point for that process is provided entirely by the analytical method of Ghosh & Bagchi, measurements must be performed in six independent planes in order to yield enough data for (6.13) to be solved for the six independent χ_{ij}. On the other hand, recourse to the informed guesswork afforded by observation of molecular structure might fix the values of some off-diagonal coefficients like χ'_{13} in (6.15) and so obviate the need for measurements in as many planes as the 'minimum' six. In §6.3.5 we review an example of this approach. Ultimately for crystals with very small anisotropies, even tactics like these will fail to distinguish very differently oriented tensors derived from different subsets of the complete data in the first stage and the whole technique will fail. The same would undoubtedly be true of any of the earlier methods proposed for the measurement of the susceptibilities of triclinic crystals.

6.3.4 Longitudinal magnetic field

Most magnetometers and Faraday balances employing superconducting solenoids suffer the disadvantage of providing for the measurement of only that susceptibility oriented parallel to the sample suspension. If $\{\xi, \eta, \zeta\}$ is now taken as the set of direction cosines of the longitudinal magnetic field with respect to the 'old' cartesian reference frame, we find, in contrast to (6.13),

$$\chi_{\|(\xi,\eta,\zeta)} = \chi_{11}\xi^2 + \chi_{22}\eta^2 + \chi_{33}\zeta^2 + 2\chi_{12}\xi\eta + 2\chi_{13}\xi\zeta + 2\chi_{23}\eta\zeta. \quad (6.17)$$

At least six different suspension directions are required to provide six different values of $\chi_{\|(\xi,\eta,\zeta)}$ from which the complete susceptibility tensor may be derived by simultaneous solution of the equations (6.17), linear in χ_{ij}, for known ξ, η, ζ. However, the differences between various values of $\chi_\|$ will generally be greater than those between the corresponding values of $(\chi_{max} + \chi_{min})$ in the planes perpendicular to the suspension directions. Values of $\chi_{\|(\xi,\eta,\zeta)}$ from longitudinal, magnetic field measurements should therefore provide a better estimate of the tensor components than would the associated values of $(\chi_{max} + \chi_{min})$. This is fortunate because the second stage of the procedure described using transverse magnetic fields is not available here. Note once again, that measurements with respect to six axes using a longitudinal field setup will, at best, yield all six tensor components χ_{ij} with no least-squares optimization: six similar crystal orientations used with a transverse field magnetometer provide twelve pieces of information – χ_{max} and χ_{min} in each plane.

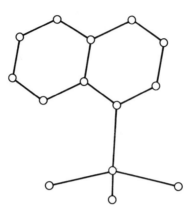

Fig. 6.1. The molecular geometry of $[Ni(quinoline)Br_3]^-$ ions.

6.3.5 An example

The complex, tribromoquinoline nickel(II) crystallizes in the space group $P\bar{1}$ with only one molecule in the asymmetric unit. The ionic geometry is shown in figure 6.1. Citations and a discussion of the results of a study of this system within ligand-field theory are presented in §12.1. Susceptibility measurements[3,9] were made on crystals weighing 1–4 mg using the transverse field Faraday balance[12] and techniques described above. Crystal axes were identified by X-ray rotation and Weissenberg photographic methods, together with computational aids devised by Davies. Values for the maximum and minimum susceptibilities in planes normal to eight different suspension axes were determined throughout the temperature range 20–300 K. They are illustrated for our present purposes by the (typical) results at 95 K in table 6.1. All values in tables 6.1–6.4 have been interpolated from the raw experimental data and corrected for a mean molar diamagnetism $\bar{\chi}_M^{dia} = -398 \times 10^{-6}$ CGS emu, calculated from Pascal's constants.[13] Values of the principal crystal susceptibilities and moments, together with their orientations, derived from (6.16) using the least-squares procedure with all 16 pieces of data, are given at that same representative temperature in table 6.2. The following observations have been made[3,9] upon the progress of these measurements.

(i) χ_{max} and χ_{min} were measured for each axis of suspension of the sample, so providing 16 pieces of data from which the six independent tensor elements were determined by least squares. Back-calculation from the refined tensor elements, of which those in table 6.2 are quite typical, yields values of χ_{max} and χ_{min} which may be compared with the raw experimental

Table 6.1. *Tetra-n-butylammonium   tribromo(quinoline)   nickelate(II): experimental values[9] of maximum and minimum susceptibilities in planes prependicular to the zone axes [u, v, w]. Values in CGS emu × 10⁴, interpolated at 95 K from raw data*

zone axis	[111]	[001]	[110]	[100]	[101]	[112]	[111]	[010]
χ_{max}	215	170	176	211	200	205	146	144
χ_{min}	108	105	125	130	106	102	125	111

Table 6.2.   *Tetra-n-butylammonium   tribromo (quinoline) nickelate(II): principal susceptibilities and moments,[9] and their orientations with respect to the orthogonalized crystal axes (a, c* $\wedge$ a, c*) at 95 K. Values corrected for $\chi_M^{dia} = -398 \times 10^{-6}$ CGS emu*

χ (CGS × 10⁴)	μ_{eff} (Bohr magnetons)	orientation (degrees)		
214	4.03	89.8	26.0	116.0
130	3.14	79.6	64.5	27.9
103	2.79	10.4	94.7	99.3

Table 6.3. *Progress of χ measurements[3]: χ values in CGS emu × 10⁴*

	data from first five axes			data from all eight axes		
Temp/K	χ	orientation[a]		χ	orientation[a]	
295	62	79.4	79.3	15.2	62	81.0	81.0	12.8
	47	69.6	158.5	83.5	47	75.1	163.7	83.5
	42	23.2	71.6	103.7	42	17.5	76.5	101.0
25	915	68.9	76.0	25.7	901	77.0	81.2	15.8
	360	83.1	165.9	77.7	328	89.0	171.2	81.2
	184	22.3	88.2	112.2	185	13.1	91.0	103.0

[a]with respect to a molecular frame in which z lies parallel to Ni—N.

data. Indices calculated as in (6.14) for data over the complete temperature range varied from $R = 0.012$ to 0.055 with an average of 0.020.

(ii) The first stage, using (6.13), with $\bar{\chi}_\perp$ values from the first six suspension axes, failed to produce a suitable estimate of the susceptibility

Table 6.4. *Comparison*[3] *of constrained and unconstrained* χ *tensors (values in* CGS emu $\times 10^4$)

Temp/K	χ^a	χ^b
295	54	62
	53	47
	43	42
25	992	901
	323	328
	86	185

[a]using the first five axes with χ_{max} constrained to lie parallel to Ni—N.
[b]using data from all eight axes, without constraint.

tensor from which the least-squares, second stage could refine. This was due to a combination of inevitable experimental errors and to an unfortunate choice of suspension axes in that four of the six directions lay in the same plane. Even with data from seven planes, the least-squares process failed to converge at some temperatures. Hence the need for eight separate crystal mountings despite the theoretical 'minimum' of six discussed above.

(iii) It is interesting to compare the tensor finally obtained from the complete data set with that deduced from five planes plus the initializing assumption that one principal susceptibility lies parallel to the Ni—N (quinoline) vector. This assumption was only used in the first stage to help provide a reasonable first estimate of the direction cosines to be used in (6.16) *et seq.*: the values shown in table 6.3, for experiments at both ends of the temperature range, show that very close agreement can be obtained, so emphasizing the problem with the slight variation in $\bar{\chi}_\perp$ values sometimes being unable to provide a satisfactory beginning for the least-squares stage, rather than an intrinsic lack of accuracy of the individual values of χ_{max} and χ_{min}.

(iv) The tensor orientation given in table 6.2 correspond[3,9] to the largest susceptibility being oriented some 13° from the Ni—N (quinoline) at 295 K, or 16° at 25 K – probably equal within experimental error – so making comprehensible the success of the initializing simplification just discussed. It is to be emphasized, however, that a *fixed* assumption that one principal susceptibility lies parallel to the Ni—N vector is not at all satisfactory for the calculation of the magnitudes of the principal susceptibilities. Thus, in table 6.4 are compared the susceptibility tensors calculated (a) with the complete set of data from eight planes, and (b) with the same five axes as in (iii) but now maintaining, throughout, the fiction that one principal susceptibility lies along Ni—N. It is clear that this second procedure is quite unacceptable, exemplifying the assertion made in §1.4.

(v) Principal sections of the 'observed' susceptibility magnitude ellipsoid for this complex at 25 K are shown in figure 12.2, providing a readily assimilable overview of results more formally presented in tabular form.

6.4 Monoclinic crystals

More compounds have been observed to crystallize in the monoclinic system than in any other. It is fortunate from a magnetochemical point of view that measurement of the susceptibilities of these crystals usually presents only a little more difficulty than that of orthorhombic materials and is much easier than for the triclinic case. However, while much of our discussion of monoclinic crystals can be presented by reference to the higher-symmetry systems, the general treatment required for the triclinic case can also be helpful.

The susceptibility tensor of a monoclinic crystal is characterized by one principal susceptibility, conventionally χ_3, lying parallel to the unique monoclinic b axis, so that $\chi_b \equiv \chi_3$. Symmetry does not define the orientation of χ_1 and χ_2 in the crystal ac plane, however. By convention, $\chi_1 < \chi_2$ and ϕ is the angle between a and χ_1 measured from the positive a axis towards, or through, the positive c axis, as shown in figure 4.1. Using a transverse field Faraday balance or similar magnetometer, suspension of a monoclinic crystal along b leaves the two principal susceptibilities χ_1 and χ_2 in the horizontal measurement plane. This single experimental orientation thus directly yields values for χ_1 and χ_2. As discussed in the last chapter, it is not simple to measure the absolute orientation of an in-plane principal susceptibility with satisfactory accuracy. Nevertheless, by using a small mirror glued to the suspension or by reflecting light from a vertical face, a rough estimate for this orientation can be made and hence an approximate value for ϕ determined. The same remarks hold for a similar crystal orientation using a torsion balance from which the principal anisotropy,

$$\Delta\chi_b = \chi_2 - \chi_1, \qquad (6.18)$$

is measured.

Returning to the transverse field Faraday experiment; a second orientation of the crystal parallel to *any* axis in the *ac* plane provides for the measurement of χ_b which is necessarily perpendicular to the *ac* plane, and the susceptibility along the mutual normal to *b* and the suspension axis. In principle, the value of the susceptibility in this latter direction furnishes the means to calculate ϕ or, equivalently, the off-diagonal susceptibility component χ_{21} in table 4.1. Suppose, for example, that the second crystal is mounted parallel to the crystal *a* axis. The maximum and minimum susceptibilities in the plane traversed by the magnet are χ_b and $\chi_{c'}$, where c' is normal to both *a* and *b*. Now, by the conventions above and in figure 4.1, the orientations of the principal susceptibilities with respect to the orthogonal crystal axes *a*, *b*, c' are given by the direction cosines

$$
\begin{array}{cccc}
 & \chi_1 & \chi_3 & \chi_2 \\
a & \begin{pmatrix} \cos\phi & 0 & -\sin\phi \\ 0 & 1 & 0 \\ \sin\phi & 0 & \cos\phi \end{pmatrix} & & \equiv \mathbf{u},
\end{array}
\qquad (6.19)
$$

so that the crystal magnetic tensor referred to that frame (a, b, c') is given by

$$
\begin{bmatrix} \chi_{aa} & 0 & \chi_{ac'} \\ 0 & \chi_{bb} & 0 \\ \chi_{ac'} & 0 & \chi_{c'c'} \end{bmatrix} = \mathbf{u} \begin{bmatrix} \chi_1 & 0 & 0 \\ 0 & \chi_2 & 0 \\ 0 & 0 & \chi_3 \end{bmatrix} \mathbf{u}^\dagger
\qquad (6.20)
$$

Therefore, the susceptibilities measured parallel to *a*, *b* and c' are

$$\chi_{aa} = \chi_1 \cos^2\phi + \chi_2 \sin^2\phi, \qquad (6.21a)$$

$$\chi_b = \chi_3, \qquad (6.21b)$$

$$\chi_{c'c'} = \chi_1 \sin^2\phi + \chi_2 \cos^2\phi, \qquad (6.21c)$$

and so from two experimental crystal orientations parallel to *b* and, say, *a* we obtain χ_1, χ_2, χ_3 and ϕ. For example, from (6.21c), we find

$$\cos^2\phi = \frac{\chi_{c'c'} - \chi_1}{\chi_2 - \chi_1}, \qquad (6.22)$$

but it is important to recognize that this expression for ϕ, involving a ratio of anisotropies, is particularly subject to experimental error. There are several ways of improving this situation.

Firstly, it is clear that ϕ values calculated by way of (6.22) should agree, at least approximately, with those determined by direct observation, as described above. Indeed, this qualitative equality is used to decide whether

$\chi_{\max}$ in the bc' plane is to be assigned to χ_3 or to $\chi_{c'c'}$. Secondly, measurements can be performed for crystals oriented parallel to other axes in the ac plane. For example, if the crystal is mounted along c', measurements yield values for χ_b (as always) plus χ_{aa}, and so, from (6.21a) we have

$$\sin^2 \phi = \frac{\chi_{aa} - \chi_1}{\chi_2 - \chi_1}. \tag{6.23}$$

Of course, the same objection about this function being a ratio of susceptibility differences can be raised once more, but we might at least hope to improve the statistical accuracy of our determination of ϕ by this further experiment. In passing, note too that the sign ambiguity in (6.22) is removed by (6.23), or *vice versa*. Similar expressions can be derived for crystals oriented along any direction in the ac plane, although the simplest way to treat the more general cases is to redefine the crystal a and c axes which, because they are not defined by symmetry may be done with impunity. A third, though lengthy, procedure for improving the estimate of ϕ is afforded by treating the monoclinic system as a special case of the triclinic one, as follows.

As noted in the previous section, a convenient way of monitoring progress in the collection and processing of data from triclinic crystals is to presume knowledge of the direction of one of the principal susceptibilities, for then equations (6.13) may be solved for four unknowns using data from only four planes. In the case of triclinic crystals such a procedure is based, at best, on approximate molecular symmetry and is therefore unreliable, at least in the first stage. On the other hand, in application of the same technique to monoclinic crystals, *no* approximation is made when identifying the direction of one principal susceptibility – χ_3 – as parallel to b. Measurement of the average susceptibilities in four different planes will yield a description of the complete monoclinic susceptibility tensor, provided that the appropriate equations (6.13) are linearly independent. Further, if independent measurements of $\chi_{\max}$ and $\chi_{\min}$ in each of these planes are made, the second, iterative, stage of the triclinic procedures may be adopted, but with inclusion of the (rigorous) constraints that $\chi_{13} = \chi_{23} = 0$. Altogether, for the cost of time and patience in measuring $\chi_{\max}$ and $\chi_{\min}$ in four different crystal planes, one of which should well be perpendicular to b, one obtains eight pieces of data with which to define the four independent elements of the monoclinic susceptibility tensor. The less troublesome alternative is to perform experiments on crystals aligned along only two directions – b and one other perpendicular – so obtaining reasonably reliable values for the three principal crystal susceptibilities χ_1, χ_2, χ_3 but a rather less exact estimate of the angle ϕ.

The traditional procedure for the determination of monoclinic susceptibilities using a torsion balance involves the measurement of the principal crystal anisotropies which, for crystals oriented along a, b, and c' and using (6.21), are given by

$$\left. \begin{aligned} \Delta\chi_a &= \chi_1 \sin^2\phi + \chi_2 \cos^2\phi - \chi_3, \\ \Delta\chi_b &= \chi_2 - \chi_1, \\ \Delta\chi_{c'} &= \chi_1 \cos^2\phi + \chi_2 \sin^2\phi - \chi_3. \end{aligned} \right\} \tag{6.24}$$

Analogous expressions may be derived for crystals oriented parallel to a', c or any other direction in the ac plane, though, again, the most direct approach here is to use (6.24) and redefine the directions of the crystal a and c axes. As usual, the principal susceptibilities are not derivable from the anisotropies alone, for these are related amongst themselves by

$$\Delta\chi_a + \Delta\chi_{c'} = \chi_1 + \chi_2 - 2\chi_3. \tag{6.25}$$

If average, powder, susceptibilities χ are available – perhaps from a Gouy or powder Faraday balance, for example – the principal susceptibilities may be obtained by inversion of (6.26):

$$\begin{pmatrix} 3\bar{\chi} \\ \Delta\chi_b \\ \Delta\chi_{c'} + \Delta\chi_a \end{pmatrix} = \begin{pmatrix} 1 & 1 & 1 \\ -1 & 1 & 0 \\ 1 & 1 & -2 \end{pmatrix} \begin{pmatrix} \chi_1 \\ \chi_2 \\ \chi_3 \end{pmatrix}. \tag{6.26}$$

The ϕ angle is once more given by an expression which is the ratio of anisotropies,

$$\cos 2\phi = \frac{\Delta\chi_a - \Delta\chi_{c'}}{\Delta\chi_b}, \tag{6.27}$$

and so is not generally well determined. Indeed, it is not infrequent to find that quite reasonably accurate estimates of the principal susceptibilities themselves are associated with the condition $|\Delta\chi_a - \Delta\chi_{c'}| > \Delta\chi_b$ even though this is theoretically impossible.

This completes our review of the procedures and techniques used to measure crystal susceptibilities but we have reached a convenient point at which to discuss some general experimental results to do with the monoclinic ϕ angle.

6.5 Variation of the monoclinic ϕ angle with temperature

There is no connection between temperature and the various features of monoclinic crystal symmetry and so there is no obvious reason to expect that the angle in figure 4.1 should remain constant while the magnitudes of the principal crystal susceptibilities vary markedly with temperature, as they certainly do for paramagnets. Nevertheless, as a matter of empirical observation, the commonest circumstance of all for monoclinic systems is to find that ϕ values vary virtually insignificantly over wide ranges of temperature. A notable exception occurs for iron(II)phthalocyanine where ϕ has been observed[14] to change dramatically, as shown in figure 6.2(a). The present section is concerned, therefore, to rationalize why essentially temperature-independent ϕ angles are the norm and also to identify the apparently rare conditions in which behaviour like that shown in figure 6.2(a) might obtain. I am indebted to Dr R.F. McMeeking in what follows for many discussions and prior communication of original material.

Beginning with a limiting case, we observe immediately that temperature-independent ϕ values will occur if Curie's law is obeyed by each principal crystal susceptibility. Thus we note from (6.22), (6.23) or (6.27) that trigonometric functions of ϕ are proportional to the ratios of various anisotropies so that, if $\chi_i = C_i/T$, $(\chi_i - \chi_j)/(\chi_k - \chi_l)$ is independent of temperature; and hence so also is ϕ. On the other hand, if the various crystal susceptibilities follow a Curie–Weiss law, $\chi_i = C_i/(T + \theta_i)$, ϕ will be independent of temperature generally only if all Weiss constants θ_i take the same value, and there seems to be no obvious reason for that to occur. Further, if we try to map the behaviour of ϕ in the general case, we have to consider a very complicated multivariable problem involving molecular and crystal magnetic rhombicity, molecular orientation, and temperature. However, McMeeking has shown[15] that the problem may be simplified, at least to the extent that the crystal and molecular rhombicities may each be represented by a single variable, so far as their effects on ϕ are concerned.

Let the principal molecular susceptibilities $\{K_i\}$ lie parallel to the axes $\{X, Y, Z\}$ for $i = 1, 2, 3$, respectively, and be oriented with respect to the orthogonalized monoclinic frame $\{a, b, c'\}$ according to the direction cosines,

$$
\begin{array}{cccc}
 & K_1 & K_2 & K_3 \\
a & \begin{pmatrix} \alpha_1 & \alpha_2 & \alpha_3 \\ \beta_1 & \beta_2 & \beta_3 \\ \gamma_1 & \gamma_2 & \gamma_3 \end{pmatrix} \\
b & & & \\
c' & & &
\end{array}
\qquad (6.28)
$$

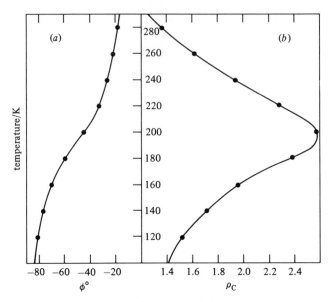

Fig. 6.2. Experimental data for iron(II)phthalocyanine, (*a*) marked variation of ϕ with temperature and, (*b*) the variation of the crystal rhombicity, $\rho_C \equiv (\chi_2 - \chi_3)/(\chi_2 - \chi_1)$, with temperature.

Figure 6.3 shows these frames, together with $\{X', Y', Z'\}$, representing the orientations of the projections $\{K_i'\}$ of the $\{K_i\}$ onto the *ac* plane. Let the angles subtended by the pairs of axes $\{X, X'\}$, $\{Y, Y'\}$, $\{Z, Z'\}$ be ψ, ψ', ψ'' and those by *a* and X', Y', Z' be ζ, ζ', ζ'' respectively. Projections of all molecules on the *ac* plane in the monoclinic system are magnetically equivalent, of course, so we need only consider the one orientation as given in (6.28) for our present purposes. Consider the crystal susceptibility $\chi(\theta)$ parallel to a vector in the *ac* plane at angle θ with respect to the crystal *a* axis, again shown in figure 6.3. We have

$$\chi(\theta) = K_1' \cos^2(\theta - \zeta) + K_2' \cos^2(\theta - \zeta') + K_3' \cos^2(\theta - \zeta''), \quad (6.29)$$

and the condition that the angle θ corresponds to the monoclinic ϕ angle as defined in figure 4.1 in that $\chi(\theta)$ takes the maximum value for $0 < \theta < 2\pi$ in the *ac* plane; that is, that $d\chi(\theta)/d\theta = 0$. Therefore,

$$
\begin{aligned}
0 &= K_1' \cos(\theta - \zeta)\sin(\theta - \zeta) + K_2' \cos(\theta - \zeta')\sin(\theta - \zeta') \\
&\quad + K_3' \cos(\theta - \zeta'')\sin(\theta - \zeta'') \\
&= K_1' \sin(2\theta - 2\zeta) + K_2' \sin(2\theta - 2\zeta') + K_3' \sin(2\theta - 2\zeta'') \\
&= \sin 2\theta(K_1' \cos 2\zeta + K_2' \cos 2\zeta' + K_3' \cos 2\zeta'') \\
&\quad - \cos 2\theta(K_1' \sin 2\zeta + K_2' \sin 2\zeta' + K_3' \sin 2\zeta'') \quad (6.30)
\end{aligned}
$$

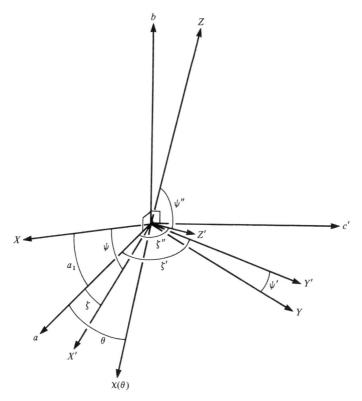

Fig. 6.3. X', Y', Z' and $\chi(\theta)$ all lie in the ac' plane.

and hence,

$$\tan 2\theta = \frac{K_1' \sin 2\zeta + K_2' \sin 2\zeta' + K_3' \sin 2\zeta''}{K_1' \cos 2\zeta + K_2' \cos 2\zeta' + K_3' \cos 2\zeta''}. \qquad (6.31)$$

Now, by construction

$$K_1' = K_1 \cos^2 \psi, \quad K_2' = K_2 \cos^2 \psi', \quad K_3' = K_3 \cos^2 \psi'' \qquad (6.32)$$

and ψ and ζ are related to α_1, β_1, γ_1, of (6.28) as follows.

In figure 6.4 is shown that part of figure 6.3 that relates to the first axis of the molecular sets. By inspection we have

$$\cos \psi = \mathrm{OX}'/\mathrm{OX}, \quad \cos \zeta = \mathrm{OA}/\mathrm{OX}', \quad \cos a_1 = \mathrm{OA}/\mathrm{OX}, \qquad (6.33)$$

so that

$$\cos a_1 = \cos \zeta \cos \psi = \alpha_1 \qquad (6.34)$$

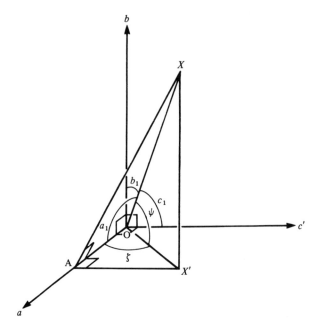

Fig. 6.4. The molecular X axis and its projection onto the ac' plane.

and, by similar constructions with respect to the remaining axes, we have the analogous relationships

$$\cos a_2 = \cos \zeta' \cos \psi' = \alpha_2 \quad \text{and} \quad \cos a_3 = \cos \zeta'' \cos \psi'' = \alpha_3. \quad (6.35)$$

Since $\psi = 90 - b_1$, $\cos \psi = \sin b_1$ and so (6.34) may also be written

$$\alpha_1 = \cos \zeta \sin b_1 \quad (6.36)$$

and there follow the relationships

$$\cos^2 \zeta = \alpha_1^2/(\alpha_1^2 + \gamma_1^2), \quad \sin^2 \zeta = \gamma_1^2/(\alpha_1^2 + \gamma_1^2), \quad \sin \zeta \cos \zeta = \alpha_1 \gamma_1/(\alpha_1^2 + \gamma_1^2) \quad (6.37)$$

and hence,

$$\cos 2\zeta = (\alpha_1^2 - \gamma_1^2)/(\alpha_1^2 + \gamma_1^2), \quad \sin 2\zeta = 2\alpha_1 \gamma_1/(\alpha_1^2 + \gamma_1^2). \quad (6.38)$$

Substitution of (6.38) and (6.32) into (6.31) gives

$$\tan 2\theta = \frac{2(\alpha_1 \gamma_1 K_1 + \alpha_2 \gamma_2 K_2 + \alpha_3 \gamma_3 K_3)}{(\alpha_1^2 - \gamma_1^2)K_1 + (\alpha_2^2 - \gamma_2^2)K_2 + (\alpha_3^2 - \gamma_3^2)K_3}, \quad (6.39)$$

which expresses θ for maximum or minimum in-plane crystal susceptibility in terms of the principal molecular susceptibilities and their orientations.

The equation has two roots, say 2θ and $(2\theta + \pi)$, whence $\phi = \theta$ or $\theta + \pi/2$. As both numerator and denominator of (6.39) are known, the ambiguity is apparently removed. However, as (6.39) corresponds to $\chi(\theta)$ of (6.29) being *either* maximal or minimal, the ambiguity remains and must be resolved by recourse to (6.29) explicitly.

Now (6.39) provides a basis for our present discussion despite its unpromising appearance. Firstly, consider an analogous expression for a similarly oriented, molecular magnetic tensor, but one whose principal values $\{L_i\}$ each differ from those above by a constant amount. Thus, suppose $L_i = K_i + A$: then the corresponding expression (6.39) for the θ value in the new system, say θ', is

$$\tan 2\theta' = \frac{2(\alpha_1\gamma_1 K_1 + \alpha_2\gamma_2 K_2 + \alpha_3\gamma_3 K_3) + 2A(\alpha_1\gamma_1 + \alpha_2\gamma_2 + \alpha_3\gamma_3)}{\begin{array}{c}[(\alpha_1^2 - \gamma_1^2)K_1 + (\alpha_2^2 - \gamma_2^2)K_2 + (\alpha_3^2 - \gamma_3^2)K_3] \\ + A[(\alpha_1^2 - \gamma_1^2) + (\alpha_2^2 - \gamma_2^2) + (\alpha_3^2 - \gamma_3^2)]\end{array}}$$

(6.40a)

$$= \tan 2\theta,$$

(6.40b)

the last equality resulting from the orthonormality of the matrix (6.28) which ensures that

$$\alpha_1\gamma_1 + \alpha_2\gamma_2 + \alpha_3\gamma_3 = 0$$

(6.41a)

and

$$(\alpha_1^2 - \gamma_1^2) + (\alpha_2^2 - \gamma_2^2) + (\alpha_3^2 - \gamma_3^2) = (\alpha_1^2 + \alpha_2^2 + \alpha_3^2) - (\gamma_1^2 + \gamma_2^2 + \gamma_3^2) = 0.$$

(6.41b)

We conclude from (6.40b), therefore, that the replacement of the real molecular tensor by one whose trace differs in this prescribed way has no effect on the expression (6.39) for θ, and hence ϕ. A special case is to write $L_1 = C\sin\rho_M$, $L_2 = C\cos\rho_M$, $L_3 = 0$ when (6.39) takes the form

$$\tan 2\theta = \frac{2(\alpha_1\gamma_1 \sin\rho_M + \alpha_2\gamma_2 \cos\rho_M)}{(\alpha_1^2 - \gamma_1^2)\sin\rho_M + (\alpha_2^2 - \gamma_2^2)\cos\rho_M},$$

(6.42)

an expression for θ in terms of the fixed molecular orientation $\{\alpha_i, \gamma_i\}$, and ρ_M relating to the molecular anisotropies: C drops out. Thus we observe how the *single* variable ρ_M can represent almost all details of the molecular rhombicity, *so far as a calculation of the crystal ϕ angle is concerned.* As above, we must supplement this parameter with a check via (6.29) to establish whether θ is maximal, and hence equal to ϕ, or not. Even so, this clever construction[15] allows one to investigate the behaviour of (6.39) in a much more economical way than might have been expected initially.

Examples of the behaviour of the crystal angle ϕ with respect to the

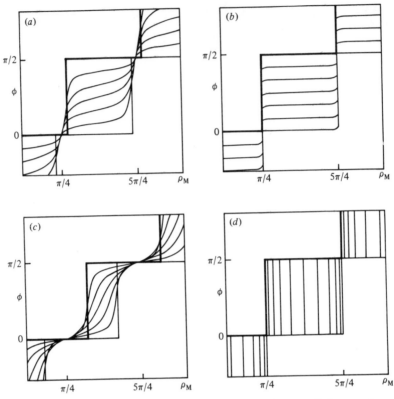

Fig. 6.5. Examples of the relationship between the monoclinic crystal
ϕ angle and the molecular rhombicity parameter ρ_M.

molecular rhombicity parameter ρ_M are shown in figure 6.5. Consider first
a limiting case in which a principal molecular susceptibility, say K_1, lies
in the ac plane. As the definition of any axis in this crystal plane is arbitrary,
in that symmetry in that plane is no determinant, a may be freely chosen:
hence, without further specialization, we take K_1 parallel to a. There follow
special conditions on the direction cosines (6.28) such that $\alpha_1 = 1$ while
$\beta_1 = \gamma_1 = \alpha_2 = \alpha_3 = 0$, whence by (6.39), $\tan 2\theta = 0$ giving $\phi = 0$ or $\pi/2$.
This simple step function, corresponding to the limiting case of K_1 lying
in the crystal ac plane, is emphasized in each part of the figure. That ϕ
takes the values 0, $\pi/2, \ldots$, rather than some other arbitrary values
separated by intervals of $\pi/2$, is just an artefact of our choice of origin
which was established by taking K_1 parallel to a. At certain values of the
rhombicity parameter ρ_M, the calculated ϕ angle changes abruptly from
one constant value to another and the corresponding vertical lines in the

figures occur when the in-plane crystal anisotropy $\Delta\chi_b$ vanishes and ϕ is undefined. The limiting values of ϕ differ by $\pi/2$ corresponding to the *interchange* in the roles of χ_1 and χ_2 before and after the zero value of the anisotropy $(\chi_2 - \chi_1)$. The *curved* traces in figures 6.5(a) and (c) are, between them, representative of all general molecular orientations in the lattice. The different families of curves in (a) and (c) are associated with different orientations of K_3 with respect to the crystal b axis; that is, with different values of β_3 in (6.28). Curves within a given family refer to various orientations of the molecule with respect to rotation of K_1 and K_2 about the direction of K_3. Inevitably, one such orientation places K_1 parallel to a and another places K_1 parallel to c: the traces for these two cases correspond, respectively, to the thick and thin sets of straight lines as shown.

Axial molecular tensors occur when the rhombicity parameter $\rho_M = \pi/4$, $5\pi/4$, etc., for then $L_1 = C/\sqrt{2} = L_2$. Hence

$$\tan 2\theta = \frac{2(\alpha_1\gamma_1 + \alpha_2\gamma_2)}{(\alpha_1^2 - \gamma_1^2) + (\alpha_2^2 - \gamma_2^2)} = \frac{2\alpha_3\gamma_3}{(\alpha_3^2 - \gamma_3^2)} \tag{6.43}$$

and, since the a axis may be chosen arbitrarily as above, we construct the curves in figure 6.5 for K_3 lying in the bc' plane: then $\alpha_3 = 0$ and $\phi = 0$ or $\pi/2$. So we observe all curves intersecting at $\rho_M = \pi/4$ or $5\pi/4$, when ϕ is 0 or $\pi/2$. The most special circumstances of all occur for the particular orientation in which K_3 lies parallel to b, as in figures 6.5(b) and (d). Firstly, having the axial molecular tensor oriented along the crystal b axis ensures that the condition $\Delta\chi_b = 0$, when ϕ is totally undefined, coincides with the case for molecular axiality, $\rho_M = \pi/4$, etc. The various members of the family of step functions in figure 6.5(b) merely correspond to a rotation of the molecule about $K_3 \| b$, so negating the otherwise fixed definition of the orientation of K_1 as parallel to a. Thus the circumstances depicted in figure 6.5(b) reflect one special case of which those in 6.5(a) and (c) are the more general. The remaining special, limiting case is shown in figure 6.5(d) for K_3 oriented perpendicular to b: once again, the family of step functions now corresponds to a set of molecular rotations about the K_3 axis. While the intervals between the members of *this* set, corresponding to uniform rotational increments of the molecular tensor, need not be and are not evenly spaced, those in the set shown in 6.5(b) are so spaced because, in effect, they merely correlate with a new definition of the crystal a axis.

We return now to the original query with which we began this section; namely, the frequent occurrence of virtually temperature-independent ϕ

values in practice, but without denying the occasional possibility of the rather extreme behaviour shown in figure 6.2(a). Obviously, there are two kinds of reason why ϕ may change with temperature: variations in either the molecular tensor orientation or in the molecular rhombicity parameter ρ_M. Either may ultimately be caused by each of two circumstances or, of course, a combination of them. The orientation of the molecular species themselves may vary with temperature, or a change may occur in their internal geometrical structure in some way: such alterations will probably be detectable by diffraction analysis. Finally, the quantum-mechanical behaviour of even rigid molecular structures in a magnetic field might give rise to variations in both orientation and magnetic rhombicity with temperature. Now, of course, these statements on their own leave us back at the beginning, merely providing wide-ranging reasons why ϕ may depend on temperature. However, the modelling provided by figure 6.5 helps to sharpen the discussion. Thus, for most systems, actual molecular reorientations occur rather slowly with changing temperature (unless, of course, a phase change takes place but that will normally be obvious); and variations resulting from the quantum-mechanical, or electronic, situation are likely to vary reasonably slowly also, depending upon the extent to which the system approaches pure Curie-law behaviour or even Curie–Weiss with 'isotropic' Weiss constants. Therefore, unless the molecular magnetic rhombicity parameter ρ_M takes values rather close to those corresponding to the vertical lines throughout figure 6.5, variations of a few degrees in either ρ_M or in the molecular tensor orientation – from any cause – will be associated with only a few degrees change in ϕ over the typical experimental temperature range. Empirical variations in ϕ by less than, say, 5° are hard to quantify and would often be regarded as almost negligible within experimental error. However, should a system happen to fall into a region of these curves near to the vertical lines, then we find ϕ is a very sensitive function of ρ_M and molecular orientation, so that small changes in these quantities could have a marked effect upon the setting angle of a monoclinic crystal freely suspended parallel to *b* and perpendicular to a magnetic field. In the most general terms, therefore, we have discovered why ϕ angles most frequently vary only a little over an experimental temperature range – the curves in figure 6.5 are mostly fairly flat. However, when ϕ does vary, a more detailed analysis of the problem can be rewarding and McMeeking[15] has developed the following views as part of his analysis of the magnetic anisotropy in iron(II)-phthalocyanine.

We may consider a quantity $\rho_C = (\chi_2 - \chi_3)/(\chi_2 - \chi_1)$ to represent the

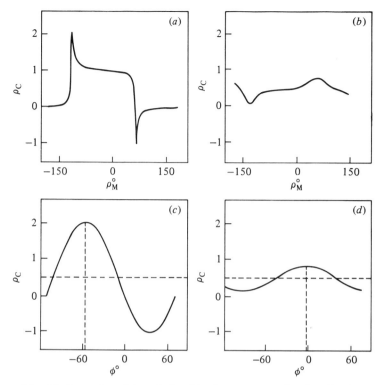

Fig. 6.6. Examples of the relationships between crystal and molecular rhombicity parameters and the sinusoidal relationship between ρ_C and ϕ. (a) and (c) illustrate the more acute cases, while (b) and (d) represent the most common circumstances.

quality of rhombicity for the *crystal* magnetic tensor. Note that, being a ratio of principal anisotropies, ρ_C neglects and retains the same kinds of information that ρ_M does and is unchanged by an overall scale factor or by a shift in the trace of χ. In figure 6.6(a) and (b) are shown relationships between the crystal and molecular rhombicity parameters for two choices of the molecular orientation, the situation in (a) illustrating a system rather near to one represented by a vertical line in figure 6.5(a) or (c), while that in 6.6(b) shows the less acute variations of the more common arrangements. Alternatively, we may plot ρ_C versus ϕ for these same molecular orientations, as shown respectively in figures 6.6(c) and (d). These latter curves are sinusoidal with the form

$$\rho_\theta = 0.5 + A \sin(2\phi + \varepsilon), \tag{6.44}$$

as may be found empirically as shown[15], after some tedious algebra,

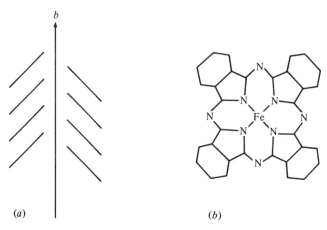

Fig. 6.7. Iron(II)phthalocyanine: (*a*) herringbone stacking in the crystal; the angle between molecular planes related by the symmetry axis is close to 90°. (*b*) the near four-fold symmetry of the macrocyclic complexes.

analytically. The curves are just two of a family of such relationships; each characterized by an amplitude A and phase ε, one for each choice of molecular orientation, and so forming the basis for further analysis as now exemplified by the case of iron(II)phthalocyanine.

Molecules of iron(II)phthalocyanine stack in the monoclinic space group $P2_1/a$ as shown in figure 6.7(*a*), their normals being inclined at about 45° to the unique b axis of the lattice. The molecules essentially possess four-fold symmetry, as shown in figure 6.7(*b*), but this is not exact, as would be defined crystallographically. Both crystal rhombicity, ρ_C, and ϕ have been measured[14] for this system and are shown in figure 6.2 as functions of temperature in the range 90–300 K. The relationship between these observed quantities, uncorrected[tt] for the known[16] diamagnetic anisotropy of metal-free phthalocyanine, is shown in figure 6.8 to follow the expected sinusoidal form quite well. Departures from this behaviour result from experimental errors – and both ρ_C and ϕ are very sensitive functions – or from temperature-dependent molecular tensor reorientations, or both. Nevertheless, approximate estimates of both A and ε can be made here.

Now recall that because of the $\approx 45°$ orientation of K_3 with respect to b, the ϕ versus ρ_M relationships will be given by curves like those in

[tt] As the present analysis is independent of any ligand-field or other quantum-mechanical model, we may proceed using corrected or uncorrected susceptibility data with equal validity.

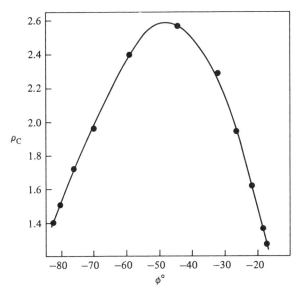

Fig. 6.8. The experimental relationship between ρ_C and ϕ for crystals of iron(II)phthalocyanine approximates the sinusoidal, theoretical curves of figure 6.6.

figure 6.5(*c*), say, rather than by figure 6.5(*b*) or (*d*). However, if the approximate four-fold structural symmetry of these molecules were taken to imply a similar rotational symmetry of the molecular magnetic tensor, such that $K_1 \approx K_2$, then $\rho_M \approx \pi/4$ and we would *not* expect to observe any marked sensitivity of the ϕ value to ρ_M or molecular tensor orientation. This point is worth emphasizing. From the analysis so far given, it seems most unlikely that the observed large variation in ϕ with temperature can be ascribed to realistic reorientations of axially symmetric molecular tensors. Instead, we should conclude, with a high probability, that the observed behaviour of the crystal anisotropies demonstrates the presence of a significant in-plane molecular anisotropy so that, notwithstanding the structural pseudosymmetry, the electronic system is rhombic, at least so far as the bulk magnetism is concerned. In fact, as McMeeking has shown[15], a formal, orbital-doublet state is likely either to be the ground state or to lie close to an orbital-singlet ground state: and then, structurally very small effects (even, perhaps, due to second-nearest neighbours) can establish an in-place anisotropy with the principal susceptibilities lying in virtually any directions. Once again, therefore, it is to be emphasized that assumptions regarding the symmetry of electronic properties may be quite unsound in the absence of the corresponding *exact* crystallographic

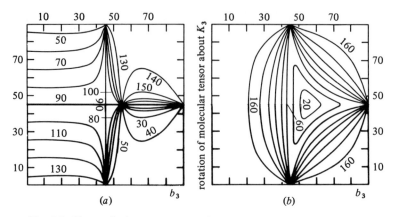

Fig. 6.9. Theoretical contour maps for (*a*) the phase ε of (6.44), and (*b*) arctan ρ_C, as functions of molecular orientations. Definitions of absolute orientations to which these refer are given by McMeeking.[15]

structural symmetry. Given the proposition that iron(II)phtholocyanine molecules posess rhombic molecular magnetic symmetry, it is possible, in principle, to exploit the foregoing analysis and estimate the orientation of the principal in-place molecular susceptibilities.

We pursue this aim, in outline at least, with the aid of either of the diagrams in figure 6.9 which give contour maps of (*a*) the phase ε of (6.44) or of (*b*) arctan ρ_C each as functions of molecular orientation. The inverse-tangent function is chosen so that the case of crystal axiality when $\Delta\chi_b = 0$ can be accommodated without infinite arguments. It is not possible to establish the complete molecular tensor orientation but an estimate for the directions of the principal in-place molecular susceptibilities can be made if that for the out-of-plane susceptibility K_3 may be presumed. This is necessary in order to establish the absolute values of ϕ and hence the phases ε, shown in figures 6.6(*c*) and (*d*) which, as described above, are related to the arbitrary value chosen for ζ in figure 6.3. However, fixing ϕ by fixing ζ in this way, simultaneously fixes the value on the abscissae of figure 6.9(*a*) and (*b*). Then, using either figure, we may determine the remaining orientation as that value of the ordinate in these diagrams at which the abscissa value and either ε or arctan ρ_C coincide: the same result should be found from either comparison and so a check of sorts is provided. Of course, the efficacy of the process may depend critically upon the accuracy with which ρ_C and ϕ are determined experimentally; but equally so, upon whether the fitting regions obtained in figure 6.9 occur in particularly sensitive regions or not. We shall not pursue the analysis[15] of this phthalocyanine complex in more detail here, however, being content to have reviewed an informative approach to a complex detail of the magnetism of monoclinic crystals.

References

[1] Mitra, S., *Transition Metal Chemistry*, **7**, 183 (1972).
[2] Mitra, S., *Prog. Inorg. Chem.*, **22**, 309 (1977).
[3] Hanton, L.R., PhD Thesis, University of Cambridge, 1980.
[4] Krishnan, K.S. & Mookerji, A., *Phys. Rev.*, **50**, 860 (1936).
[5] Krishnan, K.S. & Mookerji, A., *Phys. Rev.*, **54**, 533 (1938).
[6] Krishnan, K.S. & Mookerji, A., *Phys. Rev.*, **54**, 841 (1938).
[7] Boyd, P.D.W., Davies, J.E. & Gerloch, M., *Proc. Roy. Soc.*, **A360**, 191 (1978).
[8] Davies, J.E. & Gerloch, M., in press.
[9] Gerloch, M. & Hanton, L.R., *Inorg. Chem.*, **19**, 1692 (1980).
[10] Ghosh, U.S. & Bagchi, R.N., *Ind. J. Phys.*, **36**, 538 (1962).
[11] Ghose, J.K., *Ind. J. Pure Appl. Phys.*, **2**, 94 (1964).
[12] Cruse, D.A. & Gerloch, M., *J. Chem. Soc. Dalton Trans.*, p. 152 (1977).
[13] Mabbs, F.E. & Machin, D.J., *Magnetism and Transition Metal Complexes*, Chapman and Hall, London, 1973.
[14] Barraclough, C.G., Martin, R.L., Mitra, S. & Sherwood, R.C., *J. Chem. Phys.*, **53**, 1643 (1970).
[15] McMeeking R.F., personal communication.
[16] Barraclough, C.G., Martin, R.L. & Mitra, S., *J. Chem. Phys.*, **55**, 1426 (1972).

7

Quantum theory and magnetic susceptibilities

━━━

Any account of para- and diamagnetic susceptibilities eventually proceeds to the quantum-mechanical version of the Langevin–Debye formula that is usually referred to as the Van Vleck equation. In preparation for this and, indeed for a *generalized* susceptibility equation by which the magnetic properties of molecules of any geometry or symmetry may be calculated, we require matrix elements of the magnetic moment operator. The first part of this chapter sketches the origins of this operator and, in particular, shows how the attribute of spin, the numerical magnitude of the electron magnetic moment, and the phenomenon of spin-orbit coupling arise from a juxtaposition of quantum theory and special relativity. First, however, we construct Schrödinger's spinless wave equation by transforming the Hamiltonian function that we developed in chapter 3 into the equivalent Hamiltonian operator of quantum mechanics: relativity and spin follow close behind.

7.1 The spinless Schrödinger equation

In §3.19, we derived the classical Hamiltonian function $\mathscr{H}$ for a system of n molecular electrons[††] exposed to electric and magnetic fields; namely,

---

[††] Note again that, as in §3.19 *et seq.*, we specialize to the Hamiltonian function – and now operator – of the *electrons only* and so factors like $(\mathbf{p}_j + e\mathbf{A}_j)$ occur here with a positive sign inside because of the convention that the charge on an electron is $-e$. Appropriately changed signs occur in other derivations where the symbol e_j represents the charge on an *arbitrary* particle (j) and the Hamiltonian is to be used to compute both electron and nuclear properties.

(3.134) which we write here in the more condensed vector form,

$$\mathcal{H} = \sum_j^n \frac{1}{2m_e}(\mathbf{p}_j + e\mathbf{A}_j)^2 + V - \sum_j^n e\phi_j, \tag{7.1}$$

where, as before, m_e is the mass of the electron, V involves the electron–nuclear attraction and electron–electron repulsion energies as in (3.127), ϕ describes the usual scalar potential from all other sources of electric field, and $\mathbf{A}$ is the magnetic vector potential. The corresponding Hamiltonian *operator*, H, is obtained by making the substitution $\mathbf{p} \to \hat{\mathbf{p}}$ or $\mathbf{p} \to -i\hbar\mathrm{grad}$; that is,

$$\mathcal{H}(p_1, \ldots; q_1, \ldots) \to H\left(-i\hbar\frac{\partial}{\partial q_1}, \ldots; q_1, \ldots\right)$$

and so,

$$H = \sum_j^n \frac{1}{2m_e}(\hat{\mathbf{p}}_j + e\mathbf{A}_j)^2 + V - \sum_j^n e\phi_j$$

$$= \sum_j^n \left[-\frac{\hbar^2}{2m_e}\nabla_j^2 + \frac{e\hbar}{2im_e}(\mathbf{A}_j \cdot \mathrm{grad}_j + \mathrm{div}_j\mathbf{A}_j) + \frac{e^2}{2m_e}\mathbf{A}_j^2 - e\phi_j\right] + V. \tag{7.2}$$

Given that this Hamiltonian operator will act upon a wavefunction ψ within the Schrödinger equation, we may use the usual rules[1] of vector calculus to write

$$\mathrm{div}\,(\mathbf{A}\psi) = \psi\,\mathrm{div}\,\mathbf{A} + \mathbf{A} \cdot \mathrm{grad}\,\psi \tag{7.3}$$

and hence obtain (7.2) as

$$H = \sum_j^n \left[-\frac{\hbar^2}{2m_e}\nabla_j^2 + \frac{e\hbar}{im_e}\mathbf{A}_j\mathrm{grad}_j + \frac{e\hbar}{2im_e}\mathrm{div}_j\mathbf{A}_j + \frac{e^2}{2m_e}\mathbf{A}_j^2 - e\phi_j\right] + V. \tag{7.4}$$

As in §3.19, *we* are interested here only in the situation defined by an external time-independent, uniform magnetic field and therefore choose the potential

$$\mathbf{A}_j = \tfrac{1}{2}\mathbf{B}\wedge\mathbf{r}_j, \tag{7.5}$$

noting that those components in (3.135) are a special case when $\mathbf{B}$ is oriented parallel to z. $\mathrm{Div}\,\mathbf{A} = 0$ for the potential in (7.5) so this choice corresponds with the selection of the Coulomb gauge, as with (3.5). One immediate consequence of using this gauge is that the third term in (7.4) vanishes. It would be inelegant to write the new Hamiltonian as

$$H = \sum_{j}^{n}\left[-\frac{\hbar^2}{2m_e}\nabla_j^2 + \frac{e\hbar}{im_e}A_j\text{grad}_j - \frac{e^2}{2m_e}A_j^2 - e\phi_j\right] + V, \quad (7.6)$$

however, for we must always accompany such an equation with the statement that 'it is appropriate only for a vector potential chosen to satisfy the Coulomb gauge'. Since we have made our choice of potential – (7.5) – we should more tidily substitute this everywhere into (7.4) and henceforth claim validity for our expressions with respect to the choice (7.5). Further, as fields themselves are gauge invariant, we need not continue to qualify the equations with a caveat about the vector potential. For the second term we use the identity

$$\mathbf{B} \wedge \mathbf{r}\,\text{grad} \equiv \mathbf{B}.\,\mathbf{r} \wedge \text{grad}, \quad (7.7)$$

and so find that

$$\frac{e\hbar}{im_e}A_j\text{grad}_j = \frac{e\hbar}{2im_e}\mathbf{B}.\,\mathbf{r}_j \wedge \text{grad}_j. \quad (7.8)$$

For the term in A_j^2, we use the identity[1]

$$(\mathbf{P} \wedge \mathbf{Q}).\,(\mathbf{R} \wedge \mathbf{S}) \equiv (\mathbf{P}.\,\mathbf{R})(\mathbf{Q}.\,\mathbf{S}) - (\mathbf{P}.\,\mathbf{S})(\mathbf{Q}.\,\mathbf{R}) \quad (7.9)$$

and therefore write

$$A_j^2 \equiv \mathbf{A}_j.\,\mathbf{A}_j = \tfrac{1}{4}(\mathbf{B} \wedge \mathbf{r}_j.\,\mathbf{B} \wedge \mathbf{r}_j)$$
$$= \tfrac{1}{4}[(\mathbf{B}.\,\mathbf{B})(\mathbf{r}_j.\,\mathbf{r}_j) - (\mathbf{B}.\,\mathbf{r}_j)^2]$$
$$= \tfrac{1}{4}(1 - \mathbf{rr})\mathbf{B}^2, \quad (7.10)$$

where $(1 - \mathbf{rr})$ is a diagonal matix with elements $1 - x^2, 1 - y^2, 1 - z^2$; and $1 = x^2 + y^2 + z^2$: the abbreviation afforded by the last line of (7.10) is readily verified by direct substitution. Inserting the results of (7.8) and (7.10) into (7.4), we obtain

$$H = \sum_{j}^{n}\left[-\frac{\hbar^2}{2m_e}\nabla_j^2 + \frac{e\hbar}{2im_e}\mathbf{B}.\,\mathbf{r} \wedge \text{grad}_j + \frac{e^2}{8m_e}(1 - \mathbf{rr})\mathbf{B}^2 - e\phi_j\right] + V \quad (7.11)$$

as the Hamiltonian for atomic electrons in time-independent, uniform magnetic and electric fields.

We now *define* an orbital magnetic moment operator, μ_j^o, by

$$\mu_j^o = -\frac{e\hbar}{2im_e}\mathbf{r}_j \wedge \text{grad}_j \quad (7.12)$$

and finally write Schrödinger's equation for a system of n spinless electrons in uniform electric and magnetic fields as

$$\left\{ \sum_j^n \left[-\frac{\hbar^2}{2m_e} \nabla_j^2 - \mathbf{B} \cdot \boldsymbol{\mu}_j^0 + \frac{e^2}{8m_e} (1 - \mathbf{rr}) \mathbf{B}^2 - e\phi_j \right] + V \right\} \psi = E\psi. \quad (7.13)$$

Specializing further to the case of a central field, μ^0 is written, by analogy with (3.54), as

$$\mu^0 = -\beta_e \mathbf{l} / / \hbar, \quad (7.14)$$

where $\beta_e = eh/2m_e$ (in the SI system, with units A^2), which is called the Bohr magneton; and $\mathbf{l}$ is the usual orbital angular momentum operator.

The expression (7.13) is probably more familiar in the form taken for an applied magnetic field oriented parallel to the z axis (and for no electric field): direct substitution of $B_x = B_y = 0$ in (7.5) and (7.10), together with $\phi = 0$, leads to (7.11) being replaced by

$$H^z = \sum_j^n \left[-\frac{\hbar^2}{2m_e} \nabla_j^2 + \frac{eh}{2im_e} B_z \left(x_j \frac{\partial}{\partial y_j} - y_j \frac{\partial}{\partial x_j} \right) + \frac{e^2}{8m_e} B_z^2 (x_j^2 + y_j^2) \right] + V,$$
$$(7.15)$$

whose form could also have been deduced by direct substitution of (7.2) into (3.136).

7.2 The Pauli–Schrödinger equation

It is, of course, well known that a great part of molecular quantum mechanics – not only the spectral and magnetic properties of atoms and molecules in magnetic and electric fields – cannot be understood unless we recognize the existence of an internal degree of freedom for the electron (and for other particles) which we usually call 'spin'. There are, however, no terms in the Hamiltonian of (7.13) which show any *explicit* dependence upon it. The logical position is that we must rework the fundamental theory covered explicitly or implicitly so far, in order to provide a sound basis for this phenomenon. We know from a chemists' background of quantum theory and the electronic structures of atoms and molecules that spin does not represent a small correction factor in the main bulk of theory but an inherent and fundamental building block in the conventional edifice. Of course, the manifestations of spin with which *we* are especially concerned – magnetic moments and spin-orbit coupling – *are* generally small 'corrections' viewed against the vastly greater energies arising from the kinetic and other potential energy terms in the usual Hamiltonian.

The largest contributions to atomic and molecular energies that arise from the existence of 'spin', however, manifest themselves in the application of the antisymmetry and exclusion principles which give rise to exchange terms and all that ensues therefrom. It is perfectly possible to construct[2] molecular quantum mechanics without the notion of spin where, instead, we must focus on the appropriate permutation symmetry of the Schrödinger Hamiltonian. In this sense, the conventional Schrödinger equation can be considered to include, or to be made to include, the major effects of 'spin' by appropriate definition of the rules governing the operation of the Hamiltonian to many-electron wavefunctions: 'spin' might then be considered as *implicit* within a Schrödinger-like equation. However, spin is more usually considered explicitly and the appropriate theory was so developed. We shall look at that theory shortly, but we begin, by way of introduction, review and custom, by 'grafting' spin onto the theory we have so far, in a totally *ad hoc* way that has come to be known as 'phenomenological'.

We start from the recognition, first required – and hence discovered – by the experimental facts of atomic spectroscopy, that an electron possesses not only rest-mass and charge, but also a magnetic moment which may be considered as deriving from a spin angular momentum of magnitude $\frac{1}{2}\hbar$. We *define* a spin magnetic moment operator, $\boldsymbol{\mu}^{S}$, for the electron by

$$\boldsymbol{\mu}^{S} = -g_{e}\beta_{e}\mathbf{s}/\hbar, \tag{7.16}$$

where $\mathbf{s}$ is a spin angular momentum operator and g_{e}, called the electron gyromagnetic or magnetochemical ratio, is an intrinsic factor of magnitude 2.0023, as determined experimentally: for most purposes, and certainly for all those to do with magnetic susceptibilities, $g_{e} \approx 2$ with sufficient accuracy. Appropriate modification of the spinless Schrödinger equation (7.13) yields

$$\left\{ \sum_{j}^{n} \left[-\frac{\hbar^{2}}{2m_{e}}\nabla_{j}^{2} - \mathbf{B} \cdot (\boldsymbol{\mu}_{j}^{O} + \boldsymbol{\mu}_{j}^{S}) + \frac{e^{2}}{8m_{e}}(1 - \mathbf{rr})\mathbf{B}^{2} - e\phi_{j} \right] + V \right\}\psi = E\psi \tag{7.17}$$

known as the *Pauli–Schrödinger equation*. Its solutions, ψ, are functions of both space and spin coordinates, of course. The significance of the various terms in (7.17) are made plainer by writing

$$[H^{(0)} + H^{(1)} + H^{(2)} + H^{(3)}]\psi = E\psi, \tag{7.18}$$

where

$$H^{(0)} = -\sum_{j}^{n}\frac{\hbar^{2}}{2m_{e}}\nabla_{j}^{2} + V, \tag{7.19}$$

is the usual Schrödinger Hamiltonian for a system of electrons in the absence of external electric and magnetic fields; and

$$H^{(1)} = -\sum_{j}^{n} \mathbf{B} \cdot (\mu_j^O + \mu_j^S), \tag{7.20a}$$

$$H^{(2)} = \frac{e^2}{8m_e} \sum_{j}^{n} (1 - \mathbf{rr})\mathbf{B}^2, \tag{7.20b}$$

and

$$H^{(3)} = -e \sum_{j}^{n} \phi_j. \tag{7.20c}$$

For problems involving only magnetic fields, the last term vanishes; and (7.20a) and (7.20b) are associated with para- and diamagnetism respectively. In atoms, we may use the relationships (7.14) and (7.16) to rewrite (7.20a) as

$$H^{(1)} = \hbar^{-1}\beta_e \mathbf{B} \cdot \sum_{j}^{n} (\mathbf{l}_j + g_e\mathbf{s}_j) \tag{7.21a}$$

$$\approx \hbar^{-1}\beta_e \mathbf{B} \cdot \sum_{j}^{n} (\mathbf{l}_j + 2\mathbf{s}_j). \tag{7.21b}$$

7.3 Pauli spin matrices

As familiar as (7.16) are the spin angular momentum commutation relationships,[3]

$$[\mathbf{s}^2, s_x] = [\mathbf{s}^2, s_y] = [\mathbf{s}^2, s_z] = 0 \tag{7.22a}$$

and

$$[s_x, s_y] = i\hbar s_z, \quad [s_y, s_z] = i\hbar s_x, \quad [s_z, s_x] = i\hbar s_y; \tag{7.22b}$$

and the operations

$$\mathbf{s}^2|\tfrac{1}{2}, \pm\tfrac{1}{2}\rangle = \tfrac{3}{4}\hbar^2|\tfrac{1}{2}, \pm\tfrac{1}{2}\rangle \tag{7.23a}$$

and

$$s_z|\tfrac{1}{2}, \pm\tfrac{1}{2}\rangle = \pm\tfrac{1}{2}\hbar|\tfrac{1}{2}, \pm\tfrac{1}{2}\rangle, \quad s_x|\tfrac{1}{2}, \pm\tfrac{1}{2}\rangle = \tfrac{1}{2}\hbar|\tfrac{1}{2}, \mp\tfrac{1}{2}\rangle,$$
$$s_y|\tfrac{1}{2}, \pm\tfrac{1}{2}\rangle = \pm\tfrac{1}{2}i\hbar|\tfrac{1}{2}, \mp\tfrac{1}{2}\rangle, \tag{7.23b}$$

exemplifying the close parallels between spin and orbital angular momentum operators.

Now an angular momentum **s** with a square $\mathbf{s}^2$ which takes *only the one value*, as in (7.23a), has certain algebraic properties which are not shared by general angular momenta. These are exhibited[7,8] most clearly if we consider an associated quantity $\boldsymbol{\sigma}$, defined by

$$\mathbf{s} = \tfrac{1}{2}\hbar\boldsymbol{\sigma} \tag{7.24}$$

for, by combining (7.23) and (7.24), we observe how σ_x^2, σ_y^2 and σ_z^2 each has the same effect on each of the eigenfunctions $|\tfrac{1}{2}, \pm\tfrac{1}{2}\rangle$ as the number 1. Hence we may write the *operator equations*

$$\sigma_x^2 = \sigma_y^2 = \sigma_z^2 = 1. \tag{7.25}$$

We can similarly verify, on substitution of (7.24) into (7.23), that pairs of components of $\boldsymbol{\sigma}$ *anticommute*; that is,

$$\left. \begin{aligned} \sigma_x\sigma_y + \sigma_y\sigma_x &= 0, \\ \sigma_y\sigma_z + \sigma_z\sigma_y &= 0, \\ \sigma_z\sigma_x + \sigma_x\sigma_z &= 0. \end{aligned} \right\} \tag{7.26}$$

Finally, the behaviour of the conventional spin operators in (7.23) may be reconstructed using a matrix representation of $\boldsymbol{\sigma}$, such that

$$\sigma_x \equiv \begin{pmatrix} 0 & 1 \\ 1 & 0 \end{pmatrix}, \quad \sigma_y \equiv \begin{pmatrix} 0 & -i \\ i & 0 \end{pmatrix}, \quad \sigma_z \equiv \begin{pmatrix} 1 & 0 \\ 0 & -1 \end{pmatrix}, \tag{7.27}$$

called *Pauli's spin matrices*, with $|\pm\tfrac{1}{2}\rangle$ kets being written as column vectors;

$$|\tfrac{1}{2}\rangle \equiv \begin{pmatrix} 1 \\ 0 \end{pmatrix}, \quad |-\tfrac{1}{2}\rangle \equiv \begin{pmatrix} 0 \\ 1 \end{pmatrix}. \tag{7.28}$$

For example,

$$s_x|\tfrac{1}{2}\rangle \equiv \tfrac{1}{2}\hbar \begin{pmatrix} 0 & 1 \\ 1 & 0 \end{pmatrix}\begin{pmatrix} 1 \\ 0 \end{pmatrix} = \tfrac{1}{2}\hbar \begin{pmatrix} 0 \\ 1 \end{pmatrix} \equiv \tfrac{1}{2}\hbar|-\tfrac{1}{2}\rangle. \tag{7.29}$$

It is simple to check that the matrices in (7.27) satisfy the commutation conditions

$$[\sigma_x, \sigma_y] = 2i\sigma_z, \quad [\sigma_y, \sigma_z] = 2i\sigma_x, \quad [\sigma_z, \sigma_x] = 2i\sigma_y, \tag{7.30}$$

the anticommutation rules (7.26), and the operator equations (7.25).

7.4 A review of spin theory

7.4.1 The Klein–Gordon equation

Schrodinger's equation,

$$\left(-\frac{\hbar^2}{2m_e}\nabla^2 + V - i\hbar\frac{\partial}{\partial t} \right)\psi(\mathbf{x}, t) = 0, \tag{7.31}$$

which we write here in its time-dependent form, suffers one major drawback: space and time do not appear to the same power and so it follows immediately that it is not invariant under Lorentz transformations and hence does not explicitly conform to the principle of relativity. An early attempt to put this right began with the relativistic expression[9] for the energy of a free particle with rest-mass m_0 and momentum $\mathbf{p}$,

$$\tilde{E} = c\sqrt{(\mathbf{p}^2 + m_0^2 c^2)}. \tag{7.32}$$

It is not helpful simply to replace the classical quantities by their quantum counterparts,

$$x_i \to \hat{x}_i, \quad p_i \to i\hbar \frac{\partial}{\partial x_i}, \quad \tilde{E} \to i\hbar \frac{\partial}{\partial t}, \tag{7.33}$$

for there would result a non-uniquely defined square root of a linear operator. It was proposed, therefore, to square (7.32), yielding

$$\tilde{E}^2 = c^2(p_1^2 + p_2^2 + p_3^2) + m_0^2 c^4, \tag{7.34}$$

which has the quantum-operator equivalent,

$$-\hbar^2 \frac{\partial^2}{\partial t^2} = -c^2 \hbar^2 \nabla^2 + m_0^2 c^4. \tag{7.35}$$

The corresponding eigenvalue equation

$$\left(\nabla^2 - \frac{1}{c^2} \frac{\partial^2}{\partial t^2} - \frac{m_0^2 c^2}{\hbar^2} \right) \psi(\mathbf{x}, t) = 0, \tag{7.36a}$$

often written as

$$(\Box^2 - \kappa^2)\psi(\mathbf{x}, t) = 0, \tag{7.36b}$$

is known as the *Klein–Gordon equation*: the quantity,

$$\Box^2 \equiv \frac{\partial^2}{\partial x_1^2} + \frac{\partial^2}{\partial x_2^2} + \frac{\partial^2}{\partial x_3^2} - \frac{1}{c^2} \frac{\partial^2}{\partial t^2} = \nabla^2 - \frac{1}{c^2} \frac{\partial^2}{\partial t^2} \tag{7.37}$$

is called the d'Alembertian operator, and

$$\kappa \equiv \frac{m_0 c}{\hbar}. \tag{7.38}$$

The Klein–Gordon expression *is* invariant under Lorentz transformations, as it should be given the starting point (7.32), and as looks reasonable when we compare the structure of the d'Alembertian with the metric[9] between events in the four-dimensional space–time continuum,

$$s^2 = ct^2 - x_1^2 - x_2^2 - x_3^2. \tag{7.39}$$

While the Klein–Gordon equation meets the exigencies of special relativity, it fails to meet several of quantum theory. The problems are thoroughly reviewed elsewhere[6] and we confine ourselves here to a brief mention of just one or two of them. Firstly, (7.36) involves a second-order partial derivative with respect to time. According to the Copenhagen interpretation of the wavefunction ψ, the state of a system at a given time is completely specified by ψ at that time, and so the future development of the system can be calculated by solving the wave equation and finding $\psi(t)$ at a future time. By contrast, solutions of the Klein–Gordon equation are specified only after setting initial conditions upon *both* ψ and $\partial\psi/\partial t$ and the latter has no direct physical interpretation. A second series of problems is associated with the fact that (7.36) gives rise to both positive- and negative-energy solutions. Just one of several – though by no means obvious – consequences of this result is that no reasonable expressions for the probability density can be constructed which do not necessarily have values which may be negative. Finally, when extended to the case of an electron in an external electromagnetic field, and to the hydrogen atom in particular, the Klein–Gordon equation fails yet again by yielding results for the spectral fine structure that are in marked disagreement with experiment.

Though succeeding in the primary aim of establishing a relativistically invariant eigenvalue equation to replace the conventional Schrödinger equation, the construction of the Klein–Gordon equation fails to achieve the quantum-mechanical validity of its predecessor. Lest it seem perverse to review this apparent disaster – 'apparent' because the equation has since found an honoured place in quantum-field theory in relation to spinless particles – we note that its failures essentially stem from its second-order character. The purpose of this brief review, therefore, has been to identify the need for a *linear* Lorentz-invariant wave equation and so provide the historical starting point for Dirac's approach, which we now describe.

7.4.2 The Dirac equation

As for the Klein–Gordon equation, the point of departure for Dirac's work[10] is the relativistic energy expression (7.32) but, instead of squaring the equation to remove the square root, one can *propose* that it may be executed formally by linearizing the expression: let

$$\sqrt{(p_1^2 + p_2^2 + p_3^2 + m_0^2 c^2)} = \alpha_1 p_1 + \alpha_2 p_2 + \alpha_3 p_3 + \beta m_0 c, \qquad (7.40)$$

where the character and attributes of the α_i and β are to be determined. Put the other way round: we need to determine whether and how these multipliers may be characterized so that (7.40) is valid. We begin by squaring (7.40), to get

$$p_1^2 + p_2^2 + p_3^2 + m_0^2 c^2 = \alpha_1^2 p_1^2 + \alpha_2^2 p_2^2 + \alpha_3^2 p_3^2 + \beta^2 m_0^2 c^2$$
$$+ (\alpha_1 \alpha_2 + \alpha_2 \alpha_1) p_1 p_2 + \ldots$$
$$+ (\alpha_1 \beta + \beta \alpha_1) p_1 m_0 c + \ldots \tag{7.41}$$

and then, comparing coefficients on both sides, we find that the operators α_i and β must obey the algebraic relationships,

$$\left. \begin{array}{l} \alpha_1^2 = \alpha_2^2 = \alpha_3^2 = \beta^2 = 1, \\[6pt] \alpha_i \alpha_j + \alpha_j \alpha_i = 0 \text{ for } i, j = 1, 2, 3 \text{ and } i \neq j, \\[6pt] \alpha_i \beta + \beta \alpha_i = 0 \text{ for } i = 1, 2, 3. \end{array} \right\} \tag{7.42}$$

Provided algebraic quantities with these properties exist, the linearizing process (7.40) is valid and we may proceed by substituting (7.40) into (7.32) and obtain the linearized relativistic energy equation for a free particle,

$$\tilde{E} - c \sum_{i=1}^{3} \alpha_i p_i - \beta m_0 c^2 = 0. \tag{7.43}$$

The equivalent quantum-mechanical eigenvalue equation is now obtained after transformation using (7.33) to get

$$\left(-i\hbar \frac{\partial}{\partial t} - ic\hbar \sum_{j=1}^{3} \alpha_j \frac{\partial}{\partial x_j} + \beta m_0 c^2 \right) \psi(\mathbf{x}, t) = 0: \tag{7.44}$$

this is *Dirac's equation*. A *formal* similarity with Schrödinger's equation emerges if we write the Dirac Hamiltonian as

$$H^{\mathrm{D}} \equiv -ic\hbar \sum_{j=1}^{3} \alpha_j \frac{\partial}{\partial x_j} + \beta m_0 c^2, \tag{7.45}$$

when (7.44) takes the form

$$\left(-i\hbar \frac{\partial}{\partial t} + H^{\mathrm{D}} \right) \psi = 0. \tag{7.46}$$

It can be shown[10] that the Dirac equation is Lorentz invariant, as we had hoped given (7.32) as our point of departure. Unlike the Klein–Gordon equation, Dirac's equation is linear in all coordinates. Space and time coordinates appear on the same footing, in contrast to those in the non-relativistic Schrödinger equation. The equivalent standing of the space and time coordinates is made especially clear if we transpose the Dirac equation into a more symmetric form, as follows. We introduce new γ operators, defined by the linear transformation,

$$\left.\begin{array}{l} \gamma_j \equiv -i\beta\alpha_j, \quad j = 1, 2, 3, \\ \gamma_0 \equiv \beta. \end{array}\right\} \tag{7.47}$$

On substitution into (7.44) and multiplication by (β/ch), we obtain the completely symmetrical form[6,7] of the Dirac equation,

$$\left(\gamma_0 \frac{\partial}{\partial x_0} + \gamma_1 \frac{\partial}{\partial x_1} + \gamma_2 \frac{\partial}{\partial x_2} + \gamma_3 \frac{\partial}{\partial x_3} + \kappa\right)\psi = 0, \tag{7.48a}$$

by putting[tt] $\partial/\partial x_0 = -(i/c)(\partial/\partial t)$; recalling that $\beta^2 = 1$ from (7.42); and where κ is again given by (7.38). In an even more brief notation,

$$(\gamma_\mu \partial_\mu + \kappa)\psi = 0, \tag{7.48b}$$

where $\partial_\mu \equiv \partial/\partial x_\mu$ and the Einstein summation convention has been used – but here over all four dimensions of the space–time continuum. Analogous to (7.42) we now require, following the transformation (7.47), that the γ_μ satisfy the conditions

$$\left.\begin{array}{l} \gamma_\mu^2 = 1, \quad \mu = 0, 1, 2, 3, \\ \gamma_\mu\gamma_\nu + \gamma_\nu\gamma_\mu = 0, \quad \mu \neq \nu; \quad \mu, \nu = 0, 1, 2, 3, \end{array}\right\} \tag{7.49a}$$

or, using the usual Kronecker delta function $\delta_{\mu\nu}$,

$$\gamma_\mu\gamma_\nu + \gamma_\nu\gamma_\mu = 2\delta_{\mu\nu}, \quad \mu, \nu = 0, 1, 2, 3. \tag{7.49b}$$

7.4.3 The emergence of spin

We turn now to consider what the γ_μ (and hence α_i and β) operators might be. By way of introduction, consider Dirac's equation in the form (7.44) or (7.48). We observe from its derivation via (7.41) that the multipliers α_i, β, and hence γ_μ, are independent of the momenta $-p_i$ in (7.41) and $(-i\hbar\partial/\partial x_i)$ subsequently. Further, since we consider here the case with no external fields, all points in space–time are equivalent so that the operator in Dirac's equation does not involve the four coordinates x_i ($i = 0, 1, 2, 3$). Therefore the α_i, β and γ_μ are also independent of the x_i and so *commute*

[tt] We write the four coordinates of space–time as x_0, x_1, x_2, x_3, where $x_0 = ict$. From this it follows that $(\partial/\partial x_0) = -(i/c)(\partial/\partial t)$ and that the 'distance' or metric between 'points' or events is $\sum_{j=0}^{3} x_j^2$ as given in (7.39).

with all the momenta and the coordinates. They therefore describe a new degree of freedom, associated with some *internal* motion for the electron. Thus we see how quickly and naturally the notion of electron spin begins to emerge from the relativistic linear wave equation.

Confidence that the new degree of freedom does indeed refer to spin grows when we recognize the similarity of the algebraic conditions (7.42) or (7.49) with (7.25) and the anticommutation properties (7.26) of Pauli's σ matrices, described earlier. They are, in fact, the same except that the index range for the σ matrices is three, and for the γ operators, four. A convenient Hermitian matrix representation of the γ_μ that may be verified by straightforward substitution into (7.49) is,

$$\gamma_1 = \begin{pmatrix} 0 & 0 & 0 & -i \\ 0 & 0 & -i & 0 \\ 0 & i & 0 & 0 \\ i & 0 & 0 & 0 \end{pmatrix}, \quad \gamma_2 = \begin{pmatrix} 0 & 0 & 0 & -1 \\ 0 & 0 & 1 & 0 \\ 0 & 1 & 0 & 0 \\ -1 & 0 & 0 & 0 \end{pmatrix},$$

$$\gamma_3 = \begin{pmatrix} 0 & 0 & -i & 0 \\ 0 & 0 & 0 & i \\ i & 0 & 0 & 0 \\ 0 & -i & 0 & 0 \end{pmatrix}, \quad \gamma_0 = \begin{pmatrix} 1 & 0 & 0 & 0 \\ 0 & 1 & 0 & 0 \\ 0 & 0 & -1 & 0 \\ 0 & 0 & 0 & -1 \end{pmatrix}.$$

$$(7.50)$$

Reference to the Pauli matrices of (7.27), provides for the alternative condensed form

$$\gamma = \left(\begin{array}{c|c} 0 & -i\sigma \\ \hline i\sigma & 0 \end{array} \right), \quad \gamma_0 = \left(\begin{array}{c|c} \mathbf{I} & 0 \\ \hline 0 & -\mathbf{I} \end{array} \right) \tag{7.51}$$

for the first three and the last respectively; $\mathbf{I}$ is the 2×2 unit matrix. Of course, we may always transpose back to the α_i, β operators, when that is more convenient, by the inverse transformation to (7.47); namely,

$$\beta = \gamma_0, \quad \alpha_j = i\gamma_0\gamma_j. \tag{7.52}$$

That the internal degree of freedom does describe intrinsic spin does not rest merely upon the above analogy. Thus we observe that, unlike the non-relativistic operator in the Schrödinger equation (7.31), the Dirac Hamiltonian H^D does not commute with components of the orbital angular momentum operator: for example,

$$[H^D, l_1] = -ch^2 \left(\alpha_2 \frac{\partial}{\partial x_3} - \alpha_3 \frac{\partial}{\partial x_2} \right) \neq 0. \tag{7.53}$$

The orbital angular momentum $\mathbf{l}$ is not, therefore, a constant of the motion;

however, if we describe a new angular momentum **j**, such that

$$\mathbf{j} = \mathbf{l} + \mathbf{s}, \tag{7.54}$$

appropriate characteristics of **s** can be chosen to ensure that (i) **s** satisfies the usual commutation laws of angular momentum,

$$[s_1, s_2] = i\hbar s_3, \quad \text{etc.,} \tag{7.55}$$

(ii) **s** commutes with **l**, so validating the usual 'triangle rule' by which **j** may be constructed from **l** and **s**, and (iii) **j** commutes with H^D and is, therefore, a constant of the motion. It is a straightforward matter to verify that the only solution satisfying all these requirements is that

$$\mathbf{s} = \tfrac{1}{2}\hbar\tilde{\boldsymbol{\sigma}}, \tag{7.56}$$

where

$$\tilde{\sigma}_1 = -i\gamma_2\gamma_3, \quad \tilde{\sigma}_2 = -i\gamma_3\gamma_1, \quad \tilde{\sigma}_3 = -\gamma_1\gamma_2, \tag{7.57}$$

or an equivalent representation. For $\tilde{\sigma}_3$, for example, this result yields

$$\tilde{\sigma}_3 = \begin{pmatrix} 1 & 0 & 0 & 0 \\ 0 & -1 & 0 & 0 \\ 0 & 0 & 1 & 0 \\ 0 & 0 & 0 & -1 \end{pmatrix}, \tag{7.58}$$

and hence eigenvalues for s_3 are $\pm\tfrac{1}{2}\hbar$, *twice each*: the same holds for the other components of **s**. The overall result, whose proof we have only sketched, is that the Dirac equation describes particles of spin $\tfrac{1}{2}$. To emphasize: the consequences of the juxtaposition of quantum theory and special relativity for a free particle is that a description of its motion must include a quantity we call spin because of attributes like (7.54) and (7.55), and that, for a free Dirac particle, this intrinsic spin must take the value $\tfrac{1}{2}\hbar$ and *no other*. In passing, we note that a description of particles with other values of spin angular momentum is not to be found in Dirac's model.

7.4.4 Electrons and positrons

The eigenvalues $\pm\tfrac{1}{2}\hbar$ are each doubly degenerate in the Dirac model. Indeed, the fact that the γ operators may be represented as 4×4 matrices requires the objects of their action to comprise four separate components. The ket wavefunctions in Dirac's equation are therefore written as column vectors,

$$|\psi(\mathbf{x}, t)\rangle = \begin{pmatrix} \psi_1(\mathbf{x}, t) \\ \psi_2(\mathbf{x}, t) \\ \psi_3(\mathbf{x}, t) \\ \psi_4(\mathbf{x}, t) \end{pmatrix}, \tag{7.59}$$

and the bra functions as row vectors. The notation is illustrated by the equation,

$$\langle \phi | \gamma_3 | \psi \rangle \equiv (\phi_1 \phi_2 \phi_3 \phi_4) \begin{pmatrix} 0 & 0 & -i & 0 \\ 0 & 0 & 0 & i \\ i & 0 & 0 & 0 \\ 0 & -i & 0 & 0 \end{pmatrix} \begin{pmatrix} \psi_1 \\ \psi_2 \\ \psi_3 \\ \psi_4 \end{pmatrix} \tag{7.60}$$

Note how the γ operators scramble the components of the wavefunction on which they act, while the differential operators ∂_μ of (7.48) act discretely upon the separate parts. The components of the Dirac vector (7.59) are functions of space–time variables only, of course, for it is the operation of relativity, as it were, that confers upon such a space–time function, the attributes of spin. The wavefunction (7.59) is also known as a four-component *spinor*. Altogether, Dirac's equation (7.48) comprises, therefore, a simultaneous set of four linear differential equations, one representation of which, using the matrices in (7.50) is, in full:

$$\left. \begin{array}{l} +\partial_0 \psi_1 \quad -i\partial_1 \psi_4 \quad -\partial_2 \psi_4 \quad -i\partial_3 \psi_3 \quad +\kappa \psi_1 = 0, \\ +\partial_0 \psi_2 \quad -i\partial_1 \psi_3 \quad +\partial_2 \psi_3 \quad +i\partial_3 \psi_4 \quad +\kappa \psi_2 = 0, \\ -\partial_0 \psi_3 \quad +i\partial_1 \psi_2 \quad +\partial_2 \psi_2 \quad +i\partial_3 \psi_1 \quad +\kappa \psi_3 = 0, \\ -\partial_0 \psi_4 \quad +i\partial_1 \psi_1 \quad -\partial_2 \psi_1 \quad -i\partial_3 \psi_2 \quad +\kappa \psi_4 = 0, \end{array} \right\} \tag{7.48c}$$

The question arises: what physical significance attaches to the four independent components of these Dirac vectors? As we have seen so far, and shall find in more detail shortly, Dirac's approach provides a theoretical origin for spin in a beautiful way: but the orbitals used in the Pauli–Schrödinger theory – and throughout all chemistry, for example – are *two*-component entities, not four.

If one analyses the Dirac equation applied, for illustration, to a one-dimensional motion only, one finds[6] that the eigen*values* are

$$\tilde{E} = \pm \sqrt{[(cp)^2 + (m_0 c^2)^2]}, \tag{7.61}$$

where p is the momentum in the chosen direction, and that these *positive*- and *negative*-energy solutions are associated with two pairs of the

Dirac vector components, such that *each* of the positive and negative energies belong to each of the spin solutions $\pm\frac{1}{2}\hbar$. Perhaps it seems as if no progress has been made for here we have negative-energy solutions of the Dirac equation just as we had for the Klein–Gordon equation. There certainly are difficulties, but these do *not*[5] include the problem which followed for the Klein–Gordon approach, namely that of 'negative-electron density'. In 1930, Dirac proposed his *hole theory* in an attempt to interpret the negative-energy solutions to his equation. The first problem which had to be solved was to explain why electrons should not simply adopt the lowest-energy solutions – in this case the unphysical negative energies predicted by the theory. Dirac proposed that the normal ground state of nature involves fully occupied, negative-energy states by the spin $\frac{1}{2}$ electrons, obeying Pauli's exclusion principle as usual. The observed stability of positive-energy electrons follows from their inability to fall into the negative-energy states and thence violate Pauli's principle. However, Dirac predicted, if sufficient energy is provided – by a γ-ray with $E \geqslant 2m_0c^2$, for example – an electron could be promoted from a negative eigenvalue to a positive one. The result would be the creation of a normal, positive-energy electron, together with a hole in the otherwise filled, negative-energy states: and, of course, the energy of a hole in a negative-energy state is positive. The behaviour of the hole, therefore – prior to its inevitably imminent decay as a positive-energy electron falls into it with the release of a γ-ray – is just like that of a normal positive-energy electron, but one having a positive charge. The name 'positron' was coined and the process just described may be summarized by the notion that a suitably energetic γ-ray can create an electron–positron pair; a reaction that is clearly reversible.

This clever theory does not meet with total support, however. Just two points which arise immediately are: that its character is alien to that of the premise from which it arose and, more concretely perhaps, that it is clearly not a one-electron or one-particle theory as was originally supposed to be the case: it is concerned with a sea of negative-energy particles and is therefore an 'infinitely-many-particle' theory. In fact the Dirac equation cannot really be considered as the relativistic generalization of the Schrödinger equation of a particle but is,[6] rather, a field equation like the Klein–Gordon or Maxwell equations. The problems we have very briefly reviewed disappear within quantum-field theory, but that is another story. Having pointed to the difficulties with Dirac's approach, it is only right to say that his 'alien' hole-theory prediction of the creation of electron–positron pairs received experimental confirmation by Anderson in 1932 and was recognized with a Nobel prize in 1933. By bravely ignoring the inconsistencies in the theory as first presented, we can now reasonably readily rediscover what else Dirac's equation has to offer our subject.

7.4.5 An electron in an external electromagnetic field

The quickest and perhaps most elegant way to introduce the effects of external electric and magnetic fields into the Dirac equation is by working with the form which most transparently reveals the symmetry of space–time; namely, (7.48). Then, analogous to the replacement

$$\hat{p}_j \to \hat{p}_j + eA_j, \quad j = 1, 2, 3,$$ (7.62a)

we write

$$\partial_\mu \to \partial_\mu + \frac{ie}{\hbar} A_\mu, \quad \mu = 0, 1, 2, 3,$$ (7.62b)

where, as before, the positive sign arises out of our restriction to the case of an electron, with charge $-e$. The analogy is not altogether obvious, perhaps, for the dimensionality implicit in (7.62a) is three, but in (7.62b) it is four. This arises because we have replaced the $\{\phi, A_x, A_y, A_z\}$ of the classical treatment by the four-vector (A_0, A_x, A_y, A_z) in the space–time continuum: the scalar potential ϕ is replaced by the zeroth element in this new potential but note that, in the SI system and using relativistic notation, we have

$$\phi = -icA_0.$$ (7.63)

One may discern here the self-consistent feel of the theory, recalling, from chapter 2, how a magnetic field arises out of the relativistic consequences of moving electric charges. The condensed adjective 'electromagnetic' field acquires a fresh depth of meaning from this viewpoint. Anyway, direct substitution of (7.62b) into (7.48) yields Dirac's equation for an electron in the presence of an electromagnetic field:

$$\left[\gamma_\mu \left(\partial_\mu + \frac{ie}{\hbar} A_\mu \right) + \kappa \right] \psi = 0,$$ (7.64a)

where ψ is once again a four-component spinor. Less elegantly, but with the same result, we make the replacement (7.62a) in (7.43) and (7.44) to get

$$\left[-(\tilde{E} + e\phi) + \sum_{j=1}^{3} c\alpha_j(\hat{p}_j + eA_j) + \beta m_0 c^2 \right] \psi = 0,$$ (7.64b)

which, by consultation with (7.33), (7.38) and (7.47) and after multiplication by $(1/\beta\hbar c)$, gives (7.64a).

From Dirac's more general equation, (7.64), we shall determine the electron spin magnetic moment, establish the ground rules for spin-orbit coupling and make some passing remarks on other phenomena to which we refer in later chapters. As chemists, however, our interest is not with positrons and we would prefer to work with two-component wavefunctions and hence use the more familiar Pauli–Schrödinger equation on a day-to-day basis. Strictly speaking, we cannot just ignore two components of the four-component spinor, for the γ of (7.50) form an *irreducible* representation. However, because the energy eigenvalues of the two pairs – representing electrons and positrons, as it were – are so different, an approximate but effectively complete separation can be made. This is possible because the creation of positrons is a very high energy process, involving (say) hard γ-rays with energies equal to twice the rest-energy of an electron or more, while chemical bonding energies, let alone energies of interaction with external magnetic fields, are very small and the two types of process are almost totally uncoupled. This sort of situation is the meat and drink of projection theory. However, as we are only concerned to review spin theory in this section, we confine ourselves to a sketch of the results: the use of projection operators and partitioning theory in a different context is discussed in rather more detail in chapter 11. Thus transformations like those of Foldy & Wouthuysen[6,11] seek to partition the Dirac four-vector into the states

$$\psi_+ \equiv \begin{pmatrix} \psi_1 \\ \psi_2 \\ 0 \\ 0 \end{pmatrix} \quad \text{and} \quad \psi_- \equiv \begin{pmatrix} 0 \\ 0 \\ \psi_3 \\ \psi_4 \end{pmatrix}, \tag{7.65}$$

corresponding to positive- and negative-energy solutions respectively. Even for arbitrary Dirac states, it turns out that in the representation in which Dirac's β operator is written,

$$\beta \equiv \begin{pmatrix} 1 & 0 & 0 & 0 \\ 0 & 1 & 0 & 0 \\ 0 & 0 & -1 & 0 \\ 0 & 0 & 0 & -1 \end{pmatrix}, \tag{7.66}$$

the operators

$$\tfrac{1}{2}(1+\beta) = \begin{pmatrix} 1 & 0 & 0 & 0 \\ 0 & 1 & 0 & 0 \\ 0 & 0 & 0 & 0 \\ 0 & 0 & 0 & 0 \end{pmatrix}, \quad \tfrac{1}{2}(1-\beta) = \begin{pmatrix} 0 & 0 & 0 & 0 \\ 0 & 0 & 0 & 0 \\ 0 & 0 & -1 & 0 \\ 0 & 0 & 0 & -1 \end{pmatrix} \tag{7.67}$$

project out functions $\tilde{\psi}_\pm$ with the *form* of those in (7.65):

$$\tilde{\psi}_\pm = \tfrac{1}{2}(1 + \beta)f\psi. \qquad (7.68)$$

In the low-energy non-relativistic limit, the operator f takes the approximate form

$$f \approx \tfrac{1}{2}\left(2 + \beta\alpha_i \frac{v_i}{c}\right), \qquad (7.69)$$

where $v_i \approx p_i/m_0$. Then, from (7.68),

$$\tilde{\psi}_\pm \approx \tfrac{1}{4}\left(2 \pm 2\beta + \beta\alpha_i \frac{v_i}{c} \pm \alpha_i \frac{v_i}{c}\right)\psi \qquad (7.70)$$

and, by using the commutation relations (7.42), it can be shown[6] that

$$\frac{\tilde{\psi}_-}{\tilde{\psi}_+} \approx \tfrac{1}{2}\frac{\beta\alpha_i(v_i/c)\psi}{[\beta + \alpha_i(v_i/c)]\psi} \qquad (7.71)$$

and, since β is of order one, $\tilde{\psi}_- \ll \tilde{\psi}_+$ in this non-relativistic limit. The point of the last few equations, merely reviewed here, is to indicate how the operators (7.67), even with arbitrary Dirac vectors, almost completely decouple the functions with positive and negative energies, provided we consider only circumstances near the non-relativistic limit. Of course, the separation is not total: overall, however, we may talk of 'large' and 'small' components of ψ.

It is helpful to consider the notion of large and small components from another point of view. For this, we return to the form (7.44) of Dirac's equation. Using (7.51) and (7.52), we find

$$\alpha_j = \left(\begin{array}{c|c} 0 & \sigma_j \\ \hline \sigma_j & 0 \end{array}\right) \quad \text{for} \quad j = 1, 2, 3,$$

and

$$\beta = \left(\begin{array}{c|c} 1 & 0 \\ \hline 0 & -1 \end{array}\right). \qquad (7.72)$$

Substitution of (7.76) into (7.44), and using (7.33) and

$$p_0 = \tilde{E}/c = i\frac{\hbar}{c}\frac{\partial}{\partial t} = -\hbar\frac{\partial}{\partial x_0}, \qquad (7.73)$$

gives Dirac's equation in the form

$$\left[-p_0\left(\begin{array}{c|c} 1 & 0 \\ \hline 0 & 1 \end{array}\right) + \boldsymbol{\sigma}\cdot\mathbf{p}\left(\begin{array}{c|c} 0 & 1 \\ \hline 1 & 0 \end{array}\right) + m_0 c\left(\begin{array}{c|c} 1 & 0 \\ \hline 0 & -1 \end{array}\right)\right]\left(\begin{array}{c} \Phi_1 \\ \Phi_2 \end{array}\right) = 0, \qquad (7.74a)$$

in which the scalar product extends only over the three space dimensions; and the blocked form of matrix indicates, as in (7.51), the two-fold nature of the partitioned functions Φ_1 and Φ_2 of the four-vector ψ:

$$\psi \equiv \begin{pmatrix} \Phi_1 \\ \Phi_2 \end{pmatrix} \equiv \begin{pmatrix} \psi_1 \\ \psi_2 \\ \psi_3 \\ \psi_4 \end{pmatrix}. \tag{7.75}$$

Multiplying out (7.74a), we have

$$\left. \begin{array}{l} -(p_0 - m_0 c)\Phi_1 + \boldsymbol{\sigma} \cdot \mathbf{p}\, \Phi_2 = 0, \\ -(p_0 + m_0 c)\Phi_2 + \boldsymbol{\sigma} \cdot \mathbf{p}\, \Phi_1 = 0, \end{array} \right\} \tag{7.74b}$$

for Dirac's equation in the absence of an electromagnetic field.

By similar reasoning, or by use of (7.62), Dirac's wave equation *in the presence* of external fields is,

$$\left. \begin{array}{l} \left(p_0 - m_0 c + \dfrac{e}{c}\phi \right)\Phi_1 - \boldsymbol{\sigma} \cdot (\mathbf{p} + e\mathbf{A})\Phi_2 = 0, \\[2mm] \left(p_0 + m_0 c + \dfrac{e}{c}\phi \right)\Phi_2 - \boldsymbol{\sigma} \cdot (\mathbf{p} + e\mathbf{A})\Phi_1 = 0, \end{array} \right\} \tag{7.76}$$

and can be shown to be relativistically[10] and gauge[8] invariant. In preparation for effecting a separation between the 'large' two-component spinor Φ_1, associated with normal electrons, and the 'small' spinor Φ_2, we now multiply (7.76) by c, replace cp_0 by the energy as in (7.73), and redefine our energy origin as being relative to the electronic rest-energy $m_0 c^2$:

$$E = \tilde{E} - m_0 c^2 \tag{7.77}$$

and there results,

$$\left. \begin{array}{l} (E + e\phi)\Phi_1 - c\boldsymbol{\sigma} \cdot (\mathbf{p} + e\mathbf{A})\Phi_2 = 0, \\ (E + 2m_0 c^2 + e\phi)\Phi_2 - c\boldsymbol{\sigma} \cdot (\mathbf{p} + e\mathbf{A})\Phi_1 = 0. \end{array} \right\} \tag{7.78}$$

Elimination of Φ_2 from (7.78) gives

$$\left[\frac{1}{2m_0}\boldsymbol{\sigma} \cdot (\mathbf{p} + e\mathbf{A})k\boldsymbol{\sigma} \cdot (\mathbf{p} + e\mathbf{A}) - e\phi - E \right]\Phi_1 = 0, \tag{7.79}$$

where

$$k = 2m_0 c^2/(E + 2m_0 c^2 + e\phi). \tag{7.80}$$

Equation (7.79) is an eigenvalue equation for the energies associated with the two-component spinor Φ_1. That it corresponds closely with the Pauli–Schrödinger equation emerges as we investigate its non-relativistic limit; for when we consider normal 'chemical' electron energies and scalar potentials, the denominator in (7.80) is dominated by m_0c^2. This has the effect of making Φ_2 small compared with Φ_1 in (7.78) so that we confirm the earlier description of Φ_1 and Φ_2 as being 'large' and 'small' components respectively. We now examine this limit in more detail.

7.4.6 The non-relativistic approximation and the spin magnetic moment

In the non-relativistic limit, when $(E + e\phi) \ll 2m_0c^2$, k in (7.80) will be very nearly one except very near a field source, and may be expanded as

$$k = 1 - \frac{E + e\phi}{2m_0c^2} + \cdots \cdot \tag{7.81}$$

The lowest-order approximation we may consider is to take $k = 1$ exactly, when (7.79) becomes

$$\left\{ \frac{1}{2m_0} [\sigma \cdot (\mathbf{p} + e\mathbf{A})]^2 - e\phi - E \right\} \Phi_1 = 0. \tag{7.82}$$

We now require the identity[8]

$$(\sigma \cdot \mathbf{P})(\sigma \cdot \mathbf{Q}) \equiv \mathbf{P} \cdot \mathbf{Q} + i\sigma \cdot \mathbf{P} \wedge \mathbf{Q} \tag{7.83}$$

for two general vectors $\mathbf{P}$ and $\mathbf{Q}$ which commute with σ but not necessarily with each other[tt]. We then evaluate the square in (7.82) as

$$\begin{aligned}
[\sigma \cdot (\mathbf{p} + e\mathbf{A})]^2 &= (\mathbf{p} + e\mathbf{A})^2 + i\sigma \cdot (\mathbf{p} + e\mathbf{A}) \wedge (\mathbf{p} + e\mathbf{A}) \\
&= (\mathbf{p} + e\mathbf{A})^2 + i\sigma \cdot (\mathbf{p} \wedge e\mathbf{A} + e\mathbf{A} \wedge \mathbf{p}) \\
&= (\mathbf{p} + e\mathbf{A})^2 + e\hbar\sigma \cdot \operatorname{curl} \mathbf{A},
\end{aligned} \tag{7.84}$$

where, as above (7.2), we again use the three-dimensional relationship $\mathbf{p} = -i\hbar\nabla$ and the standard vector formula,[12]

$$\operatorname{curl}(m\mathbf{P}) \equiv \nabla \wedge (m\mathbf{P}) = m\nabla \wedge \mathbf{P} - \mathbf{P} \wedge \nabla m. \tag{7.85}$$

[tt] With $\mathbf{P} \equiv \mathbf{Q}$, the last term in (7.83) involves $(\mathbf{P} \wedge \mathbf{P})$. That this need not vanish is made clear in the second line of (7.84), which latter vanishes only if $\mathbf{p}$ and $\mathbf{A}$ are coparallel: the point, of course, is that $\mathbf{P}$ is an operator here and not simply a vector.

On substitution into (7.82), there results

$$\left[\frac{1}{2m_0}(\mathbf{p} + e\mathbf{A})^2 + \frac{eh}{2m_0}\boldsymbol{\sigma}\cdot\mathbf{B} - e\phi - E \right]\Phi_1 = 0. \tag{7.86}$$

Thus far we have merely assumed that ϕ and $\mathbf{A}$ are independent of the time: so (7.86) represents the simplest approximation of the non-relativistic form of Dirac's equation for ordinary free electrons. It differs from the spinless Schrödinger equation of §7.1 in two respects: (i) the eigenfunction is the two-component spinor Φ_1, and (ii) there has appeared the additional term $(eh/2m_0)\boldsymbol{\sigma}\cdot\mathbf{B}$. By comparison with the definition of β_e under (7.10), and with (7.14) and (7.24), we find for this 'extra' term:

$$\frac{eh}{2m_0}\boldsymbol{\sigma}\cdot\mathbf{B} = h^{-1}\beta_e(2\mathbf{s}\cdot\mathbf{B}), \tag{7.87}$$

which is of precisely the form introduced into the Pauli–Schrödinger equation and thence into (7.21b) on a phenomenological or *ad hoc* basis.

The factor 2 occurring in the magnetic moment operator

$$\mu = h^{-1}\beta_e\mathbf{B}(\mathbf{l} + 2\mathbf{s}) \tag{7.88}$$

thus emerges naturally, though admittedly after some effort, from Dirac's attempt to find a relativistically invariant substitute for Schrödinger's equation; and the 'anomalous' spin magnetic moment, first introduced here in (3.55), has been explained. The reason for the departure of the free-electron g_e value from 2 to 2.0023 lies outside Dirac's theory and, like the explanation of the existence of particles with spin not equal to $\frac{1}{2}h$, belongs to the subject called quantum electrodynamics.

7.4.7 Relativistic corrections and spin-orbit coupling

So the attribute of electron spin arises from the lowest-order approximation in the non-relativistic limit of Dirac's equation. We now review how spin-orbit coupling emerges when we retain the first-order terms in the expression (7.81) for k. Spin-orbit coupling occurs in atoms and molecules even in the absence of an external magnetic field, of course, and so we remove that complication here by setting $\mathbf{A} = 0$. For the moment, we leave the scalar potential ϕ unspecified, though eventually we shall be concerned most particularly with the case of a central electric field.

In this somewhat higher-order approximation,[8] which we shall see does not correspond to the non-relativistic limit, (7.79) becomes

$$\left[\frac{1}{2m_0}(\boldsymbol{\sigma}\cdot\mathbf{p})k(\boldsymbol{\sigma}\cdot\mathbf{p}) - e\phi - E \right]\Phi_1 = 0. \tag{7.89}$$

As **p** is a differential operator, we have

$$(\boldsymbol{\sigma} \cdot \mathbf{p})k = \boldsymbol{\sigma} \cdot (\mathbf{p}k) + k\boldsymbol{\sigma} \cdot \mathbf{p}, \tag{7.90}$$

where $(\mathbf{p}k)$ indicates that **p** operates only upon k. Therefore,

$$(\boldsymbol{\sigma} \cdot \mathbf{p})k(\boldsymbol{\sigma} \cdot \mathbf{p}) = (\boldsymbol{\sigma} \cdot \mathbf{p})^2 k - (\boldsymbol{\sigma} \cdot \mathbf{p})\boldsymbol{\sigma} \cdot (\mathbf{p}k)$$

$$\equiv X_1 - X_2 \text{ say.} \tag{7.91}$$

Using (7.83),

$$(\boldsymbol{\sigma} \cdot \mathbf{p})^2 k = \mathbf{p}^2 k + i\boldsymbol{\sigma} \cdot \mathbf{p} \wedge \mathbf{p} = \mathbf{p}^2 k, \tag{7.92}$$

and so, from (7.81), we find that

$$X_1 = \mathbf{p}^2 \left(1 - \frac{E + e\phi}{2m_0 c^2} \right) \tag{7.93}$$

to first order. Were we only concerned with the term X_1, (7.89) would reduce to

$$\left(\frac{\mathbf{p}^2}{2m_0} k - e\phi - E \right) \Phi_1 = 0, \tag{7.94}$$

so that we see that the second term in (7.93) represents a relativistic correction to the momentum arising out of the recognition via (7.81) that $(E + e\phi)$ is not entirely negligible compared with $2m_0 c^2$: if it were, k would be unity again, and (7.94) collapse to

$$\left(\frac{\mathbf{p}^2}{2m_0} - e\phi - E \right) \Phi_1 = 0, \tag{7.95}$$

corresponding once more to the spinless Schrödinger equation. Herein we may put $(E + e\phi) = \mathbf{p}^2/2m_0$, so that to first order, we find

$$X_1 = \mathbf{p}^2 \left(1 - \frac{\mathbf{p}^2}{4m_0^2 c^2} \right) = \mathbf{p}^2 - \left(\frac{1}{4m_0^2 c^2} \right) \mathbf{p}^4 \tag{7.96}$$

if $\mathbf{p}^4$ is taken to represent the operation $(\mathbf{p}^2)^2$. So the term $(1/4m_0^2 c^2)\mathbf{p}^4$ in X_1 is the first relativistic correction to have emerged from (7.89). The second arises from a development of X_2 in (7.91).

With Griffith,[8] we again use (7.83) and rewrite X_2 as

$$X_2 = \mathbf{p} \cdot (\mathbf{p}k) + i\boldsymbol{\sigma} \cdot \mathbf{p} \wedge (\mathbf{p}k)$$

$$= (-i\hbar)^2 \nabla \cdot (\nabla k) + i(-i\hbar)^2 \boldsymbol{\sigma} \cdot \nabla \wedge (\nabla k) \tag{7.97}$$

$$= -\hbar^2 \nabla^2 k - \hbar^2 (\nabla k) \cdot \nabla + i(-i\hbar)^2 \boldsymbol{\sigma} \cdot [-(\nabla k) \wedge \nabla + \text{curl} \, \nabla k],$$

where in the second line we have used $\mathbf{p} = -i\hbar\nabla$ again, and in the third (7.85). The first term on the right-hand side vanishes because ϕ satisfies

Laplace's equation, $\nabla^2 \phi = 0$, and the last also because of the identity, curl grad $\equiv 0$. Therefore,

$$X_2 = -\hbar^2(\nabla k)\cdot \nabla - i(-i\hbar)^2\boldsymbol{\sigma}\cdot (\nabla k)\boldsymbol{\wedge}\nabla$$
$$= -\hbar^2(\nabla k)\cdot \nabla - \hbar\boldsymbol{\sigma}\cdot(\nabla k)\boldsymbol{\wedge}\mathbf{p}, \tag{7.98}$$

which is as far as the present approximation leads.

If now we specialize our discussion to the atomic case and require that ϕ, and hence k, be simply a function of r, for the central-field situation, we have,

$$\nabla k = \frac{1}{r}\mathbf{r}\frac{dk}{dr}, \tag{7.99}$$

and so

$$X_2 = -\frac{\hbar^2}{r}\frac{dk}{dr}\mathbf{r}\cdot\nabla - \frac{\hbar}{r}\frac{dk}{dr}\boldsymbol{\sigma}\cdot\mathbf{r}\boldsymbol{\wedge}\mathbf{p}$$

$$= -\frac{\hbar^2}{r}\frac{dk}{dr}\frac{\partial}{\partial r} - \frac{\hbar}{r}\frac{dk}{dr}\boldsymbol{\sigma}\cdot\mathbf{l}. \tag{7.100}$$

Now, from (7.81), we have

$$\frac{dk}{dr} = -\frac{e}{2m_0c^2}\frac{d\phi}{dr}, \tag{7.101}$$

and so, altogether with (7.100), (7.96), and (7.91), the higher-order equation, (7.89), becomes

$$\left[\frac{1}{2m_0}\mathbf{p}^2 - \frac{1}{8m_0^3c^2}\mathbf{p}^4 - \frac{e\hbar}{4m_0^2c^2r}\frac{d\phi}{dr}\boldsymbol{\sigma}\cdot\mathbf{l} - \frac{e\hbar^2}{4m_0^2c^2}\frac{d\phi}{dr}\frac{\partial}{\partial r} - e\phi - E\right]\Phi_1 = 0 \tag{7.102}$$

as our complete Dirac equation for an atomic electron, relativistically correct to first order and in the absence of an external magnetic field. The third term in (7.102) provides the second relativistic correlation to have emerged in this analysis and describes the spin-orbit coupling Hamiltonian, H_{SO}; for when we write $\mathbf{s} = \frac{1}{2}\hbar\boldsymbol{\sigma}$, the new contribution reads

$$H_{SO} = -\frac{e}{2m_0^2c^2r}\frac{d\phi}{dr}\mathbf{l}\cdot\mathbf{s}, \tag{7.103}$$

and in the even more particular case that $\phi = Ze/4\pi\varepsilon_0 r$ for a nuclear charge Ze (in the SI system; see (2.44)), we get

$$H_{SO} = \frac{Ze^2}{8\pi\varepsilon_0 m_0^2c^2r^3}\mathbf{l}\cdot\mathbf{s}. \tag{7.104}$$

In parenthesis, consider the language used here. The spin-orbit coupling phenomenon arises as a relativistic correction that recognizes the imperfect decoupling of Φ_1 and Φ_2 belonging to the positive- and negative-energy solutions of Dirac's equation. The attribute of spin, on the other hand, emerges directly in the non-relativistic limit: but note that this corresponds to the 'non-relativistic limit' of a relativistically invariant theory! One may observe[13] that the consequences of special relativity in molecular quantum mechanics fall into two classes: those effects with classical analogues that arise when a particle's speed becomes comparable with that of light, and those without classical parallel – like spin and the permutation symmetry of $H^{(0)}$ in (7.16) – and with no obvious dependence upon a particle's speed. Outside quantum mechanics, one might include the magnetic field in this latter category and so illuminate further the intimate connection between spin and magnetism.

7.4.8 Spin-orbit coupling matrix elements

The spin-orbit interaction Hamiltonian (7.103) is often written

$$H_{SO} = \xi \mathbf{l} \cdot \mathbf{s}, \tag{7.105}$$

where

$$\xi(r) = -\frac{e}{2m_0^2 c^2 r} \frac{d\phi}{dr} \tag{7.106}$$

for a central field; and

$$\xi(r) = \frac{Ze^2}{8\pi\varepsilon_0 m_0^2 c^2 r^3} \tag{7.107}$$

for the simple Coulomb potential, $\phi = -Ze/4\pi\varepsilon_0 r$. Within as $\{lsm_l m_s\}$ basis, matrix elements of H_{SO} may[8,14] be factorized into radial and angular parts, as

$$\langle nlsm_l m_s | H_{SO} | n'l's'm_l' m_s' \rangle$$

$$= \delta(l, l') \langle lsm_l m_s | \mathbf{l} \cdot \mathbf{s} | l's m_l' m_s' \rangle \int_0^\infty R_{nl}(r) \xi(r) R_{n'l'}(r) r^2 \, dr, \tag{7.108}$$

in which the total radial part of a ket is defined by $R_{nl}(r)$. Note in particular that the spin-orbit matrix is rigorously diagonal in l. Further, we shall be dealing in practice with 'effective' spin-orbit coupling coefficients, so that the non-diagonality of the matrix with respect to the principal quantum number is unimportant (and is, in any case, expected to contribute rather little by the usual perturbation arguments): the matter is discussed further in chapter 11. So it is convenient to obviate any need to evaluate the radial integral in (7.108) explicitly by *defining* the quantity

$$\zeta_{nl} \equiv \hbar^2 \int_0^\infty R_{nl}^2(r)\xi(r)r^2 dr \qquad (7.109)$$

as the 'one-electron, spin-orbit coupling coefficient' for an electron in an (nl) atomic subshell. For the corresponding angular part of (7.108), we use the standard relationship

$$\mathbf{j} = \mathbf{l} + \mathbf{s} \qquad (7.110)$$

and hence,

$$2\mathbf{l}\cdot\mathbf{s} = \mathbf{j}^2 - \mathbf{l}^2 - \mathbf{s}^2, \qquad (7.111)$$

so that the angular part of the spin-orbit energy is $\hbar^2[j(j+1) - l(l+1) - \frac{3}{4}]$. Altogether, the one-electron, spin-orbit interaction energy is then given by

$$\begin{aligned} E_{SO} &= \hbar^{-2}\zeta_{nl}\langle j|\mathbf{l}\cdot\mathbf{s}|j\rangle \\ &= \tfrac{1}{2}l\zeta_{nl} \quad \text{when } j = l + \tfrac{1}{2}, \\ or &-\tfrac{1}{2}(l+1)\zeta_{nl} \quad \text{when } j = l - \tfrac{1}{2}, \end{aligned} \qquad (7.112)$$

as is simply verified by substitution of the appropriate j value in the angular energy expression.

In the case of the Coulomb central-field potential, $\phi = Ze/4\pi\varepsilon_0 r$ and $\xi(r)$ given by (7.107), the one-electron spin-orbit coupling coefficient, can be evaluated as

$$\zeta_{nl} = \frac{Ze^2\hbar^2}{8\pi\varepsilon_0 m_0^2 c^2} \int_0^\infty r^{-3} R_{nl}^2(r)r^2 dr, \qquad (7.113a)$$

$$= \frac{Ze^2\hbar^2}{8\pi\varepsilon_0 m_0^2 c^2 a^3} \frac{Z^3}{n^3 l(l+\frac{1}{2})(l+1)}, \qquad (7.113b)$$

where (7.113b) follows using tables[14] of standard integrals; and the Bohr radius $a = 4\pi\varepsilon_0 \hbar^2/m_0 e^2$. Now we may substitute (7.113b) into (7.112). Consider the case for $j = l + \frac{1}{2}$:

$$E_{SO}(l+\tfrac{1}{2}) = \frac{Z^4 e^2 \hbar^2}{16\pi\varepsilon_0 m_0^2 c^2 n^3 a^3 (l+\frac{1}{2})(l+1)}. \qquad (7.114)$$

Even more specifically, let us evaluate the spin-orbit interaction energy for the case $l = 0$:

$$E_{so}(l = 0) = \frac{Z^4 e^2 \hbar^2}{32\pi\varepsilon_0 m_0^2 c^2 n^3 a^3}. \tag{7.115}$$

But this cannot be correct, surely, for the expectation value of $\mathbf{l} \cdot \mathbf{s}$ for a state characterized by $l = 0$ vanishes. Actually the problem is greater yet, for the integral in (7.113a) diverges at the origin as we can see from the fact that the screening of all electrons near there must vanish, so that ϕ really does represent the full potential due to the nuclear charge Ze and the special form of (7.113a) is then appropriate. Under these circumstances the total radial function $R_{nl}(r)$ starts as r^l, just as in hydrogen, so that the complete integral of (7.113a) goes as $1/r$ for small r when $l = 0$. Altogether, therefore, the spin-orbit energy of an s orbital ($l = 0$) involves the product of a diverging radial integral and a vanishing angular product; so the result is indeterminate by the present calculation. Let us see how the problem is resolved.

Well, the indeterminancy is really a consequence of the approximation we used right back at the beginning; namely, that $(E + e\phi) \ll 2m_0 c^2$. While the assumption is reasonable for chemical problems in most regions of space, very near the source of the field it fails totally. Near the nucleus the reverse can happen $(E + e\phi) \gg 2m_0 c^2$, and this is relevant for s orbitals by virtue of the relatively large probability of the electron being close to the nucleus in these circumstances. Therefore, we eschew the expansion of k in (7.81) and instead of (7.101) for dk/dr we have

$$\left(\frac{dk}{dr}\right)_{\text{small } r} = \frac{-2m_0 c^2 e}{(E + m_0 c^2 + e\phi)^2} \frac{d\phi}{dr}, \tag{7.116}$$

which approaches a *finite* limit near the nucleus. Meanwhile, (7.109) must be replaced by a general expression in which the approximation (7.101) is not made: thus

$$\zeta_{nl} = \hbar^2 \int_0^\infty \frac{1}{r} R_{nl}^2(r) \frac{dk}{dr} r^2 dr, \tag{7.117}$$

which is finite with (dk/dr) given by (7.116) and the potential $\phi = Ze/4\pi\varepsilon_0 r$ for the field near the nucleus. Altogether, therefore, proper application of the first-order approximation to the complete Dirac equation yields vanishing spin-orbit energies for s orbitals, rather than the quantity in (7.115). However, even that is not quite the end of the matter.

7.4.9 The Darwin term

We have not yet discussed the fourth term in the relativistically corrected equation (7.102). It does not have a simple classical parallel and is difficult to demonstrate experimentally because it does not involve angular momenta. We consider its effect in the case of the Coulomb central field, $\phi = Ze/4\pi\varepsilon_0 r$ when we find an energy shift, ΔE,

$$\Delta E = \frac{-e\hbar^2}{16\pi\varepsilon_0 m_0^2 c^2} \int_0^\infty \left(\frac{-Ze}{r^2}\right) R_{nl}(r) \frac{\partial R_{nl}(r)}{\partial r} r^2 dr$$

$$= \frac{Ze^2\hbar^2}{16\pi\varepsilon_0 m_0^2 c^2} \int_0^\infty \frac{1}{r^2} R_{nl}(r) \frac{dR_{nl}(r)}{dr} r^2 dr$$

$$= \frac{Ze^2\hbar^2}{32\pi\varepsilon_0 m_0^2 c^2} R_{nl}(0) \tag{7.118}$$

because $R_{nl}(\infty) = 0$. We note straightaway that this relativistic correction term only contributes to integrals involving s orbitals for which $R_{nl}(0) \neq 0$ uniquely. Using standard expressions[14] for radial functions, we have $R_{n0}(0) = Z^3/n^3 a^3$ and so

$$\Delta E = \frac{Z^4 e^2\hbar^2}{32\pi\varepsilon_0 m_0^2 c^2 n^3 a^3}, \tag{7.119}$$

which is precisely the quantity calculated in (7.115)! So, after all, we find a relativistic energy correction for s orbitals which is of equal magnitude to that calculated by an incorrect application of the spin-orbit formula (7.114) to s orbitals. Now Darwin had discovered (7.102) prior to the work of Dirac but *without* this extra term involving $\partial/\partial r$: the term is, in this sense, unique to Dirac's approach. Somewhat contrarily, perhaps, the extra term is known as the Darwin term. The sketch presented here makes reasonable the fact which emerges in a full analysis, that the Darwin correction achieves a magnitude comparable with spin-orbit coupling energies, even when the scalar potential assumes more complex forms in many-electron systems, and is generally only considered practically significant in heavier atoms. It is briefly referred to in §11.3.

7.4.10 Many electrons

Our purpose in this main section §7.4 has been to sketch and review spin theory because of its obviously intimate relevance to paramagnetism.

Spin-orbit coupling followed rather naturally in the development and a little detail was included because this perturbation is of particular importance to magnetic susceptibility by virtue of the fact that energy splittings caused by it are frequently comparable with thermal energies. However, our brief here is not the theory of atomic energy levels and we must bring the review to a close. We do so with just one or two remarks about the many-electron case.

Everything in §7.4 has been concerned with the theory of the electron: *pace* remarks about the positron 'hole' theory in §7.4.4., we have been concerned with a one-electron theory. In practice, of course, we are interested in the magnetic and ligand-field properties of d^n (and f^n) systems and so require an extension of the Pauli–Schrödinger equation to the many-electron case, as in (7.17). Unfortunately this is not a simple matter: indeed, the problem has still not been completely solved. Part of the difficulty in constructing a relativistically invariant equation for more than one particle concerns the incorporation of the electrostatic interaction e^2/r_{12}, for r_{12} is not a scalar in the space–time continuum. The interaction between electrons, for example, is not independent of their velocities. Further, moving charged particles generate magnetic fields and hence contribute to the vector potential **A**. Instead of the assumption of the potentials at point $\mathbf{r}_1$ due to a moving charge Q_2 at $\mathbf{r}_2$ being given by

$$\phi(\mathbf{r}_1) = Q_2/r_{12}, \quad \mathbf{A}(\mathbf{r}_1) = 0, \tag{7.120}$$

we require[15] the relativistic potentials known as the retarded or Lienard–Wiechert potentials, with the forms

$$\left. \begin{aligned} \phi(\mathbf{r}_1, t) &= Q_2/(r_{12} + \mathbf{v}_2 . \mathbf{r}_{12}/c), \\[2mm] \mathbf{A}(\mathbf{r}_1, t) &= Q_2\mathbf{v}_2/(r_{12} + \mathbf{v}_2 . \mathbf{r}_{12}/c). \end{aligned} \right\} \tag{7.121}$$

The ground covered as one pursues this line includes the Breit equation; orbit–orbit, spin–spin, and spin–other–orbit coupling; and all manner of electron–nuclear interactions.[15,4] For our purposes, however, it is sufficient to observe that we may proceed in a phenomenological way and write for the spin-orbit interaction in a many-electron atom

$$H_{SO} = \sum_i^N \xi(r_i)\mathbf{l}_i . \mathbf{s}_i \tag{7.122}$$

or, since we shall be concerned with electrons in the same subshell,

$$H_{SO} = \zeta_{nl} \sum_i^N \mathbf{l}_i . \mathbf{s}_i : \tag{7.123}$$

similarly, for the interaction with an applied magnetic field we employ the sum of magnetic moment operators as in (7.17), (7.20) and (7.21). The business of evaluating matrix elements of these two operators in a many-electron basis, with due regard to the theory of angular momentum, is discussed at some length in the next chapter.

7.5 Time-reversal symmetry and Kramers' degeneracy

7.5.1 Time inversion in classical mechanics

The major consequences or associations of spin in molecular quantum mechanics are the *aufbau* principle, the permutation symmetry of the Schrödinger equation in a many-electron basis, and exchange–coupling phenomena. Energetically much smaller effects are those we have discussed in some detail – the electron magnetic moment and spin-orbit coupling. We turn now to another phenomenon which might be considered as an attribute of spin and it is one of a group-theoretical nature. Generally we make little explicit reference to group theory in this book for, as explained in the Introduction and as is emphasized throughout the remainder of the volume, we argue that *real* ligand-field and magnetochemical studies must be concerned overwhelmingly with unsymmetrical molecules: the philosophy, therefore, is to let 'accidents of symmetry' occur as they may within a generalized computational approach. However, the phenomenon of Kramers' degeneracy[16] depends upon the number of electron spins rather than on the geometrical arrangement of atoms in a complex and so a discussion of it properly belongs here.

The idea that the equations of motion of the states of a system are invariant under the various symmetry transformations of an appropriate group is equivalent to the proposition that the Hamiltonian commutes with the operators of that group. An observable formed with those operators is a constant of the motion and so, associated with any invariance to the group operators, there arise a number of conservation laws. In turn, by taking due account of these symmetries in the given Hamiltonian, we can establish certain conclusions about the degeneracies of its eigenvalues. The group operators that we normally consider, of course, include spatial rotations and inversions. However, the equivalence of space and time that we recognize in the special theory of relativity, suggests that we should seek a further symmetry by inclusion of time transformations within these operator sets. In parallel with spatial inversion, we consider time inversion through the so-called *time-reversal* operator T which transforms t into $-t$.

Some[17] would prefer to describe the time-reversal operation simply as one involving the reversal of the direction of motion, in which all velocities are reversed. In terms of the classical Lagrangian velocities and coordinates of §3.18, time reversal replaces $\{\dot{q},q\}$ by $\{-\dot{q},q\}$; and, since momenta are linear functions of velocities, for Hamilton's equations, T replaces $\{p,q\}$ by $\{-p,q\}$. By reversing all velocities and letting time move in a reverse direction, a system runs back through its past history. That the time-reversal operation is a classically proper, *symmetry operation* follows by noting, for example, that Newton's law $\mathbf{F} = m\mathbf{a}$ involves only a *second* time derivative or, more generally, that the classical Lagrangian function L is a second-degree polynomial in the velocities with first-degree terms usually absent: then $L(\dot{q},q) = L(-\dot{q},q)$. Such is the case for isolated particles or for those subject to electric fields. On the other hand, the time-reversal operation is *not* a symmetry – even in the classical world – in the presence of a magnetic field, for this introduces terms which are linear in the velocities via the Lorentz equation (2.52).

7.5.2 The antiunitary quantum operator

Within quantum mechanics we define the time-reversal operator, T, by the transformations

$$T\mathbf{r}T^{\dagger} = \mathbf{r}, \quad T\mathbf{p}T^{\dagger} = -\mathbf{p}, \tag{7.124}$$

where the dagger denotes the hermitian conjugate, as usual: as in the classical situation, T reverses all velocities and momenta. Consider its effect on a solution of the time-independent Schrödinger equation which, for a time-independent Hamiltonian in a conservative system, may be written

$$\Psi(\mathbf{r},t) = e^{iEt/\hbar}\psi(\mathbf{r}) \tag{7.125}$$

corresponding to the solution with eigenvalue E. The effect of time reversal is

$$T(\Psi) = e^{-iEt/\hbar}T(\psi) \tag{7.126}$$

or, writing Ψ as $c(t)\psi$,

$$T(c\psi) = c^*(T\psi). \tag{7.127}$$

Such is the behaviour of a so-called *antilinear* operator[10]; to be contrasted with that for the much more common, *linear* operators[1,10] which are defined by

$$A(c\psi) = c(A\psi). \tag{7.128}$$

Corresponding to the further condition that makes a linear operator unitary,

$$\langle A\psi | A\chi \rangle = \langle \psi | \chi \rangle, \tag{7.129}$$

is the condition for antiunitarity for an antilinear operator:

$$\langle A\psi | A\chi \rangle = \langle \psi | \chi \rangle^* = \langle \chi | \psi \rangle. \tag{7.130}$$

Since T is a symmetry operation, by assumption and by construction of the quantum analogue of the classical description, it cannot change the transition probability between two states ψ and χ: therefore, the magnitude of the scalar product is conserved,

$$|\langle T\psi | T\chi \rangle| = |\langle \psi | \chi \rangle|. \tag{7.131}$$

7.5.3 Complex conjugation

Altogether, therefore, T is found to be an antilinear, antiunitary operator. The establishment of this very important property does not completely define the time-reversal operator, however. Some first clue to its nature is given by (7.127) in which the effect of time reversal upon $e^{i\alpha t}$ was seen to be equivalent to complex conjugation. We explore this idea further by considering the Schrödinger equation in its time-dependent form,

$$H\psi = i\hbar \frac{\partial \psi}{\partial t}: \tag{7.132}$$

we are explicitly ignoring spin at this stage. Note that the Hamiltonian – representing energy – is real. Now let us take the complex conjugate of (7.124) throughout, remembering the reality of H:

$$H\psi^* = -i\hbar \frac{\partial \psi^*}{\partial t} \tag{7.133a}$$

$$= i\hbar \frac{\partial \psi^*}{\partial(-t)}. \tag{7.133b}$$

The exercise[18] tells us that the state ψ^* will evolve into positive time in just the same way that ψ would have evolved into negative time. This follows because (7.133b) is exactly the same equation as (7.132) when we write

$$H\Psi = i\hbar \frac{\partial \Psi}{\partial \tau}, \tag{7.133c}$$

where $\Psi \equiv \psi^*$ and $\tau \equiv -t$. We conclude, therefore, that if $\psi(\mathbf{r}, t)$ is a solution of the spinless Schrödinger equation, then so also is $\psi^*(\mathbf{r}, -t)$. This is not to say that the time-reversal operator may be identified with that of complex conjugation, however, except in the present special circumstances. We can at least be sure that there exist close parallels between the two operations and a more precise definition of the time-reversal operator emerges from our complete knowledge of complex conjugation, as follows.

The complex conjugation operator K acts upon the product of a (complex) coefficient and a wavefunction according to

$$K(c\psi) = c^*\psi^* \equiv c^*(K\psi), \tag{7.134}$$

and is therefore antilinear. It is also antiunitary because

$$\langle K\psi | K\chi \rangle = \langle \psi^* | \chi^* \rangle = \langle \chi | \psi \rangle. \tag{7.135}$$

These properties it shares with time reversal. However, we know one further property of complex conjugation which we have not established one way or another for time reversal: it is that the vectors ψ and ψ^* are in reciprocal correspondence; that is

$$(\psi^*)^* = \psi \tag{7.136}$$

and hence $K^2 = 1$. That the equivalent result does not always hold for the time-reversal operator is the key result of the present discussion: we establish it now.

7.5.4 The square of the operator

From the definitions (7.129) and (7.130), it is evident that the product of two antiunitary operators is unitary. Let us define, therefore, a unitary operator, U, such that

$$TK = U; \tag{7.137}$$

but since $K^2 = 1$, we also have

$$T = UK, \tag{7.138}$$

which expresses the fact that the most general form that the time-reversal operator may take is that of the product of some unitary operator and the operation of complex conjugation. Now, since T is a symmetry operation, $T\psi$ can differ from ψ by no more than a phase factor; that is, by a constant of unit magnitude: the same is true of the functions $T^2\psi$ and ψ. Therefore, we can write

$$T^2\psi = (UKUK)\psi = (UK^{-1}UK)\psi = (UU^*)\psi = c\psi. \tag{7.139}$$

The unitarity of U means that, in matrix notation,

$$\mathbf{U} = c(\mathbf{U}^*)^{-1} = c\tilde{\mathbf{U}}, \tag{7.140}$$

where the tilde denotes the transpose; and then by taking the transpose throughout, we get

$$\tilde{\mathbf{U}} = c\mathbf{U} = c^2\tilde{\mathbf{U}}, \tag{7.141}$$

and hence the very important result that $c^2 = 1$: so that

$$T^2 = \pm 1. \tag{7.142}$$

In the case that $T^2 = +1$, the time-reversal operation is just that of complex conjugation[††]: when $T^2 = -1$, it is not. Our next task, therefore, is to establish the physical circumstances corresponding to these two solutions.

The simple physical operators with which we are concerned may be divided into two classes; those, like ordinary coordinate operators, that involve quantities in even powers of time; others, typified by linear and angular momentum operators, contain odd powers. Using the definitions (7.124), the first group commute with the time-reversal operator while the second group anticommute:

$$\mathbf{r}T = T\mathbf{r}, \quad \mathbf{p}T = -T\mathbf{p}. \tag{7.143}$$

We use these commutation relationships to establish analogous rules for the unitary operator U of (7.137). Thus, ignoring spin for the moment, we substitute (7.138) into (7.143) to get

$$\mathbf{r}UK = UK\mathbf{r}, \quad \mathbf{p}UK = -UK\mathbf{p},$$
$$\mathbf{r}U = UK\mathbf{r}K^{-1}, \quad \mathbf{p}U = -UK\mathbf{p}K^{-1}. \tag{7.144}$$

In a representation where $\mathbf{r}$ is real and $\mathbf{p}$ pure imaginary, we have

$$K\mathbf{r}K^{-1} = \mathbf{r}, \quad K\mathbf{p}K^{-1} = -\mathbf{p}, \tag{7.145}$$

so that

$$\mathbf{r}U = U\mathbf{r}, \quad \mathbf{p}U = U\mathbf{p}. \tag{7.146}$$

In the absence of spin, then, U commutes with all the dynamical variables and must therefore be a constant: we are free to choose it as unity. We therefore confirm that in the absence of spin and in the $\mathbf{r}$ representation,

---

[††] This is true only when complex conjugation is defined in the $\mathbf{r}$ representation: that is, when the pure imaginary i appears out in the open rather than being implicit within the base functions.[17]

T may be identified with complex conjugation K: and, of course, $T^2 = 1$ in this case. The alternative solution to (7.142) arises when we consider spin.

7.5.5 The relevance of spin

The time-reversal operator was defined at the outset to reverse all velocities and momenta: and that includes spin which, like spatial linear and angular momenta, therefore anticommutes with the time-reversal operator:

$$T\mathbf{s}T^\dagger = -\mathbf{s}, \quad T\mathbf{s} = -\mathbf{s}T. \tag{7.147}$$

Using (7.138) again, we find

$$\mathbf{s}UK = -UK\mathbf{s} \tag{7.148}$$

and hence

$$\mathbf{s}U = -UK\mathbf{s}K^{-1}, \tag{7.149}$$

whence

$$\mathbf{s}U = -U\mathbf{s}^*, \tag{7.150}$$

where the conjugate operator $\mathbf{s}^*$ is defined by the transformation[17,19]

$$\mathbf{s}^* = K\mathbf{s}K^{-1}. \tag{7.151}$$

Thus U does not commute with spin as it did with the spatial variables in (7.146): it cannot therefore be written as a constant and T is no longer to be identified with K. Now, in the usual $\{\mathbf{r}, s_z\}$ representation, s_x and s_z are real while s_y is pure imaginary as we recall from the Pauli σ matrices in (7.27). Therefore, U must anticommute with s_x and s_z but commute with s_y. The Pauli matrix σ_y has just these properties: for example,

$$s_x\sigma_y = \frac{2}{\hbar}\begin{pmatrix} 0 & 1 \\ 1 & 0 \end{pmatrix}\begin{pmatrix} 0 & -i \\ i & 0 \end{pmatrix} = \frac{2}{\hbar}\begin{pmatrix} i & 0 \\ 0 & -i \end{pmatrix}, \tag{7.152}$$

while

$$\sigma_y s_x = \frac{2}{\hbar}\begin{pmatrix} 0 & -i \\ i & 0 \end{pmatrix}\begin{pmatrix} 0 & 1 \\ 1 & 0 \end{pmatrix} = \frac{2}{\hbar}\begin{pmatrix} -i & 0 \\ 0 & i \end{pmatrix} = -s_x\sigma_y. \tag{7.153}$$

We may therefore take U as the matrix σ_y but it is even better to take it as $i\sigma_y$ which has the further property of commuting with K. Altogether then, in the usual $\{\mathbf{r}, s_z\}$ representation of wave mechanics, the time-reversal operator for a single electron can be represented by

$$T = i\sigma_y K. \tag{7.154}$$

Because $(\sigma_y)^2$ equals the unit matrix and $i\sigma_y$ commutes with K, the square of T is given by

$$T^2 = (i\sigma_y K)^2 = (i\sigma_y)^2 K^2 = -1, \tag{7.155}$$

and so the alternative solution to (7.142) is seen to be associated directly with the electron spin. The application of the time-reversal operator in an n-electron system involves the action on n-fold orbital products by the product operator

$$T = \prod_{k=1}^{n} T_k \qquad (7.156)$$

whose square is ± 1 according as n is even or odd. This result all but establishes Kramers' theorem.

7.5.6 Degeneracy

The realness of the Hamiltonian is a symmetry with interesting consequences. If ψ is an eigenfunction of the Schrödinger equation

$$H\psi = E\psi, \qquad (7.157)$$

so also is its time-conjugate function $(T\psi)$ because T is a symmetry operator and so commutes with H:

$$HT = TH \qquad (7.158)$$

and hence,

$$H(T\psi) = TH\psi = TE\psi = E(T\psi). \qquad (7.159)$$

Whether the fact that ψ and $T\psi$ are eigenfunctions of the Schrödinger equation with the same energy implies that all eigenfunctions are at least two-fold degenerate depends on whether or not these functions are independent, by which is meant orthogonal. Two cases arise depending upon whether ψ and $T\psi$ are in reciprocal correspondence with each other or not; that is, upon whether

$$T(T\psi) = \psi \qquad (7.160)$$

or not. If (7.160) holds, $T^2 = +1$ and we may take $T = K$ so that $T\psi = \psi^*$. Testing for orthogonality, we find

$$\int (T\psi)^* \psi d\tau = \int (\psi^*)^* \psi d\tau = \int \psi^2 d\tau, \qquad (7.161)$$

which does not vanish except for the trivial case that $\psi = 0$ everywhere. Hence $T\psi$ and ψ are not orthogonal and may be related to one another by some phase factor of unit modulus,

$$\psi^* = e^{i\alpha}\psi, \qquad (7.162)$$

where a is some real number. The case $T^2 = +1$ is thus without too much interest, except to note that since $T\psi$ and ψ are in reciprocal correspondence, it is always possible[19] to define *real* vectors and hence a real representation.

When $T^2 = -1$, $T\psi$ and ψ are orthogonal: the proof is simple. We require one short word of introduction, however: because confusion can result[17,19] between the two types of conjugation – T and K – with respect to the conventional Dirac bra and ket functions, we introduce the notation[17] $(A\psi, B\phi)$ to represent a scalar product in which the operators A and B act to their immediate right only. So: for any two functions ψ and ϕ, the scalar product

$$(T\psi, T\phi) = (UK\psi, UK\phi) = (K\psi, K\phi), \qquad (7.163a)$$

because U is unitary,

$$= (\psi, \phi)^* = (\phi, \psi). \qquad (7.163b)$$

Testing for the orthogonality of $T\psi$ and ψ when $T^2 = -1$, we have

$$(T\psi, \psi) = (T\psi, T^2\psi), \qquad (7.164a)$$

using (7.163),

$$= (T\psi, -\psi) = -(T\psi, \psi), \qquad (7.164b)$$

which can only be true if $(T\psi, \psi) = 0$. So in this case $T\psi$ and ψ *are* independent functions and describe a two-fold degeneracy of the Schrodinger eigenfunctions. This will be true for all solutions of an odd-electron system, but we complete the formalities by stating that if ϕ is orthogonal to two time-conjugate vectors ψ and $T\psi$, then so is its own time conjugate $T\phi$. This easily-proved result establishes that the whole basis of an odd-electron system comprises pairs of time-conjugate vectors. It also follows – indeed directly from $T^2 = -1$ – that no real vectors can exist for these systems. The doubly-degenerate eigenvectors of odd-electron systems are called *Kramers' doublets*.

7.5.7 Microreversibility and magnetic fields

We remarked earlier that the fact that time reversal is a symmetry – that T commutes with H – rested upon the even powers of linear operators like **p** appearing in it. It is possible to show[19] generally that if the law of motion of a conservative system is reversible with respect to time, the Hamiltonian is real, and *vice versa*: this is one formulation of the *principle of microreversibility*. It is an assumption, but one that has never been found to be in conflict with experiment, that any quantum system evolving in the absence of external fields satisfies this principle. Whether or not the principle is also obeyed in the presence of applied fields depends on the behaviour of those fields with respect to time reversal. An electrostatic field *is* time invariant as is apparent when we note that the source of such a field can always be viewed as an array of fixed electric charges (see §2.1) which will not be modified by time inversion. This immediately means that Kramers' degeneracy cannot be lifted by electrostatic fields – and that includes ligand fields – so that eigenfunctions of even totally un-symmetrical molecules are always at least doubly degenerate for odd-electron species. On the other hand, the sources of a static magnetic field can always be equivalenced by fixed electric *currents* whose directions *are* reversed by time inversion. As in the classical argument earlier, the Hamiltonian for a system in an applied magnetic field contains odd powers of the linear operators (**p** or **v**) and so it no longer commutes with T. Time reversal is *not* now a symmetry and so imposes no requirement upon the solutions of H that they be degenerate. Another view of the inapplicability of the microreversibility principle to systems in external magnetic fields is provided by the form of the magnetic moment operator in (7.12), for example, which is pure imaginary. In general, therefore, Kramers' degene-racy is removed in the presence of an external magnetic field. A more positive statement, perhaps, is that a Kramers' degeneracy can be removed *only* by an external *magnetic* field. Note that an internally produced magnetic field will not do this, for the reversal of the internal momenta on time inversion automatically reverses the field. Similarly, spin-orbit coupling is even under time reversal, being a product of two 'time-odd' operators.

7.5.8 Some properties of time-conjugate vectors

Some useful theorems[17] follow in the case that $T^2 = -1$:

(i) A 'time-even' operator – Coulomb, spin-orbit or ligand-field effects being examples of particular interest to us – has no matrix elements between the time conjugates of a Kramers' doublet, because

$$\langle \psi | O | T\psi \rangle = (\psi, OT\psi) = (TOT\psi, T\psi)$$

$$= -(TOT^{-1}\psi, T\psi) = -(O^{\dagger}\psi, T\psi) = -\langle \psi | O | T\psi \rangle = 0. \quad (7.165)$$

(ii) A time-even operator has the same 'diagonal' expectation value in each component of a time-conjugate pair of states, because

$$\langle \psi | O | \psi \rangle = (\psi, O\psi) = (TO\psi, T\psi) = (TOT^{-1}T\psi, T\psi)$$

$$= (O^{\dagger}T\psi, T\psi) = (T\psi, OT\psi) = \langle (T\psi) | O | (T\psi) \rangle. \quad (7.166)$$

When $T^2 = +1$, theorem (i) above is replaced by the statement that a time-odd operator has no matrix elements between a pair of time-conjugate states. This time the proof goes as

$$\langle \psi | O | T\psi \rangle = (\psi, OT\psi) = (TOT\psi, T\psi)$$
$$= + (TOT^{-1}\psi, T\psi) = - (O^{\dagger}\psi, T\psi) = - \langle \psi | O | T\psi \rangle = 0.$$

$$(7.167)$$

Now consider a time-even Hamiltonian H with a non-degenerate eigenstate ψ, necessarily implying $T^2 = +1$. Then the time conjugates $T\psi$ and ψ coincide within a phase factor as in (7.162). But (7.167) shows that matrix elements of a time-odd operator vanish between these functions. Since $T\psi(=\psi^*)$ and ψ differ by only a phase factor, this off-diagonal matrix element is proportional to the diagonal element $\langle \psi | O | \psi \rangle$. Therefore, *all* components of the magnetic moment, involving time-odd operators, vanish in non-degenerate states of time-even Hamiltonians.[17] This is a generalization of Van Vleck's theorem[20] on the quenching of orbital angular momentum. A very brief summary of that theorem is as follows. If ψ is a non-degenerate eigenstate of a spinless Schrödinger equation whose Hamiltonian can necessarily be put into a real form, then it is also real: for otherwise, if ψ were complex, of the form $\phi_1 + i\phi_2$, the reality of H would determine ϕ_1 and ϕ_2 as separate and independent solutions of the Schrödinger equation with the same energy – which is a contradiction. Now the orbital angular momentum operator $l = -i\hbar r \wedge \nabla$ is pure imaginary, so that the expectation value of any of its components over a real ψ is imaginary also. However, l is also hermitian, so that the expectation value of a non-degenerate state must be real. Both conditions can be satisfied only if the expectation value vanishes. We have reviewed the proof merely to point up the relationship between the steps required here and those encountered in the more general analysis.

7.6 Susceptibilities

7.6.1 Van Vleck's diagonal susceptibility equation

We return at last to the subject of para- and diamagnetic susceptibilities which are, of course, the main concern of magnetochemistry. In §3.20 we briefly examined the calculation of susceptibilities within classical mechanics leading to the famous formula of Langevin. Although the approach reproduced many experimental features of dia- and paramagnetism and seemed so right, it rested, as we saw in §3.21, upon a fundamentally false assumption with respect to classical statistical thermodynamics. An immediate triumph of quantum theory, however, is that, by substituting discrete for continuous distributions, the objections of van Leeuwen and others, leading to identically zero susceptibilities in the classical theory, were rendered inapplicable. We may therefore seek to construct a quantum-mechanical equation equivalent to Langevin's formula as a basis for the contemporary calculation of magnetic susceptibilities. The result is widely known as *Van Vleck's susceptibility equation*.

Knowing that the statistical thermodynamic argument of van Leeuwen loses relevance in the quantum treatment, we can base the latter upon the pseudoclassical approach implicit within Langevin's model. Consider a classical model of paramagnetism – leaving diamagnetism till later – in which we envisage an assemblage of molecules each possessing a permanent magnetic dipole moment **m**. In the absence of an externally applied magnetic field, these dipoles will be randomly oriented and the *assemblage* will thus display no corporate magnetic moment **M**. In the presence of an external magnetic field **B**, however, the dipoles tend to align parallel to it and a non-zero magnetic moment **M** of the molecular ensemble now arises out of a balance between this tendency of dipoles to align with **B**, and the exigencies of thermal randomization. Within this simple picture, **M** will vary with temperature, while the intrinsic molecular moments **m**, of course do not. The magnetic energies of individual molecular dipoles in the field vary about the original, zero-field, energy *in a symmetrical way*, ranging from $+ \mathbf{m} \cdot \mathbf{B}$, for a dipole aligned exactly antiparallel to **B**, to $- \mathbf{m} \cdot \mathbf{B}$ for one oriented exactly parallel. Of course, the mean energy of the ensemble is not the same before and after the application of the field; that result depending upon the appropriate Boltzmann weighting of the various dipole orientations. Within the model just described, the maximum and minimum energies available to the molecular magnets are directly proportional to the strength *B* of the applied field. The proportionality constant is equivalent to the slope of a plot of dipole energy versus field strength and is obviously equal to the magnitude *m* of the molecular magnetic dipole. We take this result and the necessarily symmetrical nature of the energy splitting as key features of the quantum treatment that we now examine. Again, for the moment, we confine the discussion to the case of paramagnetism.

In the quantum picture we move from the idea of each molecule in the ensemble possessing the same permanent magnetic moment to one in which a molecule populating state *i* behaves in an external magnetic field as if it acquires a magnetic moment whose magnitude, m_i, is given by the negative of the linear term in an expression relating the energy E_i and the

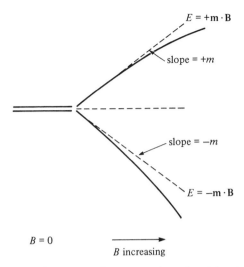

Fig. 7.1. The classical magnetic moment is replaced by the negative of the limiting slope of an energy level in a magnetic field.

field magnitude B. Thus, in figure 7.1, is shown the case of a two-fold level whose degeneracy is raised in an applied field by terms, linear and quadratic (or higher), in the field. This picture and the classical one of the last paragraph are brought into correspondence by equating the slopes of the component energies at near-vanishing field with the constant of proportionality $\mathbf{m}$ referred to earlier. We may therefore refer to the 'magnetic moment of a state' defined by

$$\mathbf{m}_i = -\frac{\partial E_i}{\partial \mathbf{B}}. \tag{7.168}$$

Now we suppose that the energy of each state of the system can be expanded as a power series[††] in the field strength B:

$$E_i = E_i^{(0)} + E_i^{(1)} B + E_i^{(2)} B^2 + \ldots, \tag{7.169}$$

and hence find the equivalent state magnetic moment to have the magnitude:

$$m_i = -E_i^{(1)} - 2E_i^{(2)} B - \ldots. \tag{7.170}$$

As in the classical Langevin treatment, we compute the macroscopic, ensemble magnetic moment (or magnetization) $\mathbf{M}$ by considering the result

†† Griffith[8] points out that this may not always be a physically sensible thing to do; it may, for example, be better to use a power series in $1/B$. He goes on to show that the same result emerges from a proof which does not use the power series expansion.

of a Boltzmann distribution of molecules amongst all possible levels at some temperature T. The magnitude of the magnetization is given as

$$M = N_A \sum_i m_i e^{-E_i/kT} \bigg/ \sum_i e^{-E_i/kT}, \qquad (7.171)$$

using the fundamental probability postulate of statistical thermodynamics: N_A is Avagadro's number, so that M refers to the molar magnetization; and k is the usual Boltzmann constant.

Before substituting the expression (7.170) into (7.171), we use the experimental fact that the sort of magnetic energy shifts illustrated in figure 7.1 are almost always tiny compared with kT and so we simplify the algebra by expanding the exponential series, writing

$$e^{-E_i/kT} = e^{-E_i^{(0)}/kT} e^{-(E_i^{(1)}B + \ldots)/kT} \approx e^{-E_i^{(0)}/kT}(1 - E_i^{(1)}B/kT). \qquad (7.172)$$

The net ensemble magnetization is then

$$M = N_A \frac{\sum_i (- E_i^{(1)} - 2E_i^{(2)}B)(1 - E_i^{(1)}B/kT)e^{-E_i^{(0)}/kT}}{\sum_i e^{-E_i^{(0)}/kT}(1 - E_i^{(1)}B/kT)}, \qquad (7.173)$$

in which terms ultimately leading to powers of B greater than one are dropped. This approximation is made partly in the spirit of the reasonable neglect of small terms but also, as in the expansion of the exponentials in (7.172), by reference to experimental fact. Under normal experimental conditions for paramagnets and before the onset of saturation effects, discussed in §4.4, the induced magnetization $\mathbf{M}$ is linear in the applied field: alternatively, the susceptibility in $\mathbf{M} = \chi\mathbf{H}$ is independent of the field strength. So terms in higher powers of B are discarded.

Next, since energy shifts engendered by typical laboratory fields (say, < 10 tesla) are rarely greater than one wavenumber, we can safely neglect $E_i^{(1)}B/kT$ in the denominator of (7.173). We cannot do so in the numerator, however, as the following evaluation shows. Multiplying out the brackets in the numerator gives

$$M = N_A \frac{\left[- \sum_i E_i^{(1)} e^{-E_i^{(0)}/kT} + \sum_i (E_i^{(1)^2} B/kT - 2E_i^{(2)}B)e^{-E_i^{(0)}/kT} \right]}{\sum_i e^{-E_i^{(0)}/kT}}, \qquad (7.174)$$

and the first term in the numerator, ultimately deriving from the number 1 in the above expansion of the exponential, vanishes identically for the following reason. Consider *any* level, degenerate or not, of the manifold eigenfunctions of the system prior to the introduction of the magnetic field. Regardless of the value of the exponential factor $e^{E_i^{(0)}/kT}$, related to the population of that level, the summation vanishes because $\sum_i E_i^{(1)}$ vanishes and each member of the originally degenerate level is multiplied by the *same* exponential factor involving the *zeroth*-order energy $E_i^{(0)}$. The sum $\sum_i E_i^{(1)}$ vanishes, either because the level is non-degenerate and there is therefore no first-order shift in a magnetic field, as described in §7.5.8; or if the level *is* originally degenerate, because the splitting in the magnetic field must by symmetrical about $E_i^{(0)}$. Sometimes this is explained by arguing that there will be no residual paramagnetism in the whole system on removing the applied field. Alternatively, as here, we appeal to the classical analogue above. So generally the trace $\sum_i E_i^{(1)}$ of a manifold under a magnetic field perturbation is zero and the final expression for the magnetization is

$$M = N_A \frac{\sum_i (E_i^{(1)^2} B/kT - 2E_i^{(2)}B)e^{-E_i^{(0)}/kT}}{\sum_i e^{-E_i^{(0)}/kT}}. \tag{7.175}$$

For an experiment where we may write $\chi = M/H$, the molar susceptibility is given by

$$\chi \approx \mu_0 N_A \frac{\sum_i (E_i^{(1)^2}/kT - 2E_i^{(2)})e^{-E_i^{(0)}/kT}}{\sum_i e^{-E_i^{(0)}/kT}}, \tag{7.176}$$

where, as discussed in §3.17, we have made the approximation (3.114).

We construct Van Vleck's susceptibility equation by identifying the multipliers $E_i^{(0)}, E_i^{(1)}$ and $E_i^{(2)}$ of the power series expansion (7.169) with the corresponding coefficients in a standard perturbation expression. We recall the usual energy expansion for a possibly degenerate level as given by perturbation theory:[3]

$$E_i = E_i^{(0)} + H_{ii}^{(1)} + \sum_j{}' \frac{H_{ij}^{(1)} H_{ji}^{(1)}}{E_i^{(0)} - E_j^{(0)}} + \dots, \tag{7.177}$$

where $H_{ij}^{(1)} \equiv \langle \psi_i^{(0)} | H^{(1)} | \psi_j^{(0)} \rangle$ for the perturbative Hamiltonian $H^{(1)}$ acting within the basis $\{\psi^{(0)}\}$ having eigenvalues $\{E^{(0)}\}$. As usual, it is assumed that appropriate zeroth-order wavefunctions have been constructed such that the first-order term in (7.177) is purely diagonal in a degenerate level: and the primed summation $\sum_j{}'$ implies that j takes all possibilities other than those labelling functions originally degenerate with i. Writing the (paramagnetic) magnetic moment operator as μ for the moment means that the perturbative Hamiltonian $H^{(1)}$ for a system subject to a magnetic field of strength B is given by

$$H^{(1)} = \mu B, \tag{7.178}$$

so that

$$H^{(1)}_{ij} = \langle \psi^{(0)}_i | \mu | \psi^{(0)}_j \rangle B, \tag{7.179a}$$

$$\equiv \langle i|\mu|j \rangle B \tag{7.179b}$$

for short. The perturbation expression (7.177) now takes the form,

$$E_i = E^{(0)}_i + \langle i|\mu|i \rangle B + \sum_j{}' \frac{\langle i|\mu|j \rangle \langle j|\mu|i \rangle}{E^{(0)}_i - E^{(0)}_j} B^2 + \dots, \tag{7.180}$$

and so, by comparison with the original power series (7.169), we have

$$E^{(1)}_i = \langle \psi^{(0)}_i | \mu | \psi^{(0)}_i \rangle, \tag{7.181a}$$

$$E^{(2)}_i = \sum_j{}' \frac{\langle \psi^{(0)}_i | \mu | \psi^{(0)}_j \rangle \langle \psi^{(0)}_j | \mu | \psi^{(0)}_i \rangle}{E^{(0)}_i - E^{(0)}_j}. \tag{7.181b}$$

These two expressions, taken together with the quantum version (7.176) of the Langevin formula, comprise the celebrated Van Vleck susceptibility equation.[20]

7.6.2 Curie's law and TIP

Except in one important respect that we discuss in the following sections, Van Vleck's equation describes the most general behaviour for a paramagnet. Of course, its use involves the evaluation, multiplication and summation of perhaps thousands of matrix elements μ_{ij} and normally one is unable to 'second-guess' the result in any particular case. We are able, however, to comment upon two simple limiting cases corresponding to situations involving energetically isolated ground states which are either degenerate or non-degenerate respectively.

Firstly, we consider an n-fold degenerate ground state so far separated energetically from all excited states that these latter are both insignificantly populated and involve such large values for $(E^{(0)}_i - E^{(0)}_j)$ in (7.181b) that $E^{(2)}_i$ in (7.176) is negligible. The situation before and after the application of a magnetic field is shown schematically in figure 7.2 and corresponds to a simple first-order Zeeman effect: first-order, because second-order terms are negligible by assumption. The expression (7.176) collapses in this limit to

$$\chi = \mu_0 N_A \sum_i^n \langle i|\mu|i \rangle^2 / kT \equiv C/T, \tag{7.182}$$

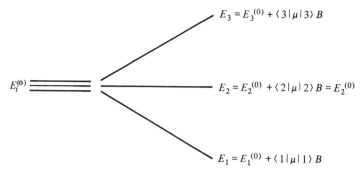

Fig. 7.2. Curie's law and the first-order Zeeman effect.

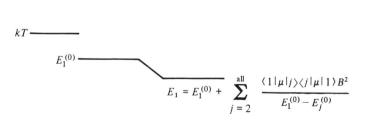

Fig. 7.3. TIP and a pure second-order Zeeman effect.

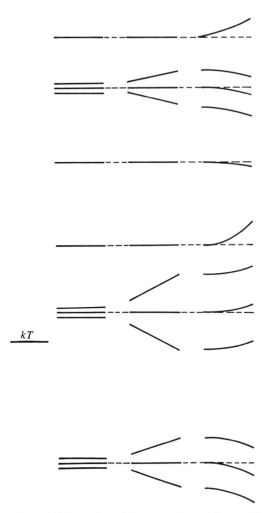

kT

Fig. 7.4. Schematic illustration of the general case: first- and second-order perturbations are shown in the middle and on the right respectively.

which is *Curie's law*. So Curie-law behaviour is characteristic of the first-order Zeeman effect within a single degenerate state.

Next we consider the situation described by a non-degenerate ground state, energetically separated from all other states sufficiently to remove the possibility of any significant population of excited states. Here $\langle 1|\mu|1 \rangle = 0$ because of (7.167) *et seq.* and $e^{E_i^{(0)}/kT} = 1$ *for* $i = 1$ and zero otherwise: therefore, (7.176) collapses to

$$\chi = -\mu_0 N_A(2E_i^{(2)}) = \text{constant.} \tag{7.183}$$

The constant is positive because the denominator in (7.181b) is necessarily negative for $i = 1$ and, of course, the numerator $|\langle i|\mu|j\rangle|^2$ is always positive: this is a general consequence for the ground state in *second*-order perturbation theory. The result (7.183), illustrated schematically in figure 7.3, describes a *para*magnetic situation (because the ground-state energy is lowered by the magnetic field) but one that is independent of temperature – essentially because there exists no manifold of levels in the field within which relative thermal populations may change with temperature. The situation illustrates the phenomenon of *temperature-independent paramagnetism* or TIP: the expression 'high-frequency terms' is also applied[20] to this contribution, referring to the large value of the denominators $(E_i^{(0)} - E_j^{(0)})$ in (7.181b).

More generally, second-order Zeeman terms, which in the case above gave rise to TIP, contribute to all susceptibilities and also there is often significant thermal occupancy of states other than ground. Figure 7.4 shows, once more schematically, the more general case: the first- and second-order Zeeman effects have been separated for heuristic reasons. Note that the second-order perturbation energies for each member of a degenerate level are equal, as follows from the appearance of zeroth-order quantities, uniquely, in (7.181b).

7.6.3 Diamagnetism

We now make brief mention of diamagnetism, represented by the term (7.20b) in the Schrödinger Hamiltonian: brief, because our subject is directed toward ligand-field analysis and, in this connection, diamagnetic effects represent no more than a correction – never negligible, but rarely important – to the leading paramagnetic term. The Hamiltonian operator H^D for the diamagnetic effect is quadratic in the magnetic field and energy shifts ΔE^D described by it take the form

$$\langle\psi_i|H^D|\psi_j\rangle = \frac{e^2}{8m_e}\langle\psi_i|\sum_k^n (x_k^2 + y_k^2)|\psi_j\rangle B_z^2 \tag{7.184}$$

for a field directed along z in an n-electron system. This quadratic term might conveniently be included in the third term of the expansion (7.169) and hence added into the $E_i^{(2)}$ term of (7.181b). It is predominantly a property of all core electrons throughout the system (which is why Pascal's additivity works, of course) and so its overall effect is adequately represented as a uniform increase in energy for all molecular states. Even where contributions from valence electrons become significant, and perhaps strongly anisotropic, the effects are essentially unchanged by permutations of the metal d or f electrons in a complex that give rise to the so-called ligand-field states. Throughout magnetochemical studies of paramagnetism, therefore, the normal procedure is to 'correct' raw susceptibility data for the gross diamagnetism of the system, and perhaps also for its diamagnetic anisotropy,[tt] by reference either to direct measurements on analogous diamagnetic materials or to Pascal's and supplementary constants.[21]

We note that a problem exists with respect to the separation of the various contributions to the quadratic term in the magnetic field expansion (7.169). Thus the diamagnetism just discussed and the 'high-frequency' paramagnetic contribution both vary as B^2: further TIP-type terms can arise even in 'non-paramagnetic' molecules and hence, generally, in association with any geometric location throughout a complex. In turn this raises the question of defining an origin with respect to which r in both linear and quadratic terms of (7.11) is defined. Certainly in any *ab initio* or semi-empirical *calculation* of diamagnetic effects, the origin dependence of models using incomplete bases can be important. Further, the definition of a spatial origin is only one aspect of the wider question of the definition of gauge. As discussed in §7.1, we have chosen a description of the uniform magnetic field (7.5) which conforms to the Coulomb gauge; but that was merely a matter of convenience and we could have done otherwise. Difficulties associated with this question of gauge[4,22] arise in connection with the comparison of observed and calculated diamagnetic susceptibilities; with a separation of high-frequency paramagnetism from 'ground-state diamagnetism' in diamagnets; and with possible relationships between components of diamagnetism, on the one hand, and electric susceptibilities on the other. Fortunately, none of these matters appears to be important in the study of paramagnetism within the ligand-field regime and so we do not pursue the business here: there is, however, an important warning for those concerned with diamagnetic materials, studied by susceptibility methods, molecular beams, or nmr spectroscopy.

[tt] Corrections for diamagnetic *anisotropies* are usually only important when paramagnetic anisotropies are very small. A particularly important group of systems in this connection are complexes of d^5 species; and we have noted that experimental correction for shape anisotropy effects can also be important here – see §5.9.

7.6.4 A generalized off-diagonal susceptibility equation

The derivation of Van Vleck's susceptibility equation in §7.6.1 assumed in effect that we were considering a magnetically isotropic system so that magnetization and applied field are coparallel: the last point is particularly relevant in deriving (7.176) from (7.175). As discussed in the Introduction and as prepared for throughout chapters 4–6, the most useful ligand-field studies of paramagnetism by far are those concerned with low-symmetry molecules for which the susceptibility tensor is anisotropic. Our next task, therefore, is to construct a generalized susceptibility equation[23] by which such susceptibility components may be computed. Before doing so, however, it is instructive to recall a little history which helps to place the study of magnetic anisotropy into context.

As described in chapter 1, the path toward crystal- and, later, ligand-field theory was via the magnetic properties of transition-metal complexes. Very early on in the story, the subject could have floundered on the question of low-symmetry ligand fields. Many magnetic criteria and diagnostics were based upon the nature of the ground terms left in octahedral (or tetrahedral) fields. The question arose, however, as to whether these simple but important rules would disappear once one took into account the energy splittings that must arise in most real complexes which, either because of low crystal symmetry or because of more complicated chemical ligation, are subject to small low-symmetry fields superimposed upon a predominantly cubic one. The 'extra' perturbations might be small from a chemical viewpoint but are expected to be very large compared with the magnetic or Zeeman effects. However, Van Vleck proved a theorem[20,8] to show that, so far as *average* susceptibilities are concerned, the effects of such low-symmetry fields are quite unimportant, provided that the ground-state energy splittings caused by them are fairly small compared with kT. This regenerates faith in the broad theoretical predictions of the day and, one way or another, the details of magnetic *anisotropies*, which are certainly not insensitive to low-symmetry fields, came to be regarded more as minutiae, providing a testing ground for detailed theory perhaps, rather than as being indirectly, but nevertheless richly, informative about the nature of the ligand field. A central theme of this book, of course, is to show how an inversion of this attitude, combined with recent advances in ligand-field theory itself, now bears fruit in a way the chemist can instantly comprehend. Let us therefore return to the fray.

We recognize that Van Vleck's susceptibility equation is, by its nature, diagonal. When computing principal molecular susceptibilities – that is, for a mole (say) of parallel oriented molecules – **M** and **H** are parallel and (7.176) provides a relationship for χ_x, χ_y, χ_z according as μ in (7.181) is taken as μ_x, μ_y, μ_z respectively. In general, however, **M** and **H** are not coparallel and, further, the absence of any special symmetry features precludes any foreknowledge of the directions along which the principal susceptibilities will lie. We therefore seek to construct a susceptibility equation that provides for the calculation of a general susceptibility tensor element χ_{ij} rather than just the diagonal elements as in (7.176). We begin by noting that the expression (7.168), providing the connection between the classical and quantum viewpoints, is actually a vector equation comprising three parts:

$$m_{i\alpha} = -\frac{\partial E_i}{\partial B_\alpha}, \quad \alpha \equiv x, y, z. \tag{7.185}$$

Of course, at this point the simple one-to-one correspondence with the classical theory breaks down unless one wriggles to contrive a classical picture of anisotropic molecular magnets. We are working in the quantum world here and simply turn our backs on that sort of problem. The component equations (7.185) are readily comprehensible in terms of the splittings in figure 7.1 being different for fields applied in different directions when the system represented by the degenerate wavefunctions generally has different properties – electron densities, potentials – along different directions. Correspondingly, the expression (7.171) for the ensemble magnetization has three components,

$$M_\alpha = -N_A \sum_i \frac{\partial E_i}{\partial B_\alpha} e^{-E_i/kT} \bigg/ \sum_i e^{-E_i/kT} \tag{7.186}$$

and we find, as expected, that because there is no general reason to suppose that the ratios between components of **M** are equal to those between components of **B**, the vector **M** will not usually lie parallel to **B**. As in (7.185), and throughout the present section, Greek letters α and β are used to represent *fixed, but arbitrarily chosen*, cartesian directions $x, y,$ or z.

In constructing a perturbation expression equivalent to (7.177), we consider the energy of the ith wavefunction in a magnetic field **B** whose orientation is taken to be quite general: the components of **B** along the chosen reference directions are B_α. As usual, we imagine having formed zeroth-order linear combinations of an original basis of the states degenerate with i, if appropriate, such that the manifold is diagonal with respect to the perturbation, $\boldsymbol{\mu} \cdot \mathbf{B} \equiv \mu_\alpha B_\alpha$ – where we use the summation convention of §4.1 here *for Greek suffixes only*. Then, using this summation convention once again,

$$E_i = E_i^{(0)} + \langle i|\mu_\alpha|i \rangle B_\alpha + \sum_j \frac{\langle i|\mu_\alpha|j \rangle \langle j|\mu_\beta|i \rangle B_\alpha B_\beta}{E_i^{(0)} - E_j^{(0)}}. \tag{7.187}$$

Differentiating (7.187) with respect to the components of the magnetic field yields,

$$\frac{\partial E_i}{\partial B_\alpha} = \langle i|\mu_\alpha|i \rangle + \sum_j{}' \frac{\langle i|\mu_\alpha|j \rangle \langle j|\mu_\beta|i \rangle B_\beta + \langle i|\mu_\beta|j \rangle \langle j|\mu_\alpha|i \rangle B_\beta}{E_i^{(0)} - E_j^{(0)}} + \dots. \tag{7.188}$$

As in the derivation of Van Vleck's equation, we consider magnetic perturbation energies to be small compared with kT and so expand the exponentials in (7.186) as

$$e^{-E_i/kT} \approx e^{-E_i^{(0)}/kT} (1 - \langle i|\mu_\alpha|i \rangle B_\alpha/kT). \tag{7.189}$$

Noting once again that paramagnets display no residual magnetization at zero field, substitution of (7.189) and (7.188) into (7.186) yields the generalized form of (7.175):

$$M_\alpha = N_A \sum_i \left[\langle i|\mu_\alpha|i \rangle \langle i|\mu_\beta|i \rangle B_\beta/kT - \sum_j{}' \frac{\langle i|\mu_\alpha|j \rangle \langle j|\mu_\beta|i \rangle + \langle i|\mu_\beta|j \rangle \langle j|\mu_\alpha|i \rangle) B_\beta}{E_i^{(0)} - E_j^{(0)}} \right]$$

$$\times e^{-E_i^{(0)}/kT} \Big/ \sum_i e^{-E_i^{(0)}/kT}. \tag{7.190}$$

To emphasize: (7.190) provides an expression for the components of the ensemble magnetization when the basis functions are diagonal with respect to the magnetic perturbation deriving from the total applied field **B**, whose orientation is fixed but arbitrary.

Recognizing that the choice of basis is a function of field direction, ω, (7.190) will be expressed in the form

$$M_\alpha = A_{\alpha\beta}(\omega) B_\beta. \tag{7.191}$$

Now, in the derivation of Van Vleck's susceptibility equation, the neglect of terms in higher powers of B was made because it guaranteed that the general expansion (4.15) of the magnetization vector in terms of the field vector terminates after the linear term, so establishing χ in (7.176) as a field-independent quantity. The same neglect of quadratic and higher terms in the field was made in the construction of (7.190) for it is obviously true that, without that neglect, the magnetization would not be linear in the field and, ultimately, that field-independent susceptibilities would not be calculated. However, the form of (7.191) apparently suggests that the magnetization is not simply linear in the field, if only because of the dependence of $A_{\alpha\beta}$ upon the field *orientation* ω. *Pro tem*, therefore, we introduce the further *assumption* that the calculated magnetization *is* linear in the field and is hence related to the magnetic field by a second-rank, field-independent susceptibility tensor. This tactic may be justified by reference to experiment in that such is the phenomenological norm: however, we justify the assumption analytically *a posteriori*. From the *ansatz* $M_\alpha = \chi_{\alpha\beta} H_\beta$, it follows that

$$\mu_0 A_{\alpha\beta}(\omega) H_\beta = \chi_{\alpha\beta} H_\beta \tag{7.192}$$

for any ω: we note that summations are implicit on each side of this expression. In order to characterize a theoretical susceptibility tensor completely, however, we require an expression for each element $\chi_{\alpha\beta}$ of the tensor rather than just a sum and so we seek conditions ω' under which some *component* $A_{\alpha\beta}(\omega')$ may be identified with some *component* $\chi_{\alpha\beta}$. For this we recognize that the significance of a tensor element $\chi_{\alpha\beta}$ is that it relates the *component* of magnetization in direction α with the *component* of field along β. Therefore, we take the special case of (7.192) in which the field is chosen to lie in direction β (that is, for $\omega = \beta$) when we establish

$$\mu_0 A_{\alpha\beta}(\beta) = \chi_{\alpha\beta} \tag{7.193}$$

as an expression for an individual tensor element. Hence, *for a basis which is diagonal with respect to* μ_β, we find

$$\chi_{\alpha\beta} = \frac{\mu_0 N_A \sum_i \left[\langle i|\mu_\alpha|i\rangle \langle i|\mu_\beta|i\rangle / kT - \sum_j{}' \left(\dfrac{\langle i|\mu_\alpha|j\rangle\langle j|\mu_\beta|i\rangle + \langle i|\mu_\beta|j\rangle\langle j|\mu_\alpha|i\rangle}{E_i^{(0)} - E_j^{(0)}} \right) \right] e^{-E_i^{(0)}/kT}}{\sum_i e^{-E_i^{(0)}/kT}}, \tag{7.194}$$

which we observe to reduce to Van Vleck's expression when applied to a diagonal element of the susceptibility tensor.

It is somewhat clumsy to use this generalized susceptibility equation as it stands, for one must rediagonalize the basis three times – once for each value of the second (say) index in $\chi_{\alpha\beta}$. We therefore seek circumstances in which the basis need not be diagonal with respect to any field direction at all and, indeed, may comprise the functions produced directly by diagonalization of a preceding ligand-field perturbation alone. We begin by considering matrices of μ_α and μ_β within a degenerate manifold which is not diagonal with respect to either. We define matrix elements of $\mathbf{A}'$ as $a'_{ik} \equiv \langle i'|\mu_\alpha|k'\rangle$ and of $\mathbf{B}'$ as $b'_{ik} \equiv \langle i'|\mu_\beta|k'\rangle$. Now there exists some unitary transformation which brings $\mathbf{B}'$ into diagonal form, $\mathbf{B}^D$:

$$\mathbf{B}^D = \mathbf{U}^{-1}\mathbf{B}'\mathbf{U}. \tag{7.195}$$

When applied to $\mathbf{A}'$, the some unitary transformation will not generally yield a diagonal matrix, of course, and we write

$$\mathbf{A} = \mathbf{U}^{-1}\mathbf{A}'\mathbf{U}. \tag{7.196}$$

Let us define

$$\mathbf{C}' = \mathbf{A}'\mathbf{B}' \tag{7.197}$$

and the transform

$$\mathbf{C} = \mathbf{U}^{-1}\mathbf{C}'\mathbf{U}, \tag{7.198}$$

and so find

$$\mathbf{C} = \mathbf{A}\mathbf{B}^{D}. \tag{7.199}$$

As the trace of a hermitian matrix is invariant under a unitary transformation, we have $\mathrm{Tr}(\mathbf{A}\mathbf{B}^{D}) = \mathrm{Tr}(\mathbf{A}'\mathbf{B}')$ and hence

$$\sum_i \sum_k a_{ik} b_{ki} = \sum_i \sum_k a'_{ik} b'_{ki}. \tag{7.200}$$

But $\mathbf{B}^{D}$ is diagonal, so that the left-hand side of (7.200) vanishes except for $k = i$, and thus

$$\sum_i a_{ii} b_{ii} = \sum_i \sum_j a'_{ik} b'_{ki}. \tag{7.201}$$

The significance of this result is that the sum $\sum_i \langle i|\mu_\alpha|i\rangle \langle i|\mu_\beta|i\rangle$, appearing in the first-order term of (7.194), and referring to a basis diagonalized with respect to μ_β, may be replaced by a double sum $\sum_i \sum_k \langle i|\mu_\alpha|k\rangle \langle k|\mu_\beta|i\rangle$ in a *completely general* basis of the degenerate manifold.

Turning now to the second-order Zeeman contribution in (7.194), we wish to show that the sum $\sum_i \sum_j'(\ldots)$ is also invariant to a unitary transformation of the degenerate manifold $\{\phi\}$ containing wavefunction i. The following proof is valid regardless of the degree of completeness of the sum $\sum_j'$. The second-order term comprises a double sum, $\sum_i \sum_j'$: we consider a typical component with respect to j, and hence the sum

$$S = \sum_i [\langle i|\mu_\alpha|j\rangle \langle j|\mu_\beta|i\rangle + \langle i|\mu_\beta|j\rangle \langle j|\mu_\alpha|i\rangle], \tag{7.202}$$

as a rediagonalized basis leaves values of $E^{(0)}$ unchanged. In fact, we need only analyse either term in (7.202) and so focus on $S' \equiv \sum_i \langle i|\mu_\alpha|j\rangle \langle j|\mu_\beta|i\rangle$. We recall that the sum $\sum_i$ is complete over the degenerate manifold that includes level i, and so recognize that the *sum S'* describes a scalar product between two vectors whose components i span the set $\{\phi\}$. By definition, a unitary transformation is one which preserves the magnitude of such scalar products and therefore S', and hence S, is invariant (for each j) under a unitary transformation of the basis $\{\phi\}$.

When we incorporate the invariances of both first- and second-order terms into (7.194), we arrive at our[23] *generalized susceptibility equation*,

$$
\chi_{\alpha\beta} = \frac{\mu_0 N_A \sum_i \left[\sum_j' \langle i|\mu_\alpha|j\rangle \langle j|\mu_\beta|i\rangle /kT - \sum_j'' \left(\dfrac{\langle i|\mu_\alpha|j\rangle \langle j|\mu_\beta|i\rangle + \langle i|\mu_\beta|j\rangle \langle j|\mu_\beta|i\rangle}{E_i^{(0)} - E_j^{(0)}} \right) \right] e^{-E_i^{(0)}/kT}}{\sum_i e^{-E_i^{(0)}/kT}},
$$

$$(7.203)$$

where the basis need not be diagonalized in any particular way: we have changed notation here, slightly, so that $\sum_j'$ means a sum in which j refers to all levels degenerate with i, and $\sum_j''$ means j labels all states which are not degenerate with i. By independent variation of α and β over the reference directions x, y and z, all nine components of the susceptibility tensor may be computed. Actually only six need be evaluated explicitly, for the susceptibility tensor is symmetric such that $\chi_{\alpha\beta} = \chi_{\beta\alpha}$. In §4.5 this symmetry was presented first as an experimental fact and then supported by a simple thermodynamic argument: on the other hand, the theoretical quantum expression (7.203) reveals this symmetry directly. It is interesting to note that the same property was not obvious in (7.194) in which the basis was explicitly diagonal with respect to μ_β.

Finally we return to the question of the assumption prior to (7.192) that our theoretical expressions conformed to the experimentally desired relationship, $M_\alpha = \chi_{\alpha\beta} H_\beta$. Verification of (7.192) would follow if

$$
\sum_j \langle i|\mu_\alpha|i\rangle \langle i|\mu_\beta|i\rangle H_\beta = \sum_i \sum_k \langle i|\mu_\alpha|k\rangle \langle k|\mu_\beta|i\rangle H_\beta \qquad (7.204)
$$

within the degenerate manifold spanned by $\{\phi\}$ and diagonal with respect to $\boldsymbol{\mu}\cdot\mathbf{H}$, the left-hand side deriving from (7.190) and the right-hand side from (7.203). Note, by the way, that the proof of (7.204) is not simply given by (7.201), for summations over $\beta = x$, y, z are implicit throughout (7.204). Now, since the basis was defined to be diagonal with respect to $\mu_\beta H_\beta$ at that stage of the proof, we have

$$
\langle k|\mu_\beta|i\rangle H_\beta = \delta_{ki} \qquad (7.205)
$$

and hence (7.204) follows directly. In other words, (7.192) is consistent with the generalized susceptibility expression (7.203) and hence the calculated expression for $\chi_{\alpha\beta}$ is not dependent upon the field orientation.

For any given temperature T, each independent component of the susceptibility tensor can be evaluated through (7.203). Subsequent numerical diagonalization of that tensor yields values of the principal theoretical susceptibilities and (from the vectors) their orientation with respect to the

fixed, but arbitrarily chosen, cartesian reference frame. Altogether, therefore, given an appropriate ligand-field model that provides the eigenvectors
and energies upon which (7.203) operates, this equation furnishes the
means for computing the magnitudes, orientations and temperature
dependencies of the principal magnetic susceptibilities of molecules
possessing any symmetry or, indeed, none whatever. The practical procedure for doing this is generally very lengthy, of course, each matrix element
$\langle i|\mu_\alpha|j\rangle$ having the form

$$\mu_{ij}^\alpha \equiv \left\langle \sum_p^m a_{ip}\phi_p \left| \sum_t^n (kl_\alpha + 2s_\alpha)_t \right| \sum_q^n a_{jq}\phi_q \right\rangle \hbar^{-1}\beta_e$$

$$= \sum_p^m \sum_q^m a_{ip}^* a_{jq} \left\langle \phi_p \left| \sum_t^n (kl_\alpha + 2s_\alpha)_t \right| \phi_q \right\rangle \hbar^{-1}\beta_e, \qquad (7.206)$$

where the functions i, j,..., produced by the ligand-field model are
expressed as linear combinations of some basic set of m kets $\{\phi\}$, often
in the form $|J, M_J\rangle$; and the operators for spin and orbital angular
momenta are summed over the n electrons of the basis; k is an orbital
reduction factor, first introduced by Stevens and discussed further in
chapter 11. The technical business of evaluating the components of such
matrix elements is described in chapter 8 and tactics for the construction
of an efficient computer program for the computation of $\chi_{\alpha\beta}$ via (7.203),
and all that follows, are discussed in chapter 10.

7.6.5 Saturation

The phenomenon of magnetic saturation was introduced in §4.4: by
definition it is associated with the non-linearity of the field-magnetization
relationship. As discussed there, it seems more direct to consider
magnetization itself under these circumstances rather than construct a
derivative quantity of doubtful utility like a field-dependent susceptibility.
In the saturation regime, which one expects to be of importance only with
larger magnetic fields at the lowest temperatures, we may no longer make
the approximations which characterized the derivation of the susceptibility
equations. Exponentials must not be approximated as in (7.172) nor terms
in higher powers of the field neglected. Instead, we must fall back on the
thermodynamic defining equation for magnetization

$$M_\alpha = -N_A \sum_i \frac{\partial E_i}{\partial B_\alpha} e^{-E_i/kT} \Big/ \sum_i e^{-E_i/kT}. \qquad (7.207)$$

A calculation of such a component of magnetization may proceed along the following lines. The energies E_i that we require in the exponential functions are determined by diagonalization of the model basis under all 'ligand-field' perturbations (Coulomb interelectron repulsion, the ligand field proper, and spin-orbit coupling) as usual, *plus* the magnetic perturbation arising from a field of strength B_α directed along the chosen direction α. Let the corresponding eigenfunctions resulting from this process be labelled $\{\Psi_i\}$. The differentials $\{\partial E_i/\partial B_\alpha\}$ we require in (7.207) are obtained as the slopes of the functions $\{\Psi_i\}$ rather than of the wavefunctions prior to the perturbation by the ligand field etc. that we require. This means that the final result will depend on the magnitude of the field appearing in the diagonalization process: that is quite proper, of course, for we recognize the saturation regime here. We compute[††] the requisite differentials as the diagonal matrix elements $\langle \Psi_i|\mu_\alpha|\Psi_i \rangle$ which are the slopes of the quantities $\langle \Psi_i|\mu_\alpha|\Psi_i \rangle B_\alpha$: note that we do *not* imply the summation of α over x, y, z on this particular occasion! Substitution of these differentials, together with exponentials accurately calculated using the $\{E_i\}$ from the diagonalization procedure, into (7.207) will provide the required theoretical value for the magnetization parallel to the applied field. The same process must be repeated at a variety of temperatures for different field strengths and for fields applied in different directions. This is obviously a time-consuming task, not least because of the new diagonalization required for each field strength and orientation. Some compensation for all this effort is that the resulting set of magnetization curves appears to provide a more exacting test of the same ligand-field model by requiring the reproduction of more independent pieces of experimental data. Experience is limited in this area, however, and it may turn out that this advantage is illusory. Much more serious an objection – and the main reason why the present section is so brief – is that, under the conditions leading to magnetic saturation effects, it is very common to observe the onset of intermolecular exchange effects which, as remarked in §5.4.2, are frequently exaggerated manifestations of chemically uninteresting, low-energy processes. While these may well be of interest to those specializing in exchange phenomena, this book is concerned with the utility of paramagnetism in ligand-field analysis: from our point of

---

[††] From a computer programming point of view, the required matrix elements would not be evaluated in quite this way, for all necessary components would have been constructed already for the previous diagonalization. Instead, one merely needs to apply the unitary transformation corresponding to the total diagonalization, to the matrix of μ_α in the original basis.

view, therefore, we simply observe that clear-cut saturation effects are too easily obscured by irrelevant exchange effects which are not easily 'corrected for'. Finally, before leaving the subject, we note that the calculation of a mean magnetization under saturation conditions does not proceed simply by forming the average of three 'principal' magnetizations. When magnetization and applied field are no longer related by a simple second-rank tensor, the mean magnetization of a sample of randomly oriented crystallites is no longer proportional to the trace of a susceptibility matrix. Instead, one is obliged to compute the magnetization over all directions and then average numerically.

7.7 Esr g values for Kramers' doublets

Consider an orbitally non-degenerate spin-doublet level characterized by $s = \frac{1}{2}, m_s = \pm\frac{1}{2}$: that is, a system with angular momentum properties equivalent to those of a free electron. The energy of the upper state in a magnetic field B_z is given, according to (7.21a), by

$$\langle\tfrac{1}{2}|H^{(1)}|\tfrac{1}{2}\rangle \equiv \langle\tfrac{1}{2}|\mu_z B_z|\tfrac{1}{2}\rangle \equiv \langle\tfrac{1}{2}|\hbar^{-1}\beta_e B g_e s|\tfrac{1}{2}\rangle = \tfrac{1}{2}g_e\beta_e B \qquad (7.208)$$

and of the lower state, by $-\frac{1}{2}g_e\beta_e B$. The splitting of these levels is therefore $g_e\beta_e B$, which is approximately $2\beta_e B$. By analogy, there has arisen the convention of defining the splitting, caused by a magnetic field, of any two adjacent levels arising from a degenerate manifold as $g\beta_e B$: g is the familiar g value determined by electron spin resonance spectroscopy. That technique is amply documented elsewhere[17,24] and we confine our attention here to the calculation of g values according to the philosophy and techniques of §7.6.

We construct the quantity D denoting the sum of the squares of the matrix elements of a degenerate manifold, diagonal in $\boldsymbol{\mu}\cdot\mathbf{B}$:

$$D = \sum_i (\langle i|\mu_\alpha|i\rangle B_\alpha)^2, \qquad (7.209)$$

noting the use of the summation convention for Greek suffixes once more. For a doubly degenerate level we therefore find that $D = \frac{1}{2}g^2\beta_e^2 B^2$ and, for a triplet, $D = 2g^2\beta_e^2 B^2$, etc. We may rewrite (7.209) in the form

$$D = d_\beta B_\beta, \qquad (7.210)$$

where

$$d_\beta = \sum_i \langle i|\mu_\alpha|i\rangle\langle i|\mu_\beta|i\rangle B_\alpha. \qquad (7.211)$$

But d_β involves exactly that sum on the left-hand side of (7.204) and so we may describe the relationship between d_β and B_β by

$$d_\beta = T_{\alpha\beta} B_\alpha, \tag{7.212}$$

where

$$T_{\alpha\beta} = \sum_{i'} \sum_{k'} \langle i' | \mu_\alpha | k' \rangle \langle k' | \mu_\beta | i' |. \tag{7.213}$$

Hence, we find

$$D = T_{\alpha\beta} B_\alpha B_\beta, \tag{7.214}$$

where, as demonstrated in §7.6.4, **T** is a symmetrical, second-rank tensor which can always be brought into diagonal form by a suitable choice of axis frame. We can then identify **T** as proportional to a $\mathbf{g}^2$ tensor, provided the splitting of a degenerate manifold is perfectly equal in the magnetic field. This might not be the case for high multiplicities but is trivially true for the most important case of isolated doublets. This *is* the most important case for us because, as emphasized repeatedly, we are concerned with *real* systems which usually possess little or no strict spatial symmetry. Esr signals are normally only observed for odd-electron systems, so we expect the usual ground states we encounter to be Kramers' doublets. In those circumstances the principal values of **T**, calculated directly from the same quantities accumulated in a corresponding susceptibility computation, are equal to $\frac{1}{2}\beta_e^2$ times the squares of the principal g values. Clearly, the procedure does not provide for the calculation of the signs of the principal g values but only their absolute magnitudes and, of course, their orientations. In low-symmetry molecules, $g_{\alpha\beta}$ may not be equal to $g_{\beta\alpha}$ and one may not then be able to diagonalize the g property.[17] The $\mathbf{g}^2$ tensor *is* symmetric and can always be diagonalized to give principal values but at the cost of sign ambiguities as we have seen.

References

[1] Margenau, H. & Murphy, G.M., *The Mathematics of Physics and Chemistry*, Van Nostrand, Princeton, New Jersey, 2nd. edn, 1956.
[2] Matson, F.A., *Adv. Quant. Chem.*, **1**, 59 (1964).
[3] Eyring, H., Walter, J. & Kimball, G.E. *Quantum Chemistry*, Wiley, New York, 1944.
[4] Davies, D.W., *The Theory of the Electric and Magnetic Properties of Molecules*, Wiley, London, 1967.
[5] Landau, L.D. & Lifshitz, E.M., *Quantum Mechanics*, vol. 3. Pergamon Press, Oxford, 2nd edn, 1965.
[6] Roman, P., *Advanced Quantum Theory*, Addison-Wesley, Reading, Mass., 1965.

[7] Avery, J.S., *The Quantum Theory of Atoms, Molecules and Photons*, McGraw-Hill, London, 1972.

[8] Griffith, J.S., *The Theory of Transition Metal Ions*, Cambridge University Press, London, 1961.

[9] Feynman, R.P., *The Feynman Lectures on Physics*, vol. 1, Addison-Wesley, Reading, Mass., 1963.

[10] Dirac, P.A.M., *The Principles of Quantum Mechanics*, Oxford University Press, 4th edn, 1958.

[11] Foldy, L.L., & Wouthysen, S.A., *Phys. Rev.*, **78**, 29 (1950).

[12] Bleaney, B.I., & Bleaney, B., *Electricity and Magnetism*, Oxford University Press, 3rd edn, 1976.

[13] Woolley, R.G., *Mol. Phys.* **30**, 649 (1975).

[14] Condon, E.U. & Shortley, G.H., *The Theory of Atomic Spectra*, Cambridge University Press, 1935.

[15] McWeeny, R. & Sutcliffe, B.T., *Methods of Molecular Quantum Mechanics* Academic Press, London, 1969.

[16] Kramers, H.A., *Koninkl. Ned. Akad. Wetenschap., Proc.*, **33**, 959 (1930).

[17] Abragam, A. & Bleaney, B., *Electron Paramagnetic Resonance of Transition Metal Ions*, Oxford University Press, 1970.

[18] Tinkham, M., *Group Theory and Quantum Mechanics*, McGraw-Hill, 1964.

[19] Messiah, A., *Quantum Mechanics*, North-Holland, 1961.

[20] Van Vleck, J.H., *The Theory of Electric and Magnetic Susceptibilities*, Oxford University Press, 1932.

[21] König, E., *Tables of Magnetic Susceptibilities*, Landolt-Börnstein, K.H. Hellwege & A.M. Hellwege, eds., Springer-Verlag, Berlin, 1965.

[22] Buckingham, A.D., *Quart. Rev.*, **13**, 183 (1959).

[23] Gerloch, M. & McMeeking, R.F., *J. Chem. Soc. Dalton. Trans.*, p. 2443, (1975).

[24] for example; Carrington, A. & McLauchlan, A.D., *Introduction to Magnetic Resonance*, Harper and Row, New York, 1967, and references therein.

PART III

——

LIGAND-FIELD THEORY

The derivation of the susceptibility equation and of corresponding expressions for esr g values in chapter 7 completed our discussions of paramagnetism. Those formulae express molecular magnetic properties in terms of the eigenvalues and eigenvectors of the system prior to the application of the magnetic field and so indicate the way in which the measurement of magnetic properties can probe the electronic structure of transition-metal complexes. Given the aim of discovering something of the bonding and electron distribution in complexes, magnetochemistry is inevitably as concerned with the electronic structure of molecules *before* the application of magnetic fields as after. We consider now, therefore, the group of models most commonly used to describe transition-metal molecular states, collectively called ligand-field theory. The susceptibility and other expressions in chapter 7 are equally appropriate for other descriptions of the molecular states, of course, including those produced by one form or other of non-empirical, *ab initio* quantum theory. It is the case, however, that such methods are not yet capable of providing eigenvalues and eigenvectors which are generally of sufficient accuracy for subsequent calculations of magnetic properties, whereas the semi-empiricism of parametric ligand-field models continues to be most rewarding.

The distinguishing character of the ligand-field regime is the recognition of the existence of a set of low-lying excitations in a transition-metal complex that is principally responsible for both the magnetic behaviour and a band of spectral features of weak intensity, and which is associated with electronic states dominated by orbital contributions of metal-ion *d*-electron parentage. Conventional ligand-field theory is confined to an explicit treatment of just the metal *d* electrons and is not, therefore, a comprehensive theory of chemical bonding in metal complexes: nevertheless, the hope is that we shall be able to extract chemically useful information about metal–ligand bonding from the ligand-field parametrization. Ligand-field theory is concerned, then, with eigensolutions of an effective Hamiltonian $\mathcal{H}'_{LF}$, spanning an energy range of a few volts above the ground state,

$$\mathcal{H}'_{LF} = \sum_{i<j}^{N} U(i,j) + \sum_{i}^{N} V_{LF}(\mathbf{r}_i) + \zeta \sum_{i}^{N} \mathbf{l}_i \cdot \mathbf{s}_i, \qquad \text{(III.1)}$$

referring, respectively, to interelectron repulsion terms, the ligand-field potential, and to spin-orbit coupling. This effective operator acts within a basis of N orbitals which are defined as *d* orbitals for transition-metal complexes or *f*, for lanthanides. It is of central importance to this book, and to the concept of ligand-field analysis, that ligand-field theory and the use of (III.1) be properly understood and recognized as a relevant and valid procedure within chemistry and bonding theory. Frequently, ligand-field texts are built upon an expression like (III.1) but have little or no space devoted to its theoretical justification within quantum chemistry at large or, indeed, to the theoretical notion of an effective operator and orbital basis. This omission is important even for those with no taste for theoretical structure for, without an *explicit definition* of the ligand-field *ansatz*, there can be little confidence in the validity of the concomitant parametrization scheme. It is crucial to meet, discuss and ultimately refute the widely-held charge that, while ligand-field models may provide convenient, and perhaps clever, means of parametrizing some spectral and magnetic properties, they have little rigorous connection with proper theory or with the concepts and quantities used elsewhere in chemistry. It would therefore seem logical to begin this section with a detailed analysis of this seminal issue but, partly because of the need for much forward reference and partly because the discussion sits best against a textured view of the subject as practised, a different development has been chosen. We reserve this important matter to the last chapter of the present section, remarking meanwhile that a satisfactory and explicit definition of ligand-field theory in general, and of the angular overlap model in particular, *can* be given;[1−3] so that the more technical and procedural matters arising with the use of (III.1) may be reviewed with confidence.

As indicated in chapter 1, our aim is to derive chemically interesting information from a wide-ranging body of ligand-field properties for virtually any type of single-centre transition-metal or lanthanide complex. In particular, we wish to study complexes with central metals having *any* d^n or f^n configuration, with variable coordination geometry involving different coordination numbers and ligand types. Further, having once established the theoretical technicalities for the appropriate calculations, we do not wish to repeat the effort or tedium such computations involve by having to construct new theoretical relationships as we change $d^n(f^n)$ configuration or coordination geometry: we require a *general* system and, ultimately, a general computer program to go with it. Should we wish to know the effect on some calculated ligand-field property of a change in one of the system parameters, we want the answer in seconds at the push of the proverbial button. If the solution to the problem entails the construction of new theory or the writing of a new computer program, we are unlikely to ask many such questions and hence unlikely to explore parameter variations as thoroughly as we should. We wish to relegate the business of finding parameter fits to the status of a technicality, however beautiful the required theory may be. Therefore, while some mention is made of the more specialized circumstances of relatively high-symmetry molecules, the greater part of the next four chapters is concerned with the most general regime in which the ligand-field properties of real (that is, actual), unsymmetrical molecules may be treated in a uniform way.

Matrix elements $\langle \psi_i | \mathcal{H}'_{LF} | \psi_j \rangle$ of the perturbation $\mathcal{H}'_{LF}$ involve, as usual, both angular and radial parts. The treatment of the angular parts of all three contributions to $\mathcal{H}'_{LF}$ in (III.1), together with those of the orbital and spin angular momentum operators required in the calculation of magnetic properties, forms the subject matter of the next chapter. We review the powerful tensor-operator theory of Racah, Wigner and others in some detail, ending the chapter with a set of 'master equations' upon which the contemporary computing package reviewed in chapter 10 is built. While simpler methods are available for calculating these various angular matrix elements, the effort of learning about the tensor-operator techniques is recompensed many times over when subsequent and repeated applications of their results can be made so simply. In fact the 'master equations' of §8.16 may be used 'blind' if desired, for an understanding of ligand-field theory and analysis does not depend upon the techniques by which the various angular matrix elements are evaluated. However, since this approach will not be to everyone's taste and yet bearing in mind the length of texts specializing on vector coupling, angular momentum and tensor-operator theory, the review given in chapter 8 might be welcome. Two matters have contributed to its relatively short length. Firstly, virtually no aspects of the theory are discussed which do not bear directly upon the ultimate significance and derivation of the 'master equations' required for the particular tasks we encounter in ligand-field analysis. Secondly, the development is restricted to free-ion-type basis functions: in particular, no mention is made of real tensorial sets and to the contributions of Griffith[4] or of Harnung & Schäffer.[5] As will be made plain in chapter 11, where the fundamental character of ligand-field theory is discussed and the use of (III.1) justified, we may reasonably work with basis orbitals specified as of d (or f) type: in which case a preference for a 'strong-field' basis, like $(t_{2g})^n(e_g)^m$, over a 'weak-field' one, employing $|J, M_J\rangle$ eigenkets, can be based only on subjective taste. Further, as is discussed in §10.5, the conventional cubic-based, strong-field functions do not offer a generally more-natural or well-adapted basis for unsymmetrical molecules than does a $|JM_J\rangle$ basis. Whether we work within a complete $d^n(f^n)$ manifold – in which case the choice of a strong- or weak-field basis is essentially irrelevant – or within a symmetry-adapted subset, as described in §10.5, all necessary computations can be carried out equally conveniently within a weak-field $|J, M_J\rangle$ basis.

The radial parts of the matrix elements of $\mathscr{H}'_{LF}$ are parametrized. In the case of the interelectron repulsion and spin-orbit coupling operators, the parameters are 'spherical'; that is, like those of the conventional theory of atomic spectroscopy, except in numerical magnitude, and we refer to effective Condon–Shortly F_k (or Racah B, C) parameters for the interelectron repulsion, and to ζ for the spin-orbit coupling. Only the ligand-field operator V_{LF} carries information on the molecular geometry, and parametrization schemes for this are many and varied. In chapter 1 we discussed reasons for our preference for the angular overlap model of the ligand-field over older, symmetry-based schemes and in chapter 9, we investigate the relationship between the AOM superposition of a ligand-field potential and the more widely-known, multipole expansion. The remainder of the chapter is concerned with the detailed application of the AOM to molecules with any, and indeed no, symmetry, using the angular momentum theory developed in chapter 8. The potentially vexatious problem of the degree of parametrization engendered by AOM analysis is considered in detail in §9.8.

Ligand-field analysis is every bit as practical a matter as a theoretical scheme. Once constructed, if a detailed theoretical and computational package is not easily exploited, the prospect of a general and successful scheme for ligand-field analysis would fade depressingly. Chapter 10, therefore, is devoted to the total analytical package and enters into much of the fine detail of how the various ligand-field properties are calculated and compared with experiment. Whilst there is great stress on computation and organization – and indeed the discussion is illustrated repeatedly with examples from the computer program[6] developed by the author and his colleagues in Cambridge – the chapter is not intended as a program handbook; nor is any program code explicitly provided. Instead there have been two aims: to provide an overview of the total computing procedure so that the reader might be convinced that thorough analysis is now possible, and to suggest ways of reducing computing cost for those who may wish to construct an appropriate system of their own. Indeed, in this chapter is a discussion of how basis functions may be 'geometry adapted' to the system under study so that basis truncation can be achieved with the optimal convenience and accuracy.

Finally, as mentioned here already, the last chapter of Part III summarizes recent work addressed to the nature of ligand-field theory and of the angular overlap model. The material is divided into two parts. In the first, the context of ligand-field theory within quantum chemistry is described within a many-electron framework: and we include here a discussion of the nature of effective operators and of their concomitantly defined bases. The angular overlap model is analysed in the second part within a one-electron basis. Particular emphasis is placed on the distinction between the AOM in ligand-field theory – as used consistently throughout this book – and the molecular-orbital version of the model whose basis has never been properly defined. It is possible, within the former, to provide a theoretical interpretation of the AOM parameters which reproduces many, but not all, of the aspirations originally expressed for the approach. An important strand of the development presented in this part of chapter 11 is the relationship between experimental observables and theoretical quantities, between 'qualitative' and *ab initio* theories, between $10Dq$ and Δ_{oct}; and the significance of molecular orbital diagrams like figure 1.1 within ligand-field theory.

References

[1] Woolley, R.G., *Mol. Phys.*, **42**, 703 (1981).
[2] Gerloch, M., Harding, J.H. & Woolley, R.G., *Structure and Bonding*, **46**, (1981).
[3] Gerloch, M. & Woolley, R.G. *Prog. Inorg. Chem.*, **31**, 371 (1983).
[4] Griffith, J.S., *The Irreducible Tensor Method for Molecular Symmetry Groups*, Prentice-Hall, Englewood Cliffs, New Jersey, 1962.
[5] Harnung, S.E. & Schäffer, C.E., *Structure and Bonding*, **12**, 201, 257 (1972).
[6] 'CAMMAG', a FORTRAN computer program by D.A. Cruse, J.E. Davies, J.H. Harding, M. Gerloch, D.J. Mackey & R.F. McMeeking.

8

Tensor-operator theory

The common currency of any ligand-field and magnetochemical analysis is the evaluation of large numbers of matrix elements of the operators of (III.1) – Coulomb, ligand-field, and spin-orbit coupling – and of the magnetic moment, within an appropriate many-electron basis. Values for all of these quantities may be determined using the simple methods described in many basic texts.[1-3] These methods are characterized by the necessity to decompose *explicitly* the various many-electron wavefunctions into even more products of one-electron functions; and, hence, many-electron matrix elements into large sums of one-electron matrix elements. The processes involved are not uniform; are very lengthy and hence rather subject to manipulative error; and prohibitively clumsy in *f*-electron computations. Tensor-operator theory exploits symmetry with great subtlety to provide an enormously powerful tool by which all these, and other, quantities may be calculated very simply. There is, of course, a trade-off between the effort saved by using the more sophisticated technique and that required to understand and utilize it: introductory texts in ligand-field theory rightly stick to the simple ways but the professional ligand-field analyst, so to speak, makes so many calculations that the use of tensor operators pays off handsomely. Even so, the present chapter is totally self-contained, its purpose being to provide a set of 'master equations' for the calculation of the required matrix elements within appropriate computer programs. The reader who prefers not to work through the present synopsis of recondite tensor-operator theory will find all the answers he requires in §8.16 and Appendix C.

314

8.1 Rotation operators

The way in which various entities transform under coordinate rotations is central to the task of calculating the angular parts of matrix elements of (III.1). Descriptions of coordinate rotations may specify either operations performed on the original entities or the relationships between the result of the rotation and the original entity. We shall describe coordinate rotations in terms of operations. In this respect, as in the general development of our material, we follow the splendid text of B. L. Silver:[4] however, the subject matter of that book is heavily pruned in the present chapter where we have retained only one path towards our own specific goals.

Throughout, we work with right-handed cartesian frames. We define a positive rotation as one involving clockwise motion viewed from the origin and looking in the positive direction of the given axis, a rotation by a positive angle θ about an axis n being written $R_n(\theta)$. It is possible to transform one cartesian frame into another by suitable sequences of rotations $R_n(\theta)$. In figure 8.1 we illustrate one such way of accomplishing the transformation from the initial frame x, y, z to a final frame x''', y''', z'''. The sequence of standard operations shown in the diagram is given as follows:

  (i) Rotate the axes by a positive angle $\alpha\,(0 \leqslant \alpha < 2\pi)$ about the z axis to give a new coordinate frame x', y', z' in which $z' \equiv z$;
  (ii) rotate the frame x', y', z' by a positive angle $\beta\,(0 \leqslant \beta < \pi)$ about the y' axis to give a further new cartesian frame x'', y'', z'' in which $y'' \equiv y'$;
  (iii) rotate x'', y'', z'' by $\gamma\,(0 \leqslant \gamma < 2\pi)$ about z'' to give the final frame x''', y''', z''' in which $z'''' \equiv z''$.

The angles α, β, γ are the so-called *Euler angles* and the sequence of operations just described is written $R_{z''}(\gamma)\,R_{y'}(\beta)\,R_z(\alpha)$, the order indicating that we operate with $R_z(\alpha)$ first, and so on. This sequence does not describe a unique pathway in transforming x, y, z into x''', y''', z''' nor need the process be carried out in three distinct steps: the two frames may be interconverted by a single rotation of one frame about a suitable axis, whose direction will not generally coincide with any of the principal cartesian axes under consideration. The foregoing recipe is one which will always work and, should figure 8.1 be difficult to comprehend visually, the reader may find it helpful to play with stick models.

As a practical consideration, it is worth noting here that calculation of the values of the Euler angles in a given system may not be particularly

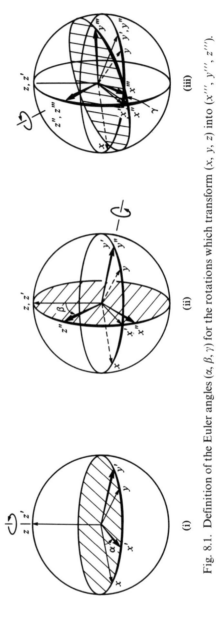

Fig. 8.1. Definition of the Euler angles (α, β, γ) for the rotations which transform (x, y, z) into (x''', y''', z''').

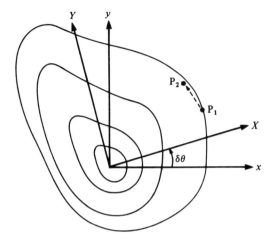

Fig. 8.2. Transformation of the scalar function f under an incremental rotation of the coordinate frame.

simple, so that while the Eulerian operators are especially convenient in the theoretical development which follows, it is worth indicating here that values for α, β, γ can be computed from matrices of direction cosines, like (4.10), which are themselves readily calculable from coordinate lists, for example: the process is described in Appendix B.

We now establish an explicit form for the rotation operators $R_n(\theta)$ and this we do by considering first the nature of the infinitesimal rotation operator. The central tactic is to define the form of the operator by its action upon a function. We take a function $f(x, y, z)$ in a coordinate frame x, y, z and consider its form in a second frame X, Y, Z obtained by rotation of the original axes by an infinitesimal positive angle about the z axis. We consider such infinitesimal rotations first in view of the presumed continuously varying nature of the function f. Let us illustrate the function and the reference frames within two dimensions, as in figure 8.2. We seek the form of $F(X, Y, Z)$, the function referred to the new coordinate frame X, Y, Z.

We choose some arbitrary point P_1 in the original system which moves, with the coordinate frame as it rotates, to the point P_2. The coordinates of P_1 are (x_1, y_1, z_1) and (X_1, Y_1, Z_1) in the old and new frames respectively. Similarly, the coordinates of P_2 are (x_2, y_2, z_2) and (X_2, Y_2, Z_2). By construction, the set of numbers $\{x_1, y_1, z_1\}$ is identical to the set of numbers $\{X_2, Y_2, Z_2\}$,

$$\{x_1, y_1, z_1\} = \{X_2, Y_2, Z_2\}. \tag{8.1}$$

As the magnitude of a function at any given point cannot depend upon the choice of coordinate frame (that is, we are dealing with a scalar field in f), the value of f at P_2 is identical with the value of F at the same point;

$$f(x_2, y_2, z_2) = F(X_2, Y_2, Z_2). \tag{8.2}$$

Since $F(X_2, Y_2, Z_2) = F(x_1, y_1, z_1)$ from (8.1), we therefore deduce from (8.2) that

$$f(x_2, y_2, z_2) = F(x_1, y_1, z_1). \tag{8.3}$$

In seeking the form of the operator which changes f into F, we must look at the geometrical relationships between the points P_1 and P_2 in some detail.

Referring to figure 8.3, we have $OP_1 = OP_2$ for the points terminating the arc subtending the small rotation $\delta\theta$ and so,

$$\begin{aligned}
OA_2/OP_1 &= OA_2/OP_2 \\
&= \cos(\theta + \delta\theta) \\
&= \cos\theta\cos\delta\theta - \sin\theta\sin\delta\theta \\
&= (OA_1/OP_1)\cos\delta\theta - (A_1P_1/OP_1)\sin\delta\theta \\
&= (x_1/OP_1)\cos\delta\theta - (y_1/OP_1)\sin\delta\theta.
\end{aligned}$$

Now, as $\delta\theta \to 0$, $\sin\delta\theta \to \delta\theta$ and $\cos\delta\theta \to 1$; therefore

$$OA_2/OP_1 \to x_1/OP_1 - (y_1/OP_1)\delta\theta$$

or

$$OA_2 \to x_1 - y_1\,\delta\theta.$$

Further,

$$CP_1 = A_2A_1 = OA_1 - OA_2 = x_1 - x_1 + y_1\,\delta\theta = y_1\,\delta\theta.$$

Similarly, we can show that $CP_2 = x_1\,\delta\theta$.

We may now express the value of the function f at P_2 in terms of its value at P_1:

$$f(x_2, y_2, z_2) = f(x_1, y_1, z_1) + y_1\,\delta\theta\left[-\frac{\partial f(x_1, y_1, z_1)}{\partial x} \right]$$
$$+ x_1\,\delta\theta\left[\frac{\partial f(x_1, y_1, z_1)}{\partial y} \right], \tag{8.4}$$

and so, from (8.3) and dropping the common suffix 1, we have

$$F(x, y, z) = \left[1 + \delta\theta\left(x\frac{\partial}{\partial y} - y\frac{\partial}{\partial x} \right) \right] f(x, y, z). \tag{8.5}$$

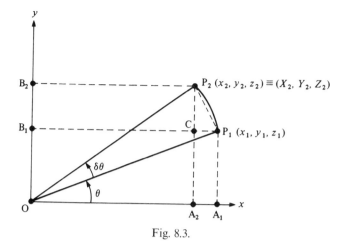

Fig. 8.3.

Now, recalling the form[5] of the quantum-mechanical operator for angular momentum about the z axis,

$$l_z = -i\hbar\left(x\frac{\partial}{\partial y} - y\frac{\partial}{\partial x}\right), \tag{8.6}$$

we may write the quantum equivalent of (8.5) as

$$F(x, y, z) = [1 + i\hbar\delta\theta l_z]\, f(x, y, z), \tag{8.7}$$

or, in atomic units of $\hbar$ for angular momentum, the quantum operator for infinitesimal rotation about the z axis, is

$$R_z(\delta\theta) = 1 + i\delta\theta l_z. \tag{8.8}$$

The equivalent operator for *finite* rotations is now constructed by noting that a finite rotation θ can be expressed as a sum of n successive rotations by θ/n. Then as $n \to \infty$, $\theta/n \to 0$ and we define the infinitesimal rotation angle $\delta\theta$ as the limit of (θ/n) as $n \to \infty$. Therefore,

$$R_z(\theta) = \lim_{n \to \infty}\ [R_z(\theta/n)]^n = \lim_{\delta\theta \to \infty}\ [R_z(\delta\theta)]^{\theta/\delta\theta}$$

$$= \lim_{\delta\theta \to 0}\ [1 + i\delta\theta l_z]^{\theta/\delta\theta}. \tag{8.9}$$

Using the identity,

$$\lim_{x \to 0}(1 + x)^{y/x} = \exp y,$$

we finally have the form of the rotation operator as

$$R_z(\theta) = \exp(i\theta l_z) \tag{8.10}$$

and, by cyclic permutation of the axis labels,

$$R_x(\theta) = \exp(i\theta l_x),$$

$$R_y(\theta) = \exp(i\theta l_y).$$

(8.11)

We have nearly reached the position of formulating the quantum-mechanical expression associated with the rotation operator $R_{z'}(\gamma)R_{y'}(\beta)$ $R_z(\alpha)$ which transforms the cartesian frame x, y, z into x''', y''', z'''. However, in order to apply (8.10), we must take note of a crucial feature of the foregoing derivation. It is that, in (8.5) the coordinates on both sides of the equation refer to a common frame. We may generalize (8.5) as

$$F(\mathbf{r}) = Rf(\mathbf{r}),$$

(8.12)

expressing the fact that $F(\mathbf{r})$ is produced when R operates on $f(\mathbf{r})$. But R must be expressed in terms of the coordinates $\mathbf{r}$ if it is to operate upon $f(\mathbf{r})$. Thus, to obtain the transformed function $F(\mathbf{r})$ we operate with the operator $R(\mathbf{r})$ on the function $f(\mathbf{r})$, the *original* function of the *new* coordinates. Accordingly,

$$R_{z''}(\gamma)R_{y'}(\beta)R_z(\alpha) = \exp(i\gamma l_{z''})\exp(i\beta l_{y'})\exp(i\alpha l_z)$$

(8.13)

and the action of this operator upon an arbitrary function $f(\mathbf{r})$ must be seen by careful recognition of the meaning of (8.12). First we have

$$f_1(\mathbf{r}') = \exp(i\alpha l_{z'})f(\mathbf{r}'),$$

(8.14)

where $f_1(\mathbf{r})$ is the original function referred to the coordinate frame produced by rotation of x, y, z by α about z. Proceeding similarly,

$$f_2(\mathbf{r}'') = \exp(i\beta l_{y''})f(\mathbf{r}'')$$
$$= \exp(i\beta l_{y''})\exp(i\alpha l_{z''})f(\mathbf{r}'')$$

(8.15)

and then,

$$f_3(\mathbf{r}''') = \exp(i\gamma l_{z'''})f_2(\mathbf{r}''')$$
$$= \exp(i\gamma l_{z'''})\exp(i\beta l_{z'''})\exp(i\alpha l_{z'''})f(\mathbf{r}''')$$

(8.16)

and hence, dropping the now common primes,

$$f_3(\mathbf{r}) = \exp(i\gamma l_z)\exp(i\beta l_y)\exp(i\alpha l_z)f(\mathbf{r}).$$

(8.17)

Note that the coordinates appearing throughout this equation, in both functions and operators, refer to the *final* frame.

The final form, therefore, of the quantum-mechanical operator which expresses the form of any function as the cartesian coordinate frame is rotated by the standard Eulerian angles, defined above, is given by

$$D(\alpha\beta\gamma) \equiv \exp(i\gamma l_z)\exp(i\beta l_y)\exp(i\alpha l_z). \tag{8.18}$$

We can define the inverse rotation operator as that which brings the system back to its original form, so that $[D(\alpha\beta\gamma)]^{-1}[D(\alpha\beta\gamma)] = 1$ and so we have,

$$[D(\alpha\beta\gamma)]^{-1} = \exp(-i\alpha l_z)\exp(-i\beta l_y)\exp(-i\gamma l_z)$$
$$= D(-\gamma - \beta - \alpha). \tag{8.19}$$

The foregoing derivation of the $D(\alpha\beta\gamma)$ rotation operator considered the effects of rotation upon a function $f(\mathbf{r})$ of ordinary three-dimensional space. We shall, of course, be concerned also with electronic spin functions. The angular momentum properties of such functions involve both spin and orbital parts and, in general, are characterized by a total angular momentum quantum number j. Although requiring explicit proof,[6] which we do not present here, it may be shown that the form of the operator (8.18) also holds under these circumstances and l_z may be replaced by either s_z or j_z as appropriate. The most general form of (8.18), therefore, is given by

$$D(\alpha\beta\gamma) \equiv \exp(i\gamma j_z)\exp(i\beta j_y)\exp(i\alpha j_z), \tag{8.20}$$

in which j is to be read as any appropriate angular momentum – l, s, or j.

8.2 Rotation matrices

Within the set of angular eigenfunctions $\{|jm\rangle\}$ the j quantum number refers to the total angular momentum and m to that angular momentum projected onto the quantization axis. If we rotate the coordinate frame, the total angular momentum of any member of the set does not change but the description of its projection upon the reference axes does. Thus, under an arbitrary coordinate rotation $D(\alpha\beta\gamma)$, the state $|jm\rangle$ transforms into a linear combination of the complete set of $(2j+1)$ states with the same j:

$$D(\alpha\beta\gamma)|jm\rangle = \sum_{m'=-j}^{j} |jm'\rangle \mathscr{D}_{m'm}^{j}(\alpha\beta\gamma), \tag{8.21}$$

in which $\mathscr{D}_{m'm}^{j}(\alpha\beta\gamma)$ are coefficients depending on the j, m' and m and on the angles α, β and γ. More fully, (8.21) describes a matrix of $\mathscr{D}$ coefficients relating the row vectors $|jm\rangle$:

$$D(\alpha\beta\gamma)(|jj\rangle,|j,j^{-1}\rangle\cdots|j,-j\rangle)=(|jj\rangle,|j,j^{-1}\rangle\cdots|j,-j\rangle)$$

$$\times \begin{pmatrix} \mathscr{D}^j_{jj}(\alpha\beta\gamma) & \mathscr{D}^j_{j,j-1}(\alpha\beta\gamma)\ldots & & \mathscr{D}^j_{j,-j}(\alpha\beta\gamma) \\ \mathscr{D}^j_{j-1,j}(\alpha\beta\gamma) & & & \\ \vdots & & & \vdots \\ \mathscr{D}^j_{-j,j}(\alpha\beta\gamma) & & & \mathscr{D}^j_{-j,-j}(\alpha\beta\gamma) \end{pmatrix}. \qquad (8.22)$$

Such $\mathscr{D}$ matrices are called Wigner rotation matrices, elements of which are evaluated by premultiplying both sides of (8.21) by $\langle jm'|$ and integrating; whence, because of the orthonormality of the $\{|jm\rangle\}$, we obtain

$$\mathscr{D}^j_{m'm}(\alpha\beta\gamma) = \langle jm'|D(\alpha\beta\gamma)|jm\rangle. \qquad (8.23)$$

On substituting the explicit form (8.20) for the rotation operator, we find

$$\begin{aligned} \mathscr{D}^j_{m'm}(\alpha\beta\gamma) &= \langle jm'|D(\alpha\beta\gamma)|jm\rangle \\ &= \langle jm'|\exp(i\gamma j_z)\exp(i\beta j_y)\exp(i\alpha j_z)|jm\rangle \\ &= \exp(i\gamma m')\langle jm'|\exp(i\beta j_y)|jm\rangle\exp(i\alpha m) \\ &= e^{i(\gamma m' + \alpha m)}d^j_{m'm}(\beta), \end{aligned} \qquad (8.24)$$

in which

$$d^j_{m'm}(\beta) = \mathscr{D}^j_{m'm}(0\beta0) = \langle jm'|\exp(i\beta j_y)|jm\rangle \qquad (8.25)$$

are coefficients independent of the α and γ Euler angles. The evaluation of the $d^j_{m'm}(\beta)$ is not trivial and we merely quote here an expression[4] for them in closed form, *viz*:

$$d^j_{m'm}(\beta) = [(j+m')!(j-m')!(j+m)!(j-m)!]^{1/2}$$

$$\times \sum_t (-1)^t \frac{(\cos\beta/2)^{2j+m-m'-2t}(-\sin\beta/2)^{2t-m+m'}}{(j+m-t)!(j-m'-t)!t!(t-m+m')!}, \qquad (8.26)$$

where t ranges over all integers leading to non-negative factorials. Tables of all $d^j_{m'm}(\beta)$ for j ranging from 0 to 7 in half-integers have been compiled by Buckmaster *et al.*[1-9] Note that throughout this book we follow the conventions of Silver[4] so that our $d^j_{m'm}(\beta)$ correspond to the $d^j_{mm'}(\beta)$ of Buckmaster. In general, the question of phase is of the utmost importance in this subject. We do not describe the relationships between our choice and those of others but we adhere to that of Silver who discusses the question at length.

Wigner rotation matrices possess a number of interesting properties, several of which are useful in the present development. Rather more complete lists of these properties than we now describe are to be found in the texts by Silver[4] (pp. 23ff) and by Brink & Satchler[10] (Appendix V). For our purposes, we note the following:

(*i*) *Unitarity*. If A is an hermitian operator, the operator $\exp(iA)$ is unitary as the following argument shows. Expanding the exponential, we have

$$B \equiv \exp(iA) = 1 + iA + i^2 A^2/2! + i^3 A^3/3! + \ldots \qquad (8.27)$$

and, by the hermiticity of A,

$$B^\dagger = \exp(-iA) = 1 - iA + i^2 A^2/2! - i^3 A^3/3! + \ldots . \qquad (8.28)$$

The inverse of B is simply,

$$B^{-1} = 1 - iA + i^2 A^2/2! - i^3 A^3/3! + \ldots = B^\dagger, \qquad (8.29)$$

and so $BB^\dagger = BB^{-1}$. Further, $BB^{-1} = 1$ and so $\exp(iA)$ is indeed unitary. Therefore $D(\alpha\beta\gamma)$ is unitary also, since it has the form $\exp(i\gamma j_z)\exp(i\beta j_y)$ $\exp(i\alpha j_z)$ in which the operators j_z and j_y are hermitian. Hence we have the relationships,

$$[\mathscr{D}^j_{m'm}(\alpha\beta\gamma)]^{-1} = \mathscr{D}^j_{m'm}(-\gamma - \beta - \alpha) = [\mathscr{D}^j_{m'm}(\alpha\beta\gamma)]^\dagger = \mathscr{D}^j_{mm'}(\alpha\beta\gamma)^*. \qquad (8.30)$$

From the unitarity of the Wigner rotation matrix, there follow the usual sum rules:

$$\sum_m \mathscr{D}^j_{m'm}(\alpha\beta\gamma)\mathscr{D}^j_{m''m}(\alpha\beta\gamma) = \delta_{m'm''}$$

and

$$\sum_m \mathscr{D}^j_{mm'}(\alpha\beta\gamma)\mathscr{D}^j_{mm''}(\alpha\beta\gamma) = \delta_{m'm''},$$

$$\qquad (8.31)$$

expressing the orthonormality of rows and columns.

There is a further, important orthogonality relationship which we quote without proof. It is that

$$\int \mathscr{D}^j_{mm'}(\alpha\beta\gamma)\mathscr{D}^J_{MM'}(\alpha\beta\gamma)\,d\Omega = \left(\frac{8\pi^2}{2j+1}\right)\delta_{iJ}\delta_{mM}\delta_{m'M'},$$

$$\qquad (8.32)$$

where $\displaystyle\int d\Omega = \int_0^{2\pi} d\alpha \int_0^{\pi} \sin\beta\, d\beta \int_0^{2\pi} d\gamma.$

The relationship bears a close similarity to the great orthogonality theorem of the theory of finite groups, but we do not explore this matter here.

(*ii*) *Symmetry.* Explicit use of (8.20), (8.21), (8.24) and (8.26) yields the following relationships:

$$d^j_{m'm}(\beta) = (-1)^{m'-m} d^j_{mm'}(\beta) = d^j_{-m,-m'}(\beta) = d^j_{mm'}(-\beta)$$
$$= (-1)^{j-m'} d^j_{m',-m}(\pi-\beta) = (-1)^{j+m} d^j_{m',-m}(\pi+\beta), \quad (8.33)$$

from which we also find

$$\mathcal{D}^j_{m'm}(\alpha\beta\gamma)^* = (-1)^{m'-m}\mathcal{D}^j_{-m',-m}(\alpha\beta\gamma) = \mathcal{D}^j_{mm'}(-\gamma-\beta-\alpha). \quad (8.34)$$

(*iii*) *Special cases.* Three useful identities which are only defined for integer j (and so written with $j \to l$) are,

$$\mathcal{D}^l_{m0}(0\beta\gamma) = (-1)^m \left(\frac{4\pi}{2l+1}\right)^{1/2} Y^l_m(\beta\gamma), \quad (8.35)$$

$$\mathcal{D}^l_{0m}(\alpha\beta0) = \left(\frac{4\pi}{2l+1}\right)^{1/2} Y^l_m(\beta\alpha), \quad (8.36)$$

$$\mathcal{D}^l_{00}(\alpha\beta\gamma) = P_l(\cos\beta) = d^l_{00}(\beta). \quad (8.37)$$

Two interesting relations for general j are:

$$d^j_{m'm}(\pi) = (-1)^{j-m'}\delta_{m,-m'}, \quad (8.38)$$

$$d^j_{m'm}(2\pi) = (-1)^{2j}\delta_{m',m}. \quad (8.39)$$

Now $d^j_{m'm}(2\pi)$ equals $+1$ or -1 depending on whether j is integral or half-integral, a relationship which presages the idea of double groups. The fact, from (8.39), that the effect of a rotation of 2π is not the same as that of one of 4π, at least for functions with half-integral j, is the reason for our convention at the beginning of this chapter for describing coordinate rotations in terms of operations.

8.3 Coupling of two angular momenta

The basis functions with which we are concerned using (III.1) comprise d spin-orbitals or Slater determinantal products of such functions. In general, therefore, we are dealing with functions in a central field which are built from two or more parts involving various spin and orbital angular momenta. If we consider the constants of motion arising for two d electrons within the Russell–Saunders coupling scheme, for example, we may trace a progression of 'good' quantum numbers (with a constant number of degrees of freedom) as we consider perturbations of, first, electrostatic origin and, then, of a magnetic (relativistic) kind:

$$\begin{Bmatrix} l_1 & s_1 & m_{l1} & m_{s1} \\ l_2 & s_2 & m_{l2} & m_{s2} \end{Bmatrix} \rightarrow \begin{Bmatrix} l_1 & s_1 & L & S \\ l_2 & s_2 & M_L & M_S \end{Bmatrix} \rightarrow \begin{Bmatrix} l_1 & s_1 & L & S \\ l_2 & s_2 & J & M_J \end{Bmatrix}$$

$$\mathscr{H} = \qquad \mathscr{H}_{\text{hydrogen-like}} \qquad + \qquad \sum_{i<j} 1/r_{ij} \qquad + \qquad \lambda \mathbf{L} \cdot \mathbf{S} \qquad (8.40)$$

It is important for us to establish relationships between bases expressed in these various ways. For the sake of generality, recognizing that spin-, orbital- and 'total'-angular momenta are of the same ilk, we now write j and m for the quantum numbers of (general) angular momentum operators $\mathbf{J}^2$ and J_z respectively. Thus, for an 'uncoupled' pair of angular momenta, which we write as

$$|j_1 j_2 m_1 m_2\rangle \equiv |j_1 m_1\rangle |j_2 m_2\rangle, \qquad (8.41)$$

we have

$$\left. \begin{aligned} \mathbf{J}_1^2 |j_1 j_2 m_1 m_2\rangle &= j_1(j_1 + 1)|j_1 j_2 m_1 m_2\rangle, \\[6pt] \mathbf{J}_{1z} |j_1 j_2 m_1 m_2\rangle &= m_1 |j_1 j_2 m_1 m_2\rangle, \end{aligned} \right\} \qquad (8.42)$$

with similar expressions for $\mathbf{J}_2^2$ and J_{2z}. After coupling, the individual z components of angular momentum are not conserved, so that the expressions for J_{1z} and J_{2z} are replaced by those for the new total angular momentum squared and its z projection, namely,

$$\left. \begin{aligned} \mathbf{J}^2 |j_1 j_2 jm\rangle &= j(j + 1)|j_1 j_2 jm\rangle, \\[6pt] J_z |j_1 j_2 jm\rangle &= m|j_1 j_2 jm\rangle, \end{aligned} \right\} \qquad (8.43)$$

in which $|j_1 j_2 jm\rangle$ are the 'coupled' eigenfunctions of the operators,

$$\left. \begin{aligned} \mathbf{J}^2 &= (\mathbf{J}_1 + \mathbf{J}_2)^2, \\[6pt] J_z &= J_{1z} + J_{2z}. \end{aligned} \right\} \qquad (8.44)$$

Now there are $(2j_1 + 1)$ states of the type $|j_1 m_1\rangle$ and $(2j_2 + 1)$ of the type $|j_2 m_2\rangle$. The conservation of the number of degrees of freedom in (8.40) reflects the fact that the product states (8.41) and the coupled eigenfunctions $|j_1 j_2 jm\rangle$ span the same space: angular momentum is conserved even though it may be partitioned differently within different schemes. There are therefore $(2j_1 + 1)(2j_2 + 1)$ coupled functions and they may be expressed as some linear combination of eigenfunctions spanning the same space.

In particular, we write

$$|j_1 j_2 jm\rangle = \sum_{m_1, m_2} C^{j_1 j_2 j}_{m_1 m_2 m} |j_1 j_2 m_1 m_2\rangle, \qquad (8.45)$$

where the C coefficients are called vector-coupling or Clebsch–Gordan coefficients. In the usual way, by premultiplication of both sides of (8.45) by $\langle j_1 j_2 m_1 m_2|$ and integrating over the orthonormal states, we find

$$C^{j_1 j_2 j}_{m_1 m_2 m} = \langle j_1 j_2 m_1 m_2 | j_1 j_2 jm\rangle. \qquad (8.46)$$

The coupled and uncoupled functions are thus connected by the unitary transformation

$$|j_1 j_2 jm\rangle = \sum_{m_1, m_2} |j_1 j_2 m_1 m_2\rangle \langle j_1 j_2 m_1 m_2 | j_1 j_2 jm\rangle \qquad (8.47a)$$

or

$$|j_1 j_2 jm\rangle = \sum_{m_1, m_2} |j_1 j_2 m_1 m_2\rangle \langle j_1 j_2 m_1 m_2 | jm\rangle \qquad (8.47b)$$

in a shorter form where the common $j_1 j_2$ are mentioned explicitly only once. Conversely, we can write,

$$|j_1 j_2 m_1 m_2\rangle = \sum_{j, m} |j_1 j_2 jm\rangle \langle j_1 j_2 jm | m_1 m_2\rangle, \qquad (8.48)$$

a relationship which reminds us that the descriptions 'coupled' and 'uncoupled' are convenient but only relative terms.

It is well known that values for the total angular momentum j must lie between the sum and difference of the uncoupled j_1 and j_2 (the limits corresponding to parallel and antiparallel alignment of j_1 and j_2) and we refer to the 'vector-triangle condition',

$$j_1 + j_2 \geqslant j \geqslant |j_1 - j_2|, \qquad (8.49)$$

and again recall the relativity of the terms 'coupled' and 'uncoupled'. Condition (8.49) therefore describes one property of the Clebsch–Gordan coefficients in (8.46). A further condition is imposed by (8.44) in that the vector-coupling coefficient vanishes unless $m = m_1 + m_2$. We have, in addition, from the orthogonality of the states $|j_1 j_2 jm\rangle$ and $|j_1 j_2 m_1 m_2\rangle$, the orthogonality between Clebsch–Gordan coefficients,

$$\sum_{m_1, m_2} \langle j_1 j_2 jm | m_1 m_2\rangle \langle j_1 j_2 m_1 m_2 | j'm'\rangle = \delta_{jj'} \delta_{mm'} \qquad (8.50)$$

and

$$\sum_{j, m} \langle j_1 j_2 m_1 m_2 | jm\rangle \langle j_1 j_2 jm | m_1' m_2'\rangle = \delta_{m_1 m_1'} \delta_{m_2 m_2'}, \qquad (8.51)$$

expressing the unitary nature of the transformations (8.47) and (8.48).

Note, however, that the sum over m in (8.51) is purely formal as the coefficients vanish unless $m = m_1 + m_2$, as above.

8.3.1 Evaluation of Clebsch–Gordan coefficients

We quote below, without proof, a formula for the calculation of Clebsch–Gordan coefficients which Racah[12] derived by making repeated use of angular momentum shift operators. That such a procedure might be expected to yield the required result, albeit after some lengthy algebra, is suggested by recalling an especially simple case, illustrated in several standard texts.[1,11] Consider the microstates arising from the electrostatic coupling of the two d electrons of the d^2 configuration. Inspection provides the unique description of the $|M_L = 4, M_S = 0\rangle$ component of the ^1G term,

$|4400\rangle$ in $|LM_L SM_S\rangle$ quantization, as $(\overset{+}{2},\overset{-}{2})$ or $(2\alpha,2\beta)$, referring to the d spin-orbitals. Application of the usual orbital lowering operator L_- yields the well-known result that

$$|4300\rangle = \sqrt{(\tfrac{1}{2})}[(\overset{+}{2},\overset{-}{1}) - (\overset{-}{2},\overset{+}{1})], \qquad (8.52)$$

in which the coefficients, $\pm\sqrt{(\tfrac{1}{2})}$ are examples of Clebsch–Gordan coefficients.

The general, powerful expression given by Racah[12] is,

$$
\begin{aligned}
\langle j_1 j_2 m_1 m_2 | j m \rangle = {}& \delta_{m_1 + m_2, m} \Delta(j_1 j_2 j) \\
& \times [(2j+1)(j_1+m_1)!(j_1-m_1)!(j_2+m_2)!(j_2-m_2)!(j+m)!(j-m)!]^{1/2} \\
& \times \sum_t (-1)^t [(j_1-m_1-t)!(j-j_2+m_1+t)!(j_2+m_2-t)! \\
& \times (j-j_1-m_2+t)!t!(j_1+j_2-j-t)!]^{-1}
\end{aligned}
\qquad (8.53)
$$

where

$$\Delta(j_1 j_2 j) = \left[\frac{(j_1+j_2-j)!(j_1+j-j_2)!(j_2+j-j_1)!}{(j_1+j_2+j+1)!} \right]^{1/2}$$

and t runs over all integer values leading to non-negative factorials.

We have emphasized that the terms 'coupled' and 'uncoupled' are relative, referring simply to different bases within the same Hilbert space. However, the three angular momenta do not appear on *exactly* the same footing: for example, j lies between the sum and difference of j_1 and j_2; and $m = m_1 + m_2$. The latter condition appears in the Kronecker function in (8.53) and the formula in general is not totally symmetric in the various j and m quantum numbers. There are, however, many symmetry properties of the Clebsch–Gordan coefficients, one such being,

$$\langle j_1 j_2 m_1 m_2 | jm \rangle = (-1)^{j_1 + j_2 - j} \langle j_2 j_1 m_2 m_1 | jm \rangle, \qquad (8.54)$$

which implies that the order in which coupled states are written can affect their phase. Should this seem strange, we may recall the example associated with (8.52) in which the signs of the coefficients depend upon our choice of 'standard order'; that is, on whether we write $(\overset{+}{2}, \bar{1})$ or $(\bar{1}, \overset{+}{2})$ etc. Other symmetry relationships are of great important in respect of saving labour. For example, it can be shown that

$$\langle j_1 j_2 m_1 m_2 | jm \rangle = (-1)^{j_2 + m_2} \left(\frac{2j+1}{2j_1+1} \right)^{1/2} \langle j_2 j, -m_2 m | j_1 m_1 \rangle, \qquad (8.55)$$

an expression which reduces the number of Clebsch–Gordan coefficients which need be calculated by (8.53) explicitly. Symmetry rules like these are not readily memorable, but Wigner invented a symbol, closely related to the Clebsch–Gordan coefficient, which possesses very simple symmetry properties. Wigner's coefficients are the so-called 3–j symbols.

8.3.2 The 3–j symbol

Wigner's 3–j symbol is related to the Clebsch–Gordan coupling coefficient by

$$\begin{pmatrix} j_1 & j_2 & j_3 \\ m_1 & m_2 & m_3 \end{pmatrix} = (-1)^{j_1 - j_2 - m_3} (2j_3 + 1)^{-1/2} \langle j_1 j_2 m_1 m_2 | j_3, -m_3 \rangle, \qquad (8.56)$$

where we have written j_3 instead of j to emphasize the equivalent footing of 'coupled' and 'uncoupled' bases: note the negative sign in the ket of the Clebsch–Gordan coefficient.

Symmetry. In addition to the vector-triangle condition that $j_1 + j_2 \leqslant j_3 \leqslant |j_1 - j_2|$, and that $-m_3 = m_1 + m_2$, as for the Clebsch–Gordan coefficients, the 3–j symbols have the following simple symmetry properties:

(i) An even permutation of columns leaves the value of the symbol unchanged:

$$\begin{pmatrix} j_1 & j_2 & j_3 \\ m_1 & m_2 & m_3 \end{pmatrix} = \begin{pmatrix} j_2 & j_3 & j_1 \\ m_2 & m_3 & m_1 \end{pmatrix} = \begin{pmatrix} j_3 & j_1 & j_2 \\ m_3 & m_1 & m_2 \end{pmatrix}. \qquad (8.57)$$

(ii) An odd permutation (non-cyclic) of columns multiplies the value of the symbol by $(-1)^{j_1+j_2+j_3}$:

$$\begin{pmatrix} j_2 & j_1 & j_3 \\ m_2 & m_1 & m_3 \end{pmatrix} = \begin{pmatrix} j_1 & j_3 & j_2 \\ m_1 & m_3 & m_2 \end{pmatrix} = \begin{pmatrix} j_3 & j_2 & j_1 \\ m_3 & m_2 & m_1 \end{pmatrix}$$

$$= (-1)^{j_1+j_2+j_3} \begin{pmatrix} j_1 & j_2 & j_3 \\ m_1 & m_2 & m_3 \end{pmatrix}. \tag{8.58}$$

By using symmetries (8.57) and (8.58), different permutations of the angular momenta corresponding to six different 3–j symbols can be related simply to one symbol evaluated via (8.53).

(iii)

$$\begin{pmatrix} j_1 & j_2 & j_3 \\ m_1 & m_2 & m_3 \end{pmatrix} = (-1)^{j_1+j_2+j_3} \begin{pmatrix} j_1 & j_2 & j_3 \\ -m_1 & -m_2 & -m_3 \end{pmatrix}. \tag{8.59}$$

From this rule, we note that a 3–j symbol with all m values equal to zero vanishes unless $j_1 + j_2 + j_3$ is even.

Orthogonality. The orthogonality condition in (8.50) can be written in terms of 3–j symbols as,

$$\sum_{m_1,m_2} \begin{pmatrix} j_1 & j_2 & j \\ m_1 & m_2 & m \end{pmatrix} \begin{pmatrix} j_1 & j_2 & j' \\ m_1 & m_2 & m' \end{pmatrix} = (2j+1)^{-1} \delta_{j'j} \delta_{m'm}. \tag{8.60}$$

As there are $(2j + 1)$ values of m, we also have

$$\sum_{m_1,m_2,m} \begin{pmatrix} j_1 & j_2 & j \\ m_1 & m_2 & m \end{pmatrix}^2 = 1, \tag{8.61}$$

a formal expression revealing one aspect of the unitarity of the expansion (8.47).

Special cases. A great many special cases and recurrence relationships for 3–j symbols have been listed.[10] A few of especial use for us are as follows:

(i)

$$\begin{pmatrix} j & j & 1 \\ m & -m & 0 \end{pmatrix} = (-1)^{j-m} m[(2j+1)(j+1)j]^{-1/2}, \tag{8.62}$$

(ii)

$$\begin{pmatrix} j_1 & j_2 & j_3 \\ 0 & 0 & 0 \end{pmatrix} = (-1)^{J/2} \left[\frac{(j_1+j_2-j_3)!(j_1+j_3-j_2)!(j_2+j_3-j_1)!}{(j_1+j_2+j_3+1)!} \right]^{-1/2}$$

$$\times \frac{(J/2)!}{(J/2-j_1)!(J/2-j_2)!(J/2-j_3)!}, \tag{8.63}$$

where $J = j_1 + j_2 + j_3$. Note from (8.59) that the symbol (8.63) vanishes if J is odd. Further, *any* symbol vanishes if J is odd *and* it has two identical columns, a result following from (8.58).

(iii)

$$\begin{pmatrix} j_1 & j_2 & 0 \\ m_1 & -m_2 & 0 \end{pmatrix} = (-1)^{j_1 - m_1} (2j_1 + 1)^{-1/2} \delta_{j_1 j_2} \delta_{m_1 m_2}. \qquad (8.64)$$

General 3–j symbols. Rotenberg *et al.*[13] have tabulated a large and essentially comprehensive list of 3–j symbols, their values being given in the form of powers of primes. This form was used in particular as a safeguard against printing errors, especially regarding the loss or misplacing of a decimal point: further, the compilations are given as photographic reproductions of computer line-printer output, again in the interests of reducing errors. While the method of using these tables is given clearly by the authors, we illustrate the system with the following example:
Suppose we require the value of the symbol

$$\begin{pmatrix} 4 & 4 & 3 \\ -2 & -1 & 3 \end{pmatrix}.$$

It is not listed explicitly in Rotenberg's tables but the symbol

$$\begin{pmatrix} 4 & 4 & 3 \\ -1 & -2 & 3 \end{pmatrix}$$

is. Here we use the symmetry rules (8.57) and (8.58) and so determine that

$$\begin{pmatrix} 4 & 4 & 3 \\ -1 & -2 & 3 \end{pmatrix}$$

is the negative of the symbol required. The tables give

$$\begin{pmatrix} 4 & 4 & 3 \\ -1 & -2 & 3 \end{pmatrix} = *0\underline{2}21, 1,$$

and the numbers on the right-hand side are to be interpreted as follows. Each digit gives the power to which prime numbers are raised, bars under them denoting negative powers. The prime numbers (greater than 1) are *all* included in the notation, in order, up to the last one required. All 3–j symbols so listed are given as squares and an asterisk at the beginning denotes the fact that the negative square root is to be taken.

In the present illustration we have

$$\begin{pmatrix} 4 & 4 & 3 \\ -1 & -2 & 3 \end{pmatrix} = -\sqrt{(2^0 3^{-2} 5^2 7^{-1} 11^{-1})} = -\sqrt{\left(\frac{25}{9 \times 7 \times 11}\right)} \quad (8.65)$$

In the case of our main interest in fitting magnetic and other ligand-field properties, we need values for many $3-j$ symbols, but not so many (as we shall see) that it is worth holding the contents of Rotenberg's tables in a computer file. It is our practice, therefore, to compute the values of $3-j$ symbols as required: Appendix C describes a FORTRAN routine for this purpose.

8.4 The Clebsch–Gordan series

The set of all coordinate rotations about a point forms the three-dimensional rotation group R_3 and the $(2j + 1)$ functions $\{|jm\rangle\}$ form bases for irreducible representations of this group. The elements of R_3, being the set of all coordinate rotations about a point, are just the rotation operators $D(\alpha\beta\gamma)$. Thus do the set of matrices $\mathscr{D}^j(\alpha\beta\gamma)$ for all α, β and γ form an irreducible representation of dimension $(2j + 1)^2$ of the group R_3. Now the wavefunctions $|j_1 m_1\rangle |j_2 m_2\rangle$ for a two-component system span a $(2j_1 + 1)(2j_2 + 1)$-fold degenerate representation of the rotation group denoted by the product $\mathscr{D}^{j_1}(\alpha\beta\gamma) \times \mathscr{D}^{j_2}(\alpha\beta\gamma)$. Reduction of this product may be achieved using the well-known result that the character χ of the direct product representation equals the product of the individual representation characters,

$$\chi(\mathscr{D}^{j_1} \times \mathscr{D}^{j_2}) = \chi(\mathscr{D}^{j_1}) \cdot \chi(\mathscr{D}^{j_2}). \quad (8.66)$$

Further, the character of a given matrix $\mathscr{D}^j(\alpha\beta\gamma)$ is independent of the coordinate frame so that we lose no generality by considering the case $\mathscr{D}^j(\alpha 00)$ when the matrix takes the diagonal form[††]

$$\mathscr{D}^j = \begin{pmatrix} e^{ij\alpha} \cdots & & & \\ \vdots & e^{i(j-1)\alpha} & & \\ & & \cdot & \\ & & & \cdot e^{-ij\alpha} \end{pmatrix} \quad (8.67)$$

with character given by

$$\chi(\mathscr{D}^j) = (e^{ij\alpha} + e^{i(j-1)\alpha} + \ldots + e^{-ij\alpha}). \quad (8.68)$$

[††] Take care not to confuse the imaginary i here with an angular momentum!

On substitution of (8.68) into (8.66), we find

$$\chi(\mathscr{D}^{j_1} \times \mathscr{D}^{j_2}) = (e^{ij_1\alpha} + e^{i(j_1-1)\alpha} + \ldots + e^{-ij_1\alpha})(e^{ij_2\alpha} + e^{i(j_2-1)\alpha} + \cdots + e^{-ij_2\alpha})$$

$$= (e^{i(j_1+j_2)\alpha} + e^{i(j_1+j_2-1)\alpha} + \ldots + e^{-i(j_1+j_2)\alpha})$$

$$+ (e^{i(j_1+j_2-1)\alpha} + e^{i(j_1+j_2-2)\alpha} + \ldots e^{-i(j_1+j_2-1)\alpha})$$

$$\vdots$$

$$+ (e^{i(|j_1-j_2|)\alpha} + e^{i(|j_1-j_2|-1)\alpha} + \ldots + e^{-i(|j_1-j_2|\alpha)})$$

$$= \chi(\mathscr{D}^{j_1+j_2}) + \chi(\mathscr{D}^{j_1+j_2-1}) + \ldots + \chi(\mathscr{D}^{|j_1-j_2|}),$$

$$(8.69)$$

or, writing the result in condensed form,

$$\mathscr{D}^{j_1} \times \mathscr{D}^{j_2} = \sum_{j=|j_1-j_2|}^{j_1+j_2} \mathscr{D}^{j}. \qquad (8.70)$$

This is, of course, just the vector-triangle rule. However, an important relationship between Wigner rotation matrices emerges if we write the reduction (8.70) explicitly:

$$\mathscr{D}^{j}_{m'm} = \langle jm' | D | jm \rangle$$

$$= \langle j_1 j_2 jm' | D | j_1 j_2 jm \rangle$$

$$= \sum_{\substack{m_1 m_2 \\ m'_1 m'_2}} \langle j_1 j_2 jm' | m'_1 m'_2 \rangle \langle j_1 j_2 m'_1 m'_2 | D | j_1 j_2 m_1 m_2 \rangle \langle j_1 j_2 m_1 m_2 | jm \rangle$$

$$= \sum_{\substack{m_1 m_2 \\ m'_1 m'_2}} \langle j_1 j_2 jm' | m'_1 m'_2 \rangle \mathscr{D}^{j_1}_{m'_1 m_1} \mathscr{D}^{j_2}_{m'_2 m_2} \langle j_1 j_2 m_1 m_2 | jm \rangle.$$

$$(8.71)$$

The converse relationship is proved similarly,

$$\mathscr{D}^{j_1}_{m'_1 m_1} \mathscr{D}^{j_2}_{m'_1 m_2} = \sum_{jmm'} \langle j_1 j_2 m'_1 m'_2 | jm' \rangle \mathscr{D}^{j}_{m'm} \langle j_1 j_2 jm | m_1 m_2 \rangle. \qquad (8.72)$$

These relationships show how rotation matrices can be built up recursively and the very approximate similarity in form for the generating formulae for $d(\beta)$ coefficients and Clebsch–Gordan coefficients, (8.26) and (8.53) should not be too surprising. The recursive series implied by (8.71) and (8.72) define the Clebsch–Gordan series. In view of a common preference for 3–j symbols over Clebsch–Gordan coefficients, we rewrite (8.72), using (8.34), as

$$\mathscr{D}^{j_1}_{m'_1 m_1} \mathscr{D}^{j_2}_{m'_2 m_2} = \sum_{j} (2j+1) \begin{pmatrix} j_1 & j_2 & j \\ m'_1 & m'_2 & m' \end{pmatrix} \begin{pmatrix} j_1 & j_2 & j \\ m_1 & m_2 & m \end{pmatrix} \mathscr{D}^{j*}_{m'm}. \qquad (8.73)$$

8.4.1 An important integral

We multiply both sides of (8.73) by $\mathscr{D}^{j_3}_{m'_3 m_3}$ and integrate over all angles α, β, γ to get

$$I = \int_0^{2\pi}\int_0^{\pi}\int_0^{2\pi} \mathscr{D}^{j_1}_{m'_1 m_1}(\alpha\beta\gamma)\mathscr{D}^{j_2}_{m'_2 m_2}(\alpha\beta\gamma)\mathscr{D}^{j_3}_{m'_3 m_3}(\alpha\beta\gamma)\,d\alpha\sin\beta\,d\beta\,d\gamma$$

$$= \int_0^{2\pi}\int_0^{\pi}\int_0^{2\pi} \sum_j (2j+1)\begin{pmatrix} j_1 & j_2 & j \\ m'_1 & m'_2 & m' \end{pmatrix}\begin{pmatrix} j_1 & j_2 & j \\ m_1 & m_2 & m \end{pmatrix}$$

$$\times\, \mathscr{D}^{j*}_{m'm}(\alpha\beta\gamma)\mathscr{D}^{j_3}_{m'_3 m_3}(\alpha\beta\gamma)\,d\alpha\sin\beta\,d\beta\,d\gamma. \qquad (8.74)$$

Now apply the orthogonality condition in (8.32) to the right-hand side of (8.74) and note that all terms vanish unless $m_3 = m(= m_1 + m_2)$ and $m'_3 = m'(= m'_1 + m'_2)$ and $j = j_3$: therefore

$$I = 8\pi^2 \begin{pmatrix} j_1 & j_2 & j_3 \\ m'_1 & m'_2 & m'_3 \end{pmatrix}\begin{pmatrix} j_1 & j_2 & j_3 \\ m_1 & m_2 & m_3 \end{pmatrix}. \qquad (8.75)$$

To obtain a special, but important, result we now use (8.36) which holds only for integer $j(\equiv l)$ values and so reduce (8.75) to give

$$\int_0^{2\pi}\int_0^{\pi} Y^{l_1}_{m_1}(\theta, \phi)\, Y^{l_2}_{m_2}(\theta, \phi)\, Y^{l_3}_{m_3}(\theta, \phi)\sin\theta\,d\theta\,d\phi$$

$$= \left[\frac{(2l_1 + 1)(2l_2 + 1)(2l_3 + 1)}{4\pi}\right]^{1/2}\begin{pmatrix} l_1 & l_2 & l_3 \\ 0 & 0 & 0 \end{pmatrix}\begin{pmatrix} l_1 & l_2 & l_3 \\ m_1 & m_2 & m_3 \end{pmatrix}, \qquad (8.76)$$

where θ, ϕ are the usual polar angles used with spherical harmonics. This integral is, of course, just the kind encountered in the evaluation of ligand-field and, when taken in product pairs, of interelectron-repulsion, or Coulomb, matrix elements. In conjunction with, say, Rotenberg's tables of 3–j symbols or with an equivalent computer routine, the evaluation of these integrals is now simple and generalized in comparison with the less elegant methods described elsewhere.[2,3]

8.5 Functions and operators

We described in chapter 4 how scalars, vectors and tensors are defined by their transformation properties and, further, how we may formalize all such entities as tensors of different rank. Our discussion there was confined to cartesian tensors. Within quantum mechanics and especially in the context of central-field problems and the relevance of the rotation group R_3, it is desirable to define equivalent structures which we call spherical tensors. In the same way that the cartesian vector $\mathbf{r} = (x, y, z)$ transforms under the 3×3 matrix of direction cosines, (4.12) for example, we may define a set of $(2k + 1)$ entities which span $\mathscr{D}^k$ in standard form as an *irreducible spherical tensor* of rank k. The transformation under rotations of such a set is given, as in (8.21), by

$$D(\alpha\beta\gamma)T_q^k = \sum_{q'} T_{q'}^k \mathscr{D}_{q'q}^k(\alpha\beta\gamma). \tag{8.77}$$

As the set T_q^k spans an *irreducible* representation $(\mathscr{D}^k)$ of R_3, it is itself irreducible, comprising no more and no less than $(2k + 1)$ elements.

Analogously, we may also define an *irreducible tensor field* of rank k to be a set of $(2k + 1)$ *functions* transforming under rotation according to

$$D(\alpha\beta\gamma)f_q^k(\mathbf{r}') = \sum_{q'} f_{q'}^k(\mathbf{r}')\mathscr{D}_{q'q}^k(\alpha\beta\gamma), \tag{8.78}$$

which, bearing in mind (8.12) and associated remarks, can be rewritten as

$$f_q^k(\mathbf{r}) = \sum_{q'} f_{q'}^k(\mathbf{r}')\mathscr{D}_{q'q}^k(\alpha\beta\gamma). \tag{8.79}$$

A typical example of a set of functions transforming in this way are the spherical harmonics $Y_m^l(\theta, \phi)$: for example, the set $Y_m^2(\theta, \phi)$ comprise an irreducible tensor field of rank 2 spanning the irreducible representation $\mathscr{D}^2$ of R_3.

Now, functions can also behave as quantum mechanical operators: for example, in the Schrödinger equation

$$(-\tfrac{1}{2}\nabla^2 + V)\psi = E\psi, \tag{8.80}$$

V, a function of the usual space coordinates $\mathbf{r}$, acts as an operator upon the wavefunction ψ (also a function of $\mathbf{r}$); again, in ligand-field theory, the ligand-field potential V_{LF} operates in the same way and is commonly expressed as a linear superposition of spherical harmonics Y_q^k also acting as operators. However, functions acting as functions are not the same as functions acting as operators, the fundamental reason being that they transform differently in the two roles. An operator, A, may be defined by its action upon some general vector $\mathbf{V}$:

$$\mathbf{V}' = A\mathbf{V}. \tag{8.81}$$

Under a rotation R, the vector $\mathbf{V}$ is transformed into $R\mathbf{V}$, and $\mathbf{V}'$ into $R\mathbf{V}'$;

therefore

$$RV' = R(AV) = RA(R^{-1}R)V = RAR^{-1}(RV); \qquad (8.82)$$

that is,

$$(RV') = B(RV), \qquad (8.83)$$

where $B = RAR^{-1}$ is the equivalent operator to A that now relates RV' and RV. Accordingly,

$$B = RAR^{-1} \qquad (8.84)$$

is the transformation rule for operators under coordinate rotations R. We may now define by analogy, an *irreducible spherical tensor operator* of rank k as a set of $(2k + 1)$ operators T_q^k which transform under coordinate rotations according to the relation,

$$D(\alpha\beta\gamma)T_q^k D^{-1}(\alpha\beta\gamma) = \sum_{q'} T_{q'}^k \mathscr{D}_{q'q}^k(\alpha\beta\gamma). \qquad (8.85)$$

This expression means that, under the coordinate rotation $D(\alpha\beta\gamma)$, the operator T_q^k transforms into a linear combination of the $(2k + 1)$-fold set $\{T_{q'}^k\}$ with expansion coefficients given by appropriate elements of the Wigner rotation matrices.

It is also possible to define and construct irreducible *cartesian* tensors, fields and operators as the quantum equivalents of corresponding classical structures, some of which were discussed in chapter 4. A cartesian tensor-operator theory corresponding to that we are now developing may be constructed but in central-field problems leads to identical physical results, of course. We shall argue, in chapter 11, that the ligand-field model may be treated adequately within a metal d^n basis of the rotation group R_3 and so have no need to explore the elegant but rather complex area of cartesian tensor operators.

8.5.1 Scalar, vector and tensor operators

We now illustrate some simple examples and properties of the irreducible tensor operators (8.85) of various rank k. With $k = 0$, $(2k + 1) = 1$ and so the set comprises the single operator T_0^0 and this transforms as,

$$DT_0^0 D^{-1} = T_0^0 \mathscr{D}_{00}^0 = T_0^0 \times 1 = T_0^0. \qquad (8.86)$$

T_0^0 is therefore unchanged under coordinate rotations and is called a *scalar operator*. As all terms appearing in a Hamiltonian operator must be invariant under all coordinate transformations, they too are scalar operators: accordingly, we shall have much more to say about zero-rank tensors.

For $k = 1$, we have $(2k + 1) = 3$ component operators, T_1^1, T_0^1, T_{-1}^1 which form a first-rank tensor transforming as

$$DT_0^1 D^{-1} = T_{-1}^1 \mathcal{D}_{-10}^1 + T_0^1 \mathcal{D}_{00}^1 + T_1^1 \mathcal{D}_{10}^1, \tag{8.87}$$

that is, like a vector. The set comprises a so-called *vector operator*. The most important examples of vector operators for our purposes are the operators for orbital and spin angular momentum, each having three components. Most commonly these components are given in cartesian form, those for $\mathbf{l}$ being l_x, l_y and l_z. Neither these, nor the equally common forms,

$$\left. \begin{array}{l} l_+ \equiv (l_x + il_y), \\ l_- \equiv (l_x - il_y), \end{array} \right\} \tag{8.88}$$

and l_z, transform correctly according to (8.85), however, simply because they are not in the correct standard form defined for spherical tensor operators, which depends in its turn upon the standard form (8.22) for the Wigner rotation matrices in which the elements range $\mathcal{D}_{jj}^j(\alpha\beta\gamma)$, $\mathcal{D}_{j,j-1}^j(\alpha\beta\gamma)\ldots\mathcal{D}_{j,-j}^j(\alpha\beta\gamma)$ in that order. Instead, and ultimately depending also on the Condon–Shortly phase convention,[1,2,11] the $\mathbf{l}$ operator in correct standard form comprises the three components, l_{+1}, l_0, l_1, where

$$l_{+1} = -2^{-1/2}l_+, \quad l_0 = l_z, \quad l_{-1} = +2^{-1/2}l_-: \tag{8.89}$$

it is important to distinguish carefully the subscripts $\pm$ from ± 1. Similar definitions are appropriate for the spin operator $\mathbf{s}$.

A second-rank spherical tensor operator comprises five operators $T_2^2, T_1^2, T_0^2, T_{-1}^2, T_{-2}^2$, the most obvious example being the set of spherical harmonics $Y_m^2(\theta, \phi)$, occurring, for instance, in the usual multipole analyses of ligand-field or Coulomb operators.

Tensor operators of higher rank are of concern to us only in the multipole expansions of electrostatic and ligand-field effective potentials.

Tensor operators with k half-integer, though not formally excluded from the definition (8.85), are irrelevant in physics for they would be associated with a process involving half-integral spin change: fermions and bosons are not interconvertible.

8.6 Tensor-operator commutators

The rotation operator $D(\alpha\beta\gamma)$ in (8.85) is, of course, that for finite rotations, but the equation is equally valid for infinitesimal rotations. Considering an infinitesimal rotation α about an axis n, we rewrite (8.85) as,

$$(1 + i\alpha J_n)T_q^k(1 - i\alpha J_n) = \sum_{q'} T_{q'}^k \mathscr{D}_{q'q}^k(\alpha_n). \tag{8.90}$$

Now, from (8.23),

$$\mathscr{D}_{q'q}^k(\alpha_n) = \langle kq'|1 + i\alpha J_n|kq \rangle = \delta_{q'q} + i\alpha \langle kq'|J_n|kq \rangle, \tag{8.91}$$

which on substitution in (8.90) and following the neglect of terms in α^2, gives

$$T_q^k - i\alpha T_q^k J_n + i\alpha J_n T_q^k = \sum_{q'} [\delta_{q'q} + i\alpha \langle kq'|J_n|kq \rangle]T_{q'}^k \tag{8.92}$$

and hence,

$$J_n T_q^k - T_q^k J_n = \sum_{q'} T_{q'}^k \langle kq'|J_n|kq \rangle. \tag{8.93}$$

Putting $J_n = J_z$ and $J_\pm$ in turn into (8.93) and using the standard relationships,[5]

$$\left. \begin{array}{l} \langle jm|J_z|jm \rangle = m, \\[2mm] \langle j, m \pm 1|J_\pm|jm \rangle = [(j \pm m + 1)(j \mp m)]^{1/2}, \end{array} \right\} \tag{8.94}$$

yields the commutation relations,

$$\left. \begin{array}{l} [J_z, T_q^k] = qT_q^k, \\[2mm] [J_\pm, T_q^k] = [(k \pm q + 1)(k \mp q)]^{1/2} T_{q \pm 1}^k, \end{array} \right\} \tag{8.95}$$

relations by which Racah originally defined[12] irreducible tensor operators. Apart from this aspect of historical interest, we make three points from (8.95).

(i) Firstly, we may use the commutation relationships of **J** with spherical tensor operators to determine the spherical equivalents of cartesian tensors. Thus, consider for the vector operator **A** that we have $[J_z, A_z] = 0$ so that $A_0 = A_z$. Further, from (8.95) we find

$$\begin{aligned} A_{\pm 1} &= \frac{1}{\sqrt{2}}[J_\pm, A_0] \\[2mm] &= \frac{1}{\sqrt{2}}([J_x, A_z] \pm i[J_y, A_z]) \\[2mm] &= \pm \frac{1}{\sqrt{2}}(A_x \pm iA_y). \end{aligned} \tag{8.96}$$

The relationships in (8.89) furnish one example of this equation.

(ii) Secondly, following Silver,[4] we may use the commutators (8.95) to clarify further the exact meanings of the expressions 'scalar operator' and 'vector operator' and, in so doing, lay some useful groundwork for a subsequent discussion of scalar operators. In the case of a vector operator, we put $k = 1$ in (8.95) to get

$$\left.\begin{array}{ll} [J_+, V_0] = 2^{1/2} V_1 & [J_-, V_0] = 2^{1/2} V_{-1}, \\ [J_+, V_{-1}] = 2^{1/2} V_0 & [J_-, V_1] = 2^{1/2} V_0, \\ [J_z, V_0] = 0 & [J_z, V_{+1}] = V_{\pm 1}. \end{array}\right\} \tag{8.97}$$

A full description of an operator obeying these commutation rules is that of a vector operator *with respect to the angular momentum* **J**. The italicized clause had not previously been made explicit and the point, emerging from Racah's form of (8.85), is essential. It is possible for a vector operator to commute with **L** – for example, the spin angular momentum operator **S** acting in a different subspace – and we therefore cannot describe **S** as a vector operator with respect to **L**. While **L** and **S** commute, each separately non-commutes with **J** according to (8.97): **L** and **S** behave as vector operators with respect to **J** but not with respect to each other.

An irreducible tensor operator which commutes with **J** is called a scalar operator *with respect to* **J**. This can only occur for a zero-rank tensor (for which $k = q = 0$). Following the discussion immediately above, **S** is a scalar operator with respect to **L** and vice versa. Now the spin-orbit coupling operator **L·S** is a scalar operator with respect to **J** as it is formed as the scalar product of two vector operators with respect to **J**: it is not scalar with respect to **L** or **S**, however.

(iii) Thirdly, we make use of the commutation rules to determine the adjoint of a tensor operator. An operator T is called hermitian if $T = T^\dagger$ or, in terms of its matrix elements, if

$$\langle \alpha | T^\dagger | \beta \rangle \equiv \langle \beta | T | \alpha \rangle^* = \langle \alpha | T | \beta \rangle. \tag{8.98}$$

A necessary and sufficient condition for hermiticity is that all eigenvalues of the operator be real. The components of **J** are hermitian. Given the operator T as an irreducible tensor operator, we ask now whether the adjoint $T^\dagger$ is also; that is, whether $T^\dagger$ transforms as (8.85) or, equivalently, whether its commutators with components of **J** are given by (8.95). For this, we need to establish a few simple relationships, as follows. We consider two operators A and B. Then

$$(AB)^\dagger = B^\dagger A^\dagger. \tag{8.99}$$

Further, if a is a complex number,

$$(aA)^\dagger = a^* A^\dagger \tag{8.100}$$

and so we have the commutator rules,

$$[A, B]^\dagger = [B^\dagger, A^\dagger] = -[A^\dagger, B^\dagger]. \tag{8.101}$$

Now, given the hermitian property of the components of $\mathbf{J}$, it follows that

$$J_\pm^\dagger = J_\mp \quad \text{and} \quad J_0^\dagger = J_0. \tag{8.102}$$

We are now ready to determine the commutators of $T^{k\dagger}$ with $\mathbf{J}$, and find:

$$[J_0, T_q^{k\dagger}] = -[J_0, T_q^k]^\dagger = -q T_q^{k\dagger},$$
$$[J_\pm, T_q^{k\dagger}] = -[J_\mp, T_q^k]^\dagger = -[(k \mp q + 1)(k \pm q)]^{1/2} T_{q\mp1}^{k\dagger}. \tag{8.103}$$

Thus we observe that $\mathbf{T}^{k\dagger}$ does not transform as does $\mathbf{T}^k$ with respect to $\mathbf{J}$. Instead $\mathbf{T}^{k\dagger}$ transforms *contragrediently* to $\mathbf{T}^k$. Compare here the effect of the rotation operator $D(\alpha\beta\gamma)$ on the state $|jm\rangle$ in (8.21),

$$|jm\rangle' = D|jm\rangle = \sum_{m'=-j}^{j} |jm'\rangle \mathscr{D}_{m'm}^j, \tag{8.21}$$

with the transformation of conjugate states

$$\langle jm|' = \langle jm|D^\dagger = \sum_{m'=-j}^{j} \mathscr{D}_{m'm}^{j*} \langle jm'|. \tag{8.104}$$

Again, the states $|jm\rangle$ and $\langle jm|$ are said to transform contragrediently: cogredience is defined by (8.21). Now one symmetry property of Wigner rotation matrices described above is,

$$\mathscr{D}_{m'm}^{j*} = (-1)^{m'-m} \mathscr{D}_{-m',-m}^j, \tag{8.34}$$

which shows that the transformation (8.104) for $\langle jm|$ is the same as that for $(-1)^m |j, -m\rangle$. This relationship therefore suggests how we may construct an operator related to $\mathbf{T}^{k\dagger}$ but which also transforms cogrediently with $\mathbf{T}^k$ and hence satisfies the definition – (8.85) or (8.95) – of an irreducible spherical tensor. We define, then, the *hermitian adjoint* of T_q^k as,

$$\bar{T}_q^k = (-1)^q T_q^{k\dagger}, \tag{8.105}$$

which may be seen to satisfy the commutation rules (8.95).

8.7 Compound irreducible tensor operators

Many operators with which we are concerned are built up as products of simpler operators. Further, the component operators frequently commute by operating in different subspaces – either the space and spin subspaces of a single particle as in the spin-orbit coupling operator $\mathbf{L \cdot S}$ or on the separate orbital spaces of two distinct particles as encountered in binary products of spherical harmonic operators used in the evaluation of matrix elements of the Coulomb operator $1/r_{ij}$. The formalism we are reviewing eventually leads to powerful expressions for the evaluation of the angular parts of a wide variety of matrix elements and its efficacy stems from the desire to maximize the utility of the symmetry content of the various species in a given calculation. We are therefore concerned to characterize both functions and operators in terms of their symmetries within the group R_3 (in the present case). But we have also seen how the coupling of function bases, or the converse decoupling process, can be determined by what are essentially group-theoretical means. By entirely analogous reasoning, it is possible to couple and decouple *operators*, for their transformation properties are determined by the same (R_3) symmetry.

If $R_{q_1}^{k_1}$ and $S_{q_2}^{k_2}$ are irreducible spherical tensors of rank k_1 and k_2, respectively, the $(2k_1 + 1)(2k_2 + 1)$ products $R_{q_1}^{k_1} S_{q_2}^{k_2}$ form a tensor spanning the representation $\mathscr{D}^{k_1} \times \mathscr{D}^{k_2}$ of R_3. The representation is reducible, by (8.70) and, analogously to (8.47) for the coupling of states, we have

$$T_q^k = \{\mathbf{R}^{k_1}(1) \times \mathbf{S}^{k_2}(2)\}_q^k = \sum_{q_1 q_2} \langle k_1 k_2 q_1 q_2 | kq \rangle R_{q_1}^{k_1}(1) S_{q_2}^{k_2}(2), \quad (8.106)$$

in which k takes the values $(k_1 + k_2), \ldots, |k_1 - k_2|$ and $q = q_1 + q_2$ and where the arguments (1) and (2) refer to the separate subspaces spanned by $\mathbf{R}^{k_1}$ and $\mathbf{S}^{k_2}$.

As an example of the use of (8.106), we consider two first-rank tensor operators, $k_1 = k_2 = 1$. Their direct product spans $k = 2, 1, 0$. For the case $k = 0$ we have, from (8.106),

$$T_0^0 = \{\mathbf{R}^1 \times \mathbf{S}^1\}_0^0 = \sum_{q_1 q_2} \langle 11 q_1 q_2 | 00 \rangle R_{q_1}^1(1) S_{q_2}^1(2)$$

$$= \sum_{q_1 q_2} (-1)^{1-1-0}(2 \times 0 + 1)^{1/2} \begin{pmatrix} 1 & 1 & 0 \\ q_1 & q_2 & 0 \end{pmatrix} R_{q_1}^1(1) S_{q_2}^1(2)$$

$$= \sum_{q_1 q_2} \begin{pmatrix} 1 & 1 & 0 \\ q_1 & q_2 & 0 \end{pmatrix} R_{q_1}^1(1) S_{q_2}^1(2). \quad (8.107)$$

Using (8.64) gives the 3–j symbol as $(-1)^{1-q_1} \delta_{q_1 q_2} 3^{-1/2}$ and, as q_1 and q_2 take the values $+1, 0, -1$, we have

$$T_0^0 = \{\mathbf{R}^1 \times \mathbf{S}^1\}_0^0$$
$$= 3^{-1/2}[R_1^1(1) S_{-1}^1(2) - R_0^1(1) S_0^1(2) + R_{-1}^1(1) S_1^1(2)]. \quad (8.108)$$

Similar calculations yield the components of $\mathbf{T}^1$:

$$
\left.\begin{aligned}
T_1^1 &= \{\mathbf{R}^1 \times \mathbf{S}^1\}_1^1 = 2^{-1/2}[R_1^1(1)S_0^1(2) - R_0^1(1)S_1^1(2)], \\
T_0^1 &= \{\mathbf{R}^1 \times \mathbf{S}^1\}_0^1 = 2^{-1/2}[R_1^1(1)S_{-1}^1(2) - R_{-1}^1(1)S_1^1(2)], \\
T_{-1}^1 &= \{\mathbf{R}^1 \times \mathbf{S}^1\}_{-1}^1 = 2^{-1/2}[R_{-1}^1(1)S_0^1(2) - R_0^1(1)S_{-1}^1(2)],
\end{aligned}\right\} \quad (8.109)
$$

and of $\mathbf{T}^2$:

$$
\begin{aligned}
T_2^2 &= \{\mathbf{R}^1 \times \mathbf{S}^1\}_2^2 = R_1^1(1)S_1^1(2), \\
T_1^2 &= \{\mathbf{R}^1 \times \mathbf{S}^1\}_1^2 = 2^{-1/2}[R_1^1(1)S_0^1(2) + R_0^1(1)S_1^1(2)], \\
T_0^2 &= \{\mathbf{R}^1 \times \mathbf{S}^1\}_0^2 = (\tfrac{2}{3})^{-1/2}[R_0^1(1)S_0^1(2)] \\
&\quad + 6^{-1/2}[R_1^1(1)S_{-1}^1(2) + R_{-1}^1(1)S_1^1(2)], \\
T_{-1}^2 &= \{\mathbf{R}^1 \times \mathbf{S}^1\}_{-1}^2 = 2^{-1/2}[R_{-1}^1(1)S_0^1(2) + R_0^1(1)S_{-1}^1(2)], \\
T_{-2}^2 &= \{\mathbf{R}^1 \times \mathbf{S}^1\}_{-2}^2 = R_{-1}^1(1)S_{-1}^1(2).
\end{aligned}
\quad (8.110)
$$

8.7.1 Scalar operators

Only scalar operators can appear in a Hamiltonian, so a particularly important case of the general expression (8.106) is

$$
\begin{aligned}
T_0^0 &= \{\mathbf{R}^{k_1}(1) \times \mathbf{S}^{k_2}(2)\}_0^0 = \sum_{q_1 q_2} \langle k_1 k_2 q_1 q_2 | 00 \rangle R_{q_1}^{k_1}(1)S_{q_2}^{k_2}(2) \\
&= \sum_{q_1 q_2} (-1)^{k_1 - k_2} \begin{pmatrix} k_1 & k_2 & 0 \\ q_1 & -q_2 & 0 \end{pmatrix} R_{q_1}^{k_1}(1)S_{q_2}^{k_2}(2), \quad (8.111)
\end{aligned}
$$

which describes how a zero-rank tensor operator is built from two commuting operators of any rank. However, the 3–j symbol, given by (8.64) vanishes unless $k_1 = k_2$, which is to say that scalar operators can only be constructed in this way from products of terms of the same (arbitrary) rank. Equation (8.111) then collapses, through (8.64), to

$$
T_0^0 \equiv \{\mathbf{R}^k(1) \times \mathbf{S}^k(2)\}_0^0 = \sum_q (-1)^{k-q}(2k+1)^{-1/2} R_q^k(1)S_{-q}^k(2) \quad (8.112)
$$

and, looking ahead, we begin to recognize the form of the usual binary products of spherical harmonic operators appearing in the calculation of matrix elements of the Coulomb operator.[2,11] An alternative form of (8.112) is in common use, namely

$$
(\mathbf{R}^k(1) \cdot \mathbf{S}^k(2)) = \sum_q (-1)^q R_q^k(1)S_{-q}^k(2), \quad (8.113)
$$

by which the 'dot product' emphasizes the scalar nature of the result, but confusion can arise out of the differing definitions: we note, therefore, that

$$(\mathbf{R}^k \cdot \mathbf{S}^k) = (-1)^k (2k+1)^{1/2} \{\mathbf{R}^k \times \mathbf{S}^k\}_0^0. \tag{8.114}$$

8.7.2 Product tensors from spherical harmonics

We have seen how spherical harmonics acting as operators describe irreducible spherical tensors and we also recall how products of such operators occur in the Coulomb operator $1/r_{ij}$; we may now see more fully why this is so. In what follows, it is less troublesome to use modified spherical harmonics defined by the equation,

$$C_q^k = \left(\frac{4\pi}{2k+1}\right)^{1/2} Y_q^k. \tag{8.115}$$

In figure 8.4 we define vectors and coordinates for two electrons relative, say, to the nucleus of an atom. The Coulomb operator e^2/r_{12} is used to calculate the electron–electron interaction energy and is written as $1/r_{12}$ in atomic units. In calculating matrix elements of this operator, our problem devolves around the explicit operation of a two-electron operator upon wavefunctions normally expressed as products of two one-electron functions (or orbitals). The normal procedure is to express the Coulomb operator as (a sum of) products of one-electron operators. This involves two steps. In the first we express the (scalar) distance between the electrons,

$$1/r_{12} = (r_1^2 + r_2^2 - 2r_1 r_2 \cos \omega)^{-1/2} \tag{8.116}$$

as a superposition[5] of Legendre polynomials,

$$1/r_{12} = \sum_k (r_<^k / r_>^{k+1}) P_k(\cos \omega), \tag{8.117}$$

where $r_<$ and $r_>$ label the lesser and greater of the distances r_1 and r_2. Note that the scalar nature of $1/r_{12}$ remains intact throughout (8.117) in that $\cos \omega = \mathbf{r}_1 \cdot \mathbf{r}_2$: that is, ω and r_{12} are the only quantities in the figure which are independent of the choice of axes.

The second step is to use the so-called *spherical harmonic addition theorem* which expresses a Legendre polynomial as a superposition of products of spherical harmonics,

$$P_k(\cos \omega) = \sum_q C_q^{k*}(\theta_1, \phi_1) C_q^k(\theta_2, \phi_2). \tag{8.118}$$

The theorem arises as follows. Consider modified spherical harmonics of

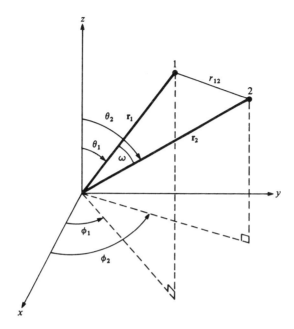

Fig. 8.4. ω and r_{12} are independent of the absolute orientation of the reference frame.

the polar coordinates of the electrons 1 and 2 in figure 8.4, $C_q^k(\theta_1, \phi_1)$ and $C_q^k(\theta_2, \phi_2)$, from which we may form the scalar product, as in (8.113):

$$\mathbf{C}^k(1) \cdot \mathbf{C}^k(2) = \sum_q (-1)^q C_q^k(\theta_1, \phi_1) C_{-q}^k(\theta_2, \phi_2). \qquad (8.119)$$

This is, of course, invariant with respect to rotation of the coordinate frame. Therefore, the scalar product must also be expressible as a function of the angle ω between the directions[††] $\hat{r}_1, (\theta_1, \phi_1)$ and $\hat{r}_2, (\theta_2, \phi_2)$, which angle is the only angular invariant in the system, as pointed out above. If we choose new axes, so that (θ_2, ϕ_2) becomes the new z axis,

$$C_q^k(\theta_2, \phi_2) \to C_q^k(0, 0) = \delta_{q,0} \qquad (8.120)$$

and

$$C_0^k(\theta_1, \phi_1) \to P_k(\cos \theta_1) = P_k(\cos \omega). \qquad (8.121)$$

Hence,

$$\mathbf{C}^k(1) \cdot \mathbf{C}^k(2) = P_k(\cos \omega), \qquad (8.122)$$

which is the spherical harmonic addition theorem (8.118), when we

---

[††] We occasionally use the rotation $\mathbf{r} \equiv (r, \hat{r})$ where $r = |\mathbf{r}|$ and $\hat{r}$ defines the polar angles of $\mathbf{r}$.

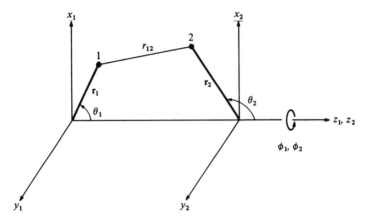

Fig. 8.5. Bipolar harmonics are used for two-centre expansions.

recognize that

$$(-1)^q C_q^k(\theta_1, \phi_1) = C_q^{k*}(\theta_1, \phi_1). \tag{8.123}$$

The Coulomb operator thus takes the form,

$$1/r_{ij} = \sum_k (r_<^k / r_>^{k+1})(\mathbf{C}_1^k \cdot \mathbf{C}_2^k), \tag{8.124}$$

and we shall use this expression later.

The foregoing addition theorem and overall (Neumann) expansion of $1/r_{ij}$ was referred to a single centre. There exists a corresponding expansion with respect to two centres which we shall use in §11.10.2. The corresponding geometrical details are shown in figure 8.5. Now, if $\mathbf{u}_1$ and $\mathbf{u}_2$ define unit vectors parallel to $\mathbf{r}_1$ and $\mathbf{r}_2$, we may construct so-called *bipolar harmonics* $\mathbf{B}^k$ as tensors formed from $C_{q_1}^{k_1}(\mathbf{u}_1)$ and $C_{q_2}^{k_2}(\mathbf{u}_2)$: most generally, we have,

$$B_q^k(k_1 k_2) = \sum_{q_1 q_2} \langle k_1 k_2 kq | q_1 q_2 \rangle C_{q_1}^{k_1}(\mathbf{u}_1) C_{q_2}^{k_2}(\mathbf{u}_2). \tag{8.125}$$

Bipolar harmonics are orthogonal for integration over the angles $\hat{u}_1$ and $\hat{u}_2$ of $\mathbf{u}_1$ and $\mathbf{u}_2$,

$$\int d\mathbf{u}_1 \int d\mathbf{u}_2 B_q^{k*}(k_1 k_2) B_{q'}^{k'}(k_1' k_2') = 16\pi^2 \frac{\delta_{k_1 k_1'} \delta_{k_2 k_2'} \delta_{kk'} \delta_{qq'}}{(2k_1 + 1)(2k_2 + 1)}, \tag{8.126}$$

which expression[10] is analogous to (8.32). Bipolar harmonics also possess the closure property,

$$\sum_{kq} |B_q^k(k_1 k_2)|^2 = 1. \tag{8.127}$$

When $k = q = 0$, (8.125) reduces to a spherical harmonic addition theorem relative to two centres:

$$B_0^0(k_1 k_2) = (-1)^{k_1} \delta(k_1 k_2) P_{k_1} (\mathbf{u}_1 \cdot \mathbf{u}_2)/(2k_1 + 1). \tag{8.128}$$

Our interest in bipolar harmonics emerges when we consider the electrostatic perturbation of metal d orbitals by the electron density of a ligand which lies, of course, at a different site.

8.8 The Wigner–Eckart theorem

The whole purpose of tensor-operator theory is to exploit symmetry to the fullest extent so that much lengthy computation may be avoided. So far we have discussed the transformation properties of both functions and operators in the three-dimensional rotation group R_3 and made connection with the coupling of two angular momenta. These ideas are now brought together, quite simply, into what is perhaps the central and most useful theorem of the subject. It is a commonplace use of symmetry to assert that a matrix element $\langle \psi_i | O | \psi_j \rangle$ vanishes unless the triple direct product of the representations – $\Gamma(\psi_i) \times \Gamma(O) \times \Gamma(\psi_j)$ – contains the totally symmetric representation, Γ_1. We should note two points about this procedure; firstly, that it cannot be carried out until both functions and operator are classified according to the symmetry of the relevant group; and secondly, that the conclusion, though valuable, is meagre. The Wigner–Eckart theorem we now prove still requires the symmetry classification of both functions and operator, of course, but is much more informative. It states that the matrix elements of an irreducible tensor operator are proportional to Clebsch–Gordan coupling coefficients,

$$\langle jm | T_{m_2}^{j_2} | j_1 m_1 \rangle \propto \langle j_1 j_2 m_1 m_2 | jm \rangle. \tag{8.129}$$

8.8.1 Proof

We address ourselves to the general matrix elements $\langle \alpha' j' m' | T_q^k | \alpha j m \rangle$ in which α, α' describe any non-rotational properties required to complete the description of the states in bra and ket: they refer to radial properties of the functions, principal quantum numbers being one example of these labels. First consider the effect of the operator T_q^k upon $|\alpha j m\rangle$. The operator generally may have both radial and angular properties. The radial property will transform $|\alpha..\rangle$ into $|\beta..\rangle$, a change (which could be no change, of course) which will apply equally to all $|jm\rangle$ states as, by definition, the radial property is spherically symmetric. The angular part of T_q^k changes the angular momentum description, $|jm\rangle$ of the initial state. The way in which this occurs becomes clear when we recall the equivalent symmetry properties of irreducible spherical tensors and angular momentum states. Thus, $T_q^k|\alpha jm\rangle$ may be viewed as a product state vector which transforms according to the representation $\mathscr{D}^k \times \mathscr{D}^j$ of R_3. Before reducing this representation, let us be quite sure why $T_q^k|\alpha jm\rangle$ may be so regarded.

Under a general rotation $D(\alpha\beta\gamma)$, which we write as D to avoid any confusion between αs, the product $T_q^k|\alpha jm\rangle$ transforms as,

$$D[T_q^k|\alpha jm\rangle] = DT_q^k D^{-1} D|\alpha jm\rangle$$

$$= \sum_{q'} \mathscr{D}_{q'q}^k T_{q'}^k \sum_{m'} \mathscr{D}_{m'm}^j |\alpha jm'\rangle$$

$$= \sum_{q'm'} \mathscr{D}_{q'q}^k \mathscr{D}_{m'm}^j T_{q'}^k |\alpha jm'\rangle, \qquad (8.130)$$

where we have used (8.21) and (8.85). Hence we see that the $(2k + 1)(2j + 1)$ products implied by $T_q^k|\alpha jm\rangle$ do indeed transform under rotations as bases for the direct product representation $\mathscr{D}^k \times \mathscr{D}^j$. This space can be decomposed into irreducible spaces $\mathscr{D}^J$ in which, by (8.70), J ranges $(k + j)$ to $|k - j|$: that is to say that $T_q^k|\alpha jm\rangle$ may be written as a linear combination of the states $|\beta JM\rangle$,

$$T_q^k|\alpha jm\rangle = C_{J_1 M_1}|\beta J_1 M_1\rangle \dots + C_{J_n M_{n'}}|\beta J_n M_{n'}\rangle \dots \qquad (8.131)$$

The expansion coefficients are just Clebsch–Gordan coupling coefficients, of course, for we may use (8.48) to get

$$T_q^k|\alpha jm\rangle = \sum_{JM} \langle kjqm|JM\rangle|\beta JM\rangle. \qquad (8.132)$$

The matrix elements we wish to simplify are then obtained by taking the scalar product of both sides of (8.132) with $\langle \alpha'j'm'|$:

$$\langle \alpha'j'm'|T_q^k|\alpha jm\rangle = \sum_{JM} \langle kjqm|JM\rangle\langle \alpha'j'm'|\beta JM\rangle. \qquad (8.133)$$

Now the scalar product $\langle \alpha'j'm'|\beta JM\rangle$ vanishes unless $J = j'$ and $M = m'$ and this condition upon J and M reduces the summation in (8.133) to a single term:

$$\langle \alpha'j'm'|T_q^k|\alpha jm\rangle = \langle kjqm|j'm'\rangle\langle \alpha'j'm'|\beta j'm'\rangle. \qquad (8.134)$$

The inner product of state vectors $\langle \alpha'j'm'|\beta j'm'\rangle$ is independent of m' for it involves the quantity $e^{-im'\phi} \cdot e^{im'\phi} = 1$, for any m'. We could therefore rewrite (8.134) as,

$$\langle \alpha'j'm'|T_q^k|\alpha jm\rangle = \langle kjqm|j'm'\rangle\langle \alpha'j' \cdot |\beta j' \cdot \rangle \qquad (8.135)$$

and in this unusual form we have the essence of the Wigner–Eckart theorem given in (8.129): that is, that the general matrix element on the left-hand side is proportional to a coupling coefficient, the constant of proportionality, which we have temporarily written $\langle \alpha'j' \cdot | \beta j' \cdot \rangle$, being independent of any magnetic quantum number. The reason why this is such a powerful result will be clarified further very shortly. Meanwhile, however, we must discuss something of the nature of the proportionality constant. It involves α' and j' which appeared in the original matrix element; β which appeared nowhere originally; and fails to mention either α or j which did occur at the beginning! The point here, of course is that the effect of the operator T_q^k upon $|\alpha jm\rangle$ is to produce $|\beta JM\rangle$ where $J = j'$ and $M = m'$ (ultimately). Therefore, a better symbol for the proportionality constant would explicitly mention not only α' and j' but also α and j, together with the operator which gave rise to the changes described. We use the symbol

$$\langle \alpha'j' \| \mathbf{T}^k \| \alpha j \rangle \equiv \langle \alpha'j' \cdot | \beta j' \cdot \rangle \tag{8.136}$$

and this double-barred matrix element is called a *reduced matrix element* (RME). The Wigner–Eckart equation now takes the form,

$$\langle \alpha'j'm' | T_q^k | \alpha jm \rangle = \langle kjqm|j'm' \rangle \langle \alpha'j' \| \mathbf{T}^k \| \alpha j \rangle. \tag{8.137}$$

One further word about why the RME notation is sensible. In equations like (8.137), the RME is to be used in conjunction with appropriate coupling coefficients, as shown. Now these involve, not only the rotational descriptions of bra and ket of the matrix elements to be evaluated, but also the rank of the tensor operator involved. The rank k of the tensor must therefore be mentioned in the RME symbol. At the same time something more specific must be included in the description concerning $\mathbf{T}^k$ as the possible natures of α and α' leading to non-zero results depend upon this: thus, the symbol $\langle \alpha'j' \| k \| \alpha j \rangle$ would not suffice. Of course, the mere act of writing down the proportionality constant as the RME in (8.136) does not define its value. While that is a matter we discuss later, for the moment we are really only concerned with definitions and, above all, with the Wigner–Eckart 'factorization', (8.137).

The theorem may be expressed in terms of 3–j symbols, so that (8.137) becomes

$$\langle \alpha'j'm' | T_q^k | \alpha jm \rangle = (-1)^{k-j+m'} (2j'+1)^{1/2} \begin{pmatrix} j' & k & j \\ -m' & q & m \end{pmatrix} \langle \alpha'j' \| \mathbf{T}^k \| \alpha j \rangle. \tag{8.138}$$

However, this form is not used: instead we have

$$\langle\alpha'j'm'|T_q^k|\alpha jm\rangle = (-1)^{j'-m'}\begin{pmatrix} j' & k & j \\ -m' & q & m \end{pmatrix}\langle\alpha'j'\|\mathbf{T}^k\|\alpha j\rangle, \quad (8.139)$$

and the missing factor of $\pm(2j'+1)^{1/2}$ is incorporated within the RME symbol. There is no problem here as the RMEs are only proportionality constants which have yet to be evaluated anyway. On the other hand it is essential to make sure that sources of RMEs are compatible with the form of the Wigner–Eckart theorem being employed.

8.8.2 Discussion

Being based only upon the transformation properties of functions and operators, the Wigner–Eckart theorem possesses great generality. The symbol $|jm\rangle$ can refer to spatial or spin arguments associated with a single-electron or a many-electron function, provided that the quantum numbers j and m define the angular momenta of the whole function.

The value of the RME depends upon the detailed forms of both operator and states and, as yet, we have not discussed any ways in which RMEs may be computed. On the other hand, the coupling coefficients (or 3–j symbols) depend only upon the rotational symmetry properties. The Wigner–Eckart theorem thus 'factorizes' matrix elements into physical and geometrical parts – the RME and 3–j symbols respectively. This factorization is the source of great saving of labour in the calculation of complete matrices. For example, if we wish to compute the matrix between d orbitals under the spherical harmonic operators $\mathbf{Y}^2$, we have $(2j+1)(2k+1)(2j'+1) = 125$ matrix elements to evaluate, corresponding to the total number of combinations of magnetic quantum numbers appearing in the functions and operator. All 125 matrix elements, however, are proportional to the same RME and we have,

$$\langle\alpha d_m|Y_q^2|\alpha'd_{m'}\rangle = (-1)^{2-m}\begin{pmatrix} 2 & 2 & 2 \\ -m & q & m' \end{pmatrix}\langle\alpha 2\|\mathbf{T}^2\|\alpha'2\rangle. \quad (8.140)$$

On the question of the calculation of RMEs we make three points. Firstly that those appropriate for many-electron systems – here, that is more than two electrons – are rather difficult to evaluate; the usual method, apart from reference to tables, being to use so-called coefficients of fractional parentage. Next we note that in any problem, an RME, which is only a proportionately constant, may be evaluated by computing one matrix element by some other means (obviously we choose the simplest) which may then be compared with the same element calculated by (8.139). In parenthesis, we note here the basis of the method of operator equivalents[15] within the Wigner–Eckart theorem. Similarly, we recall Stevens' orbital reduction factor[16] k occurring in the magnetic moment operator $\hat{\mu}_i = \beta_0(k\hat{l}_i + 2\hat{s}_i)$: it is no coincidence that all elements of the matrix of d orbitals under kl_i are simply k times those under l_i, for the relationships amongst the matrix elements are determined simply by the coupling coefficients. Finally, we may illustrate the comparative method by computing the reduced matrix element in (8.140). Thus, in (8.76) we derived an expression for the general matrix element $\langle Y^{l_1}_{m_1} | Y^{l_2}_{m_2} | Y^{l_3}_{m_3} \rangle$. Take the particular case for $m_1 = m_2 = m_3 = 0$,

$$\langle Y^{l_1}_0 | Y^{l_2}_0 | Y^{l_3}_0 \rangle = \left[\frac{(2l_1 + 1)(2l_2 + 1)(2l_3 + 1)}{4\pi} \right]^{1/2} \begin{pmatrix} l_1 & l_2 & l_3 \\ 0 & 0 & 0 \end{pmatrix}^2. \quad (8.141)$$

Using the Wigner–Eckart theorem instead, we have

$$\langle l_1 0 | Y^{l_2}_0 | l_3 0 \rangle = (-1)^{l_1} \begin{pmatrix} l_1 & l_2 & l_3 \\ 0 & 0 & 0 \end{pmatrix} \langle l_1 \| \mathbf{Y}^{l_2} \| l_3 \rangle \quad (8.142)$$

and, by equating these two expressions, we find

$$\langle l_1 \| \mathbf{Y}^{l_1} \| l_3 \rangle = (-1)^{l_1} \left[\frac{(2l_1 + 1)(2l_2 + 1)(2l_3 + 1)}{4\pi} \right]^{1/2} \begin{pmatrix} l_1 & l_2 & l_3 \\ 0 & 0 & 0 \end{pmatrix} \quad (8.143)$$

or, in terms of the modified spherical harmonics, defined in (8.115),

$$\langle l_1 \| \mathbf{C}^k \| l_3 \rangle = (-1)^{l_1} [(2l_1 + 1)(2l_3 + 1)]^{1/2} \begin{pmatrix} l_1 & k & l_3 \\ 0 & 0 & 0 \end{pmatrix}. \quad (8.144)$$

We shall find this result useful in chapter 9. Note that the 3–j symbol, and hence the RME, vanishes when $l_1 = l_3$ unless k is even: this is also a well-known parity rule.

8.9 Coupling of three angular momenta

The counting scheme of some rude tribes today and, no doubt, of all our ancestors – namely 'one, two, many' – surely identifies a basic truth: that while 'two' is different from 'one', 'many' possesses a quality quite different again. Our understanding of angular momentum broadens significantly when we turn from the coupling of two angular momentum vectors to three. Thereafter, further complexities are more along the lines of theme and variations.

Analogous to the discussion in §8.3, we write the product of three angular momentum vectors $|j_1 m_1\rangle$, $|j_2 m_2\rangle$ and $|j_3 m_3\rangle$ as $|j_1 j_2 j_3 m_1 m_2 m_3\rangle$. This uncoupled product is an eigenfunction of the operators $\mathbf{j}_1^2$, $\mathbf{j}_2^2$, $\mathbf{j}_3^2$, j_{1z}, j_{2z}, j_{3z}, of course. After coupling we get a state with quantum numbers J and M labelling eigenstates of $\mathbf{J}^2 = (\mathbf{j}_1 + \mathbf{j}_2 + \mathbf{j}_3)^2$ and of $J_z = j_{1z} + j_{2z} + j_{3z}$. However, the latter is no longer unique and we require a further quantum number. The central question, and the one which distinguishes the many-vector case from the two-vector one, is the *order* of coupling. For coupling, by its very name of course, can only involve two entities at a time: others must take their turn. With three vectors, there are clearly three possible coupling sequences. Suppose, first, that we couple $\mathbf{j}_1$ and $\mathbf{j}_2$ to form $\mathbf{j}_{12}$, subsequently adding $\mathbf{j}_3$ vectorially to give $\mathbf{J}$. Using the coupling equation (8.47), the first process is accomplished by

$$|j_1 j_2 j_{12} m_{12}\rangle = \sum_{m_1 m_2} |j_1 j_2 m_1 m_2\rangle \langle j_1 j_2 m_1 m_2 | j_{12} m_{12}\rangle \qquad (8.145)$$

and the second by

$$|a\rangle \equiv |(j_1 j_2) j_{12} j_3 J M\rangle = \sum_{m_{12} m_3} |j_1 j_2 j_{12} j_3 m_{12} m_3\rangle \langle j_{12} j_3 m_{12} m_3 | J M\rangle.$$
$$(8.146)$$

This state is an eigenfunction of $\mathbf{j}_1^2$, $\mathbf{j}_2^2$, $\mathbf{j}_3^2$, $\mathbf{J}^2$ and also of $\mathbf{j}_{12}^2 = (\mathbf{j}_1 + \mathbf{j}_2)^2$: j_{12} provides the additional quantum number to complete the state specification.

A second, alternative, coupling sequence would be to couple $\mathbf{j}_2$ and $\mathbf{j}_3$ to give $\mathbf{j}_{23}$, followed by vectorial addition of $\mathbf{j}_1$ to give the final resultant $\mathbf{J}$. Corresponding to (8.145) and (8.146), we have

$$|j_2 j_3 j_{23} m_{23}\rangle = \sum_{m_2 m_3} |j_2 j_3 m_2 m_3\rangle \langle j_2 j_3 m_2 m_3 | j_{23} m_{23}\rangle \qquad (8.147)$$

and

$$|b\rangle \equiv |j_1 (j_2 j_3) j_{23} J M\rangle = \sum_{m_1 m_{23}} |j_1 j_2 j_3 j_{23} m_1 m_{23}\rangle \langle j_1 j_{23} m_1 m_{23} | J M\rangle$$
$$(8.148)$$

and this state is an eigenfunction of $\mathbf{j}_{23}^2 = (\mathbf{j}_2 + \mathbf{j}_3)^2$.

The third coupling sequence would define states $|c\rangle \equiv |(j_1 j_3) j_{13} j_2 J M\rangle$ as eigenfunctions of $\mathbf{j}_{13}^2 = (\mathbf{j}_1 + \mathbf{j}_3)^2$. Now the three sets of $\{|a\rangle\}$, $\{|b\rangle\}$, $\{|c\rangle\}$ describe three different representations of the same states, namely the results of coupling together the original three one-momentum vectors. These three representations are clearly not independent for they span the same subspace. They must be related by a linear transformation.

$$|a\rangle = \sum_b |b\rangle\langle b|a\rangle, \text{etc.,} \tag{8.149}$$

in which the expansion coefficients $\langle b|a\rangle$ refer to the relationships between the various coupling schemes. To be explicit, the transformation coefficients between the first and second sequences above are written

$$|(j_1 j_2)j_{12}j_3 JM\rangle = \sum_{j_{23}} |j_1(j_2 j_3)j_{23} JM\rangle\langle j_1(j_2 j_3)j_{23}J|(j_1 j_2)j_{12}j_3 J\rangle.$$
$$\tag{8.150}$$

Note that, by the orthogonality of $\langle\ldots M|\ldots M'\rangle$, the summation here only extends over j_{23} and that, as in the definition (8.135), the surviving expansion coefficients are independent of M. Like the RME discussed above, these transformation, or recoupling, coefficients are therefore scalars.

Racah[12] introduced his W function,

$$\langle j_1(j_2 j_3)j_{23}J|(j_1 j_2)j_{12}j_3 \, J\rangle = [(2j_{12} + 1)(2j_{23} + 1)]^{1/2} W(j_1 j_2 Jj_3; j_{12}j_{23}),$$
$$\tag{8.151}$$

with a normalization chosen to simplify its properties. Later Wigner defined the 6–j symbol, being closely related to Racah's W coefficient, but possessing even more symmetry, as we shall discuss. The 6–j symbol is defined by

$$\begin{Bmatrix} j_1 & j_2 & j_{12} \\ j_3 & J & j_{23} \end{Bmatrix} = (-1)^{j_1+j_2+j_3+J} W(j_1 j_2 Jj_3; j_{12}j_{23})$$

$$= (-1)^{j_1+j_2+j_3+J}[(2j_{12} + 1)(2j_{23} + 1)]^{-1/2}$$

$$\times \langle (j_1 j_2)j_{12}j_3 J|j_1(j_2 j_3)j_{23}J\rangle. \tag{8.152}$$

Note the convention in which 6–j symbols are written within braces, while 3–j symbols appear in round brackets. While in algebraic form this differentiation is unnecessary – for the bottom line of 3–j symbols involve magnetic quantum numbers – in numerical form, wholly or partly, the distinction is essential.

We began this section by drawing attention to the qualitative change involved when we couple three angular momenta rather than two. The new concept introduced thereby is not so much the 6–j symbol, although it is a most convenient entity, but rather the need for it; that is to say, the new physical element is the emergence of a multiplicity of coupling sequences once we pass from 'two' to 'many'. As if to emphasize this view we follow through a simple process by which W (and hence $6{-}j$) coefficients may be expressed in terms of (Clebsch–Gordan) coupling coefficients: ultimately, of course, this means that the $6{-}j$ symbols could be dispensed with, even though the number of permutations of two angular momenta amongst n would remain. The expansion of the W coefficients proceeds as follows. We expand both sides of (8.150) in the uncoupled representation $|j_1 j_2 j_3 m_1 m_2 m_3\rangle$ using (8.145) and (8.146) in the left-hand side and (8.147) and (8.148). On the left-hand side we get,

$$|(j_1 j_2) j_{12} j_3 JM\rangle = \sum_{m_{12} m_3} \sum_{m_1 m_2} |j_1 j_2 j_3 m_1 m_2 m_3\rangle \langle j_1 j_2 m_1 m_2 | j_{12} m_{12}\rangle$$

$$\times \langle j_{12} j_3 m_{12} m_3 | JM\rangle, \tag{8.153}$$

and, on the right-hand side,

$$|j_1 (j_2 j_3) j_{23} JM\rangle = \sum_{m_1 m_{23}} \sum_{m_2 m_3} |j_1 j_2 j_3 m_1 m_2 m_3\rangle \langle j_2 j_3 m_2 m_3 | j_{23} m_{23}\rangle$$

$$\times \langle j_1 j_{23} m_1 m_{23} | JM\rangle. \tag{8.154}$$

On substitution of (8.153) and (8.154) into (8.150) and equating coefficients, we obtain

$$\langle j_1 j_2 m_1 m_2 | j_{12} m_{12}\rangle \langle j_{12} j_3 m_{12} m_3 | JM\rangle$$

$$= \sum_{j_{23}} \langle j_2 j_3 m_2 m_3 | j_{23} m_{23}\rangle \langle j_1 j_{23} m_1 m_{23} | JM\rangle \langle j_1 (j_2 j_3) j_{23} J | (j_1 j_2) j_{12} j_3 J\rangle.$$

$$\tag{8.155}$$

Temporarily we introduce some simplifying notation and write a, b, c, d, e, f for the vectors and α, β, γ, δ, ε, ϕ for the corresponding z components. Rewriting (8.155) in this notation, we have

$$\langle ab\alpha\beta | e, \alpha + \beta\rangle \langle ed, \alpha + \beta, \gamma - \alpha - \beta | c\gamma\rangle$$

$$= \sum_f \langle bd\beta, \gamma - \alpha - \beta | f, \gamma - \alpha\rangle \langle af\alpha, \gamma - \alpha | c\gamma\rangle$$

$$\times [(2e+1)(2f+1)]^{1/2} W(abcd; ef). \tag{8.156}$$

We multiply both sides by $\langle bd\beta, \gamma - \alpha - \beta | f', \gamma - \alpha\rangle$ and sum over β to obtain

$$\langle af\alpha, \gamma - \alpha | c\gamma\rangle [(2e+1)(2f+1)]^{1/2} W(abcd; ef)$$

$$= \sum_\beta \langle ab\alpha\beta | e, \alpha + \beta\rangle \langle ed, \alpha + \beta, \gamma - \alpha - \beta | c\gamma\rangle \langle bd\beta, \gamma - \alpha - \beta | f, \gamma - \alpha\rangle$$

$$\tag{8.157}$$

and, proceeding similarly, we finally obtain

$$[(2e + 1)(2f + 1)]^{1/2} W(abcd;ef)\delta_{cc'}\delta_{\gamma\gamma'}$$

$$= \sum_{\alpha\beta} \langle (ab\alpha\beta|e,\alpha + \beta\rangle\langle ed,\alpha + \beta,\gamma - \alpha - \beta|c\gamma\rangle$$

$$\times \langle bd\beta, \gamma - \alpha - \beta|f,\gamma - \alpha\rangle\langle af\alpha,\gamma - \alpha|c'\gamma'\rangle, \qquad (8.158)$$

in which summation, γ is held constant. Thus Racah's W function appears as a scalar invariant obtained by contraction of four vector coupling coefficients. Equation (8.158) obviously points the way to a general formula for the evaluation of W coefficients and Racah has derived from it and (8.53) the general expression,

$$W(abcd;ef) = \Delta(abc)\Delta(acf)\Delta(bdf)\Delta(cde)$$

$$\times \sum_{z}(-1)^{z}(a + b + c + d + 1 - z)![z!(e + f - a - d + z)!$$

$$\times (e + f - b - c + z)!(a + b - e - z)!(c + d - e - z)!$$

$$\times (a + c - f - z)!(b + d - f - z)!]^{-1}, \qquad (8.159)$$

where

$$\Delta(abc) = \left[\frac{(a + b - c)!(a + c - b)!(b + c - a)!}{(a + b + c + 1)!}\right]^{1/2}.$$

As for 3–j symbols, Rotenberg *et al.*[13] have tabulated a great many 6–j symbol values, again in the format of powers of primes discussed in §8.3.2. Once again, for our purposes it is not worth storing all these values in a computer file and we therefore compute the values of 6–j symbols as required: Appendix C describes a FORTRAN routine for this purpose.

8.9.1 Properties of the recoupling coefficients

A rather complete list of orthogonality and symmetry properties of the W coefficients is given in Appendix II of Brink & Satchler.[10] The symmetry properties of the recoupling coefficients are more memorable using Wigner's 6–j symbols so we list here a few general properties, symmetry and others, of 6–j symbols, including an interesting graphical representation of some of these fascinating properties.

Symmetry. The 6–j symbol is unchanged by interchange of any two

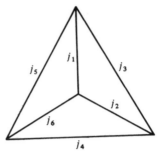

Fig. 8.6. Illustrating the symmetry of the 6–j symbol.

columns or by interchange of the upper and lower members of any two columns, e.g.,

$$\begin{Bmatrix} j_1 & j_2 & j_3 \\ j_4 & j_5 & j_6 \end{Bmatrix} = \begin{Bmatrix} j_4 & j_5 & j_3 \\ j_1 & j_2 & j_6 \end{Bmatrix} = \begin{Bmatrix} j_2 & j_1 & j_3 \\ j_5 & j_4 & j_6 \end{Bmatrix}. \qquad (8.160)$$

The symmetry of the 6–j symbol is shown most dramatically, however, by diagrams introduced by Levinson.[17] In figure 8.6 the edges of a tetrahedron represent the various js. The 6–j symbol symmetry is then given by noting that any of the 24 rotations and reflections which transform the tetrahedron into itself leave the 6–j symbol unchanged in value. For example, the first and third symbols in (8.160) are related in the diagram by a reflection in a plane perpendicular to the page and containing j_6. Further, if the sides of the figure are made proportional to the respective j values, a closed solid can only be obtained if the triangle condition holds for each face which, as we see from the general expression (8.159), is indeed a general condition for the 6–j symbol not to vanish. Pictorially the Δ functions in (8.159) may be illustrated as follows:

$$\left| \begin{matrix} \end{matrix} \right\} \quad \left| \begin{matrix} \end{matrix} \right\} \quad \left| \begin{matrix} \end{matrix} \right\} \quad \left| \begin{matrix} \end{matrix} \right\} \qquad (8.161)$$

Although we do not pursue the matter here, it is worth mentioning that diagrams like those above are two of many which may be used, not only to represent symmetries, but also coupling schemes and even to calculate matrix elements.[10]

Orthogonality and special cases. By rewriting (8.158) in terms of 6–j

symbols, we can express the 6–j symbol in terms of 3–j symbols:

$$\begin{Bmatrix} j_1 & j_2 & j_3 \\ j_4 & j_5 & j_6 \end{Bmatrix} = \sum_{\text{all } n, m_1, m_2} (-1)^{j_4+j_5+j_6+n_1+n_2+n_3}(2j_3+1)\begin{pmatrix} j_1 & j_2 & j_3 \\ m_1 & m_2 & m_3 \end{pmatrix}$$

$$\times \begin{pmatrix} j_1 & j_5 & j_6 \\ m_1 & n_5-n_6 \end{pmatrix}\begin{pmatrix} j_4 & j_2 & j_6 \\ -n_4 & m_2 & m_6 \end{pmatrix}\begin{pmatrix} j_4 & j_5 & j_3 \\ m_4 & -n_5 & m_3 \end{pmatrix},$$

$$(8.162)$$

in which each 3–j symbol corresponds to one side of the tetrahedron in figure 8.6: the triangle conditions on the faces, in the Δ symbols of (8.159), and of the 3–j symbols in (8.162) are equivalent.

The orthogonality relationship for 6–j symbols is

$$\sum_{j_3} (2j_3+1)\begin{Bmatrix} j_1 & j_2 & j_3 \\ j_4 & j_5 & j_6 \end{Bmatrix}\begin{Bmatrix} j_1 & j_2 & j_3 \\ j_4 & j_5 & j_6' \end{Bmatrix} = \delta_{j_6 j_6'}(2j_6+1). \quad (8.163)$$

Three special cases of 6–j symbols are of interest for us:
(i) a 6–j symbol with one zero collapses:

$$\begin{Bmatrix} j_1 & j_2 & 0 \\ j_4 & j_5 & j_6 \end{Bmatrix} = (-1)^{j_1+j_4+j_6}[(2j_1'+1)(2j_4+1)]^{-1/2}\delta_{j_1 j_2}\delta_{j_4 j_5}, \quad (8.164)$$

in which the δ functions are obvious if $j_3 = 0$ in figure 8.6.
(ii)

$$\begin{Bmatrix} L & L & 1 \\ S & S & J \end{Bmatrix} = (-1)^{S+L+J}\frac{J(J+1)-S(S+1)-L(L+1)}{[S(2S+1)(2S+2)L(2L+1)(2L+2)]^{1/2}}$$

$$(8.165)$$

(iii)

$$\begin{Bmatrix} j_1 & j_2 & j_3 \\ 1 & j_3 & j_2 \end{Bmatrix} =$$

$$= (-1)^{j_1+j_2+j_3+1}\frac{2[j_2(j_2+1)+j_3(j_3+1)-j_1(j_1+1)]}{[2j_2(2j_2+1)(2j_2+2)2j_3(2j_3+1)(2j_3+2)]^{1/2}}. \quad (8.166)$$

8.10 Coupling of four angular momenta

While no new principles are involved when we move from a consideration of the coupling of three angular momenta to four, the question is not without interest or utility. When four states $|jm\rangle$ are coupled together, there result $(2j_1+1)(2j_2+1)(2j_3+1)(2j_4+1)$ orthonormal components

$|JM\rangle$. The coupling processes may be carried out in many different sequences, or 'coupling schemes', being interrelated by unitary transformations. Consider the relationship between just two such schemes:

$$\begin{aligned} \mathbf{j}_1 \text{ with } \mathbf{j}_2 &\to \mathbf{j}_{12} \searrow \\ \mathbf{j}_3 \text{ with } \mathbf{j}_4 &\to \mathbf{j}_{34} \nearrow \end{aligned} \mathbf{J}$$

$$\begin{aligned} \mathbf{j}_1 \text{ with } \mathbf{j}_3 &\to \mathbf{j}_{13} \searrow \\ \mathbf{j}_2 \text{ with } \mathbf{j}_4 &\to \mathbf{j}_{24} \nearrow \end{aligned} \mathbf{J} \tag{8.167}$$

To be concrete, such a pair of coupling schemes might represent the Russell–Saunders and '*j*–*j*' coupling schemes which we would write analogously as

$$\begin{aligned} \mathbf{l}_1 \text{ with } \mathbf{l}_2 &\to \mathbf{L} \searrow \\ \mathbf{s}_1 \text{ with } \mathbf{s}_2 &\to \mathbf{S} \nearrow \end{aligned} \mathbf{J} \text{ Russell–Saunders}$$

$$\begin{aligned} \mathbf{l}_1 \text{ with } \mathbf{s}_1 &\to \mathbf{j}_1 \searrow \\ \mathbf{l}_2 \text{ with } \mathbf{s}_2 &\to \mathbf{j}_2 \nearrow \end{aligned} \mathbf{J} \quad \text{'}j\text{–}j\text{' coupling.} \tag{8.168}$$

The states in the first scheme of (8.167) are eigenfunctions of $\mathbf{j}_{12}^2 = (\mathbf{j}_1 + \mathbf{j}_2)^2$ and of $\mathbf{j}_{34}^2 = (\mathbf{j}_3 + \mathbf{j}_4)^2$ while those in the second scheme are eigenstates of $\mathbf{j}_{13}^2$ and $\mathbf{j}_{24}^2$. The two sets of states $|JM\rangle$ are related by the unitary transformation

$$|(j_1 j_2) j_{12} (j_3 j_4) j_{34} JM\rangle = \sum_{j_{13} j_{24}} \langle (j_1 j_2) j_{12} (j_3 j_4) j_{34} J | (j_1 j_3) j_{13} (j_2 j_4) j_{24} J \rangle$$
$$\times |(j_1 j_3) j_{13} (j_2 j_4) j_{24} JM\rangle, \tag{8.169}$$

the expansion coefficients of which correspond to Fano's X coefficients.[18] Again somewhat more symmetrical coefficients have been constructed by Wigner, the 9–*j* symbol being defined as

$$\langle (j_1 j_2) j_{12} (j_3 j_4) j_{34} j | (j_1 j_3) j_{13} (j_2 j_4) j_{24} j \rangle$$

$$= [(2j_{12} + 1)(2j_{34} + 1)(2j_{13} + 1)(2j_{24} + 1)]^{1/2} \begin{Bmatrix} j_1 & j_2 & j_{12} \\ j_3 & j_4 & j_{34} \\ j_{13} & j_{24} & j \end{Bmatrix} \tag{8.170}$$

Smith[19] and Howell[20] have provided listings of many 9–*j* symbols, but we shall not require them. Our main reason for introducing these symbols is so as to be able to draw on a most convenient general property of the symbol and upon a special case, as follows.

In (8.162) we saw how a 6–j symbol may be expressed in terms of 3–j symbols. It can be shown that a 9–j symbol may be expressed in terms of 6–j symbols and thence, via (8.162), in terms of 3–j symbols. There emerges the beautifully symmetrical result:

$$
\begin{Bmatrix} j_{11} & j_{12} & j_{13} \\ j_{21} & j_{22} & j_{23} \\ j_{31} & j_{32} & j_{33} \end{Bmatrix} = \sum_{\text{all } m} \begin{pmatrix} j_{11} & j_{12} & j_{13} \\ m_{11} & m_{12} & m_{13} \end{pmatrix} \begin{pmatrix} j_{21} & j_{22} & j_{23} \\ m_{21} & m_{22} & m_{23} \end{pmatrix}
$$

$$
\times \begin{pmatrix} j_{31} & j_{32} & j_{33} \\ m_{31} & m_{32} & m_{33} \end{pmatrix} \begin{pmatrix} j_{11} & j_{21} & j_{31} \\ m_{11} & m_{21} & m_{31} \end{pmatrix} \begin{pmatrix} j_{12} & j_{22} & j_{32} \\ m_{12} & m_{22} & m_{32} \end{pmatrix}
$$

$$
\times \begin{pmatrix} j_{13} & j_{23} & j_{33} \\ m_{13} & m_{23} & m_{33} \end{pmatrix} \tag{8.171}
$$

If one j in the 9–j symbol is zero, we find two 3–j symbols within (8.171) with j and its corresponding m equal to zero (the latter because m can equal zero only if j is zero). Thus for the symbol

$$
\begin{Bmatrix} j_{11} & j_{12} & j_{13} \\ j_{21} & j_{22} & j_{23} \\ j_{31} & j_{32} & 0 \end{Bmatrix}, \text{ there arise } \begin{pmatrix} j_{31} & j_{32} & 0 \\ m_{31} & m_{32} & 0 \end{pmatrix} \text{ and } \begin{pmatrix} j_{13} & j_{23} & 0 \\ m_{13} & m_{23} & 0 \end{pmatrix}.
$$

From 8.64, we have

$$
\begin{pmatrix} j_{31} & j_{32} & 0 \\ m_{31} & m_{32} & 0 \end{pmatrix} = \delta_{j_{31} j_{32}} \delta_{m_{31}, -m_{32}} (-1)^{j_{31} - m_{31}} (2j_{31} + 1)^{-1/2}. \tag{8.172}
$$

Therefore if the 9–j symbol with $j_{33} = 0$ is not to vanish, we have the condition that $j_{31} = j_{32}$ and $j_{13} = j_{23}$, giving the form

$$
\begin{Bmatrix} j_{11} & j_{12} & j \\ j_{21} & j_{22} & j \\ j' & j' & 0 \end{Bmatrix}.
$$

Making use of (8.162), we find,

$$
\begin{Bmatrix} j_{11} & j_{12} & j \\ j_{21} & j_{22} & j \\ j' & j' & 0 \end{Bmatrix} = (-1)^{j_{12} + j_{21} + j + j'} [(2j + 1)(2j' + 1)]^{-1/2} \begin{Bmatrix} j_{11} & j_{12} & j \\ j_{22} & j_{21} & j' \end{Bmatrix}.
$$

$$
\tag{8.173}
$$

8.11 Reduced matrix elements of compound irreducible tensor operators

We are rapidly approaching the construction of 'master' equations, suitable for automatic or electronic computation, for the various terms in the general perturbation Hamiltonian (III.1) and for the orbital and spin angular momentum operators within the magnetic moment operator. Generally, we wish to evaluate the angular parts of matrix elements of the form $\langle \psi_i | O | \psi_j \rangle$. The operator O for the Coulomb perturbation consists, as we saw in (8.124), of pairwise products of spherical harmonic operators acting on the spatial coordinates of one electron at a time: the operator for ligand-field potentials involves similar spherical harmonics but taken singly: spin-orbit coupling involves the scalar product of spin and orbital angular momentum operators acting on the spin and space parts of individual electronic wavefunctions: finally, the components of the magnetic moment operator act upon individual electronic orbitals with respect to their space or spin parts. Altogether, while we have developed the basic theoretical apparatus with which to solve our problem, we have not yet assembled the pieces in a form suitable for automatic computation. Thus, on the one hand, some of the operators we shall use are commonly expressed as simple irreducible tensor operators while others appear as (scalar) products; and, on the other, standard expressions usually appear with respect to different basis quantizations. It is really too clumsy to have to change from bases defined as products of $|lsm_l m_s\rangle$ to $|LSM_L M_S\rangle$ to $|LSJM_J\rangle$ as we consider the various perturbations of interest to us. As will transpire, we work in $|LSJM_J\rangle$ quantization throughout and we therefore wish to develop 'master equations' which all pertain to this basis. Finally, we still have the problem of the calculation of RMEs. We begin our review of this nest of problems by deriving an expression for the reduced matrix element of a compound tensor operator in terms of those for simple tensor operators.

The most general operator T_Q^K we need consider can have any rank K and be built from two operators acting upon two independent sets of variables, labelled 1 and 2. The fact that our basis functions will generally involve n d electrons $(1 \leqslant n \leqslant 9)$, or even $(1 \leqslant n \leqslant 13)$ for f-electron problems, is irrelevant at this stage, for one- and two-particle operators only act on one or two electrons at a time. The operator T_Q^K, therefore, may also be written, *via* (8.106), as

$$T_Q^K = \sum_{q_1 q_2} R_{q_1}^{k_1}(1) S_{q_2}^{k_2}(2) \begin{pmatrix} k_1 & k_2 & K \\ q_1 & q_2 & -Q \end{pmatrix} (2K+1)^{1/2} (-1)^{k_1 - k_2 + Q}, \qquad (8.174)$$

which we consider to operate upon states $|j_1j_2JM\rangle$ built from two independent component states $|j_1m_1\rangle$ and $|j_2m_2\rangle$ by the usual expansion (8.47),

$$|j_1j_2JM\rangle = \sum_{m_1m_2} |j_1j_2m_1m_2\rangle \begin{pmatrix} j_1 & j_2 & J \\ m_1 & m_2 & -M \end{pmatrix} (2J+1)^{1/2}(-1)^{j_1-j_2+M}.$$

(8.175)

Within this basis, the general matrix element under T_Q^K is given by

$$\langle j_1j_2JM|T_Q^K(1,2)|j_1'j_2'J'M'\rangle = (-1)^{J-M} \begin{pmatrix} J & K & J' \\ -M & Q & M' \end{pmatrix}$$

$$\times \langle j_1j_2J \| \mathbf{T}^K(1,2) \| j_1'j_2'J'\rangle, \quad (8.176)$$

after application of the Wigner–Eckart theorem, (8.137).

This same matrix element can be evaluated after expanding both states and operator, using (8.175) and (8.174), to give

$$\langle j_1j_2JM|T_Q^K(1,2)|j_1'j_2'J'M'\rangle = \sum_{m_1m_2} \sum_{m_1'm_2'} \sum_{q_1q_2} (-1)^{k_1-k_2+Q}$$

$$\times (-1)^{j_1-j_2+M}(-1)^{j_1'-j_2'+M'}[(2J'+1)(2J+1)(2K+1)]^{1/2}$$

$$\times \begin{pmatrix} j_1 & j_2 & J \\ m_1 & m_2 & -M \end{pmatrix} \begin{pmatrix} j_1' & j_2' & J' \\ m_1' & m_2' & -M' \end{pmatrix} \begin{pmatrix} k_1 & k_2 & K \\ q_1 & q_2 & -Q \end{pmatrix}$$

$$\times \langle j_1j_2m_1m_2|R_{q_1}^{k_1}(1)S_{q_2}^{k_2}(2)|j_1'j_2'm_1'm_2'\rangle. \quad (8.177)$$

Now the operators $R_{q_1}^{k_1}$ and $S_{q_2}^{k_2}$ act on variables 1 and 2, respectively, which variables appear in separated form in bra and ket; therefore we can write,

$$\langle j_1j_2m_1m_2|R_{q_1}^{k_1}(1)S_{q_2}^{k_2}(2)|j_1'j_2'm_1'm_2'\rangle = \langle j_1m_1|R_{q_1}^{k_1}|j_1'm_1'\rangle\langle j_2m_2|S_{q_2}^{k_2}|j_2'm_2'\rangle.$$

(8.178)

Now using the Wigner–Eckart theorem on both parts of (8.178) and substituting the result into (8.177), we obtain,

$$\langle j_1j_2JM|T_Q^K(1,2)|j_1'j_2'J'M'\rangle = \sum_{m_1m_2} \sum_{m_1'm_2'} \sum_{q_1q_2} (-1)^{k_1-k_2+Q}$$

$$\times (-1)^{j_1-j_2+M}(-1)^{j_1'-j_2'+M'}(-1)^{j_1-m_1}(-1)^{j_2-m_2}$$

$$\times [(2J+1)(2J'+1)(2K+1)]^{1/2} \begin{pmatrix} j_1 & j_2 & J \\ m_1 & m_2 & -M \end{pmatrix} \begin{pmatrix} j_1' & j_2' & J' \\ m_1' & m_2' & -M' \end{pmatrix}$$

$$\times \begin{pmatrix} k_1 & k_2 & K \\ q_1 & q_2 & -Q \end{pmatrix} \begin{pmatrix} j_1 & k_2 & j_1' \\ -m_1 & q_1 & m_1' \end{pmatrix} \begin{pmatrix} j_2 & k_2 & j_2' \\ -m_2 & q_2 & m_2' \end{pmatrix}$$

$$\times \langle j_1 \| \mathbf{R}^{k_1} \| j_1'\rangle\langle j_2 \| \mathbf{S}^{k_2} \| j_2'\rangle. \quad (8.179)$$

The right-hand sides of the two expressions (8.179) and (8.176) may now be equated, each multiplied by

$$(-1)^{J-M}\begin{pmatrix} J & K & J' \\ -M & Q & M' \end{pmatrix}$$

and then summed over M, M' and Q. Thus, for (8.176), we get

$$\sum_{M,M',Q}[(-1)^{J-M}]^2\begin{pmatrix} J & K & J' \\ -M & Q & M' \end{pmatrix}^2\langle j_1j_2J\|\mathbf{T}^K(1,2)\|j_1'j_2'J'\rangle,$$

but, using (8.61), this reduces to $\langle j_1j_2J\|\mathbf{T}^K(1,2)\|j_1'j_2'J'\rangle$. Altogether, therefore, we arrive at the result,

$$\langle j_1j_2J\|\mathbf{T}^K(1,2)\|j_1'j_2'J'\rangle = \sum_{M,M',Q}\sum_{m_1m_2}\sum_{m_1'm_2'}\sum_{q_1q_2}$$

$$\times(-1)^{k_1-k_2+j_1'-j_2'+Q+M'-m_1-m_2+J}\begin{pmatrix} j_1 & j_2 & J \\ m_1 & m_2 & -M \end{pmatrix}$$

$$\times\begin{pmatrix} j_1' & j_2' & J' \\ m_1' & m_2' & -M' \end{pmatrix}\begin{pmatrix} k_1 & k_2 & K \\ q_1 & q_2 & -Q \end{pmatrix}\begin{pmatrix} j_1 & k_1 & j_1' \\ -m_1 & q_1 & m_1' \end{pmatrix}$$

$$\times\begin{pmatrix} j_2 & k_2 & j_2' \\ -m_2 & q_2 & m_2' \end{pmatrix}\begin{pmatrix} J & K & J' \\ -M & Q & M' \end{pmatrix}[(2J+1)(2J'+1)(2K+1)]^{1/2}$$

$$\times\langle j_1\|\mathbf{R}^{k_1}\|j_1'\rangle\langle j_2\|\mathbf{S}^{k_2}\|j_2'\rangle, \tag{8.180}$$

which expresses the relationship between the RME of a compound irreducible tensor operator and those of its components. Actually, this fearsome expression can be written in a very neat way when we recognize that the sum over six 3–j symbols appears in the definition of the 9–j symbol (8.171) once suitable symmetry transformations of the 3–j symbols have been made to remove the phase factor: ultimately we get,

$$\langle j_1j_2J\|\mathbf{T}^K(1,2)\|j_1'j_2'J'\rangle = \langle j_1\|\mathbf{R}^{k_1}(1)\|j_1'\rangle\langle j_2\|\mathbf{S}^{k_2}(2)\|j_2'\rangle$$

$$\times[(2J+1)(2J'+1)(2K+1)]^{1/2}\begin{Bmatrix} j_1 & j_1' & k_1 \\ j_2 & j_2' & k_2 \\ J & J' & K \end{Bmatrix}. \tag{8.181}$$

Note, as a check, that all quantities appearing in this expression are scalars. This very important equation is a generator of several useful expressions.

8.11.1 Scalar compound operators

In (III.1) we are concerned with scalar operators. If we put $K=0$ in the 9–j symbol in (8.181) we find from (8.173) and the associated discussion that the symbol vanishes unless $k_1 = k_2$ and $J = J'$. This defines selection rules for scalar (compound) tensor operators:

(i) any zero-rank operator is diagonal in the total angular momentum J, and

(ii) a zero-rank compound irreducible tensor operator must be constructed from two operators of the same rank.

With $K = 0$, we further note that the $9-j$ symbol collapses to a $6-j$ symbol *via* (8.173) and so we arrive at a general expression for the reduced matrix elements of zero-rank compound operators in terms of those for general-rank, simple operators,

$$\langle j_1 j_2 J \| \mathbf{T}^0(1,2) \| j_1' j_2' J' \rangle = \langle j_1 j_2 J \| \{ \mathbf{R}^k(1) \times \mathbf{S}^k(2) \}^0 \| j_1' j_2' J' \rangle$$
$$= (-1)^{k+j_2+J+j_1}(2k+1)^{-1/2}(2J+1)^{1/2}$$
$$\times \begin{Bmatrix} j_1 & j_2 & J \\ j_2' & j_1' & k \end{Bmatrix} \langle j_1 \| \mathbf{R}^k \| j_1' \rangle \langle j_2 \| \mathbf{S}^k \| j_2' \rangle,$$

(8.182)

or, written in the more conventional 'scalar product' form (8.114),

$$\langle j_1 j_2 J \| (\mathbf{R}^k(1) \cdot \mathbf{S}^k(2)) \| j_1' j_2' J \rangle = (-1)^{j_1+j_2+J}(2J+1)^{1/2} \begin{Bmatrix} j_1 & j_2 & J \\ j_2' & j_1' & k \end{Bmatrix}$$
$$\times \langle j_1 \| \mathbf{R}^k(1) \| j_1' \rangle \langle j_2 \| \mathbf{S}^k(2) \| j_2' \rangle. \quad (8.183)$$

We have therefore arrived at a very important equation for matrix elements of compound scalar operators:

$$\langle j_1 j_2 J M | (\mathbf{R}^k(1) \cdot \mathbf{S}^k(2)) | j_1' j_2' J' M' \rangle = \delta_{JJ'} \delta_{MM'}(-1)^{j_1+j_2+J} \begin{Bmatrix} j_1 & j_2 & J \\ j_2' & j_1' & k \end{Bmatrix}$$
$$\times \langle j_1 \| \mathbf{R}^k(1) \| j_1' \rangle \langle j_2 \| \mathbf{S}^k(2) \| j_2' \rangle.$$

(8.184)

8.11.2 Single-variable operators acting on coupled states

An important case arises when we consider an operator $\mathbf{R}^{k_1}(1)$ acting on a single set of variables 1 within a coupled state $|j_1 j_2 JM\rangle$. To be concrete, the operator may be the spin or the orbital angular momentum operator within the Russell–Saunders states $|JM\rangle$: we have in mind here the constituent parts of the magnetic moment operator. We proceed generally by putting $\mathbf{S}^{k_2} = \mathbf{1}$ in our master generator (8.181), whence $k_2 = 0$. Once again the $9-j$ symbol collapses and the conditions $j_2' = j_2$ and $k = k_1$ are imposed as in §8.11.1, so we get

$$\langle j_1 j_2 J \| \mathbf{R}^{k_1}(1) \| j_1' j_2' J' \rangle = (-1)^{j_1 + j_2 + J' + k_1} [(2J' + 1)(2J + 1)]^{1/2}$$

$$\times \begin{Bmatrix} j_1 & J & j_2 \\ J' & j_1' & k_1 \end{Bmatrix} \langle j_1 \| \mathbf{T}^{k_1}(1) \| j_1' \rangle. \qquad (8.185)$$

If we are interested in $\mathbf{S}^{k_2}(2)$, similar reasoning implies $\mathbf{R}^{k_1} = 1$, $k_1 = 0$, $j_1' = j_1$, $K = k_2$ and hence,

$$\langle j_1 j_2 J \| \mathbf{S}^{k_2}(2) \| j_1' j_2' J' \rangle = (-1)^{j_1 + j_2 + J + k_2} [(2J + 1)(2J' + 1)]^{1/2}$$

$$\times \begin{Bmatrix} J & k_2 & J' \\ j_2' & j_1 & j_2 \end{Bmatrix} \langle j_2 \| \mathbf{S}^{k_2}(2) \| j_2' \rangle. \qquad (8.186)$$

8.12 Decoupled operators

In the most recent sections we have discussed how reduced matrix elements of some product tensors are related to the RMEs of component, simple, tensor operators. Shortly we shall investigate the basis of evaluating RMEs for many-electron systems using the concept of coefficients of fractional parentage. An important strand of that theory is the progressive expression of such RMEs ultimately in terms of one-variable RMEs and in that connection we shall introduce the concept of unit tensors. In our final computing package we shall draw on published tables of such unit tensors and our 'master equations', for the various perturbations of interest to us are all cast in a way compatible with RMEs of unit tensors. As part of the final preparation of these equations, we now introduce the idea of *double-tensor operators*.

Exploitation of the Wigner–Eckart theorem in the evaluation of matrix elements like $\langle j_1 m_1 | T_q^k | j_2 m_2 \rangle$ relies on both operator and states being classified in the same way, that is, with respect to the same angular momenta. Should we wish to work with bases of uncoupled states, for example $|l m_l s m_s \rangle$, we can proceed only when the operator has been cast into a compatible form. Now, the process of classifying states and functions has been effected throughout by reference to their transformation properties within the rotation group R_3. In the same way that states are decoupled, we may decouple operators to give *simple* products of the kind $R_q^k(1) S_v^u(2)$ which, in their separate parts, operate selectively upon the variable sets 1 and 2. It is common to write such double tensor operators as T_{qv}^{ku}. One example of decoupling an operator is for the spin-orbit coupling perturbation: we may write

$$\mathbf{l} \cdot \mathbf{s} = l_0 s_0 - (l_{+1} s_{-1} + l_{-1} s_{+1}), \tag{8.187}$$

by using the phases in (8.89), or in a formal double-tensor form

$$\mathbf{l} \cdot \mathbf{s} = \sum_q (-1)^q (ls)^{11}_{q, -q}. \tag{8.188}$$

The matrix elements of a double-tensor operator within a decoupled state basis are evaluated by applying the Wigner–Eckart theorem twice:

$$\langle j_1 j_2 m_1 m_2 | T^{ku}_{qv} | j'_1 j'_2 m'_1 m'_2 \rangle \equiv \langle j_1 j_2 m_1 m_2 | R^k_q(1) S^u_v(2) | j'_1 j'_2 m'_1 m'_2 \rangle$$

$$= \langle j_1 m_1 | R^k_q(1) | j'_1 m'_1 \rangle \langle j_2 m_2 | S^u_v(2) | j'_2 m'_2 \rangle$$

$$= (-1)^{j_1 - m_1} \begin{pmatrix} j_1 & k & j'_1 \\ -m_1 & q & m'_1 \end{pmatrix} \langle j_1 \| \mathbf{R}^k \| j_1 \rangle$$

$$\times (-1)^{j_2 - m_2} \begin{pmatrix} j_2 & u & j'_2 \\ -m_2 & v & m'_2 \end{pmatrix} \langle j_2 \| \mathbf{S}^u \| j'_2 \rangle$$

$$\equiv (-1)^{j_1 + j_2 - m_1 - m_2} \begin{pmatrix} j_1 & k & j'_1 \\ -m_1 & q & m'_1 \end{pmatrix} \begin{pmatrix} j_2 & u & j'_2 \\ -m_2 & v & m'_2 \end{pmatrix}$$

$$\times \langle j_1 j_2 \| \mathbf{T}^{ku} \| j'_1 j'_2 \rangle, \tag{8.189}$$

the last line of which serves as a definition of the RME of a double-tensor operator.

8.13 Coefficients of fractional parentage

The state functions appearing in our bases must, of course, be anti-symmetric with respect to an exchange of the coordinates of any two electrons. The simplest and perhaps most common way of achieving this property is to construct the bases in terms of Slater determinants. Although convenient for many purposes, this method does not provide a ready relationship between antisymmetrized wavefunctions with different numbers of electrons and, as we shall see, a representation with this property can be very useful. An alternative method, originated by Goudsmit & Bacher[21] but first studied systematically and applied to the theory of atomic spectra by Racah,[22,23] involves the concept of fractional parentage.

In general it is not possible to write an antisymmetrized wavefunction for n electrons as a simple product of an antisymmetrized function for $(n-1)$ electrons times a wavefunction for the nth. It *is* possible, however, to write it as a linear combination of such products. Let us illustrate the idea with a Slater determinant for three electrons and three spin-orbitals, a, b, c. We expand the determinant part-way:

$$|a(1)b(2)c(3)| \equiv \frac{1}{\sqrt{(3!)}} \begin{vmatrix} a(1) & b(1) & c(1) \\ a(2) & b(2) & c(2) \\ a(3) & b(3) & c(3) \end{vmatrix}$$

$$= \frac{\sqrt{(2!)}}{\sqrt{(3!)}} [a(1)|b(2)c(3)| - b(1)|a(2)c(3)| + c(1)|a(2)b(3)|].$$

(8.190)

Note that the two-electron determinants like $|b(2)c(3)|$ are antisymmetrized functions of $(n-1) = 2$ electrons, while the simple products, like $a(1)|b(2)c(3)|$, though three-electron functions, are not antisymmetrized: however, the linear combination within the square brackets of (8.190) is antisymmetrized, for we obtained it from the Slater determinant. Suppose we had already available the two-electron, antisymmetrized products and wished to proceed in a forward direction to the construction of the three-electron antisymmetric product. The required function is obviously given as a linear combination of the various single products (or 'parents' of the final result) and we refer to the coefficients in this expansion – $+\sqrt{(\frac{1}{3})}$, $-\sqrt{(\frac{1}{3})}$, $+\sqrt{(\frac{1}{3})}$ in the present illustration – as *coefficients of fractional parentage* (cfp). As a further preliminary to setting up a formal nomenclature for this expansion, we must notice too that the way in which the final result is achieved is not unique. The expansion (8.190), for example, could have been done as follows:

$$|a(1)b(2)c(3)| \equiv \frac{1}{\sqrt{(3!)}} \begin{vmatrix} a(1) & b(1) & c(1) \\ a(2) & b(2) & c(2) \\ a(3) & b(3) & c(3) \end{vmatrix}$$

$$= \frac{\sqrt{(2!)}}{\sqrt{(3!)}} [a(3)|b(1)c(2)| - b(3)|a(1)c(2)| + c(3)|a(1)b(2)|]. \quad (8.191)$$

Therefore, our nomenclature must describe not only the n-electron antisymmetric product being constructed, but also the $(n-1)$-electron, antisymmetric product (which need not be in Slater determinantal form, of course) and the orbital of the nth electron.

We write a general expression for the sort of expansions in (8.190) suitable under R_3, as

$$|l^n \alpha J M\rangle = \sum_p |l^{n-1}\alpha_p J_p \cdot lj; J M\rangle \langle l^{n-1}\alpha_p J_p lj| \} l^n \alpha J\rangle, \quad (8.192)$$

which is to be read as follows: we are constructing an antisymmetrized product of n electrons within some shell l (say 2 for d electrons), the whole function being characterized by the total angular momentum quantum numbers J and M; α denotes any additional quantum number required to define the state. The expansion (8.192) is over the component 'parent', simple or 'dot' products p, characterized by the l configuration $(n-1)$, together with appropriate J and α quantum numbers of this $(n-1)$-electron, antisymmetric function, times the nth electron (in shell l) having a total angular momentum labelled by j: and, after the semicolon, is mentioned the fact that the two parts of the simple product are to be coupled to give a part of the resulting function with angular momentum J, M. In fact, the nomenclature for the parent state means that, anti-symmetric or not, the correct angular momentum properties are to be established; this is done *via*,

$$|l^{n-1}\alpha_p J_p \cdot lj; JM\rangle = \sum_m |l^{n-1}\alpha_p J_p, M - m\rangle |ljm\rangle_n$$
$$\times \langle J_p j, M - m, m | JM \rangle. \qquad (8.193)$$

Finally, the expansion coefficients (cfp) in (8.192) similarly identify the $(n-1)$-electron, antisymmetrized function and single $(n$th), electron function within the pth parent which contributes to the final state $|l^n\alpha J\rangle$: the components of the parent, dot product appear to the left of the symbol $|\}$, the final state to the right.

In summary, then, the coefficients of fractional parentage describe how the state $|l^n\alpha JM\rangle$ may be built from its possible parent states obtained by the removal of one electron. (In passing, we note here the relevance to the theory of photoelectron spectroscopy.) Equation (8.193) simply describes the vector coupling of the two parts to the correct total angular momentum. The orthogonality of the states in the expansion (8.192) means that the cfp obey the relationship,

$$\sum_{\alpha_p J_p l_j} \langle l^n\alpha J\{|l^{n-1}\alpha_p J_p lj\rangle \langle l^{n-1}\alpha_p J_p lj|\}l^n\alpha'J'\rangle = \delta_{\alpha\alpha'}\delta_{JJ'}. \qquad (8.194)$$

General values of fractional parentage coefficients. Although we shall not use cfp *explicitly* in this book, we note that extensive tables have been prepared for p^n, d^n and f^n configurations by Nielson & Koster:[24] as in Rotenberg *et al.*[13]'s tables of 3–j and 6–j symbols, the numerical values are presented in the form of powers of primes.

8.14 Reduced matrix elements in many-electron systems

Consider first the action of a one-electron operator within a system of n electrons. We have

$$F_q^k = \sum_i^n f_q^k(i), \qquad (8.195)$$

in which $f_q^k(i)$ operates upon the ith electron. Matrix elements of this operator between two antisymmetric n-electron states are given by

$$\langle\psi|F_q^k|\psi'\rangle = \langle\psi|\sum_i^n f_q^k(i)|\psi'\rangle = n\langle\psi|f_q^k(i)|\psi'\rangle \qquad (8.196)$$

and, as i labels any of the n electrons, we can put $i = n$ without loss of generality, to get

$$\langle\psi|F_q^k|\psi'\rangle = n\langle\psi|f_q^k(n)|\psi'\rangle. \qquad (8.197)$$

Just as we have expanded the operator in (8.195), we now expand the state functions, here using the concept of fractional parentage. We let

$$\psi = |l^n LSJM_J\rangle = \sum_{J''} \langle l^{n-1}J''lJ|\}l^nJ\rangle|l^{n-1}J''\cdot l(n)LSJM_J\rangle, \qquad (8.198)$$

where the nth electron has been labelled in the dot product to stress that it falls to the right of the dot. On substitution of (8.198) with (8.197), we have

$$\langle\psi|F_q^k|\psi'\rangle = n \sum_{J'',J'''} \langle l^{n-1}J''lJ|\}l^nJ\rangle\langle l^{n-1}J'''lJ'|\}l^nJ'\rangle$$
$$\times \langle l^{n-1}J''\cdot l(n)LSJM_J|f_q^k(n)|l^{n-1}J'''\cdot l(n)L'S'J'M'_J\rangle. \qquad (8.199)$$

In this matrix element $f_q^k(n)$ acts only upon the variables of the nth electron: this is important in two ways. Firstly, by orthogonality, it means that J'' must equal J''' if the matrix element on the left-hand side is not to vanish. Secondly, we may exploit the relationship (8.186) which, in conjunction with the Wigner–Eckart theorem, yields, for the matrix element on the right-hand side,

$$\langle l^{n-1}J''\cdot l(n)LSJM_J|f_q^k|l^{n-1}J''\cdot l(n)L'S'J'M'_J\rangle$$

$$= (-1)^{J-M_J}\begin{pmatrix} J & k & J' \\ M_J & q & -M'_J \end{pmatrix}\langle l^{n-1}J''\cdot l(n)LSJ\|\mathbf{f}^k\|l^{n-1}J''\cdot l(n)L'S'J'\rangle$$

$$= (-1)^{J-M_J}\begin{pmatrix} J & k & J' \\ M_J & q & -M'_J \end{pmatrix}(-1)^{J''+l+J+k}\begin{Bmatrix} l & l & k \\ J' & J & J'' \end{Bmatrix}$$

$$\times [(2J+1)(2J'+1)]^{1/2}\langle l\|\mathbf{f}^k\|l\rangle. \qquad (8.200)$$

But, from (8.198), we also have,

$$\langle\psi|F_q^k|\psi'\rangle = \langle l^nLSJM_J|F_q^k|l^nL'S'J'M'_J\rangle$$

$$= (-1)^{J-M_J}\begin{pmatrix} J & k & J' \\ M_J & q & -M'_J \end{pmatrix}\langle l^nLSJ\|\mathbf{F}^k\|l^nL'S'J'\rangle,$$

$$(8.201)$$

and so, by comparing (8.201) with (8.200) and (8.199), we obtain the very important result

$$\langle l^n LSJ \| \mathbf{F}^k \| l^n L'S'J' \rangle = n \sum_{J''} \langle l^{n-1} J'' lJ | \} l^n J \rangle \langle l^{n-1} J'' lJ' | \} l^n J' \rangle$$

$$\times (-1)^{J''+l+J+k} \begin{Bmatrix} l & l & k \\ J' & J & J'' \end{Bmatrix} [(2J+1)(2J'+1)]^{1/2} \langle l \| \mathbf{f}^k \| l \rangle.$$

$$(8.202)$$

We have here the expression for the RME in a many-electron basis in terms of a linear superposition of one-electron RMEs.

Now, we spend much of our time dealing with operators which act only upon the spatial coordinates of functions (for example, in the Coulomb and ligand-field operators) or only upon the spin coordinates. In the case of spin-independent operators, we may construct an equivalent expression to (8.202); one in which, not only is the operator classified with respect to spatial coordinates, but also the functions. There results,

$$\langle l^n LS \| \mathbf{F}^k \| l^n L'S \rangle = n \sum_{L''S''} \langle l^{n-1} L''S'' l | \} l^n LS \rangle \langle l^{n-1} L''S'' l | \} l^n L'S \rangle$$

$$\times (-1)^{L''+l+L+k} \begin{Bmatrix} l & l & k \\ L' & L & L'' \end{Bmatrix} [(2L+1)(2L'+1)]^{1/2} \langle l \| \mathbf{f}^k \| l \rangle.$$

$$(8.203)$$

8.15 Unit tensor operators

Even if one-electron RMEs like $\langle l \| \mathbf{f}^k \| l \rangle$ were readily to hand, the process of calculating corresponding many-electron RMEs involves the summation over products of cfp in (8.203), which is rather time consuming. It would be convenient, therefore, to have available tables of the many-electron RMEs themselves. However, as things stand, we would require one such set of tables for each operator that we might ever use, of a given rank; and in each case the computation in (8.203) would have been repeated in their preparation.

RMEs, however, are just scalefactors in the Wigner–Eckart theorem (8.129, 8.137) and, in the form introduced in §8.8, have already been 'rescaled' to correspond with conventions deriving from other proofs, as shown by (8.139). Consider the Wigner–Eckart theorem for one-electron operators in a one-electron function basis. We can always choose to factorize the RME scalefactors into two parts in such a way that all operators may be regarded as multiples of a unique *unit tensor operator*:

$$\langle \alpha j m | f_q^k | \alpha' j' m' \rangle = (-1)^{j-m} \begin{pmatrix} j & k & j' \\ m & q & -m' \end{pmatrix} \langle \alpha j \| \mathbf{u}^k \| \alpha' j' \rangle A_f, \quad (8.204)$$

$$\langle \alpha j m | g_q^k | \alpha' j' m' \rangle = (-1)^{j-m} \begin{pmatrix} j & k & j' \\ m & q & -m' \end{pmatrix} \langle \alpha j \| \mathbf{u}^k \| \alpha' j' \rangle A_g. \quad (8.205)$$

For the operator f_q^k in (8.204), the one-electron RME has been expressed as a product of a unit tensor RME and a numerical factor A_f: in (8.205), a different numerical factor A_g goes with the operator g_q^k. By this device, the particular and physical character of the operator in question contributes mainly to the numbers A_f or A_g: the unit tensor RME merely retains the rank of the operator and identifies the functional basis. This unit tensor operator is so called because we define its RME to be unity, writing generally

$$\langle \alpha j \| \mathbf{u}^k \| \alpha' j' \rangle = \delta_{jj'} \delta_{\alpha\alpha'}. \quad (8.206)$$

We may therefore use the computational procedures in (8.203) in a way which is independent of the precise nature of the operator in question. Thus, if we define a many-electron unit tensor operator of rank k by

$$U_q^k = \sum_i u_q^k(i), \quad (8.207)$$

its RME is given, from (8.203), as

$$\langle l^n LS \| U^k \| l^n L'S \rangle = n \sum_{L'',S''} \langle l^{n-1} L''S'' l | \} l^n LS \rangle \langle l^{n-1} L''S'' l | \} l^n L'S \rangle$$

$$\times (-1)^{L''+l+L+k} \begin{Bmatrix} l & l & k \\ L' & L & L'' \end{Bmatrix} [(2L+1)(2L'+1)]^{1/2}, \quad (8.208)$$

an expression defining many-electron RMEs in terms of cfp and 6-j symbols. Reduced matrix elements of the many-electron unit tensor operators U^k have been compiled for configurations p^n, d^n, f^n by Nielson & Koster,[24] and their utility in conjunction with the Wigner–Eckart theorem derives from the relationships,

$$\langle \cdot \| \mathbf{F}^k \| \cdot \rangle = \langle \cdot \| \sum_i \mathbf{f}^k(i) \| \cdot \rangle = \langle \cdot \| \sum_i A_f \mathbf{u}^k(i) \| \cdot \rangle$$

$$= A_f \langle \cdot \| \sum_i \mathbf{u}^k(i) \| \cdot \rangle = A_f \langle \cdot \| \mathbf{U}^k \| \cdot \rangle. \quad (8.209)$$

The 'physical' multiplier A_f of (8.209) and (8.204) is given, by definition, as

$$A_f = \langle j \| \mathbf{f}^k \| j' \rangle. \quad (8.210)$$

Unit double tensor operators are defined in a similar manner, Racah[22] having defined a particular subset by the relationship,

$$\mathbf{V}^{1k} = \sum_i \mathbf{s}(i)\mathbf{u}^k(i), \tag{8.211}$$

in which the rank of $\mathbf{s}$ is unity, indicated by the superscript in $\mathbf{V}$. Although Racah lists values of $\mathbf{V}^{11}$ and $\mathbf{V}^{12}$, we only require the concept of double tensor operators for spin-orbit coupling and hence for the operator $\mathbf{l} \cdot \mathbf{s}$. The orbital angular momentum operator is of rank 1 and so we shall only require values of $\mathbf{V}^{11}$. Indeed, *we* hardly need the concept of unit double tensors at all, being concerned with but the one double tensor operator, but the listing[24] of values of $\mathbf{V}^{11}$ by Nielson & Koster renders the idea convenient and we shall refer to these quantities in §8.16.4.

8.16 MASTER EQUATIONS ✦

Having characterized operators and states by their transformation properties under the three-dimensional (spherical) rotation group R_3; and having introduced 3–*j*, 6–*j* symbols and reduced matrix elements, and their roles within the central Wigner–Eckart theorem, we now construct 'master equations' for the evaluation of matrix elements under the various perturbations (III.1) of interest to us. In conjunction with the tables of Rotenberg *et al.*[13] for 3–*j* and 6–*j* symbols and by Nielson & Koster[24] for reduced matrix elements, these equations may be used 'blind'; that is, the reader does not need to understand the significance of the components of these equations or to have read any of the foregoing sections of this chapter in order to evaluate the various integrals.

8.16.1 The Coulomb interaction: (M.1)

Interaction repulsion effects are dealt with using the Coulomb operator e^2/r_{ij} where i, j label all pairs of electrons. For the simple case of one pair of electrons, configuration l^2, we have from (8.124), and (8.184),

$$\langle l^2 L S M_L M_S | e^2/r_{12} | l^2 L' S' M'_L M'_S \rangle$$

$$= \sum_k \langle l^2 L S M_L M_S | (\mathbf{C}_1^k \cdot \mathbf{C}_2^k) | l^2 L' S' M'_L M'_S \rangle$$

$$\times e^2 \int_0^\infty \int_0^\infty \frac{r_<^k}{r_>^{k+1}} R_l^2(1) R_l^2(2) r_1^2 dr_1 r_2^2 dr_2$$

$$= e^2 \sum_k (-1)^{2l+L} \begin{Bmatrix} l & l & k \\ l & l & L \end{Bmatrix} \langle l \| \mathbf{C}^k \| l \rangle^2 \delta_{SS'} \delta_{M_S M'_S} \delta_{LL'} \delta_{M_L M'_L} \cdot F^k, \tag{8.212}$$

where F^k are the usual Condon–Shortley radial integrals. For general configurations of the form $l^n (n \geqslant 2)$, we require appropriate expressions for the squares of the reduced matrix elements in (8.212). However, the whole problem, in this quantization, has been solved once and for all, Nielson & Koster[24] listing many-electron reduced matrix elements for use in the simple relationship:

✦
$$\left\langle \alpha l^n L S J M_J \left| \sum_{i<j} \frac{1}{r_{ij}} \right| \alpha' l^n L' S' J' M'_J \right\rangle$$

$$= \sum_k \langle \alpha l^n L S \| \mathbf{F}^k \| \alpha' l^n L' S' \rangle F^k \delta_{LL'} \delta_{SS'} \delta_{JJ'} \delta_{M_j M'_j}, \qquad \textbf{(M.1)}$$

which we shall refer to as our master equation for the Coulomb interaction. Note that the various Kroneker δ functions in both (M.1) and (8.212) express the fact that (i) the Coulomb operator is spin independent, and (ii) the angular expressions do not involve J and hence give identical results for all $(2J+1)$ components of Russell–Saunders term, ^{2S+1}L. As discussed elsewhere,[2] it is conventional to use the renormalized parameters F_k, related to the F^k by

$$F_0 = F^0, \quad F_2 = \tfrac{1}{49} F^2, \quad F_4 = \tfrac{1}{441} F^4. \qquad (8.213)$$

8.16.2 The ligand-field potential: (M.2) and (M.3)

The ligand-field perturbation in (III.1) represents the action of the set of ligands upon the ith electron of a metal function as that of an effective charge distribution with an effective potential $V_{LF}(i)$. It is well known that this potential may be expanded in a series of multipoles so that a matrix element of the ligand-field potential may be written as

$$\langle \psi | V_{LF} | \psi' \rangle = \sum_k \sum_{q=-k}^{k} c_{kq} \langle \psi | Y_q^k | \psi' \rangle, \qquad (8.214)$$

where the expansion coefficients $\{c_{kq}\}$ subsume radial integrals and have units of energy. All ligand-field models, including the angular overlap model we discuss in the following chapter, may be brought into this form. We shall therefore need to evaluate the angular matrix elements $\langle \psi | Y_q^k | \psi' \rangle$.

In the case of bases defined as one-electron functions, or orbitals, the solution is already available for, combining (8.143) with the Wigner–

Eckart theorem, we have

✦
$$\langle l, m_l | Y_q^k | l, m_l' \rangle = (-1)^{m_l}(2l+1)\left(\frac{2k+1}{4\pi}\right)^{1/2}\begin{pmatrix} l & k & l \\ -m_l & q & m_l' \end{pmatrix}\begin{pmatrix} l & k & l \\ 0 & 0 & 0 \end{pmatrix}.$$

$$(\text{M.2})$$

The case for many-electron bases, quantized as $|LSJM_J\rangle$, is only a little more complex. First we use the Wigner–Eckart theorem to get

$$\langle \alpha l^n LSJM_J | Y_q^k | \alpha' l^n L'S'J'M_J' \rangle$$

$$= (-1)^{J-M_J}\begin{pmatrix} J & k & J' \\ -M_J & q & M_J' \end{pmatrix}\langle \alpha l^n LSJ \| \mathbf{Y}^k \| \alpha' l^n L'S'J' \rangle \quad (8.215)$$

and then, recognizing that the spherical harmonic operator only acts upon the spatial part of the function, we use (8.185) for the RME,

$$\langle \alpha l^n LSJ \| \mathbf{Y}^k \| \alpha' l^n L'S'J' \rangle = (-1)^{L+S+J'+k}[(2J+1)(2J'+1)]^{1/2}$$

$$\times \begin{Bmatrix} L & J & S \\ J' & L' & k \end{Bmatrix}\langle \alpha LS \| \mathbf{Y}^k \| \alpha' L'S' \rangle \delta_{SS'}.$$

$$(8.216)$$

Finally, from (8.209) and (8.210), we have

$$\langle \alpha LS \| \mathbf{Y}^k \| \alpha' L'S' \rangle = \langle l \| \mathbf{Y}^k \| l \rangle \langle \alpha LS \| \mathbf{U}^k \| \alpha' L'S' \rangle, \quad (8.217)$$

in which

$$\langle l \| \mathbf{Y}^k \| l \rangle = (-1)^l(2l+1)\left(\frac{2k+1}{4\pi}\right)^{1/2}\begin{pmatrix} l & k & l \\ 0 & 0 & 0 \end{pmatrix} \quad (8.143)$$

Combining (8.215), (8.216), (8.217), (8.143), we obtain our third 'master equation',

✦
$$\langle \alpha l^n LSJM_J | Y_q^k | \alpha' l^n L'SJ'M_J' \rangle$$

$$= (-1)^{J+J'+L+S-M_J+k+l}[(2J+1)(2J'+1)]^{1/2}(2l+1)\left[\frac{2k+1}{4\pi}\right]^{1/2}$$

$$\times \begin{pmatrix} l & k & l \\ 0 & 0 & 0 \end{pmatrix}\begin{pmatrix} J & k & J' \\ -M_J & q & M_J' \end{pmatrix}\begin{Bmatrix} L & J & S \\ J' & L' & k \end{Bmatrix}\langle \alpha l^n LS \| \mathbf{U}^k \| \alpha' l^n L'S \rangle.$$

$$(\text{M.3})$$

We shall discuss the use of (**M.3**) within a computer package in chapter 10. Meanwhile, however, we note that such matrix elements are very simply evaluated with this equation, by recourse to the tables of 3–j, 6–j and RME coefficients.

8.16.3 The magnetic moment operator: (M.4)

Here we are concerned with the angular part of the magnetic moment operator μ_α which we write as

$$\mu_\alpha = kL_\alpha + 2S_\alpha, \tag{8.218}$$

in which $\alpha \equiv x, y, z$. The presence of Stevens' orbital reduction factor[16] k means that we wish to keep the spin and orbital parts separate and so evaluate matrix elements $\langle \psi | L_\alpha | \psi' \rangle$ and $\langle \psi | S_\alpha | \psi' \rangle$. This is very conveniently done within $|LSJM_J\rangle$ quantization. We note immediately that $\mathbf{L}$ is diagonal in L, and $\mathbf{S}$ diagonal in S, from the equations defining these operators. Further, L_z is diagonal in M_L; S_z in M_S; and hence μ_z in M_J.

When considering a matrix element under $\mathbf{L}$ we proceed as in §8.16.2 for Y_q^k, using the Wigner–Eckart theorem together with (8.185) in recognition that $\mathbf{L}$ operates on only one part of the function:

$$\langle \alpha l^n LSJM_J | L_q^1 | \alpha l^n LSJ'M_J' \rangle$$

$$= (-1)^{J-M_J} \begin{pmatrix} J & 1 & J' \\ -M_J & q & M_J' \end{pmatrix} \langle \alpha l^n LSJ \| \mathbf{L}^1 \| \alpha l^n LSJ' \rangle \tag{8.219}$$

and

$$\langle \alpha l^n LSJ \| \mathbf{L}^1 \| \alpha l^n LSJ' \rangle = (-1)^{L+S+J'+1} [(2J+1)(2J'+1)]^{1/2}$$

$$\times \begin{Bmatrix} L & J & S \\ J' & L & 1 \end{Bmatrix} \langle L \| \mathbf{L}^1 \| L \rangle. \tag{8.220}$$

It is unnecessary to use unit tensors here, for the simplicity of the $\mathbf{L}$ operator furnishes a readily calculable value for the RME, as follows.

The matrix element

$$\langle LL | L_z | LL \rangle \equiv \langle LL | L_0^1 | LL \rangle$$

$$= (-1)^0 \begin{pmatrix} L & 1 & L \\ -L & 0 & L \end{pmatrix} \langle L \| \mathbf{L}^1 \| L \rangle \tag{8.221}$$

is known from traditional methods to equal L. Therefore,

$$\langle L \| \mathbf{L}^1 \| L \rangle = \begin{pmatrix} L & 1 & L \\ -L & 0 & L \end{pmatrix}^{-1} L = \begin{pmatrix} L & L & 1 \\ L & -L & 0 \end{pmatrix}^{-1} L$$

$$= (-1)^{L-L} \frac{1}{L} [(2L+1)(L+1)L]^{1/2} L$$

$$= [(2L+1)(L+1)L]^{1/2}, \tag{8.222}$$

where we have used (8.62).

Combining (8.222), (8.220) and (8.219), we have the result,

$$\langle \alpha l''LSJM_J|L_q^1|\alpha l''LSJ'M_J'\rangle = (-1)^{J-M_J+L+S+J'+1}\begin{pmatrix} J & 1 & J' \\ -M_J & q & M_J' \end{pmatrix}$$

$$\times \begin{Bmatrix} L & J & S \\ J' & L & 1 \end{Bmatrix}[(2J+1)(2J'+1)]^{1/2}[L(L+1)(2L+1)]^{1/2}. \quad (8.223)$$

The corresponding equation for the spin operator using (8.186), is

$$\langle \alpha l''LSJM_J|S_q^1|\alpha l''LSJ'M_J'\rangle = (-1)^{J-M_J+L+S+J+1}\begin{pmatrix} J & 1 & J' \\ -M_J & q & M_J' \end{pmatrix}$$

$$\times \begin{Bmatrix} J & 1 & J' \\ S & L & S \end{Bmatrix}[(2J+1)(2J'+1)]^{1/2}[S(S+1)(2S+1)]^{1/2}. \quad (8.224)$$

We require matrix elements under the magnetic moment operator expressed with respect to a cartesian frame, however, and so now we use (8.96), or (8.88) and (8.89): explicitly we have

$$\left.\begin{aligned} L_z &= L_0, \\ L_x &= \frac{1}{\sqrt{2}}(L_{-1}-L_{+1}), \\ L_y &= \frac{i}{\sqrt{2}}(L_{-1}+L_{+1}), \end{aligned}\right\} \quad (8.225)$$

and so, combining (8.225), (8.224), (8.223) and (8.218), we obtain our fourth 'master equation':

✦
$$\langle \alpha l''LSJM_J|\mu_z|\alpha l''LSJ'M_J\rangle = (-1)^{J-M_J}\begin{pmatrix} J & 1 & J' \\ -M_J & 0 & M_J \end{pmatrix}\cdot[A],$$

✦
$$\langle \alpha l''LSJM_J|\mu_x|\alpha l''LSJ'M_J'\rangle$$
$$=\frac{1}{\sqrt{2}}(-1)^{J-M_J}\left[\begin{pmatrix} J & 1 & J' \\ -M_J & -1 & M_J' \end{pmatrix}-\begin{pmatrix} J & 1 & J' \\ -M_J & 1 & M_J' \end{pmatrix}\right]\cdot[A],$$

✦
$$\langle \alpha l''LSJM_J|\mu_y|\alpha l''LSJ'M_J'\rangle$$
$$=\frac{i}{\sqrt{2}}(-1)^{J-M_J}\left[\begin{pmatrix} J & 1 & J' \\ -M_J & -1 & M_J' \end{pmatrix}+\begin{pmatrix} J & 1 & J' \\ -M_J & 1 & M_J' \end{pmatrix}\right]\cdot[A],$$

(cont.) (M.4)

where

$$[A] = (-1)^{L+S+J'+1}[(2J+1)(2J'+1)]^{1/2}[L(L+1)(2L+1)]^{1/2}$$

$$\times \begin{Bmatrix} J & 1 & J' \\ L & S & L \end{Bmatrix} \cdot k + 2(-1)^{L+S+J+1}[(2J+1)(2J'+1)]^{1/2}$$

$$\times [S(S+1)(2S+1)]^{1/2} \begin{Bmatrix} J & 1 & J' \\ S & L & S \end{Bmatrix}.$$

Note that k appearing here is Stevens' orbital reduction factor and *not* a tensor rank!

8.16.4 Spin-orbit-coupling: (M.5)

The spin-orbit coupling operator is a scalar compound operator: using (8.184) we therefore write,

$$\langle \alpha l^n LSJM_J | (\mathbf{L}^1(1) \cdot \mathbf{S}^1(2)) | \alpha' l^n L'S'J'M'_J \rangle$$

$$= \delta_{JJ'} \delta_{M_J M'_J} (-1)^{L'+S+J} \begin{Bmatrix} L & S & J \\ S' & L' & 1 \end{Bmatrix} \langle \alpha l L \| \mathbf{L}^1 \| \alpha' l L' \rangle \langle S \| \mathbf{S}^1 \| S \rangle. \tag{8.226}$$

With (8.189) we identify the product of RMEs in (8.226) with the RME of the double tensor operator:

$$\langle LS \| (\mathbf{LS})^{11} \| L'S' \rangle = \langle L \| \mathbf{L}^1 \| L' \rangle \langle S \| \mathbf{S}^1 \| S' \rangle, \tag{8.227a}$$

$$= \sum_i^n \langle l \| \mathbf{1} \| l' \rangle \langle s \| \mathbf{s} \| s' \rangle \mathbf{u}^1(i)\mathbf{s}(i), \tag{8.227b}$$

$$= [l(l+1)(2l+1)]^{1/2}\sqrt{(\tfrac{3}{2})}\langle LS \| \mathbf{T}^{11} \| L'S' \rangle, \tag{8.227c}$$

using (8.209), (8.210) and (8.211) to get (8.227b); and (8.222) and (8.211) for (8.227c). Putting (8.227c) and (8.226) together, we establish our master equation for spin-orbit coupling within a given configuration as

$$\langle \alpha l^n LSJM_J | \sum_i^n \mathbf{l}_i \cdot \mathbf{s}_i | \alpha' l^n L'S'J'M'_J \rangle = \delta_{JJ'} \delta_{M_J M_{J'}} (-1)^{L'+S+J}$$

$$\times \begin{Bmatrix} L & S & J \\ S' & L' & 1 \end{Bmatrix} [l(l+1)(2l+1)]^{1/2} \langle \alpha l^n LS \| \mathbf{V}^{11} \| \alpha' l^n L'S' \rangle, \tag{M.5}$$

where the constant $\sqrt{(\tfrac{3}{2})}$ in (8.225c) has been subsumed by the RMEs of $\mathbf{V}^{11}$ to correspond with their original definition by Racah[22] (theory of complex spectra III, equation (25)) and the listings of Nielson & Koster.[24]

References

[1] Ballhausen, C.J., *Introduction to Ligand-Field Theory*, McGraw-Hill, New York, 1962.

[2] Gerloch, M. & Slade, R.C., *Ligand-Field Parameters*, Cambridge University Press, 1973.

[3] Figgis, B.N., *Introduction to Ligand Fields*, Interscience, New York, 1966.

[4] Silver, B.L., *Irreducible Tensor Methods*, Academic Press, New York, 1976.

[5] Eyring, H., Walter, J. & Kimball, G.E., *Quantum Chemistry*, Wiley, New York, 1944.

[6] Messiah, A., *Quantum Mechanics*, North-Holland, Amsterdam, 1961.

[7] Buckmaster, H.A., *Can. J. Phys.*, **13**, 386 (1964)

[8] Buckmaster, H.A., *Can. J. Phys.*, **44**, 2525 (1966).

[9] Buckmaster, H.A., Chatterjee, R. & Shing, Y.H., *Phys. Stat. Sol.*, **13A**, 9 (1972).

[10] Brink, D.M. & Satchler, G.R., *Angular Momentum*, Oxford University Press, 1968.

[11] Condon, E.U. & Shortley, G.H., *The Theory of Atomic Spectra*, Cambridge University Press, 1935.

[12] Racah, G. *Phys. Rev.*, **62**, 438 (1942).

[13] Rotenberg, M., Bivens, R., Metropolis, N. & Wooten, J.K., Jr, *The 3–j and 6–j Symbols*, The Technology Press, MIT, Cambridge, Mass., 1959.

[14] Cotton, F.A., *Chemical Applications of Group Theory*, Interscience, New York, 1963.

[15] Stevens, K.W.H., *Proc. Phys. Soc.*, **65**, 209 (1952).

[16] Stevens, K.W.H., *Proc. Roy. Soc.*, **A219**, 542 (1954).

[17] Levinson, J.B., *Trudy fiz.-tekh. Inst., Ashkhabad*, **2**, 17 and 31 (1957); and *Liet. TSR Mokslu Akad. Darb.*, **B4**, 3 (1957).

[18] Fano, U., *National Bureau of Standards Rep. No.* **1214**, 45 and 109 (1951).

[19] Smith, K., *Table of Wigner 9–j symbols for Integral and Half-Integral Values of the Parameters*, ANL-5860 Argonne National Lab., Chicago, Ill., 1958.

[20] Howell, K.M., *Revised Tables of 6–j Symbols*, Res. Rep. 59–1. University of Southampton, Maths. Dept., Southampton, England, 1959.

[21] Goudsmit, S. & Bacher, R.F., *Phys. Rev.*, **46**, 948 (1934).

[22] Racah, G., *Phys. Rev.*, **63**, 367 (1943).

[23] Racah, G. *Phys. Rev.*, **76**, 1352 (1949).

[24] Nielson, C.W., & Koster, G.F., *Spectroscopic Coefficients for p^n, d^n and f^n Configurations*, MIT Press, Cambridge, Mass., 1963.

9

The ligand field

While all three terms in the effective Hamiltonian (III.1) relate to a transition model or lanthanide ion within a complex, it is the second, involving the so-called ligand-field potential, that carries information about the geometry of the complex and provides the most direct association with the bonding in the molecule as a whole. It is here that the essential chemical content of ligand-field theory is to be found and the approach we shall exploit to reveal it is the cellular construction we call the angular overlap model. Before doing so, however we must review some aspects of the ligand-field potential itself and derive several general relationships between ligand-field matrix elements that serve as a basis for all models of this perturbation.

In ligand-field theory we examine how a chosen basis of d or f electrons associated with the central metal ion in a complex is perturbed by an effective potential arising from all other electrons in the system. While leaving the question of a *precise* definition of ligand-field theory till chapter 11, we emphasize from the start that the theory is concerned only with the restricted regime of 'd' or 'f' electrons and that the orbital basis and the ligand-field potential are co-defined and inextricably linked together. This important characteristic guides the brief review of the well-known multipole expansion of the ligand-field with which we begin.

9.1 Multipole expansions

Ligand-field potentials are complicated functions of all spatial coordinates in a complex. Various technical methods of handling calculations in ligand-field theory involve different ways of partitioning such potentials. Some are simpler to handle mathematically while others are easier to visualize and relate more directly to general chemical concepts: the different types of division or expansion are generally not equally successful in these two respects. Here we examine the multipole expansion[††] of the total or global ligand field in a complex.

We might contemplate using the expansion

$$V_{\mathrm{LF}}(\mathbf{r}) = \sum_{k=0}^{\infty} \sum_{q=-k}^{k} a_{kq}\rho_k(r)Y_q^k(\theta, \phi), \tag{9.1}$$

but recognition of the interdependent nature of the ligand-field potential and the object of its application suggests that a better approach is to expand *matrix elements* of the potential, for it must never be forgotten that it is only through these – as parameters – that contact with physical observables is made. So we consider instead the multipole expansion,

$$\langle \psi | V_{\mathrm{LF}}(\mathbf{r}) | \psi' \rangle = \sum_{k} \sum_{q=-k}^{k} a_{kq}\langle \psi | \rho_k(r)Y_q^k(\theta, \phi) | \psi' \rangle \tag{9.2}$$

or, factorizing radial R and angular A parts of the basis functions,

$$\langle lR(r)A(\hat{r}) | V_{\mathrm{LF}}(\mathbf{r}) | lR(r)A'(\hat{r}) \rangle = \sum_{k=0}^{2l} \sum_{q=-k}^{k} a_{kq}\langle R(r) | \rho_k(r) | R(r) \rangle$$

$$\times \langle A(\hat{r}) | Y_q^k(\theta, \phi) | A'(\hat{r}) \rangle, \tag{9.3a}$$

$$= \sum_{k=0}^{2l} \sum_{q=-k}^{k} c_{kq}(r)\langle A(\hat{r}) | Y_q^k(\theta, \phi) | A'(\hat{r}) \rangle, \tag{9.3b}$$

where the expansion coeffients $c_{kq}(r)$ subsume radial integrals and have units of energy. Of course, one immediate consequence of using (9.2) rather than (9.1) is the introduction of the series terminator such that

$$2l \geqslant k \geqslant 0. \tag{9.4}$$

as determined by the vector-triangle rule, explicitly contained within the

---

[††] Effective ligand fields generally involve both local and non-local parts so that, while the multipole expansion of local terms is achieved through a superposition of spherical harmonics centred on the metal ion, the corresponding expansion of non-local contributions is accomplished with the use of bipolar harmonics, illustrated in §8.7.2. On the other hand, as will be discussed in §11.3, the terms of an expansion in bipolar harmonics occur in the same way as for the single-centre spherical harmonics, so that no new parameters are generated. For present purposes, therefore, it is sufficient to proceed with the expansion of the local contributions to the ligand field in terms of spherical harmonics alone, recognizing that the contributions from non-local parts may be sequestered within the resulting parameter set on a one-to-one basis.

3–j symbol $\begin{pmatrix} l & k & l \\ 0 & 0 & 0 \end{pmatrix}$ in both (**M.2**) and (**M.3**). Also, parity requires k to be even.

Let us comment further, though briefly, upon the interdependent nature of the definitions of the basis functions and the potential that led to our preference for the expansion (9.2): these remarks might also be reviewed after the more technical discussions in §§11.2 and 11.9. First, consider the angular parts of the multipole expansion (9.1). In the absence of the series terminator (9.4), established by reference to the object of its application, the series is infinite, of course. In this sense, therefore, we may talk of the basis as 'sampling' the complete potential. In the trivial case where we choose a basis of s orbitals (which we never do because the conditions defining a 'ligand-field regime', as in §11.8, are never met this way[††]), the effective ligand-field potential is always spherical because $k = q = 0$. By the same token, the potential sampled by a d-orbital basis differs from that experienced by an f-orbital set, simply on the grounds that the extent of the multipole expansion varies. However, it is not only with respect to the angular functions that we observe the definitions of ligand-field potential and function basis to be interdependent. A simple illustration of the point is made if we consider the potentials associated with bases defined, for example, as atomic $3d$ versus $4d$: the same values for the observables – which then means, for the ligand-field matrix elements – are ensured only if the radial parts, ρ, of V_{LF} are adjusted in step with the radial wavefunctions, R, in (9.3a). Again, we emphasize that radial properties of both bases and operator are sequestered within the expansion coefficients c_{kq} of (9.3b). The optimization of the theoretical definition of ligand-field basis functions and of the effective Hamiltonian we call the ligand-field potential are matters we discuss in detail in chapter 11. Meanwhile, for the purposes of the manipulation and use of ligand-field formalisms, it is sufficient to proceed using an expansion in the form (9.3), together (so far as transition-metal complexes are concerned) with a d-orbital basis of fixed, but *unspecified*, radial character.

Returning to the angular parts of the multipole expansions, we make two further points briefly at this stage. In addition to those series terminators established by parity and vector-triangle rule, further special relationships between the multipole expansion coefficients may be defined by the global molecular point-group symmetry. The well-known angular

---

[††] And the same goes for a p-orbital basis.

form of the potential for octahedral molecules,

$$V_{oct} = Y_0^4 + \sqrt{\left(\frac{5}{14}\right)}(Y_4^4 + Y_{-4}^4) \tag{9.5}$$

is one example of this. Similar special cases, and the group-theoretical procedures by which they may be derived, form a large part of most conventional texts on ligand-field theory. Partly for this reason, but mostly because, as discussed in chapter 1, we are interested in *real* chemical systems, most of which possess little or no symmetry, we do not enter this area of the subject. In all cases we shall proceed with the most general expansion (9.2) and allow special 'accidents', so to speak, to occur numerically as they will. Secondly, we note that within a pure l basis, by which is meant the occurrence of matrix elements only between wavefunctions of the same orbital angular momentum quantum number (d–d or f–f, say), the ensuing even parity of the potential effectively adds an inversion centre to any molecular geometry. This means that the complete ligand-field matrix of some complex ML_N is identical to the average of ML_N and its geometric, centrosymmetric inverse. While the consequences of this phenomenon upon parametrization schemes will be discussed within the angular overlap approach later, it should be realized that its origin lies directly within the angular 'purity' of the metal basis functions; and this is so regardless of the parity of the basis functions themselves.

9.2 Inversions of the multipole expansion

We now derive some useful and general relationships[1] between the expansion coefficients c_{kq} of (9.2) and matrix elements of a ligand-field potential written simply as V. Our starting point is the 'master equation' (**M.2**), derived in the preceding chapter:

$$\langle l, m_l | Y_q^k | l, m_l' \rangle = (-1)^{m_l}(2l+1)\left(\frac{2k+1}{4\pi}\right)^{1/2}\begin{pmatrix} l & k & l \\ -m_l & q & m_l' \end{pmatrix}\begin{pmatrix} l & k & l \\ 0 & 0 & 0 \end{pmatrix}. \tag{M.2}$$

In addition to the vector-triangle rule implicit in both 3–j symbols, we also recall the selection rule which requires that $-m_l + q + m_l' = 0$ or $q = m_l - m_l'$. This restricts the number of terms in the potential which can connect any two given basis functions. For example, only Y_l^2 and Y_l^4 are effective in the particular d matrix elements $\langle m_l = 2 | V | m_l' = 1 \rangle$ and

Table 9.1. *General relationships between* c_{kq} *and complex matrix elements for d orbitals*

$$c_{00} = \tfrac{2}{5}\pi^{1/2}[2\langle 2|V|2\rangle + 2\langle 1|V|1\rangle + \langle 0|V|0\rangle]$$

$$c_{20} = -\left(\frac{4\pi}{5}\right)^{1/2}[2\langle 2|V|2\rangle - \langle 1|V|1\rangle - \langle 0|V|0\rangle]$$

$$c_{21} = \left(\frac{4\pi}{5}\right)^{1/2}[6^{1/2}\langle 2|V|1\rangle + \langle 1|V|0\rangle]$$

$$c_{22} = -\left(\frac{4\pi}{5}\right)^{1/2}[2\langle 2|V|0\rangle + (\tfrac{3}{2})^{1/2}\langle 1|V|-1\rangle]$$

$$c_{40} = \tfrac{2}{5}\pi^{1/2}[\langle 2|V|2\rangle - 4\langle 1|V|1\rangle + 3\langle 0|V|0\rangle]$$

$$c_{41} = \left(\frac{4\pi}{5}\right)^{1/2}[-\langle 2|V|1\rangle + 6^{1/2}\langle 1|V|0\rangle]$$

$$c_{42} = 2\left(\frac{2\pi}{5}\right)^{1/2}[(\tfrac{3}{2})^{1/2}\langle 2|V|0\rangle - \langle 1|V|-1\rangle]$$

$$c_{43} = -2\left(\frac{7\pi}{5}\right)^{1/2}\langle 2|V|-1\rangle$$

$$c_{44} = \left(\frac{14\pi}{5}\right)^{1/2}\langle 2|V|-2\rangle$$

$\langle 1|V|0\rangle$ and we write,

$$\langle 2|V|1\rangle = A_{21}^{21}c_{21} + A_{41}^{21}c_{41}, \tag{9.6}$$

$$\langle 1|V|0\rangle = A_{21}^{10}c_{21} + A_{41}^{10}c_{41}, \tag{9.7}$$

where the quantities A are determined by (**M.2**): thus

$$A_{21}^{10} = (-1)^1(5)\left(\frac{5}{4\pi}\right)^{1/2}\begin{pmatrix}2 & 2 & 2\\ -1 & 1 & 0\end{pmatrix}\begin{pmatrix}2 & 2 & 2\\ 0 & 0 & 0\end{pmatrix}$$

$$= -5\cdot\left(\frac{5}{4\pi}\right)^{1/2}\cdot\left(\frac{1}{70}\right)^{1/2}\cdot-\left(\frac{2}{35}\right)^{1/2} = +\frac{1}{14}\sqrt{\left(\frac{5}{\pi}\right)}, \tag{9.8}$$

in which the values of the 3–*j* symbols are given, for example, by Rotenberg *et al.*[2] Equations (9.6) and (9.7) thus form a simultaneous pair for the evaluation of c_{21} and c_{41} in terms of the one-electron matrix elements $\langle 2|V|1\rangle$ and $\langle 1|V|0\rangle$, yielding the result:

$$c_{21} = \left(\frac{4\pi}{5}\right)^{1/2}[6^{1/2}\langle 2|V|1\rangle + \langle 1|V|0\rangle], \tag{9.9}$$

Table 9.2. *General relationships between c_{kq} and real matrix elements for d orbitals*

(a) Real parts of c_{kq}

$$c_{00} = \tfrac{2}{5}\pi^{1/2}[\langle x^2 - y^2|V|x^2 - y^2\rangle + \langle xy|V|xy\rangle + \langle xz|V|xz\rangle + \langle yz|V|yz\rangle + \langle z^2|V|z^2\rangle]$$

$$c_{20} = -\left(\frac{\pi}{5}\right)^{1/2}[2\langle x^2 - y^2|V|x^2 - y^2\rangle + 2\langle xy|V|xy\rangle - \langle xz|V|xz\rangle - \langle yz|V|yz\rangle - 2\langle z^2|V|z^2\rangle]$$

$$c_{21} = -\left(\frac{2\pi}{5}\right)^{1/2}[3^{1/2}\langle yz|V|xz\rangle + 3^{1/2}\langle xz|V|x^2 - y^2\rangle + \langle xz|V|z^2\rangle]$$

$$c_{22} = -\left(\frac{2\pi}{5}\right)^{1/2}[2\langle x^2 - y^2|V|z^2\rangle + (\tfrac{3}{4})^{1/2}\langle yz|V|yz\rangle - (\tfrac{3}{4})^{1/2}\langle xz|V|xz\rangle]$$

$$c_{40} = \tfrac{1}{5}\pi^{1/2}[\langle x^2 - y^2|V|x^2 - y^2\rangle + \langle xy|V|xy\rangle - 4\langle xz|V|xz\rangle - 4\langle yz|V|yz\rangle + 6\langle z^2|V|z^2\rangle]$$

$$c_{41} = \left(\frac{4\pi}{5}\right)^{1/2}[-3^{1/2}\langle xz|V|z^2\rangle + \tfrac{1}{2}\langle yz|V|xy\rangle + \tfrac{1}{2}\langle xz|V|x^2 - y^2\rangle]$$

$$c_{42} = \left(\frac{2\pi}{5}\right)^{1/2}[3^{1/2}\langle x^2 - y^2|V|z^2\rangle + \langle xz|V|xz\rangle - \langle yz|V|yz\rangle]$$

$$c_{43} = \left(\frac{7\pi}{5}\right)^{1/2}[\langle yz|V|xy\rangle - \langle xz|V|x^2 - y^2\rangle]$$

$$c_{44} = \left(\frac{7\pi}{10}\right)^{1/2}[\langle x^2 - y^2|V|x^2 - y^2\rangle - \langle xy|V|xy\rangle]$$

(b) Imaginary parts of c_{kq}

$$c_{00} = c_{20} = c_{40} = 0$$

$$c_{21} = \left(\frac{2\pi}{5}\right)^{1/2}[-3^{1/2}\langle yz|V|x^2 - y^2\rangle + 3^{1/2}\langle xz|V|xy\rangle + \langle yz|V|z^2\rangle]$$

$$c_{22} = -\left(\frac{2\pi}{5}\right)^{1/2}[-2\langle xy|V|z^2\rangle + 3^{1/2}\langle xz|V|yz\rangle]$$

$$c_{41} = \left(\frac{4\pi}{5}\right)^{1/2}[3^{1/2}\langle yz|V|z^2\rangle + \tfrac{1}{2}\langle yz|V|x^2 - y^2\rangle - \tfrac{1}{2}\langle xz|V|xy\rangle]$$

$$c_{42} = -\left(\frac{2\pi}{5}\right)^{1/2}[3^{1/2}\langle xy|V|z^2\rangle + 4\langle xz|V|yz\rangle]$$

$$c_{43} = \left(\frac{7\pi}{5}\right)^{1/2}[\langle xz|V|xy\rangle + \langle yz|V|x^2 - y^2\rangle]$$

$$c_{44} = -\left(\frac{14\pi}{5}\right)^{1/2}\langle x^2 - y^2|V|xy\rangle$$

and

$$c_{41} = \left(\frac{4\pi}{5}\right)^{1/2} [-\langle 2|V|1\rangle + 6^{1/2}\langle 1|V|0\rangle]. \tag{9.10}$$

Proceeding in a similar fashion for all c_{kq} within a one-electron d-orbital basis involves the solution of up to three simultaneous equations; and ultimately yields table 9.1. Equivalent expressions for terms with q negative may be constructed explicitly (for example, by considering $\langle 1|V|2\rangle$ and $\langle 0|V|1\rangle$ in the preceding illustration) or by recalling the definition of the hermitian adjoint of Y_q^k in (8.105) from which we require that

$$c_{kq} = (-1)^q c_{k,-q}^*, \tag{9.11}$$

a reminder that the matrix elements in the complex (natural) basis are generally complex.

It is also useful to construct a similar inversion for the multipole expansion in which the c_{kq} are expressed in terms of matrix elements of the *real* forms of the basis orbitals. For d functions, we have the relationships within the usual Condon–Shortley phase convention,

$$\left.\begin{aligned}
d_{z^2} &= Y_0^2 \\[4pt]
d_{xz} &= \frac{1}{\sqrt{2}}(|Y_{-1}^2\rangle - |Y_1^2\rangle), \\[4pt]
d_{yz} &= \frac{i}{\sqrt{2}}(|Y_{-1}^2\rangle + |Y_1^2\rangle), \\[4pt]
d_{xy} &= \frac{i}{\sqrt{2}}(|Y_{-2}^2\rangle - |Y_2^2\rangle), \\[4pt]
d_{x^2-y^2} &= \frac{1}{\sqrt{2}}(|Y_{-2}^2\rangle + |Y_2^2\rangle),
\end{aligned}\right\} \tag{9.12}$$

which, on substitution into table 9.1, yields table 9.2. Note that the expressions in both tables 9.1 and 9.2, describing relationships between multipole expansion coefficients of (9.3) and single-electron, d matrix elements, are applicable without restriction to ligand-field problems involving molecules of any, or no, symmetry: they are, in effect, inversions of the general multipole expansion (9.2) for d electrons. We make use of these expressions in the application of the angular overlap model to which we now turn.

9.3 Localized potentials: the AOM as a ligand-field model

We shall implement the angular overlap model by expressing the ligand-field potential as a superposition of non-overlapping contributions by dividing up the molecular coordination sphere into N non-overlapping regions or cells, writing

$$V_{\text{LF}} = \sum_{l=1}^{N} v^l. \tag{9.13}$$

This all-important premise expresses the additivity idea of the AOM – although first introduced in the molecular-orbital scheme, as reviewed in §1.9 – and states that the contributions to the total ligand-field potential from a given ligand arises from that ligand, and no other, in a region of space which is thereby uniquely associated with that ligand. The idea seems obvious, and, indeed is physically very sensible, but it does not emerge trivially within ligand-field theory: we argue in §§11.8 and 11.12 that the potential v^l is sufficiently screened that non-local effects will be short-range only and may be presumed negligible outside cell l. Given this additivity principle, however, the AOM formalism emerges simply and naturally within ligand-field theory, as follows.

Writing V as the matrix of the global ligand-field potential within a basis of d orbitals referred to the same global frame, we use (9.13) to decompose its elements as

$$V_{ij} \equiv \langle d_i | V_{\text{LF}} | d_j \rangle = \sum_{l=1}^{N} \langle d_i | v^l | d_j \rangle \equiv \sum_{l=1}^{N} v_{ij}^l. \tag{9.14}$$

Thus $\mathbf{v}^l$ describes a matrix of ligand-field energies arising from the contributions of ligand l. Each of these separate hermitian matrices $\mathbf{v}^l$ can be diagonalized by a different unitary transformation:

$$\mathbf{R}^l \mathbf{v}^l \mathbf{R}^{l\dagger} = \mathbf{e}^l, \tag{9.15}$$

where

$$\mathbf{R}^l \mathbf{R}^{l\dagger} = \mathbf{R}^{l\dagger} \mathbf{R}^l = \mathbf{1} \tag{9.16}$$

and $\mathbf{e}^l$ is a diagonal 5×5 matrix with diagonal elements $\{e_k^l\}$. Now define the transformed orbitals $\langle d_k^l |$ as linear combinations of the global d orbitals using the *same* unitary matrices $\mathbf{R}^l$:

$$\langle d_k^l | = \sum_i R_{ki}^l \langle d_i |. \tag{9.17}$$

Then

$$e_k^l = \langle d_k^l | v^l | d_k^l \rangle \tag{9.18}$$

are the exact energy shifts of the orbitals $|d_k^l\rangle$ caused by the potential

v^l. Accordingly, (9.14) may be rewritten as

$$V_{ij} = \sum_l^N \sum_k \sum_{k'} R_{ik}^{l\dagger} R_{k'j}^l \langle d_k^l | v^l | d_{k'}^l \rangle, \tag{9.19a}$$

$$= \sum_{l,k} R_{ik}^{l\dagger} R_{kj}^l e_k^l \quad \begin{cases} 1 \leqslant i,j,k \leqslant 5 \\ 1 \leqslant l \leqslant N \end{cases} \tag{9.19b}$$

where (9.19b) follows because (9.15) defined $\mathbf{e}^l$ as diagonal. Equation (9.19) expresses matrix elements of the total (global) ligand-field potential in terms of (local) cellular (diagonal) energies. It is important to note that this equation, which we take as the definition of what may be called a 'localized potential model', derives from the cellular decomposition of the ligand-field potential directly by formal manipulation: given the additivity principle, which we discuss in detail in §11.12, this equation belongs entirely within ligand-field theory. Writing (9.19b) as,

$$\langle d_i | V_{\text{LF}} | d_j \rangle = \overset{N\text{ cells}}{\sum_l} \overset{5\text{ modes}}{\sum_k} R_{ik}^{l\dagger} R_{kj}^l e_k^l, \tag{9.20}$$

we observe the form of the fundamental equation of the AOM as given, for example, by Schäffer.[4]

While the *form* of the fundamental AOM equation therefore derives directly from ligand-field theory, the *utility* of the model and the significance of the label 'modes' in (9.20) requires one further, simple *assumption*: namely, that a cellular decomposition (9.13) *can be made* such that the unitary matrices $\mathbf{R}^l$ *are determined by the molecular structure*. We define, for each ligand in the complex, a local coordinate frame which reflects the local M—L pseudosymmetry: for example, for a metal–pyridine interaction, we define the local ligand z axis parallel to the M—N bond, with x and y axes taken parallel and perpendicular to the pyridine plane. It is therefore usually reasonable to presume that the unitary matrix $\mathbf{R}^l$ may be identified with the *rotation* matrix relating the local frame of the M—L pseudosymmetry and the global frame in which the matrix $\mathbf{V}_{\text{LF}}$ is to be diagonalized; and that an $\mathbf{R}^l$ matrix so defined will make $\mathbf{v}^l$ diagonal, with elements

$$\langle d_{z^2}^l | v^l | d_{z^2}^l \rangle = e_\sigma^l,$$

$$\langle d_{xz}^l | v^l | d_{xz}^l \rangle = e_{\pi x}^l, \quad \langle d_{yz}^l | v^l | d_{yz}^l \rangle = e_{\pi y}^l,$$

$$\langle d_{xy}^l | v^l | d_{xy}^l \rangle = e_{\delta xy}^l, \quad \langle d_{x^2-y^2}^l | v^l | d_{x^2-y^2}^l \rangle = e_{\delta x^2-y^2}^l. \tag{9.21}$$

The suffixes σ, π, δ characterize the d^l orbitals in the local ligand frame in terms of the local pseudosymmetry and hence clarify the summation over 'modes' in (9.20). The energy shifts e_k^l in the local ligand cells are the conventional AOM e parameters, introduced in an historical context in chapter 1. The extra assumption conferring chemical relevance on the AOM is important but it does not explicitly prescribe the nature of the metal–ligand interaction: each local cellular potential must take a diagonal form and the assumption here is just that the principal axes of that diagonalized form will closely reflect the principal features of the ligand electron density. A justification for that assumption and a discussion of the chemical significance of the signs and magnitudes of the AOM e parameters is given in chapter 11. For the moment, we note that the character of the assumption is not such as to differentiate the AOM from conventional ligand-field theory in kind. *The angular overlap model falls within ligand-field theory.*

9.4 AOM rotation matrices

The AOM equation (9.20) expresses a ligand-field matrix element within the global frame of the complex in terms of local, diagonal, energy parameters and unitary rotation matrices. We require one such rotation matrix for each ligand or, because of the even parity of the total ligand-field operator, one matrix for each non-centrically related ligand. A prerequisite for any form of AOM calculation is the construction of equations expressing the relationship between the global and various ligand reference frames.

Perhaps the simplest way of describing these is in terms of direction cosines. If we label global coordinates with capitals X, Y, Z and local ones with the lower case x, y, z, the transformation of local coordinates into the global frame is accomplished by

$$\begin{pmatrix} X \\ Y \\ Z \end{pmatrix} = \mathbf{T} \begin{pmatrix} x \\ y \\ z \end{pmatrix} \tag{9.22}$$

where the transformation matrix $\mathbf{T}$ is written, in terms of direction cosines, by

$$\begin{array}{c} \\ X \\ Y \\ Z \end{array} \begin{array}{ccc} x & y & z \\ \begin{pmatrix} \alpha_1 & \beta_1 & \gamma_1 \\ \alpha_2 & \beta_2 & \gamma_2 \\ \alpha_3 & \beta_3 & \gamma_3 \end{pmatrix} \end{array} \equiv \mathbf{T}. \tag{9.23}$$

Alternatively, global and local frames may be related by a standard

Eulerian rotation, $D(\alpha, \beta, \gamma)$; note the use of subscripted Greek letters for direction cosines and unsubscripted ones for the Euler angles. As the practical matter of establishing the Euler angles required to perform the desired transformation within a given structure is most readily effected by reference to a table of equivalent direction cosines – as described in Appendix B – the first step for either form of rotation matrix is the construction of **T** in (9.23). Appendix A includes procedures by which **T** may be determined from a source-list of crystallographic atomic fractional coordinates and the unit cell.

Once furnished with the direction cosines $\{\alpha_i, \beta_i, \gamma_i\}$ or the Euler angles $\{\alpha, \beta, \gamma\}$, the rotation matrices **R** of (9.15) *et seq.* may be constructed in accordance with (9.17):

$$
\begin{pmatrix} z^2 \\ xz \\ yz \\ xy \\ x^2-y^2 \end{pmatrix} = \begin{pmatrix} R_{11} & R_{12} & R_{13} & R_{14} & R_{15} \\ R_{21} & R_{22} & R_{23} & R_{24} & R_{25} \\ R_{31} & R_{32} & R_{33} & R_{34} & R_{35} \\ R_{41} & R_{42} & R_{43} & R_{44} & R_{45} \\ R_{51} & R_{52} & R_{53} & R_{54} & R_{55} \end{pmatrix} \begin{pmatrix} Z^2 \\ XZ \\ YZ \\ XY \\ X^2-Y^2 \end{pmatrix}. \tag{9.24}
$$

Earlier expositions[4,8,9] of the AOM use (9.20) directly and hence require that these rotation matrices $\mathbf{R}^l$ be constructed explicitly. We shall adopt an alternative procedure – though one which still requires the explicit direction cosines or Euler angles above, of course – and so only need (9.24) in the formal manner given here. Explicit forms for **R** in terms of direction cosines,[11] or using Euler angles,[4] have been published, though care must be taken with the different conventions used there.

9.5 The relationship between the cellular and multipole expansions

Consider a somewhat more explicit illustration of the AOM equation (9.20), specialized to the case of a single ligand. We take a rotation matrix **R**, in either of the forms of (9.24), with respect to the d functions written in the given order; that is, in the local frame as of σ, π_x, π_y, δ_{xy} and $\delta_{x^2-y^2}$ symmetry respectively. Suppose we wish to evaluate the global matrix element $\langle d_{z^2} | V | d_{xy} \rangle$ in terms of the local e parameters: using our standard order of d functions, we have

$$
\langle d_1 | V | d_4 \rangle = R_{11} R_{14} e_\sigma + R_{21} R_{24} e_{\pi x} + R_{31} R_{34} e_{\pi y}
$$
$$
+ R_{41} R_{44} e_{\delta xy} + R_{51} R_{54} e_{\delta x^2-y^2}. \tag{9.25}
$$

The quantities $R_{ki} R_{kj}$ on the right-hand side are just numbers, of course, so that equations like (9.25) express the various global matrix elements as linear combinations of the local e parameters. Generally we may write,

$$m_i = a_{ij}e_j, \qquad (9.26)$$

where the a_{ij} are coefficients formed from the appropriate binary products of $\mathbf{R}$ matrix elements, and i labels one of the 25 (but 15 distinct) global matrix elements $\langle d|V|d'\rangle$. In other words, we express the global matrix elements in terms of the local AOM energies *via* a matrix $\mathbf{a}$. There will be one such matrix for each ligand in the complex, required by the sum over cells in (9.20). Altogether, therefore, we may rewrite (9.20) as the matrix equation

$$M_i = A_{ij}e_j, \qquad (9.27)$$

in which i runs from 1 to 15 (those d–d matrix elements not related by hermiticity) in some standard order; and the matrix $\mathbf{A}$ is formed as the sum of the matrices $\mathbf{a}$; one per ligand.

Expressed in this way, the fundamental AOM equation (9.20) takes the form of an expansion to be compared with the multipole expansion of (9.3). In the more conventional superposition, global ligand-field matrix elements are expressed in terms of global multipoles while, in the AOM, the same integrals are given as sums of locally defined quantities – the e parameters – which can, at the very least, be correlated with more general chemical concepts in an immediately obvious way. *The AOM thus brings to ligand-field theory the notion of the functional group.*

In table 9.2 we expressed the inverse of the multipole expansion (9.2), a result which can be written in the matrix form,

$$c_i = B_{ij}M_j: \qquad (9.28)$$

here j runs over the 15 independent global matrix elements, as above and i subsumes the pair of indices (k, q) in some convenient standard order. Combining (9.28) with (9.27), we arrive at an expression relating the different kinds of expansion coefficients, namely the $\{c\}$ of (9.2) and the $\{e\}$ of (9.20):

$$c_i = B_{ij}A_{jk}e_k = F_{ik}e_k. \qquad (9.29)$$

To be clear about the question of the dimensionality of the vectors and matrices appearing in this equation, suppose we are dealing with a d-orbital basis for a complex ML_N, and each metal–ligand interaction is parametrized by its own five e parameters, defined in (9.21). The multipole harmonics span orders 0–4, so that explicit consideration of only positive q values (the c_{kq} with q negative being determined by (9.11)) leads to the index i in (9.29) ranging 1–9 corresponding to the c_{kq} in table 9.2. The

Table 9.3. *Relationships between c_{kq} and e parameters for a d-electron basis within one local M—L interaction*

All c_{kq} are real; only non zero c_{kq} are listed.

$$c_{00} = \tfrac{2}{5}\pi^{1/2}(e_\sigma + e_{\pi x} + e_{\pi y} + e_{\delta xy} + e_{\delta x^2 - y^2})$$

$$c_{20} = \left(\frac{\pi}{5}\right)^{1/2} (2e_\sigma + e_{\pi x} + e_{\pi y} - 2e_{\delta xy} - 2e_{\delta x^2 - y^2})$$

$$c_{22} = c_{2-2} = \left(\frac{3\pi}{10}\right)^{1/2} (e_{\pi x} - e_{\pi y})$$

$$c_{40} = \tfrac{1}{5}\pi^{1/2}(6e_\sigma - 4e_{\pi x} - 4e_{\pi y} + e_{\delta xy} + e_{\delta x^2 - y^2})$$

$$c_{42} = c_{4-2} = \left(\frac{2\pi}{5}\right)^{1/2} (e_{\pi x} - e_{\pi y})$$

$$c_{44} = c_{4-4} = \left(\frac{7\pi}{10}\right)^{1/2} (e_{\delta x^2 - y^2} - e_{\delta xy})$$

All c_{kq} are real; only non-zero c_{kq} are listed.

index k in (9.29) ranges 1–5N if all ligands are taken to be distinct: alternatively, if, say, three of the N ligands are considered to be chemically equivalent with shared AOM parameters, k would range from 1 to 5 $(N-2)$, etc. Note that this same point means that the ranges of the indices j in (9.26) and (9.27) are generally different, of course. We note further that the number of expansion coefficients, or degrees of freedom, in the two kinds of ligand-field expansion are not generally equal. Indeed, at first sight it appears that the AOM is likely to be greatly overparametrized relative to the conventional multipole analysis. While in no way denying this problem, we merely note at this stage that the situation is actually better than it appears, reserving a full discussion of this most important practical feature of the AOM till §9.8.

9.6 The relationship within the local frame

The procedure just described involved comparison of matrix elements of the AOM and of a multipole expansion for the *complete* molecular complex. An alternative approach is to make this comparison for each local M—L interaction, followed by superposition. We establish the relationship between the two expansions in the local frame by defining, as we may for a single M—L interaction, parallel global and ligand reference frames. If these axes are chosen to reflect the local pseudosymmetry in the usual way, we arrive immediately at the local *and* global result that the ligand-field operator is diagonal with elements given by (9.21). Substitution of these identities into the expressions in table 9.2 yield the general relationships between the $\{c\}$ and $\{e\}$ expansion coefficients within a single M—L reference frame given in table 9.3. Note that the diagonality of the local perturbation ensures that all c_{kq} in table 9.3 are real. The significance of this table is that we have an expression for the multipole expansion of the ligand field associated with a single ligand in terms of any given or postulated AOM e parameters. Construction of the multipole expansion for a complete ML_N complex is then accomplished by appropriate summation of these single-ligand expansions. That summation can only be performed, however, within some common global reference frame and so we must now transform each local multipole expansion into the global frame following the procedures established in chapter 8. If $D(\alpha\beta\gamma)$ represents the Eulerian rotation transforming the local frame into the global frame then, by (8.77), we have

$$D(\alpha\beta\gamma)Y_q^k = \sum_{q'=-k}^{k} Y_{q'}^k \mathscr{D}_{q'q}^k(\alpha\beta\gamma), \qquad (9.30)$$

that is

$$Y_q^k(\text{global}) = \sum_{q'=-k}^{k} Y_{q'}^k(\text{local})\mathscr{D}_{q'q}^k(\alpha\beta\gamma). \qquad (9.31)$$

Exactly the same transformation rule then applies to the multipliers of the $Y_q^k(\text{local})$, namely the c_{kq} of table 9.3, and so we have the general procedure for obtaining the coefficients of the multipole expansion of the ligand field of a single ligand in terms of the local AOM parameters and expressed within some (arbitrary) chosen global frame. Superposition of the contributions from all other ligands is then trivial.

9.7 The choice of method

The most general ligand-field calculation we face is the evaluation of integrals of the type $\langle \Psi | V_{LF} | \Psi' \rangle$ in which Ψ and Ψ' are ligand-field *states* of the whole complex. These functions are generally many-electron wavefunctions constructed from an appropriately chosen basis – d orbitals for transition-metal systems, f orbitals for lanthanides. The theoretical treatment in chapter 8 leading to our various 'master equations' was directed to the evaluation of these integrals for states defined in $|J, M_J\rangle$ quantization. Alternative schemes, involving real rather than complex bases, have been studied by several authors, notably by Griffith[5] and, in the AOM connection, by Harnung & Schäffer.[6] They are not discussed in this book – though an excellent summary is to be found in that by Silver[7] – partly for reasons summarized at the end of chapter 10. At any rate, given the choice, as we later justify, of a *d*- (or *f*-) orbital basis, together with the need to include all three terms in the effective Hamiltonian (III.1), $|J, M_J\rangle$ quantization is a natural choice, particularly when we observe that all the necessary mathematics for handling it are to hand.

In general then, all ligand-field elements with which we are concerned will be evaluated through the 'master equation' (**M.3**),

$$\langle \alpha l^n LSJM_J | Y_q^k | \alpha' l^n L'SJ'M_J' \rangle$$

$$= (-1)^{J+J'+L+S+M_J+k+l}[(2J+1)(2J'+1)]^{1/2}(2l+1)\left[\frac{2k+1}{4\pi}\right]^{1/2}$$

$$\times \begin{pmatrix} l & k & l \\ 0 & 0 & 0 \end{pmatrix}\begin{pmatrix} J & k & J' \\ -M_J & q & M_J' \end{pmatrix}\begin{Bmatrix} L & J & S \\ J' & L' & k \end{Bmatrix}\langle \alpha l^n LS \| \mathbf{U}^k \| \alpha' l^n L'S \rangle.$$

$$\text{(M.3)}$$

Now an important assumption of all ligand-field models and, indeed, a defining property as discussed in chapter 11, is that one and the same ligand-field potential V_{LF} is simultaneously appropriate for each of the many-electron basis functions. Were this not appropriate – and in truth it cannot be better than a fair and useful approximation – ligand-field models would be characterized by a different set of parameters for each state matrix element and the whole approach would lose utility. It is really a matter of empirical observation that 'ligand-field theory works' which justifies the approximation. Given that V_{LF} may be written as a sum of identical one-electron potentials, the expansion coefficients c_{kq} of (9.2) for the one-electron case carry over to the many-electron problem, and we may write,

$$\langle \alpha l^n LSJM_J | V_{LF} | \alpha' l^n L'SJ'M_J' \rangle$$

$$= \sum_k \sum_{q=-k}^{k} c_{kq}(r)\langle \alpha l^n LSJM_J | Y_q^k(\theta, \phi) | \alpha' l^n L'SJ'M_J' \rangle, \quad (9.32)$$

in which the $\{c_{kq}\}$ are given by the one-electron calculation; that is, by (9.29) or the procedures of §9.6.

It is the power of (**M.3**) that determines our procedure of recasting the one-electron AOM model from its natural real-orbital basis into the complex basis of the spherical multipole expansion. A simple summarizing recipe is: (i) choose AOM *e* parameters, (ii) calculate the equivalent c_{kq} via

(9.29), etc., and (iii) use the 'master equation' (**M.3**) to evaluate all many-electron ligand-field matrix elements. From an aesthetic point of view, since our basis is 'spherical', that is, it forms an irreducible representation of R_3, and our major vector-coupling relationship (**M.3**) is couched in similar terms, it seems more satisfying to extricate ourselves from the real, cartesian basis of the AOM scheme as early as possible.

Somewhat similar reasoning partly lies behind our preference for the procedures of §9.6 over those of §9.5. However, beyond a feeling that the former technique is more 'natural' or 'harmonious', there are some concrete advantages in comparing the two expansions in the local frame first. Thus (i), it may be desirable to study the effects upon calculated spectral and magnetic properties of rotating any ligand about its M—L axis. Within the local frame this may be achieved by replacing $\mathscr{D}(\alpha, \beta, \gamma)$ in (9.31) by $\mathscr{D}(\alpha + \theta, \beta, \gamma)$ or, since from (8.24)

$$\mathscr{D}^k_{q'q}(\alpha\beta\gamma) = e^{i(\gamma q' + \alpha q)}d^k_{q'q}(\beta),\tag{9.33}$$

by simply multiplying the right-hand sides of the expressions in table 9.3 by $e^{iq\theta}$ and retaining the original $\mathscr{D}$ matrix. The procedures of §9.5 are much less direct. Then (ii), extension of the computational system to f-electron systems is similarly rather easier in the local frame. There we require tables, equivalent to 9.1, 9.2, and 9.3, expressing $\{c_{kq}\}$ in terms of $f-f$ matrix elements: these are discussed in Appendix D. The procedures of §9.6 may then be applied as before. On the other hand, making the sum over ligands before comparison of the types of expansion – that is, the use of (9.24) – requires the construction of an f-orbital transformation matrix, equivalent to (9.23) or (9.24), a tedium we might just as well avoid. Thirdly (iii), within the local frame it is simpler to consider the problem of 'misdirected valency' and the non-diagonality of the local AOM potential: we shall discuss this fully in §9.9.

9.8 The degree of parametrization within the AOM

The multipole expansion (9.2) is well behaved in the sense that there exists a one-to-one relationship between the matrix elements of V_{LF} and the expansion coefficients c_{kq}. For a d-orbital basis, for example, the hermitian matrix **V** has 15 $(=\frac{1}{2}n(n+1)$, where $n=5)$ independent elements. The potential spans orders $k=0$, 2 and 4 which possess a total of 15 $(=1+5+9)$ components. The 28 independent elements within an f-orbital basis are also matched by the $1+5+9+13$ expansion coefficients in the multipole expansion. The inversions in tables 9.1 and 9.2 are similarly well behaved. An alternative check on the number of independent c_{kq} in table 9.1, for example, is to observe that all coefficients with $q \neq 0$ are generally complex, with numerically independent real and imaginary parts: so the nine c_{kq} listed there give rise to $3 + (2 \times 6) = 15$ independent numbers. All this is to say that if we have some procedure involving variation of the c_{kq} for optimally reproducing the experimentally determined matrix elements, the resulting 'fit' can be unique. Actually, of course, the spherical term in the multipole expansion is not assessable (V_{oct} in (9.5) specifically excludes the Y_0^0 term, by convention) in a theory concerned with energy *splittings*. That is to say, since experimental data are related to energy differences, the diagonal elements of V are determined only to within some constant Δ, corresponding to a choice of zero-reference energy.

$$V_{ij} = \bar{V}_{ij} + \Delta \delta_{ij}. \tag{9.34}$$

For the d-orbital basis then, 14 independent matrix-element differences are mapped by 14 c_{kq} coefficients.

Contrast this with the dimensionality of the AOM cellular decomposition in (9.27). Here those same 15 matrix elements are expressed in terms of a set of e parameters whose number varies according to the chemistry and coordination number of the complex. Suppose, for example, we are concerned with a complex with six different ligands, each associated with the five e parameters (9.21). Equation (9.27) then constitutes a system of 15 equations in 30 unknowns; or, conversely, 30 AOM parameters are required to reproduce 15 observables. Solutions are obviously not unique. The utility of the AOM depends crucially upon a reduction of the number of disposable parameters by recourse to physico–chemical arguments. In presenting the following discussion, I do not coyly suppose that the reader is ignorant of the background and some applications of the angular overlap model but, rather, that I may refer freely to the level of description given in chapter 1 and, at the same time, look ahead somewhat to some of the material in chapter 11.

Within its historical context as a molecular-orbital model, the AOM described metal orbital energy shifts as proportional to the squares of appropriate M—L overlap integrals. The notion that $e_\sigma > e_\pi \gg e_\delta$ followed immediately and it was not long before the practice grew of neglecting AOM contributions of δ symmetry altogether. This clearly advantageous procedure, removing two parameters ($e_{\delta xy}$ and $e_{\delta x^2 - y^2}$) from each cell, has been formalized by the simple device of renormalizing the earlier AOM parameters and writing,

$$\tilde{e}_k^l = e_k^l - e_\delta^l. \tag{9.35}$$

This, together with the convention of dropping the tilde, means that many recent AOM studies are to be read such that e_σ refers really to the difference between σ and δ contributions, e_π to that between π and δ. However, since such studies usually involve ligands for which $e_{\pi x} \neq e_{\pi y}$, it must be presumed that $e_{\delta xy} \neq e_{\delta x^2 - y^2}$ also; in which case, the renormalization (9.35) is less effective than one in which we also presume that $e_{\delta xy}$ and $e_{\delta x^2 - y^2}$ are both negligible. This author would argue that renormalizations like (9.35) are in the nature of *ex post facto* window dressing for what has been taken as a very simple assumption right from the beginning; namely, that contributions to an M—L interaction from δ bonding are negligible. Within the original molecular-orbital formulation of the AOM such an assumption is effectively always and obviously sensible, simply on grounds of orbital overlap. From the point of view of the classical point-charge model, the assumption appears less well founded, for the repulsive interaction between metal electrons in a $d_{x^2 - y^2}$ orbital (δ) and a negative charge placed along the z axis apparently need not be totally negligible relative, say, to that between the $d_{xz}(\pi)$ and the ligand charge. Similarly, given that the AOM is *really* a ligand-field model, not properly modelled by molecular orbitals, and describes d-orbital energy shifts under an *effective* ligand-field potential, doubts must arise about total neglect of δ-type interactions. It is obvious, however, that we wish to retain this neglect in the interests – usually overwhelming interests – of reducing the degree of parametrization, so it is with some relief that we note here that convincing, if somewhat protracted, arguments are available to support a neglect of δ contributions from first principles: these arguments are detailed in chapter 11. We shall henceforth parametrize any given M—L interaction in a complex with a maximum of three AOM variables – $e_\sigma, e_{\pi x}, e_{\pi y}$. Even so, for a complex with, say, six dissimilar ligands, each parametrized in this way, there result too many (18) AOM parameters and the ensuing ambiguity in fitting experimental data will be manifest in correlation between the system variables. The degree of parametrization must be reduced further.

One obvious way to do this, of course, is merely to study complexes with a lower coordination number! Another is to analyse systems with one or more so-called 'linear ligators' for which $e_{\pi x} \equiv e_{\pi y}$ by symmetry. Monatomic ligands like halogens fall into this class: so do naturally cylindrical ligands (bonded to the metal 'end- on'), like—CO,—CN etc; and also effectively cylindrical ligands, like —P(C$_6$H$_5$)$_3$, in which the higher-than-twofold symmetry of the M—L moiety defines a cylindrical potential so far as the metal d_π orbitals are concerned. Altogether, therefore, some degree of linear ligation in a complex is not uncommon. In passing, we note here that the major departure of the AOM formalism from the early point-charge (and point-dipole) models of the ligand field, is the possibility of distinguishing between local M—L x and y directions in the AOM: such differences may be related to the relative facility of M—L π bonding in the two directions; for example, parallel and perpendicular to a pyridine ligand, while no such distinction can be accommodated within the point-charge scheme. Conversely, the point-charge and AOM potentials *can* always be equivalenced exactly in the case of linear ligators.[††]

Probably the main avenue for minimizing the number of variables in an AOM calculation is that which arises from the assumption that two or more ligands in a complex are *chemically equivalent*. For example, in a complex $MABCD_3$ it *may* be appropriate to assume that the three D ligands present similar local potentials to the central metal: while this assumption would surely require virtually equal M—D bond lengths as a minimum condition for its acceptance, the relative positions of these ligands around the metal – the various D—M—D and D—M—L bond angles – may well be of little concern. While this type of assumption is used wherever possible in AOM calculations it must be realized that the assumption can only be exactly valid in molecules with overall symmetry of some kind. It would be totally correct in an MA_3B_3 molecule possessing exact crystallographic three-fold symmetry, for example. We argue shortly, however, that the AOM technique is generally more successful in almost or completely *un*symmetrical molecules and in these cases the assumption of chemical equivalence between ligands may or may not be very reliable. The point here is that the AOM e parameters are not simply properties of the ligands alone: they monitor metal–ligand *inter*actions within the complete complex. Thus, the acceptor function of a metal orbital interacting with one ligand of type D may differ from that of another interacting with a second ligand of type D, due to the possible asymmetry of the interactions with the other ligands in the complex. Hence, while the two reference ligands may be of the same type (that is, chemical formulae) and even lie at the same distance from the metal, the two M—L interactions may not be equivalent also. A well-known illustration of this idea is the *trans* effect, of course. Should we choose to equivalence ligands which may be subject to this kind of effect, we ought, at least, to be aware of the nature of our assumption and not expect that the 'best fit' parameter values will necessarily yield exact averages for the various members of the group.

[††] ... provided *both* σ and π AOM perturbations are included.

Lest the reader become disillusioned by the nature and quantity of the assumptions made in AOM calculations, it is worth reminding him of the nature and advantages of the approach. Being, in essence, a cellular decomposition of the ligand field, the localized potential model we call the AOM attempts no more than the division of the total ligand field in a complex into spatially discrete parts which refer to individual metal–ligand interactions, in each case further subdivided under the classification of the local M—L pseudosymmetry. The possibilities for chemistry at large, not shared by the more conventional ligand-field models, is the determination of quantities which relate, in some semi-quantitative way, to the traditional and ubiquitous concepts of the σ- and π-bonding properties of functional groups. All ligand-field models require the use of simplifying assumptions. Within schemes essentially derived from the traditional multipole expansion, the physical or chemical basis for these assumptions is frequently unclear. The nature of the approximations made within the AOM approach, on the other hand, is directly comprehensible and, as we discuss in chapter 11, reasonable in quantum-mechanical terms.

By choosing to work with molecules possessing linear ligators and/or several ligands designated as 'chemically equivalent', it is very often possible to reduce the number of AOM variables to well below the number of elements in the ligand-field matrix. Set against this, however, there are several circumstances that often arise which frustrate the situation by reducing the degrees of freedom within the total ligand-field matrix itself. They derive from the total molecular symmetry, in one form or another. Thus we note that elements of the total molecular symmetry group may define special relationships between some of the ligand-field matrix elements or, equivalently, between some of the multipole expansion coefficients. In O_h symmetry, for example, the coefficients of all terms in the potential for a d-orbital basis, other than those appearing in (9.5), vanish identically; and, further, the ratio of c_{40} to $c_{4,\pm 4}$ is fixed at $\sqrt{(\frac{5}{14})}$. The consequence, within the real d-orbital basis, is the well-known result of there being only two types of non-zero matrix elements, corresponding to the e_g and t_{2g} subsets. Ultimately, of course, an octahedral ligand field is parametrized by the single variable, Δ_{oct}. There is clearly no way in which simplifying assumptions about the degree of parametrization within the AOM can make possible a differentiation of σ- and π-bonding parameters in this case. The theoretical relationship

$$\Delta_{\text{oct}} = 3e_\sigma - 4e_\pi \tag{9.36}$$

exactly illustrates this ambiguity. Let us make the same point explicitly in only one further case; that of molecular D_{4h} symmetry, appropriate for many MA_4B_2-type complexes. Suppose we consider an ideal example in which both A- and B-type ligands are linear ligators. The AOM parameter list then reads $e_\sigma(A)$, $e_\sigma(B)$, $e_\pi(A)$, $e_\pi(B)$. On the other hand, simple, group-theoretical arguments lead to a splitting of the d-orbital manifold into four species – $a_{1g}(d_{z^2})$, $e_g(d_{yz}, d_{xz})$, $b_{2g}(d_{xy})$ and $b_{1g}(d_{z^2})$. Characterized thus, these four levels define three degrees of freedom only; that is, three energy differences and no non-zero, off-diagonal matrix elements. Once again, the ambiguous result emerges where three pieces of experimental data are unable to define four theoretical (AOM) variables. The reader will be aware of many other similar illustrations of this point. This is not to say that all types of high molecular symmetry preclude a useful AOM analysis, however. An example of a different kind would be an MX_6 molecule involving six linear ligators arranged in D_{3d} symmetry. Here values for the two AOM parameters $e_\sigma(X)$ and $e_\pi(X)$, may be determined from the two energy splittings of the three species $e_g + e_g + a_{1g}$ arising from the d basis in this symmetry. Once more, there are many similar examples of this type.

The foregoing discussion described how special relationships arise out of the molecular point-symmetry as defined by the arrangement of the ligand nuclei around the metal. Further special, and frustrating, circumstances may be defined by the other symmetry properties relating to the *electronic* variables. Jørgensen, for example, has discussed[9] the consequences of *holohedral symmetry* at some length. The most general aspect of this derives, as noted above in §9.1, from the 'purity' of the metal basis functions and leads to the equivalence of the potentials $V(x, y, z)$ and $\frac{1}{2}[V(x, y, z) + V(-x, -y, -z)]$: centrosymmetrically related ligands give rise to potentials which are equiconsequential upon the metal basis functions. Consider the equivalent ligand fields associated with the two idealized, illustrative molecules in figure 9.1. When constructing matrices like **A** in (9.27), separate contributions from centrosymmetrically related ligands need not be calculated separately – all that is required is a multiplicative factor of two. In the present context, however, we find that the contributions of different ligands, or similar ones at different M—L distances, which lie on exactly opposite sides of the central metal atom, are not distinguishable. This is obvious when we note the identical potentials arising from the arrangements in figure 9.2. Worse still, from the point of view of the amount of obtainable information, is the case shown in figure 9.3(a) where, for simplicity, we consider σ contributions only. The interaction of the metal orbital with ligand A results in an energy shift of $e_\sigma(A)$; with ligand B, of $e_\sigma(B)$. The total shift $(e_\sigma(A) + e_\sigma(B))$

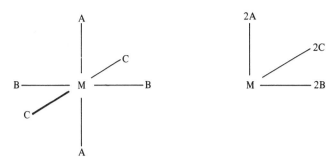

Fig. 9.1. The ligand-field potential set up for a pure-parity basis (d^n or f^n, say) by these two geometrical arrangements are equal.

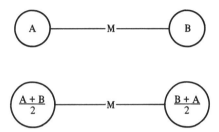

Fig. 9.2. Holohedral symmetry equivalences these two structures.

is what is observable, however, so that a differentiation between the two ligand interactions is not possible. An even more general situation is shown in figure 9.3(b) for these two ligands placed in *arbitrary* positions: here the total, observable, shift is the energy of the d orbital in some combination of the e_σ parameters, say $(c_A e_\sigma(A) + c_B e_\sigma(B))$; and at this stage in the argument, the reader might be forgiven for believing that no differentiation of individual ligands and their bonding modes would ever be possible. Rescue comes, of course, in the recognition that these same ligands simultaneously interact with, in general, all of the metal orbitals and so there arises a set of simultaneous equations, solution of which may yield the individual e parameters. However, this will only be possible if there are enough such *independent* equations. The determinant of the coefficients (like c_A in the above) of the e parameters will be singular, or effectively so, under two types of circumstances. The first is evident in cases where the exact molecular symmetry is sufficiently high and the second, approximate, case arises when the particular values of some AOM parameters in the actual geometrical arrangement happen to cause, more-or-less exactly, two or more of the simultaneous equations to be mutually dependent. While the problems of high molecular symmetry should be evident *a priori*,

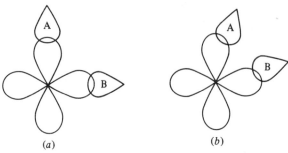

(a)   (b)

Fig. 9.3. The effects of ligands A and B upon a single metal orbital are inseparable whatever the detailed coordination.

those arising from approximate accidents might not. When fitting experimental data, underdetermination of this kind will manifest itself in a greater or lesser degree of correlation between the various system variables. In the practical application of the AOM, therefore, it is essential to study wide regions of parameter space to check for the uniqueness of a 'fit'. A discussion of some of the tactics one may use in such searches is presented in the following chapter.

Altogether, it should be apparent that the greatest opportunities for a successful application of the angular overlap model will occur in studies of molecules with little or no symmetry. This is a happy circumstance, of course, for it is precisely under those conditions where the more conventional ligand-field models, based upon symmetry-defined subdivisions of the multipole expansion, lose utility. This point is sufficiently important to bear repetition. The cellular decomposition we call the AOM promises contact with general chemical concepts in a way that the symmetry-based, multipole analyses do not. However, in the more highly symmetric complexes, for which the older methods were devised, the AOM is weakest, while the problems engendered by low or non-existent symmetry for the traditional models become advantages for the AOM. And then we must recall: the vast majority of real, interesting chemical complexes possess little or no symmetry.

One final point remains in connection with the degree of parametrization in AOM analysis. It is this: even if the circumstances are such that the reasonably reduced number of AOM e parameters is less than or equal to the number of independent ligand-field matrix elements, the values of these matrix elements may not all be determinable by a given experiment. Most of the early AOM studies involved the analysis of optical, $d-d$ spectra and even in those cases employing polarized single-crystal spectroscopy, the well-known problems usually remain. We refer here to the frequent difficulty of resolving split spectral peaks; of correct assignment; of the sufficiently accurate determination of band origins; of the frequent obscuring of $d-d$ transitions beneath intense charge-transfer, or ligand, spectra; and the typical loss of electronic spectral information below, say, $4000\,\mathrm{cm}^{-1}$ in energy. While it is true that some spectra have been obtained which are so rich and well resolved that an unambiguous ligand-field or AOM analysis has been possible, it is much more common, particularly for low-symmetry molecules, involving a wide variety of ligand types, to find that insufficient experimental data is forthcoming upon which to base a satisfactory AOM analysis.

Historically, a common response to this situation has been to reduce the degree of parametrization by fixing the values of one or more of the model parameters by reference to molecules of a related type. In essence, the factorizability of the spectrochemical series and the transferability of ligand-field parameters have been regarded as sufficiently valid that the parameters for M—A in MABCD, determined in one study, may be 'borrowed' or transferred to M—A of MAEFG in another. Occasionally, as discussed in §12.1, rather odd conclusions have emerged from studies based on this procedure. That is hardly surprising for we should always remember that these concepts of transferability and spectrochemical factorizability are only approximate and, furthermore, based upon 'averaged' results derived, in effect, via the 'law of average environment'. After all, the mutual interaction of a metal with a ligand must depend upon the remaining ligands: indeed, if the whole of the AOM has any point at all, it is to study just that sort of dependence. Many of the results described in chapter 12 illustrate that and it is this author's firm conviction that a shortfall in the amount of experimental data for one system cannot be remedied by recourse to another.

Surely the proper course to take in the face of too limited a data base from a single technique is to refer to observations from the other complementary methods of ligand-field theory. Many recent AOM analyses owe their success to the simultaneous investigation of spectral and magnetic phenomena; the latter often involving both bulk susceptibility and esr g-value determinations. Such magnetic measurements are especially valuable when derived from studies of single crystals, for then we have the task of reproducing the temperature variations, absolute values, and directions of the principal susceptibilities. The equations derived in §7.6 provide the means by which all these features may be calculated from the eigenvectors and eigenvalues of, say, an AOM representation of the ligand field. Susceptibilities and g values thus probe the same ligand-field model as spectral transition energies. It is therefore reasonable to hope that some combination of these various properties should yield sufficient information from which we may derive unique AOM parameters in any particular case. Because of the indirect nature of the calculations of magnetic properties, it is generally not possible to predict how successful the whole process will be and the final efficacy of any analysis can be assessed only after a thorough and wide-ranging sampling of the whole parameter space. Nevertheless, our inability to express the conditions for parameter definition analytically does not invalidate or prevent the success of contemporary ligand-field studies of this kind. It should by now be obvious that those studies which are most likely to succeed involve low-symmetry species analysed through the simultaneous application of several ligand-field techniques. The complementarity of spectral and magnetic methods is also manifest in the way that magnetism, being a property predominantly of the ground and low-lying states in a molecule, will tend to furnish information on some parameters more than others. This is especially true for those parameters associated with the other terms in the complete effective Hamiltonian (III.1), in that estimates of inter-electron-repulsion parameters derive, essentially, exclusively from spectral measurements, and of spin-orbit coupling coefficients from magnetism. These points are illustrated in several of the applications described in chapter 12.

9.9 Non-diagonal local potentials

In §9.3 we showed how the equation (9.19) derives directly from the cellular decomposition of the ligand-field potential. The utility of the fundamental AOM equation (9.20) follows on the assumption that the unitary matrices **R** are defined by the molecular structure. The subsequent classification of the e_λ parameters by reference to the local M—L pseudosymmetry assumed in effect that that local symmetry approximated C_{2v} or higher with respect to the M—L axis. The pyridine group in figure 9.4(*a*) illustrates this position. There may be real chemical circumstances, however, where such a classification is inappropriate. Figure 9.4(*b*) illustrates the case of 'misdirected valency' in which the local influence of the donor atom D may not be directed exactly towards the central metal, perhaps because of the constraints of chelation: the diagram illustrates this in terms of a misdirected σ orbital. Should this description be at all appropriate in a given system, the local ligand-field potential would not be

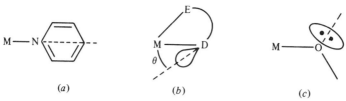

Fig. 9.4. (a) Local C_{2v} symmetry in the M—L interaction. (b) 'Misdirected valency' in chelate complexes. (c) Local off-diagonal parameter required to represent the interaction of the metal orbitals with the electron lone pair of an oxygen donor.

diagonal with respect to the M—D vector. In molecular-orbital terms (but see chapter 11) the Dσ orbital can overlap with both d_{z^2} and d_{xz} orbitals, say, with the result that the matrix element $\langle z^2|V|xz\rangle$ in the local frame will take a non-zero value. By inspection of table 9.2, table 9.3 must then be modified to include the non-zero terms

$$
\left.
\begin{aligned}
c_{21} &= -\left(\frac{2\pi}{5}\right)^{1/2} e_{\pi\sigma}, \quad \text{real,} \\
c_{41} &= -\left(\frac{12\pi}{5}\right)^{1/2} e_{\pi\sigma}, \quad \text{real,}
\end{aligned}
\right\}
\tag{9.37}
$$

where we have defined the local, non-diagonal matrix element $\langle z^2|V|xz\rangle$ as $e_{\pi\sigma}$. Transformation into the global frame may proceed as before. The contributions to $e_{\pi\sigma}$ can include the effects of local Dπ orbitals interacting with both d_{z^2} and d_{xz} also, so that ultimate chemical interpretations of the 'cross-parameters' $e_{\pi\sigma}$ may well be confused. The formalism described is nevertheless valid. Note that we have freedom to choose the plane in which to place the 'cross-interaction'. While the choice of the matrix element $\langle z^2|V|yz\rangle$ would be equally possible, there is no need to have both.

Thus far, the argument is somewhat theoretical and we might enquire whether experiment or practice can indicate that the circumstances of figure 9.4(b) actually occur. To date there is little evidence for it in this form but the example shown in figure 9.4(c), however, is of interest. AOM analyses described in chapter 12 have determined non-zero values for both $e_{\pi x}$ and $e_{\pi y}$ associated with some trigonal oxygen donors. The non-zero e_{π} value associated with the direction perpendicular to the diagram is to be expected, being associated no doubt with the lone pair of electrons in the unhybridized oxygen p orbital. No π interaction was initially expected in the plane of the formal sp^2 hybrids, however, but a freely varied parameter for it took on a substantial, positive value. A more appropriate description would be in terms of a 'cross-parameter' $e_{\pi\sigma}$ representing the simultaneous interaction of the sp^2 lone pair on the oxygen with metal σ and πx orbitals. The systems are more fully described in chapter 12. For the moment we observe that there are occasions in which non-diagonal local potentials should be recognized and the foregoing provides the procedure by which they may be incorporated within a general AOM program.

9.10 Empty cells

Throughout the discussion so far we have implicitly associated one cell with each discrete ligand or donor atom in a complex. Circumstances exist where it is appropriate also to assign cells, with their associated AOM e parameters, to physical regions in a coordination shell where there are no ligands. This strange idea is particularly important for the coordinationally void regions above and below the plane of various four-coordinate, planar complexes, for example. The reasons for adopting this approach and the physico-chemical interpretation of the AOM parameters of empty cells are discussed fully in § 11.12. No new formalism is required for handling these quantities and they are mentioned in the present chapter merely for the sake of completeness, but it is appropriate to remember here that the choice of cells is not determined in any arbitrary way but rather by the requirement that the corresponding unitary matrices $\mathbf{R}^l$ diagonalize the local potential matrix $\mathbf{v}^l$.

References

[1] Gerloch, M. & McMeeking, R.F., *J. Chem. Soc. Dalton Trans.*, p. 2443 (1975).
[2] Rotenberg, M., Bivens, R., Metropolis, N. & Wooten, J.K. Jr, *The 3−j and 6−j Symbols*, The Technology Press, MIT, Cambridge, Mass., 1959.
[3] Woolley, R.G., *Mol. Phys.* **42**, 703 (1981).
[4] Schäffer, C.E., *Structure and Bonding*, **5**, 68 (1968).
[5] Griffith, J.S., *The Irreducible Tensor Method for Molecular Symmetry Groups*, Prentice-Hall, New Jersey, 1962.
[6] Harnung, S.E. & Schäffer, C.E., *Structure and Bonding*, **12**, 201, 257 (1972).
[7] Silver, B.L., *Irreducible Tensor Methods*, Academic Press, London, 1976.
[8] Gerloch, M. & Slade, R.C., *Ligand Field Parameters*, Cambridge University Press, 1973.
[9] Jørgensen, *Modern Aspects of Ligand Field Theory*, North-Holland, Amsterdam, 1971.

10

Techniques for parametric models

—

10.1 Aims

Having assembled theoretical and experimental techniques for the study of magnetic and spectroscopic properties of single-centre transition-metal and lanthanide complexes, we turn to the practical question of how these skills may be integrated into a complete package for routine investigations. As pointed out in the introduction to Part III, we need to construct a computational and procedural system which facilitates the study of a wide variety of chemical systems in a way that makes the connections between experimental data and chemical conclusions as direct and routine as possible. While a ligand-field analysis should be comprehensive, detailed and correct, it should not intrude unduly on the central purpose of the study which is to learn something of the nature of the bonding and electron distribution in these systems.

It is obvious from the nature of the ligand-field theory that it is impossible to calculate quantities relating to structure and bonding *directly* from observable quantities like spectral transition energies or magnetic suscepti- bilities. Instead, we must calculate such properties from the model and subsequently compare the results with experiment. The package we review in this chapter therefore involves these two broad features: (i) the computation of eigenvalues and eigenvectors under the parametrized perturbation (III.1), together with the consequent crystal and molecular magnetic susceptibilities, their orientations and temperature dependencies and the magnitudes and directions of the principal molecular g values; (ii) means for systematically comparing any or all of these quantities with experimentally determined values, together with tactical procedures by which the various model parameters may be optimized for 'best fit'. The discussion is based on a FORTRAN computer program system,[1] called CAMMAG, which has been developed in Cambridge by the author and his colleagues over several years. It is inappropriate to present a full coding of this program here, nor it is suggested that CAMMAG provides the best or only way of achieving the desired aims in ligand-field analysis. At the time of writing, however, it is the only such comprehensive system known to the author and, in any case, the present discussions are directed more towards the structure of such a package rather than to a detailed prescription or handbook. The material in this chapter nevertheless deals with many of the minutiae of ligand-field analysis and we are obliged to consider aspects of theory and format concurrently. In §10.2 we consider the business of computing ligand-field properties, following the structure and philosophy of CAMMAG. It is frequently most direct to develop these features by illustrative examples. Means of comparing calculated and observed properties are described in §10.3 later.

10.2 Calculation of ligand field properties

The calculations to be performed fall into two main parts: the first is the diagonalization of an appropriate basis under the perturbation (III.1)

$$\mathcal{H}'_{\mathrm{LF}} = \sum_{i<j}^{N_d} U(i,j) + \sum_{i}^{N_d} V_{\mathrm{LF}}(\mathbf{r}_i) + \zeta \sum_{i}^{N_d} \mathbf{l}_i \cdot \mathbf{s}_i, \qquad (\text{III}.1)$$

and the second is the ensuing computation of various ligand-field properties to be compared with experiment. Each part of (III.1) is parametrized, as also is the magnetic moment operator used in the second part of the calculations. It is sensible and possible to divide the calculations into one part which needs to be performed only once and one which must be repeated as we investigate the consequences of changing the various system parameters: we shall refer to these two program segments as SETUP and RUN respectively. In SETUP we compute everything that is independent of the parameter values referring to (III.1), and of the orbital reduction factor in the magnetic moment operator. We therefore construct a series of separate matrices for the various perturbations in (III.1), or for parts thereof, and store the results on file. In RUN, copies of each matrix are multiplied by a current value of an appropriate parameter and then summed: the total matrix is then diagonalized numerically and the various ligand-field properties calculated from the resulting eigenvalues and eigenvectors.

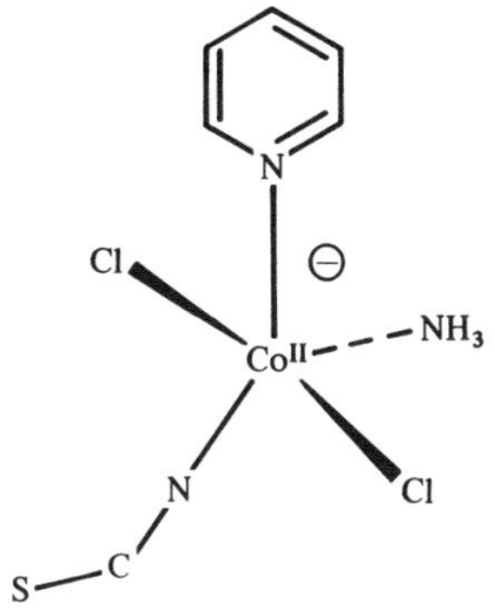

Fig. 10.1. A fictitious molecule to illustrate the parametrization scheme.

10.2.1 SETUP

Input data for this program segment comprises a description of the molecular geometry, together with ligand-field parameter-type assignment, and of the basis which is ultimately to be diagonalized under (III.1). We reserve until §10.2.2 a description of the precise way in which these data are communicated to the computer program. In addition to this information referring to the particular problem at hand, we require a body of standard data which includes the various reduced matrix elements featuring in the 'master equations' of §8.16.

The *basis* is specified as free-ion terms or states. If we consider the hypothetical complex ion in figure 10.1 as a high-spin d^7 species, it may be sufficient for many purposes to work with a basis (in J, M_J quantization) spanning the terms of maximum spin degeneracy arising from the d^7 configuration; namely, the 40 functions $^4P + {}^4F$. More refined, but obviously lengthier, calculations may be deemed necessary later in the analysis, in which case the basis might be enlarged to span the complete 120-fold d^7 manifold. We should arrange matters so that choices of this kind may be made and changed easily.

Considering now the *molecular geometry*, the example in figure 10.1 has been chosen to illustrate many, though not all, of the facilities required of an AOM calculation. We assume, for generality, that the ion possesses no strict point-group symmetry. An AOM calculation requires definitions of the orientation of the various local, ligand reference frames relative to some global, metal reference frame. In addition, therefore, to a data list of various donor-atom coordinates relative to the metal atom, standard input data for SETUP must include information defining our choice of

orientation of coordinate frames on both metal and ligands. With ligand
z axes taken to lie along the metal–donor-atom vector, we choose local
x and y axes which reflect the local metal–ligand pseudosymmetry. For
the pyridine group in the example, we might take x and y ligand axes to
lie in, and perpendicular to, the plane of the heterocycle: subsequently $e_{\pi x}$
and $e_{\pi y}$ for this ligand may be labelled $e_{\pi \parallel}$ and $e_{\pi \perp}$, respectively, and
we might hope to see the analysis demonstrate, in due course, that $e_{\pi \parallel} \approx 0$.
Similarly, for the thiocyanate ligand, we would choose to place local x and
y axes parallel and perpendicular to the Co—N—C—S plane, though
here we might not necessarily expect that $e_{\pi \parallel} \approx 0$. Both chlorine and
ammonia ligands will be considered to interact with the metal in a
cylindrically symmetric fashion (linear ligators) so that $e_{\pi x} \equiv e_{\pi y}$. For these
ligands, the choice of local x and y axes is arbitrary. However, since our
program should treat all ligands on an equal footing, AOM rotation
matrices will be constructed for the general case: therefore, a definite,
though arbitrary, choice for the local x and y axes will be required, the
linear ligation being recognized by the *subsequent*, fixed relationship
$e_{\pi x} = e_{\pi y}$ in RUN. As discussed in §9.8, it may be desirable to consider
the two chlorine ligands as chemically equivalent in the interests of
reducing the overall degree of parametrization. If so, this may be achieved
in two ways. We may regard the two chlorines as distinct in SETUP and
arrange to set equal parameter values in RUN later: or we may define
them as equivalent from the beginning and sum their contributions to the
process we now describe.

Given that we choose to work within a basis defined in terms of $|J, M_J \rangle$
eigenkets, a choice we discuss further in §10.5, the machinery of tensor-
operator theory summarized in the 'master equations' of §8.16 provides
the obvious and powerful tool for the evaluation of all required matrix
elements. In particular, those for the ligand-field operator are most
conveniently evaluated, via (**M.3**), in terms of the components of the
conventional multipole expansion of the ligand-field potential. Barring
special and rather exceptional cases discussed in §10.4, we prefer to
parametrize the ligand field within the AOM and so an important segment
of SETUP is a conversion, in effect, from the cellular decomposition of
the AOM to the global expansion of the conventional multipole description. The relationships between the AOM parameters and the multipole
expansion coefficients expressed within a given local ligand frame were
given for a d-orbital basis in table 9.3 which, after dropping the c_{00} term
and ignoring δ-bonding contributions, takes the form of the matrix

$$
\begin{array}{c|ccc}
 & e_\sigma & e_{\pi x} & e_{\pi y} \\
\hline
c_{20} & \left(\dfrac{4\pi}{5}\right)^{1/2} & \left(\dfrac{\pi}{5}\right)^{1/2} & \left(\dfrac{\pi}{5}\right)^{1/2} \\
c_{2,\pm 2} & 0 & \left(\dfrac{3\pi}{10}\right)^{1/2} & -\left(\dfrac{3\pi}{10}\right)^{1/2} \\
c_{40} & \tfrac{6}{5}\pi^{1/2} & -\tfrac{4}{5}\pi^{1/2} & -\tfrac{4}{5}\pi^{1/2} \\
c_{4,\pm 2} & 0 & \left(\dfrac{2\pi}{5}\right)^{1/2} & -\left(\dfrac{2\pi}{5}\right)^{1/2}
\end{array}
\tag{10.1}
$$

Now, furnished with input data specifying the coordination geometry of the complex, together with suitable orientations for each ligand reference frame, we may compute the geometrical relationship between each ligand coordinate frame and the fixed (but arbitrary) metal frame in terms of a set of Euler angles $(\alpha\beta\gamma)$: the process is detailed in Appendices A, B. Then, using (9.31), we construct a matrix, equivalent to (10.1) which relates that ligand's AOM e parameters to *global* c_{kq} coefficients, as in (9.29). In general, each order (k even) of the $\{c_{kq}\}$ will give rise to a combination of the complete set of q belonging to that order, so that 14 different c_{kq} will arise (for d^n systems) in the global frame. After a similar process is followed for each ligand in the complex – the identity of each being established by a different set of Euler angles – we shall have constructed a matrix **F** of the form shown in (10.2). Note that the list of global c_{kq} should include those with

$$
\begin{array}{c|c|c|c|c}
 & e_\sigma(1)e_{\pi x}(1)e_{\pi y}(1) & e_\sigma(2)e_{\pi x}(2)e_{\pi y}(2) & e_\sigma(3)\ldots & \ldots \\
\hline
c_{20} & & & & \\
c_{21} & & & & \\
c_{22} & & & & \\
c_{40} & & & & \\
c_{41} & \mathbf{F}(1) & \mathbf{F}(2) & \mathbf{F}(3) & \ldots \\
c_{42} & & & & \\
c_{43} & & & & \\
c_{44} & & & &
\end{array}
\tag{10.2}
$$

negative q values, but these need not be stored explicitly, for they are related to those with positive q by (9.11). The matrix **F** has dimensions of 8 for the global c_{kq} (or 14 if we include the implicit negative q) by $3n$, where n is the number of discrete ligand *types*. For the complex ion in figure 10.1, n is five; or four, if we designate the two chlorine atoms to be equivalent. In this latter case, the two submatrices $F(Cl1)$ and $F(Cl2)$, differing because of the different ligand reference frames, are summed element-for-element in the second of the two methods discussed at the end of the preceding paragraph. The same procedure is adopted in all cases of equivalent ligands (genuine or by simplifying assumption). In the special case of centrosymmetric molecules, the Euler matrices for inversion-related ligands can be taken as identical (for handedness presents no distinction in the pure parity basis) and so some of the labour of the construction of (10.2) can be avoided simply by consideration of the asymmetric moiety, followed by multiplication of the resulting F matrix by two: this is, of course, a manifestation of holohedral symmetry.

Although the matrix (10.1) in the local frame is real, the F matrix (10.2) is generally complex, as may be seen directly from the form (8.24) of the rotation matrices appearing in (9.31). If, however, the complex under study possesses two-fold rotation symmetry with respect to the global y axis, all components of the global ligand-field potential will be identically real: proof is given in Appendix E. Altogether, the F matrix contains all the geometric information about the complex required by the study. In the second main program segment, RUN, it provides the recipe by which a given set of AOM parameter values may be converted into an equivalent set of multipole expansion coefficients. One task of SETUP, therefore, is to store F on a permanent file to be used in all subsequent executions of the RUN segment.

Turning now to the other task of the SETUP segment, which is the construction of matrices under all perturbations of the basis except insofar that particular parameter values are involved, we consider first the separate matrices referring to different parameters involved in (III.1). If M is the total matrix to be diagonalized in any one calculation in the RUN segment, we write

$$
\begin{aligned}
M = F_2 R_2 &+ F_4 R_4 \\
&+ c_{20} L_{20} + c_{21} L_{21} + c_{2-1} L_{2-1} + \cdots \\
&+ \zeta P,
\end{aligned}
\tag{10.3}
$$

in which the matrices R_2 and R_4 refer to interelectron-repulsion energies, the $\{L\}$ to the ligand-field potential, and P to spin-orbit coupling. These component matrices are independent of the particular parameter values – the Condon–Shortley F_2 and F_4, the multipole expansion coefficients $\{c_{kq}\}$

and the spin-orbit coupling coefficient ζ – and so need to be evaluated only once; and then stored on file. Subsequently, in RUN, copies of each component matrix are multiplied by current parameter values, summed to give $\mathbf{M}$ which is then diagonalized numerically. The component matrices are square and of dimension given by the chosen basis.

The interelectron-repulsion matrices $\mathbf{R}_2$ and $\mathbf{R}_4$ (and $\mathbf{R}_6$ for lanthanide systems) are written down directly in terms of the RME of (**M.1**), taking account of the renormalizing definitions (8.213). These RMEs, taken from the tabulations of Nielson & Koster,[2] form part of the fixed data base for SETUP. Consider the example in figure 10.1. The electrostatic matrix for d^7 is the same as that for d^3 and, ignoring terms in F^0 we find, from Nielson & Koster, the RMEs in table 10.1. Allowing for the definitions of F_k in terms of F^k in (8.213), the $\mathbf{R}_2$ matrix for the complete d^7 matrix given in the order in table 10.1 will have diagonal elements: zero for the first 12 elements, -15 for the next 28, -6 for the next six, and so on. Off-diagonal elements occur only for the repeated (^2D) free-ion terms and, for the $\mathbf{R}_2$ matrix, for example, will comprise ten elements of $3\sqrt{(21)}$ lying on a diagonal parallel to the principal diagonal in accordance with the Kronecker delta functions in (**M.1**). The completed matrices $\mathbf{R}_2$ and $\mathbf{R}_4$ are stored[††] on permanent file, to be used in the RUN segment of the program.

The spin-orbit coupling matrix $\mathbf{P}$ is evaluated in the same basis, using (**M.5**). For this we require a value for l ($= 2$ in the present example) which is taken from the given free-ion configuration as data; various 6–j symbols are computed as desired, as in Appendix C; and RMEs of the unit tensor operator $\mathbf{V}^{11}$, taken from the fixed data basis from Nielson & Koster.[2] The following relationships[2] are required to complete the coverage of the explicit RMEs:

$$\langle \alpha'L'S' \| \mathbf{V}^{11} \| \alpha LS \rangle = (-1)^{L'-L+S'-S} \langle \alpha LS \| \mathbf{V}^{11} \| \alpha'L'S' \rangle, \qquad (10.4a)$$

$$\langle \alpha'L'S' \| \mathbf{V}^{11}(4l + 2 - n) \| \alpha LS \rangle = -\langle \alpha'L'S' \| \mathbf{V}^{11}(n) \| \alpha LS \rangle : \qquad (10.4b)$$

we note, for example, that elements for d^7 are the negatives of those provided for d^3.

The ligand-field matrices $\{\mathbf{L}\}$ of (10.3) are constructed using the 'master equation' (**M.3**). Again an l value is taken from input data and various 3–j and 6–j symbols calculated as required via Appendix C; values of

---

[††] The interelectron-repulsion matrices are stored either in this form or, being very sparse, as linear arrays with zeros omitted, together with appropriate addresses. Although we do not concern ourselves with computer program details of that kind, it is worth recalling that such details are often suggested by the problem itself.

Table 10.1. *Electrostatic matrix elements for d^3*

^4P	^4P		$-\frac{1}{3}F^4$
^4F	^4F	$-\frac{15}{49}F^2$	$-\frac{8}{49}F^4$
^2P	^2P	$-\frac{6}{49}F^2$	$-\frac{4}{147}F^4$
$^2D(1)$	$^2D(1)$	$\frac{1}{7}F^2$	$+\frac{1}{7}F^4$
	$^2D(2)$	$\frac{3}{49}\sqrt{(21)}F^2$	$-\frac{5}{147}\sqrt{(21)}F^4$
$^2D(2)$	$^2D(2)$	$\frac{3}{49}F^2$	$-\frac{19}{147}F^4$
^2F	^2F	$\frac{9}{49}F^2$	$-\frac{29}{147}F^4$
^2G	^2G	$-\frac{11}{49}F^2$	$+\frac{13}{441}F^4$
^2H	^2H	$-\frac{6}{49}F^2$	$-\frac{4}{147}F^4$

RMEs of the unit tensor operators $\mathbf{U}^k$ are taken from Nielson & Koster.[2] Within the 'pure' configuration bases we use, that is, d^n or f^n, necessary phase relationships are obtained from the equations[2]

$$\langle\alpha'L'S'\|\mathbf{U}^k\|\alpha LS\rangle = (-1)^{L'-L}\langle\alpha LS\|\mathbf{U}^k\|\alpha'L'S'\rangle, \qquad (10.5a)$$

$$\langle\alpha'L'S'\|\mathbf{U}^k(4l+2-n)\|\alpha LS\rangle = -\langle\alpha'L'S'\|\mathbf{U}^k(n)\|\alpha LS\rangle. \qquad (10.5b)$$

Some computational effort can be saved if the ligand-field point-group symmetry is such as to cause some c_{kq} to vanish identically, firstly because the corresponding $\mathbf{L}_{kq}$ matrices need not then be constructed in SETUP and, secondly, because they need not subsequently be multiplied by the corresponding parameter values in RUN. However, while such special group-theoretical relationships are easily determined in a calculation performed on paper, so to speak, it is simpler in the general computer package we describe here to recognize appropriately near-zero values of c_{kq} in the $\mathbf{F}$ matrix of (10.12) by numerical inspection.

Altogether we note that all component matrices of interelectron-repulsion and spin-orbit coupling are real, as also are the parameters F_2, F_4 and ζ. Therefore, the total matrix $\mathbf{M}$ is only complex for ligand-field symmetries which lack a diad parallel to the global y axis.

The final task for the SETUP program segment concerns the magnetic moment operator. The computation of both susceptibility and g values, described in chapter 7, involves evaluation of matrix elements of the magnetic moment operator:

$$\mu_\alpha = kl_\alpha + 2s_\alpha; \quad \alpha = x, y, z \qquad (10.6)$$

within the eigenfunctions resulting from the diagonalization of $\mathbf{M}$ in (10.3). Equation (7.206) expressed the fact that these matrix elements involve simple, if lengthy, sums of the corresponding matrix elements of μ_α within the original basis used throughout the construction of the component matrices we have just discussed. A great deal of computational effort can therefore be saved if matrices of μ_α, or better still of l_α and s_α separately, within the original basis are constructed once and for all, rather than their calculation being repeated for each variation of Stevens' orbital reduction factor k and for each combination of the parameters of (III.1). Accordingly, six matrices – for l_α and s_α; $\alpha = x, y, z$ – are constructed and filed within the SETUP program using the appropriate parts of the 'master equation' (**M.4**). Only $3-j$ and $6-j$ symbols are required here and these are computed as required: the single special RME of $\mathbf{L}^1$ was evaluated in (8.148) and incorporated in (**M.4**).

10.2.2 SETUP input

It cannot be emphasized too strongly that the technical business of performing calculations within a ligand-field analysis should be made as automatic as possible so that tedium should be no determinant of the scope or detail of any given study. One facet of this practical philosophy, and we shall discuss others throughout this chapter, is the provision of a simple and versatile interface between user and computer. So we turn to the question of the form of data input for CAMMAG. No attempt is made to be comprehensive, nor is it claimed that the system is unique or exclusive. We aim merely to illustrate the kind of tactics which may be adopted to simplify analysis. The scheme is actually much more important in the RUN segment which is performed many times but a description of some aspects of SETUP input serves to introduce the format.

As elsewhere in this chapter, it is convenient to illustrate the procedure specifically but we must preface our description with some remarks about the examples we use. We should like to illustrate the input features of basis and molecular geometry; and the output, referring to spectral transition energies and assignments, magnetic susceptibilities and their orientations, and molecular g tensors. However, as will be apparent in the review of recent applications of the method presented in chapter 12, no one system serves to illustrate all these features clearly. It happens, and it does not seem possible to predict when, that some analyses are established more by the electronic spectrum than by the paramagnetism, some by the esr g values, others by various combinations of these properties: sometimes parameter values are sensitively determined for complexes with few distinct ligand types but occasionally not. Rather than illustrate this chapter, therefore, with a single example worked right

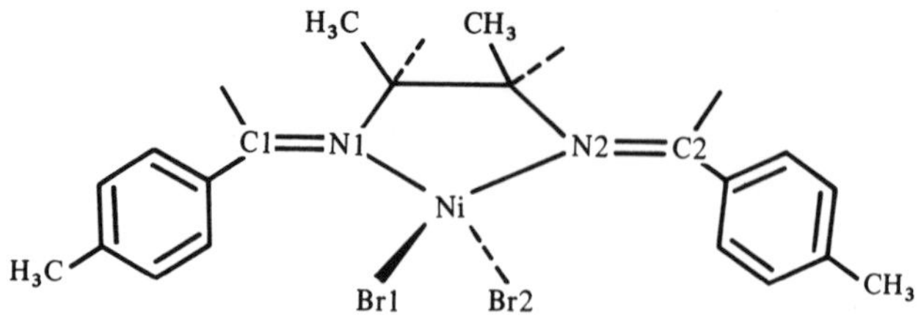

Fig. 10.2. Dibromo [cis-endo-*N*, *N'*-di (4-methylbenzylidene)-meso-2, 3-butanediamine]-nickel(II).

through, we shall refer to a series of systems, each of which serves to illuminate the particular point in question.

Consider, then, the 'tetrahedral' complex of nickel(II), shown in figure 10.2. The X-ray structure analysis[3] describes a molecular geometry close to C_{2v}, so that we shall consider the coordination to comprise a pair of chemically equivalent imine groups and a pair of equivalent bromine atoms. Let us suppose our interest is confined to the analysis of a solution absorption spectrum for which a basis of spin triplets is sufficient. The input to the SETUP segment of CAMMAG takes the following form:

Card		
1.	TITL	Example 10.2
2.	CELL	15.584 11.325 14.135 90 122.53 90
3.	CONF	2 8
4.	BASE	3F 3P
5.	NI	0 0.0259 0.25
6.	N1	0.0547 − 0.0139 0.1581
7.	N2	− 0.0605 0.1637 0.1492
8.	BR1	0.1428 0.0998 0.4174
9.	BR2	− 0.1395 − 0.1001 0.1943
10.	C1	0.0940 − 0.1146 0.1496
11.	C2	0.1126 − 0.2222 0.2113
12.	MID	0.0016 − 0.0001 0.3058
13.	XREF	8 1 2
14.	LGND	1 2 1 6
15.	LGND	1 3 1 7
16.	LGND	2 4 1 5
17.	LGND	2 5 1 4
18.	MULT	1
19.	END	

Card 3 describes the configuration as d^8: for f^5 the arguments would be 3 5. The basis is defined on card 4 as $^3F + {}^3P$: for the complete 45-fold basis of d^8, we would specify 3F 3P 1D 1G 1S. On cards 5–12 are given fractional coordinates from the X-ray analysis,[3] referred to the unit cell dimensions on card 2, for all atoms required to define metal and ligand reference frames. The global, metal frame may be defined arbitrarily, but in the present example is given by card 13 as having the z axis parallel to the vector joining 'atoms' 8 and 1 in the preceding list: that is, joining the nickel atom to the midpoint of the two bromine atoms: z is thus chosen to approximate the idealized molecular diad. The global y axis is defined by the XREF card to lie perpendicular to the plane defined by the three 'atoms' 1, 2, 8. The x axis is constructed, as a vector product, to be orthogonal to z and y and to form a right-handed coordinate frame: details are given in Appendix A. Cards 14–17 describe ligand types and axis orientations. Thus, card 14 defines the imine with donor atom N1 as of type 1, with local z axis parallel to the vector joining Ni and N1, and local y axis perpendicular to the plane defined by Ni, N1, C1. This reference frame reflects the local pseudosymmetry and we might hope the subsequent analysis would show that $e_{\pi \parallel} \equiv e_{\pi x} \approx 0$. Our choice of local y axis for the type 2 ligand on card 16, for example, is arbitrary, for we shall later take $e_{\pi x} = e_{\pi y}$ for bromine, but it must nevertheless be defined here. Finally, card 18 defines the molecule as non-centric. MULT 2 would signify a centrosymmetric molecule: as described earlier, explicit coordinates for such symmetry-related ligands are not required; the program simply arranges that the **F** matrix is multiplied by two.

Execution of SETUP with these data produces a file containing the **F** matrix of (10.2), the various matrices of (10.3), and the six matrices referring to spin and orbital angular momentum. These are all required by the second main segment of the CAMMAG package, namely RUN, which we now describe.

10.2.3 RUN

The core of the RUN segment of CAMMAG is the diagonalization of the basis under the effective Hamiltonian (III.1), a process which is performed for many trial variations of the various system parameters. Our discussion begins with the formal description of these parameter values. The appropriate part of RUN input for the complex in figure 10.2 might be as follows:

1. B   700 800 950
2. ZETA 250

3.	ESIG	1 4000 5000 500 *
4.	EPIX	1 0
5.	EPIY	1 − 500 500 500 *
6.	ESIG	2 3000 5000 1000 *
7.	EPI	2 0 500 250 *

Card 1 states that values for the Racah B parameter[††] will successively take values 700, 800, 950 cm⁻¹. The spin-orbit coupling coefficient will be assigned only one value for the moment, as given by card 2. The AOM parameters for ligand type 1 (the imines) are given on cards 3–5, referring to $e_\sigma, e_{\pi x}, e_{\pi y}$ respectively. The asterisk notation provides an alternative notation when parameter values are to be varied by fixed intervals: thus card 3 assigns e_σ values from 4000 to 5000 cm⁻¹ in steps of 500 cm⁻¹. For linear ligators, we have the facility shown on card 7 of maintaining the equality $e_{\pi x} = e_{\pi y}$ throughout the variation here of 0, 250, 500 cm⁻¹. On execution, RUN will construct and diagonalize **M** of (10.3) for all combinations of these parameter values: in the present example, 243 separate diagonalizations will be performed. It may be that we do not require calculations to be made for all combinations of these parameter values. For example, we may have some reason to consider $e_\sigma(Br) = 3000$ cm⁻¹ only, when $e_\sigma(imine) = 4000$ cm⁻¹; $e_\sigma(Br) = 4000$ cm⁻¹ when $e_\sigma(imine) = 4500$ cm⁻¹; and so on. CAMMAG does indeed have facilities for limiting parameter combinations in this way. It is similarly possible to require some relationship between parameter values, like $e_\pi = 0.25 e_\sigma$ (see LINK, for example, in §10.3.1).

In execution, RUN systematically selects one combination of the system parameters and performs all required calculations before varying a parameter and repeating the whole process. Having established the current set of parameter values, the first step is to compute those c_{kq} parameters which correspond to the current AOM e_λ parameters: this is done by reference to the **F** matrix of (10.2) taken from the file prepared by SETUP. At this stage, use is also made of any information provided by SETUP regarding the absence of any L_{kq} matrices as determined by higher molecular symmetry: there will, of course, be no such saving in the present example of an unsymmetrical molecule. Copies of the 11 component

---

[††] CAMMAG normally involves the use of Condon–Shortley F_k parameters. However, within bases like the spin triplets of d^8 where only one interelectron repulsion parameter is necessary, the use of Racah's B parameter is allowed. CAMMAG employs the tactic of using the **R**₂ and **R**₄ matrices of (10.3) and of setting $F_2 = B$ with $F_4 = 0$. As only *relative* energies are required, this substitution is permissible in this case.

matrices of (10.3) are taken from the file, appropriate action taken for the $\mathbf{L}_{kq}$ with negative q values (equation (9.11)), each multiplied by the corresponding current parameter value (several of the c_{kq} being complex) and summed to give the complete matrix $\mathbf{M}$. This is generally a complex, hermitian matrix of order given by the basis (30 in the present example) and is diagonalized numerically by standard algorithms to give (30) eigenvalues and associated eigenvectors.

Standard CAMMAG output of eigenvalues takes one of two forms, depending upon specific coded requests. Examples are shown in figure 10.3 for all (spin-triplet) eigenvalues of the current example, corresponding to the first parameter value in each of the cards above. All energies are given relative to the ground state as the nature of ligand-field theory, let alone the neglect of the 'spherical terms' F_0 and c_{00}, precludes any knowledge of absolute energies, of course. The 'full output' in (*a*) lists all eigenvalues, together with spatial and spin characteristics of their associated eigenvectors. These supplementary data are intended to aid spectral assignments, of course, and the philosophy is as follows. Where polarized electronic spectra of a strictly low-symmetry complex are available, new or already within the literature, it is often possible to perform an electric-dipole or vibronic polarization analysis to assign transitions as between certain representations of a more-or-less appropriate, idealized, higher-symmetry point group. In the current example, C_{2v} would probably be a good choice. Now the eigenvectors emerging from CAMMAG calculations in general, and from the present example in particular, will not transform exactly according to representations of such an idealized group; partly because the molecular geometry is actually somewhat less than ideal and partly because of mixing under spin-orbit coupling. Rather than perform subsidiary calculations for a corresponding ideal molecular geometry, together with a neglect of spin-orbit coupling, we construct a series of projection operators $P(\Gamma)$ which project out from the actual eigenvector ψ the proportion of the representation Γ of the chosen ideal group. The projector takes the form

$$P(\Gamma) = \sum_S \chi(\Gamma, S) \langle \psi | S\psi \rangle / h, \tag{10.7}$$

where S is a symmetry operation of the group which latter has order h. The present version of CAMMAG performs this for only two point groups – C_{2v} and D_2 – a limitation which represents a compromise between programming effort, execution time and usage. Thus, in the few studies of high-symmetry species undertaken, we can often make assignments on the basis of calculated degeneracies and the computed behaviour

(a) PARAMETER VALUES:—

	F2	F4	ZETA	LIGAND TYPE	SIGMA	PIX	PIY	PIS
	700.0	0.0	250.0	1	4000.0	0.0	−500.0	0.0
				2	3000.0	0.0	0.0	0.0

EIGENVALUES

			PROJECTIONS OF REPRESENTATIONS				TRANSFORMATIONS ⟨X : SX⟩			SPIN PROJECTIONS
			(D2)							TRIPLET
			(C2V)							
LEVEL	ENERGY	RELATIVE ENERGY	A1 / A1	B1 / A2	B2 / B1	B3 / B2	C2Z	C2Y	C2X	
30	6097.52	18451	0.0	0.9	0.1	0.0	0.9	−0.9	−1.0	1.0
29	6082.47	18436	0.0	0.9	0.1	0.0	0.9	−0.9	−1.0	1.0
28	6071.99	18425	0.0	1.0	0.0	0.0	−1.0	−1.0	−1.0	1.0
27	5603.44	17957	0.0	0.0	1.0	0.0	−0.9	0.9	−0.9	1.0
26	5575.96	17929	0.0	0.1	0.9	0.0	−0.9	0.8	−1.0	1.0
25	5563.23	17917	0.0	0.1	0.9	0.0	−0.9	0.9	−1.0	1.0
24	5059.04	17412	0.0	0.0	0.0	1.0	−1.0	−1.0	−1.0	1.0
23	5049.36	17403	0.0	0.0	0.0	1.0	−1.0	−0.9	0.9	1.0
22	5043.50	17397	0.0	0.0	0.0	1.0	−1.0	−0.9	0.9	1.0
21	−1834.14	10519	0.0	0.9	0.0	0.0	0.9	−0.9	−1.0	1.0
20	−1843.38	10510	0.0	0.9	0.0	0.1	0.9	−0.9	−1.0	1.0
19	−1859.14	10494	0.0	1.0	0.0	0.0	−1.0	−1.0	−1.0	1.0
18	−2775.28	9578	0.0	0.0	0.8	0.1	−0.9	0.7	−0.7	1.0
17	−2808.39	9545	0.0	0.0	0.8	0.1	−0.9	0.7	−0.7	1.0
16	−2838.50	9515	0.0	0.0	0.8	0.1	−1.0	−0.7	0.7	1.0
15	−3728.72	8625	0.0	0.0	0.1	0.8	−1.0	−0.7	0.7	1.0
14	−3733.63	8620	0.0	0.0	0.1	0.9	−1.0	−0.7	0.7	1.0
13	−3764.22	8589	0.0	0.0	0.0	0.8	0.5	0.5	1.0	1.0
12	−6368.57	5985	0.7	0.0	0.0	0.3	0.4	0.4	−1.0	1.0
11	−6380.87	5973	0.7	0.0	0.0	0.3	0.9	0.9	0.9	1.0
10	−6431.91	5921	1.0	0.0	0.0	0.0	−0.9	−0.9	−1.0	1.0
9	−6577.85	5776	0.3	0.0	0.0	0.7	−0.5	−0.5	−1.0	1.0
8	−6644.52	5709	0.3	0.0	0.0	0.7	−0.4	−0.4	−1.0	1.0
7	−6648.50	5705	0.0	0.0	0.0	0.0	1.0	1.0	−1.0	1.0
6	−7437.18	4916	0.0	0.0	0.0	0.0	1.0	1.0	−1.0	1.0
5	−7454.46	4899	0.0	1.0	0.0	0.0	−1.0	−1.0	−1.0	1.0
4	−7459.09	4894	0.0	1.0	0.0	0.0	1.0	1.0	−1.0	1.0
3	−12351.63	2	0.0	0.0	1.0	0.0	−1.0	1.0	−1.0	1.0
2	−12353.10	0	0.0	0.0	1.0	0.0	−1.0	1.0	−1.0	1.0
1	−12353.40	0	0.0	0.0	1.0	0.0	−1.0	1.0	−1.0	1.0

(b)

ENERGY	C2V SPECIES
18436	3A2
17933	3B1
17403	3B2
10507	3A2
9545	3B1
8610	3B2
5958	3A1
5729	3B2
4902	3A2
0	3B1

Fig. 10.3. Examples of CAMMAG output for the molecule shown in figure 10.2. Parameter values are given first and then either (*a*) a full description of the eigenvalues and some details of the associated eigenvectors, or (*b*) a condensed form for ready comparison with an observed spectrum.

of eigenvalues with respect to 'key' parameters. For lower symmetry systems, those which approximate some ideal point group, and hence those for which spectral assignments may be meaningful, frequently resemble the C_{2v} or D_{2d} groups. Further, calculations for these groups, using (10.7) are essentially identical, involving rotation of eigenfunctions by π about the three cartesian axes: inversions are irrelevant here, so all that is required to distinguish the results for these two groups is a relabelling of the representations. Should even lower-symmetry groups, like C_2, be appropriate, assignments can be made by inspection of the quantities listed to the right in output (*a*): these are the integrals $\int \psi^* S \psi \, d\tau$ where S signifies a two-fold rotation about one of the cartesian axes. From figure 10.3(*a*), we observe that the eigenvector (number 7) with eigenvalue $5705 \, \text{cm}^{-1}$, for example, transforms 30% as A_1 in the C_{2v} point group and 70% as B_2. Note that none of the supplementary data in the figure is significant if the molecule under study does not approximate *mmm* symmetry (or some subset thereof) or if the XREF card in SETUP has not been prepared to reflect the appropriate ideal symmetry in a manner compatible with the standard group tables. If XREF in the present example had placed global axes parallel to the orthogonalized crystal axes, for instance, and arbitrarily relative to the molecular structure, the supplementary output would be incorrect: this will not affect the eigenvectors themselves, however, which are then merely referred to a non-standard coordinate frame, so the calculation of various magnetic properties to follow will remain sound. Finally, in figure 10.3(*b*) we show a 'summary output' in which eigenvalues are collected together in groups according to their spin multiplicity, determined in output (*a*) by the spin-projection operator

$$P_S = |\psi_S\rangle\langle\psi_S| \qquad (10.8)$$

and assigned in a preselected point group (here C_{2v}) on the basis of majority speciation. The form (*b*) of output is obviously simpler to read but (*a*) is more informative. For example, some eigenvectors may be intimate mixtures of parts transforming as two or more different representations of the idealized point group: output type (*b*) will merely pick out the largest species.

10.2.4 Susceptibilities and g values

If so requested, RUN will continue after the diagonalization step and calculate magnetic susceptibilities and esr g values. Two further data cards are required for susceptibilities: for example,

8. K   1.0 0.8 0.6
9. TEMP   300 200 100 50 20 10 5

referring to values for Stevens' orbital reduction factor in the magnetic moment operator (10.6) and to the temperature (in °K) for which we wish to compute susceptibilities. The second card is irrelevant for g values, of course.

Consider the calculation of molecular magnetic susceptibilities via the generalized susceptibility equation (7.203),

$$K_{\alpha\beta} = \mu_0 N_A$$

$$\frac{\sum_i \left\{ \dfrac{\sum_j' \langle i|\mu_\alpha|j\rangle\langle j|\mu_\beta|i\rangle}{kT} - \sum_j'' \left[\dfrac{\langle i|\mu_\alpha|j\rangle\langle j|\mu_\beta|i\rangle + \langle i|\mu_\beta|j\rangle\langle j|\mu_\alpha|i\rangle}{E_i^{(0)} - E_j^{(0)}} \right] \right\} e^{-E_i^{(0)}/kT}}{\sum_i e^{-E_i^{0}/kT}}.$$

$$(7.203)$$

This can be an extremely time-consuming process, especially if we perform many such calculations in the sort of extensive explorations of parameter space which characterize a thorough magnetochemical analysis. It is therefore essential to optimize the computational procedure and so we now review several tactics which serve to avoid wasteful repetition.

Firstly, note that the numerator product in the first-order Zeeman term, being independent of temperature, should be evaluated only once, so that temperature variation should take place in the outermost loop of the program routine. Next, recall that $\sum_j'$ indicates a sum over all eigenvalues degenerate with E_i^0 and hence includes the contribution for $j = i$, while $\sum_j''$ in the second-order term refers to all $E_j^0 \neq E_i^0$. Therefore sums like those in the square brackets are generally also implicit within the first-order term. In essence, we may carry out virtually the same process for both first- and second-order Zeeman terms, accumulating the inner sums separately according to whether the eigenvalues of the ith and jth eigenvectors are deemed degenerate or not. Within the regime appropriate for the derivation of the susceptibility equation, it is usually satisfactory to consider eigenvalues as degenerate if they differ by less than $1\,\mathrm{cm}^{-1}$. Thirdly, note that we must perform these calculations for all six independent components $K_{\alpha\beta}$ of the susceptibility tensor. In the evaluation of K_{xy}, for example, we encounter the matrix elements $\langle i|\mu_x|j\rangle$ and $\langle j|\mu_y|i\rangle$ in the numerator products of (7.203), while K_{xz} involves $\langle i|\mu_x|j\rangle$ and $\langle j|\mu_z|i\rangle$, and so on. It is obvious, therefore, that simultaneous computation of quantities required for all six $K_{\alpha\beta}$ can be organized so as to avoid unnecessary duplication of effort. Fourthly, a particularly valuable saving can be made with respect to variation of the orbital reduction factor. At first sight it might appear that the whole calculation of susceptibilities must be performed for each variation of Stevens' k factor, but this is not so. Writing the numerator products for either first- or second-order terms in full, we get

$$\langle i|\mu_\alpha|j\rangle\langle j|\mu_\beta|i\rangle \equiv \langle i|kl_\alpha + 2s_\alpha|j\rangle\langle j|kl_\beta + 2s_\beta|i\rangle$$

$$= k^2\langle i|l_\alpha|j\rangle\langle j|l_\beta|i\rangle$$

$$+ k[\langle i|l_\alpha|j\rangle\langle j|2s_\beta|i\rangle + \langle i|2s_\alpha|j\rangle\langle j|l_\beta|i\rangle]$$

$$+ \langle i|2s_\alpha|j\rangle\langle j|2s_\beta|i\rangle. \tag{10.9}$$

It is therefore sensible to perform the inner summations of (7.203) in three separate parts which are subsequently multiplied by k^2, k or 1. In this way, the orbital reduction factor, appearing in the penultimate computation loop, is rendered a 'computationally cheap' parameter in much the same way as temperature. Note, however, that this procedure is wholly appropriate only if the orbital reduction factor takes a common value for μ_x, μ_y and μ_z. If only in the interests of reducing the degree of parametrization in the ligand-field analyses described in this book, we consistently maintain such an assumption (see also §11.7).

Summarizing: we simultaneously accumulate six separate sets of totals for each of the six independent components of the molecular magnetic susceptibility tensor. These comprise the sums

$$\left.\begin{array}{l} \displaystyle\sum_i\sum_j{}' \langle i|l_\alpha|j\rangle\langle j|l_\beta|i\rangle, \\[1em] \displaystyle\sum_i\sum_j{}' [\langle i|l_\alpha|j\rangle\langle j|2s_\beta|i\rangle + \langle i|2s_\alpha|j\rangle\langle j|l_\beta|i\rangle], \\[1em] \text{and}\quad \displaystyle\sum_i\sum_j{}' \langle i|2s_\alpha|j\rangle\langle j|2s_\beta|i\rangle \end{array}\right\} \tag{10.10}$$

for the first-order Zeeman effect, and

$$\left.\begin{array}{l} \displaystyle\sum_i\sum_j{}'' [\langle i|l_\alpha|j\rangle\langle j|l_\beta|i\rangle + \langle i|l_\beta|j\rangle\langle j|l_\alpha|i\rangle]/(E_i^0 - E_j^0), \\[1em] \displaystyle\sum_j\sum_j{}'' [\langle i|l_\alpha|j\rangle\langle j|2s_\beta|i\rangle + \langle i|2s_\alpha|j\rangle\langle j|l_\beta|i\rangle \\[0.5em] \qquad + \langle i|l_\beta|j\rangle\langle j|2s_\alpha|i\rangle + \langle i|2s_\beta|j\rangle\langle j|l_\alpha|i\rangle]/(E_i^0 - E_j^0), \\[1em] \text{and}\quad \displaystyle\sum_i\sum_j{}'' [\langle i|2s_\alpha|j\rangle\langle j|2s_\beta|i\rangle + \langle i|2s_\beta|j\rangle\langle j|2s_\alpha|i\rangle]/(E_i^0 - E_j^0) \end{array}\right\} \tag{10.11}$$

for the second-order contributions. These totals are accumulated for all levels (that is, all i) which are significantly populated. The summation limit for the outer sums $\sum_i$ is determined by the requirement that $\exp\{-(E_i^0 - E_1^0)/kT\}$ be negligible: again, experience suggests $(E_i^0 - E_1^0) > 4.5kT_{max}$, where T_{max} is the largest temperature given on the data card, is a suitable criterion for this. The summation limit for the first-order term $-\sum_j'-$ is, of course, determined by the effective degeneracy of the populated level. On the other hand, the corresponding limit for j in $\sum_j''$ for the second-order terms should, in principle, be determined by the size of the basis. However, experience suggests that convergence is safely satisfactory if contributions are ignored for those terms with $(E_i^0 - E_j^0) > 12\,000\,\text{cm}^{-1}$. It is rarely worth taking the trouble to try and reduce this value, for the computing economy made thereby tends to be annulled by the optimization process itself!

Two final observations can be made about the economic computation of susceptibilities. Firstly, consider the quantity

$$\langle i|l_\alpha|j\rangle\langle j|l_\beta|i\rangle + \langle i|l_\beta|j\rangle\langle j|l_\alpha|i\rangle \equiv A + B$$
$$= A + A^* = 2 \times (\text{real part of } A). \qquad (10.12)$$

Simple manipulations of this kind can be exploited for all the quantities requiring evaluation and, although not worth detailing further here, should clearly be incorporated within the program coding. Secondly, and as presaged by our remarks in §10.2.1., considerable economies may be made by the prior calculation of matrices of the original basis functions under the l_α and s_α operators. Since, for example,

$$\langle i|l_\alpha|j\rangle = \sum_J \sum_I a_{iI}^* a_{jJ}\langle \Phi_I|l_\alpha|\Phi_J\rangle, \qquad (10.13)$$

where the $\{a\}$ are coefficients of the *basis* functions $\{\Phi\}$ appearing in the eigenvectors of (III.1), if matrices of the $\{\Phi\}$ under l_α etc. are calculated beforehand, the evaluation of a matrix element like $\langle i|l_\alpha|j\rangle$ immediately simplifies to the calculation of sums of products like the right-hand side of (10.13). Furthermore, the first parts of such products, namely quantities like $a_{iI}^* a_{jJ}$, are common to all matrix elements of the kind $\langle i|O|j\rangle$, where O stands for $l_{x,y,z}$ or $s_{x,y,z}$: thus, there is opportunity here for further saving of computing time. Incidentally, it should now be clear why *separate* matrices for spin and orbital angular momenta were prepared in the SETUP segment.

We have reviewed the sort of detail which is of importance in the writing of an effective and economical computer program for susceptibility calculations. These details *are* important, for without them thorough examinations of the variations in a polyparameter model would be too tedious and expensive to be feasible. The successful application of contemporary ligand-field studies leading to significant and useful measures of the various σ- and π-bonding interactions in complexes generally involve molecular systems of low symmetry. As described throughout this book, we have both theoretical and experimental facilities available for such studies: but they can only be exploited properly given the equally vital third part of the synthesis, namely an efficient computing package. Of course, once that package has been constructed, our interest should become that of the user and so none of the foregoing detail should be visible: firstly, we then don't need to see it and, secondly, we don't wish to be distracted by it. So we move on.

Having calculated all independent components of the molecular susceptibility tensor for a given temperature, numerical diagonalization yields the principal molecular susceptibilities (the eigenvalues) and their orientation with respect to the global coordinate frame as defined by the XREF card in SETUP (the eigenvectors). These quantities, as well as the tensor prior to diagonalization may be output to a line printer, if desired. If the complex under study crystallizes in a triclinic space group, the computation would stop here, for the calculated molecular and crystal properties are the same by identity. For molecules in crystals of higher symmetry, however, we must perform the appropriate tensor summation over all symmetry-related molecules in the unit cell to obtain the corresponding crystal susceptibilities and their orientation, as described in chapter 6. All of these calculations would be performed within both temperature and k programming loops. Some typical output from CAMMAG is shown in figure 10.4.

If it is desired to calculate molecular g values, this is also most economically done within the k loop (but at only one (arbitrary) temperature, of course). Most of the required computations were completed when forming the sums in (10.10). As discussed in §§7.5 and 7.7, by far the most common circumstance in odd-electron systems is that involving a simple two-fold Kramers' degeneracy in the ground state. Accordingly, esr calculations in CAMMAG are performed under this assumption, using (7.213). The six independent elements of the $\mathbf{g}^2$ tensor are computed in this way, using the sums (10.10) accumulated for the susceptibility calculations. An ensuing diagonalization yields the directions and principal values of the $\mathbf{g}^2$ tensor and so, after taking square roots, we lose information regarding the signs of the g values, as discussed in §7.7.

MOLECULAR SUSCEPTIBILITY DATA

T	K	PRINCIPAL CHI VALUES	PRINCIPAL DIRECTIONS OF CHI ALPHA	BETA	VALUE GAMMA	AVERAGE CHI VALUES	PRINCIPAL MU VALUES	AVERAGE MU VALUES
295.00	0.80	3638 3686 12309	144.4 77.0 57.5	59.9 101.6 32.7	72.7 17.6 87.2	6545	2.930 2.949 5.389	3.929
195.00	0.80	4540 4600 21166	144.6 77.6 57.5	59.8 101.2 32.7	73.4 16.9 87.2	10102	2.661 2.678 5.745	3.969
125.00	0.80	5265 5334 36304	144.8 77.8 57.5	59.7 101.0 32.7	73.7 16.6 87.2	15635	2.294 2.309 6.024	3.953
75.00	0.80	5662 5736 63479	144.7 77.6 57.5	59.7 101.1 32.7	73.5 16.6 87.2	24959	1.843 1.855 6.170	3.860
45.00	0.80	5752 5827 106862	144.6 77.5 57.5	59.8 101.2 32.7	73.4 16.9 87.2	39480	1.439 1.448 6.201	3.769
25.00	0.80	5758 5833 192434	144.6 77.5 57.5	59.8 101.2 32.7	73.3 16.9 87.2	68008	1.073 1.080 6.203	3.687

CRYSTAL SUSCEPTIBILITY DATA

MONOCLINIC CRYSTAL CLASS

T	K	CHI (1)	CHI (2)	CHI (3)	ANISOTROPIES (2)—(3)	(1)—(3)	(2)—(1)	CHIBAR
295.0	0.80	3680	6170	9784	−3614	−6194	2489	5345
195.0	0.80	4592	9191	16323	−6932	−11731	4799	10102
125.0	0.80	5326	14316	27262	−12945	−21936	8991	15673
75.0	0.80	5727	22517	46634	−24117	−40997	16799	24958
45.0	0.80	5817	35222	77402	−42180	−71584	29484	70480
25.0	0.80	5824	60101	138041	−77880	−132217	54277	65870

Fig. 10.4. Illustrating the form of CAMMAG output for paramagnetic susceptibilities. The orientations of the molecular magnetic tensor are referred to the global frame defined by the XREF card.

10.3 Comparison with experiment

Sometimes it may be sufficient merely to calculate spectral and magnetic properties for a given complex or molecular geometry in order to investigate the response of the system to various, or perhaps just 'key', parameters. Most usually, however, our purpose is the analysis of experimentally determined ligand-field properties so that we might learn something of the bonding and electron distribution in a complex. In these circumstances, we must carry out the calculations just described for a large variety of parameter values and ultimately obtain a so-called 'best fit'. As we are obliged to proceed in a forward direction – from model to experiment – the process inevitably takes the form of 'trial and error'; and this may range from the haphazard, through the systematic, to the nominally analytical framework of a 'least-squares' scheme. At first sight, the most appealing approach would be that offered by least-squares analysis. More-or-less standard applications of the method are frequently used in fitting ligand-field optical spectra and a scheme for the more difficult task of reproducing magnetic susceptibilities has been proposed[4] also. However, the least-squares scheme is really a technique for refinement rather than search and we are concerned just as much with the business of finding a fit in the first place as with optimizing it later. Further, it is not uncommon to find ambiguous fits between theory and experiment, either in the form of *discrete* parameter sets affording good reproduction of the observed property or in terms of greater or lesser degrees of *correlation* between two or more parameters. The systematic trial-and-error, mapping process we describe below reveals both of those circumstances when they occur: a least-squares analysis might only locate one discrete 'fit' in a 'search mode' and so might have to be initiated at several points in polyparameter space if no important 'fits' are to be missed. Further, a mapping process can help suggest a search pattern: the purely numerical nature of the least-squares scheme frequently disguises 'signposts' and, by minimizing the degree of user intervention, decreases the number of opportunities for experience, intuition and flair. Above all, the mechanical process of searching parameter space for a fit can be informative in itself, not only in the identification of correlations but as an indicator of what is and is not possible in a ligand-field analysis. We begin our description of the mapping procedure as it is applied to crystal and molecular magnetic susceptibilities.

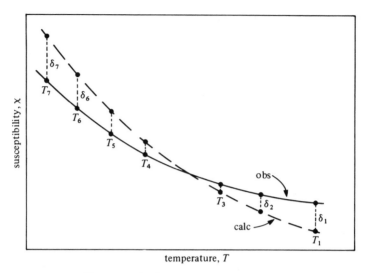

Fig. 10.5. The basis of the residual R_{mag}.

10.3.1 A mapping technique

Suppose figure 10.5 represents the observed temperature variation of some susceptibility (either principle, average, or anisotropy for either molecule or crystal), together with the corresponding property calculated within our model for a particular set of parameter values $\{p_i\}$. At a number, N_t, of representative temperatures, we may compute differences δ between the observed and calculated quantity, and define a 'figure-of-merit' R_{mag} as

$$R_{\text{mag}} = 100 - \left[\sum_i^{N_t} \frac{\delta_i^2}{\chi_i^2} \middle/ \sum_i^{N_t} \frac{|\delta_i|}{|\chi_i|} \right]^{1.7}, \tag{10.14}$$

constructed in such a way that perfect agreement gives $R_{\text{mag}} = 100$ and an average discrepancy $(\delta/\chi_{\text{obs}})$ of 10% gives $R_{\text{mag}} \approx 90$. By this empirical device, the figure-of-merit responds roughly linearly with 'error' in a parameter region 'close' to perfect fit. Any non-linear response outside that region is unimportant for there can be little objectivity in preferring one bad fit over another. Let it be emphasized straightway that we make no pretence of a mathematical foundation of the recipe (10.14): its purpose is solely as a *guide* in searching for, and refining, a 'fit'. In general, we may wish to fit several experimental quantities simultaneously: for example, all three principal crystal susceptibilities χ_1, χ_2, χ_3; or $\chi_\parallel, \chi_\perp$ for a tetragonal crystal. A corresponding figure-of-merit for the simultaneous fitting of several quantities (deriving from the same quantum-mechanical model) to (10.14) is

$$\bar{R}_{mag} = 100 - \left[\sum_{i}^{n} \frac{w_i^2 \delta_i^2}{\chi_i^2} \bigg/ \sum_{i}^{n} \frac{w_i |\delta_i|}{|\chi_i|} \right]^{1.7}, \qquad (10.15)$$

where n is the product of N_t and the number of quantities, and where the weighting factors w_i are empirically assigned factors[††] representing the relative confidence in each experimental quantity (in most real circumstances, there is no good reason for the w_i to differ from unity). Altogether, using (10.15), or any similar expression which the reader might prefer, as a recipe for a figure-of-merit we are able to attach a single 'fidelity' number to the set $\{p_i\}$ of parameter values used to perform these particular calculations. Similar numbers may be computed for each parameter variation we consider and then plotted on a polydimensional grid, there being one dimension per model parameter (temperature is therefore not considered as a parameter in this context). Taking two-dimensional slices of this polydimensional solid allows the drawing of contour maps so as to identify regions of good fit and, at the same time, any parameter correlation which may occur. The idealized sketches in figure 10.6 illustrate the sort of maps which may be constructed in this way. In (a) is represented a single, sharp 'fit'; in (b) two regions, one being rather less sensitive than the other; and in (c) an example of strong correlation between the fitting values of two parameters. Note two points about the use of such maps, immediately. (i) There exist no criteria in these diagrams which lead us to prefer one good fit over another. It may well be in a given analysis that some parameter values are sharply defined while others are correlated as in figure 10.6(c): in such circumstances we must be satisfied with some clear conclusions rather than none. Often, of course, ambiguities may be wholly or partly removed by simultaneous reference to other experimental properties, like electronic spectra. (ii) Secondly, the relative value of a figure-of-merit of 98 over one of 94, say, may not be very great; and in any case, having used the $\bar{R}_{mag}$ values and contour maps for what they are, namely guides and aids, we should then recheck the agreement between observed and calculated quantities explicitly.

The utilization of this mapping technique within the CAMMAG system illustrates the sort of tactics employed in the analyses summarized in chapter 12. We might follow a parameter value specification in the RUN segment with the cards:

[††] The formula in (10.15) is based upon the weighting factors w_i being normalized such that $w_{max} = 1.0$.

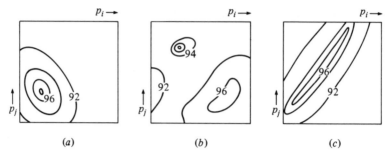

Fig. 10.6. Examples of typical types of susceptibility fitting maps. Contours are marked with $\bar{R}_{\mathrm{mag}}$ values.

9.	DATA	295 3803 4747 6354
10.	DATA	195 7128 10297 15942
11.	DATA	75 9692 16200 27860
12.	DATA	45 11860 24040 46310
13.	DATA	25 13220 37060 79750
14.	KMAP	
15.	FIT	1 1 1
16.	GRID	6 9
17.	GO	
18.	GRID	12 6
19.	GO	
20.	END	

Cards 9–13 list experimental, principal crystal susceptibilities; for example, card 9 specifies values of 3803, 4747, and 6354, for χ_1, χ_2, and χ_3 at 295 K (susceptibilities in CGSU $\times 10^6$). In the case of a triclinic system, six values would follow the temperature specification, corresponding to the six independent susceptibility tensor elements in some standard order (obviously, CAMMAG requires another card defining the crystal class). These five DATA cards are used instead of a TEMP card when we wish to compare observed and calculated magnetism rather than just list calculated quantities. Card 14 asks that a mapping procedure be performed with respect to the susceptibilities. The FIT card (15), usually defaulted, assigns equal unit weights w_i in (10.15) to each of the three principal crystal susceptibilities. Card 16 determines the way in which the polydimensional fitting map will be sliced up. The numbers on this card correspond, via a standard ordering code, to the parameters $e_{\pi y}(1)$ and $e_{\pi x}(2)$ and the meaning of card 16 is that $\bar{R}_{\mathrm{mag}}$ numbers will be plotted on two-dimensional grids with values varying with $e_{\pi y}(1)$ across the page and with $e_{\pi x}(2)$ down

the page: in the absence of further arguments on the card, the order of sampling of the polydimensional map is determined by the internal structure of the computer program. Card 17 initiates the RUN segment. On completion – that is, after diagonalization under all given parameter values, calculation of spectral assignments, of molecular and crystal magnetic susceptibilities, of $\bar{R}_{mag}$ values, and of the printing of susceptibility fitting maps – the program continues, reads cards 18 and 19 and replots the polydimensional map but sampled in an alternative way – here with respect to Stevens' k factor and $e_{\pi y}(1)$ across and down the page respectively.

The circumstance depicted in figure 10.6(c) might be seen as depressing in that seemingly any values of p_i and p_j, suitably correlated, will afford good reproduction of the experimental quantities. That may be so, but at the same time sensitivity of the system to other parameters may be quite sharp. One way of investigating this would be to hold one particular combination of p_i and p_j values fixed and then slice the 'fitting solid' in two other dimensions. On the other hand, the question of correlation between a third parameter p_k and either of p_i or p_j may be a matter of concern. Here, the following CAMMAG facilities can be helpful:

8.   LINK 9 6 1.2 150

19.   GRID 6 4

20.   GO

The LINK card controls the possible combinations of parameter values: specifically, card 8 indicates that values (in cm^{-1}) of parameter $9 (e_{\pi x}(2))$ takes the current value of parameter $6 (e_{\pi y}(1))$, multiplied by 1.2, plus $150 \, cm^{-1}$. We have, in effect, programmed the slope and intercept of the fitting 'ridge' in figure 10.6(c). The map plotted using card 19 will reveal the correlation, if any, between parameter $4 (e_\sigma(1))$, down the page, and parameters 9 and 6, which are varied together in such a way as to correspond to the best fits obtained in the earlier slice, represented in figure 10.6(c). Notice, by the way, that the EPI card introduced earlier corresponds essentially to a combination of EPIX and EPIY cards, together with a LINK between them of the form,

LINK 6 5 1.0 0

Despite an earlier assurance that the present chapter is not intended as a computer program handbook, we have looked at the LINK facility really in order to demonstrate how the necessary technical minutiae of polyparameter ligand-field analysis can be simplified: we reiterate that only when such practical methods and tactics are rationalized, simplified and easily accessed, can we be confident that thorough and standard explorations of polyparameter space have been undertaken.

The same kind of mapping technique can be employed for esr g values. Once again, a single number or figure-of-merit $\bar{R}_g$ can be attached to each combination of parameter values $\{F_2, F_4, (F_6), e_\lambda, \zeta, k\}$ and plotted as just described. The most valuable use of this facility concerns the comparison of observed and calculated molecular g tensors equivalent to the three principal g values and their orientations with respect to a given, arbitrary frame (from XREF?). The construction of $\bar{R}_g$ here is rather different from that of $\bar{R}_{mag}$ above, though it closely parallels what is required for the susceptibilities of triclinic crystals at a single temperature. The problem centres around the provision of a quantitative comparison of two ellipsoids (for the $\mathbf{g}^2$ or χ tensors), for it is clearly not satisfactory merely to compare the values of the principal property values alone. We show in Appendix F that the so-called 'difference ellipsoid', each tensor element of which is given by the difference between the corresponding calculated and observed tensor elements, provides a suitable measure for comparison and one, moreover, which is not dependent upon the frame of reference: the corresponding recipe for $\bar{R}_g$ is given there. Note that $\bar{R}_g$ refers to comparison of $\mathbf{g}^2$ tensors, while $\bar{R}_{mag}$ for triclinic susceptibilities refers to χ tensors sampled at several temperatures. Once $\bar{R}_g$ values have been computed, the presentation and use of the corresponding maps exactly parallels that described for susceptibilities: in CAMMAG, the request for a g-value fitting map is made by the codeword GMAP. A similar GRID card determines the order of 'slicing'. Occasionally we might be unsure of the experimental g-value data in one respect. In monoclinic or orthorhombic crystals, for example, it may not have been possible to establish beyond doubt with which of the crystallographically equivalent but magnetically inequivalent molecules in the unit cell the measured $\mathbf{g}$ tensor should be associated. The calculations may be used to help this 'assignment' problem, and CAMMAG has a facility to simplify this: thus

```
ESR    3.72 1.96 1.74
GDCS  0.1925  0.6678  0.7193  0.5116  0.6947  0.5071  0.8368  0.2689
      0.4772 *
```

conveys to the program the experimentally determined principal g values on the ESR card, and their direction cosines with respect to the orthogonalized crystal frame on the GDCS card. If an asterisk is included on the latter card, mapping comparisons between observed and calculated

$\mathbf{g}^2$ tensors are made for all magnetically inequivalent molecules separately, that is, all those whose direction cosines are related to those given on the GDCS card by the space-group symmetry operations. Hopefully one round of these, obviously more lengthy, calculations would be sufficient to establish or confirm the relationship between the $\mathbf{g}^2$ tensor orientation and the molecular frame: in subsequent rounds, the asterisk would be omitted.

The mapping technique we have reviewed does not employ least-squares analysis, as we have discussed. It *is* systematic, however, though not inflexible. Were the required calculations virtually cost-free, we might consider presenting RUN with enormously wide ranges for each parameter, varied by small increments. However, the calculations can be very lengthy indeed and so the CAMMAG system has evolved to provide the flexibility required by a user employing experience and tactics in searching for regions of fit. Such tactics are barely mentioned in original research papers, of course, and a book such as this is probably the only place where one might hope to find any of them written down. Even here we can only sketch the more important points which have gone to make up the lore.

10.3.2 Search tactics

The exigencies of ligand-field analysis are that we must locate and refine a 'fit' to experimental observations, check for its uniqueness or otherwise, and do it as quickly and cheaply as possible. From the beginning it must be realized that no two analyses are alike and the pursuit of cost-effectiveness can be self-defeating (it can be a long and expensive process to define the shortest and cheapest path to our goal!). So experience tells in the end and all remarks in the present section should be taken as no more than suggestions.

Firstly, if possible it is worth performing initial calculations and searches using a less-than-complete basis: for high-spin d^7 and d^8 species, for example, we would work with the 45-fold and 30-fold bases of spin quartets or spin triplets, respectively, rather than with the 120-fold or 45-fold complete d-configuration bases. It is generally not worth restricting the bases further, for the subsequent refinement process tends to take so long as to negate earlier savings. Next, the way in which we sample parameter values depends on experience and on the particular property we are studying. On the question of experience: it would be perverse not to take account of previous analysis on similar systems and use the concepts of 'transferability' of parameter values to guide our search, notwithstanding the discussion in §9.8. Provided that widely different parameter values are investigated in due course, there will be no substance to a charge of 'bias' or 'inconsistency'. We should similarly respond, within reason, to guesswork and intuition, though such a tactic must be tempered by a sense of when to quit. Now, generally, we may have somewhere between six and ten parameters to vary – sometimes more, rarely less. Flailing our arms in all directions, so to speak, is unlikely to be helpful so we might best begin by deciding to which parameters the experimental property in question is likely to be relatively insensitive. By and large, and to be sure there are many exceptions, optical spectra depend less on e_π parameters and, of course, spin-orbit coupling; while susceptibilities and g values vary relatively little with interelectron repulsion parameters and e_σ values. When searching for a fit to magnetic properties, therefore, it is a good tactic to make an intelligent guess at Racah B and all e_σ values and first study the dependence of calculated properties or the various e_π parameters. Obviously, if one or more of these are expected to be zero (for NH_3 ligands, for instance), then even though final checks should always be made, it is sensible to fix these values accordingly, *pro tem*. Since variations of Stevens' k factor are 'computationally cheap', as discussed above, it is worth calculating magnetic properties for a fairly wide spread of k values as a matter of course. Although magnetic properties can be very sensitive to spin-orbit coupling, it often transpires that this dependence is not strongly correlated with other parameters. So it is frequently possible to establish a rough value for ζ in an early round of RUN, fix the value, and move on to an investigation of other parameter dependencies, only returning to the ζ dependence from time to time.

In difficult cases, it is well to think carefully about the sort of e_π correlations which might be expected by reason of molecular geometry – as discussed in §9.8. As a desperation tactic, one could try a limited, coarse variation of all parameters from median values, two at a time. But, if the problem is that bad, generalized advice is probably without value. One approach which can be helpful with higher-symmetry molecules – or ones which approximate that situation – is to forgo the AOM and local parametrization of the ligand field in favour of the more traditional, global schemes using parameters like Dq, Ds, Dt or $Dq, D\sigma, D\tau$ for example. Alternatively, individual global orbital energies may be parametrized. These schemes are outlined in §10.4: suffice to say here that their use, either *pro tem*. or throughout the analysis, may aid a search process by reducing the degree or parametrization. Once a fit is found, a subsequent return to the AOM, if valid in terms of the relative numbers of parameters and pieces of data, might be immediately successful: in which case, the alternative parametrization scheme may be regarded merely as a guide in the whole process. A corollary is that, in such circumstances, the 'intermediate' parametrization scheme need not be pursued fully or totally consistently: we shall refer to this in §10.4. Finally, a psychological point is worth making for the particularly difficult cases: it is that confidence in one search direction occasionally only develops when certain other directions, however silly in retrospect, have first been explicitly eliminated. Following two directions in polyparameter space at the same time can sometimes lead only to a form of schizophrenic trance!

The bulk of these few observations have referred to the search for fits to magnetic properties. Reproducing optical spectral transition energies raises new issues which are discussed separately.

10.3.3 Fitting spectra

Although the *refinement* of a spectral fit can proceed by least-squares or via a mapping scheme like that described for magnetic properties, searches are best made using rather different tactics. The reasons for the different approach lie in the different natures of the experimental observations. Measurements of principal crystal susceptibilities or of **g** tensors generally provide us with a complete set of observations with values suffering something like random errors about the averages we measure. By contrast, spectral transition energies frequently correspond to a restricted set of those theoretically possible either because of instrumental limitation or due to inherent lack of suitable resolution of broad bands or of bands obscured by 'extraneous' features like charge-transfer spectra. Often the bands are not assigned with confidence from the experiment and sometimes this ignorance extends even to the spin-state identification: in many analyses, we use the calculated transition energies to assign the observed spectrum. Even when we observe a clear, resolved, unsplit peak in a spectral trace, the common large width of $d-d$ transition, especially of spin-allowed bands, implies considerable uncertainty in our knowledge of electronic transition energies when band origins are unknown.[††] Of course, some spectra are very fine indeed, perhaps because of the detailed work of the researcher, using low temperatures, single-crystal techniques, polarized light, maybe MCD methods to aid some assignments, and so

---

[††] It is probably more correct to fit to spectal peak maxima rather than origins: see the discussion in §12.2.1.

on; but also by luck, for the spectra of many complexes defy resolution under any experimental effort. Generally, we have to deal with spectra of variable quality, deriving from studies on single crystals, solutions or powders and we cannot afford to discard such information just because it is incomplete, whether for good or bad reasons. We have repeatedly emphasized the ability and need of the contemporary approach to analyse ligand-field properties of unsymmetrical molecules involving a wide range of ligand types. Spectra of these systems are rarely finely resolved and frequently involve large charge-transfer tails. We must therefore be realistic and accept the clay with which we must work.

Set against these difficulties is the fact that the information we seek to reproduce is all of the same kind: that is, transition energies. There is no question of variation with temperature (of the calculated quantities, at least) or of some data referring to magnitudes (like intensities – but what a pity!) while others refer to angles. In short, we can represent the experimental and calculated quantities in a simple way directly, without the need to resort to 'figures-of-merit'. The scheme illustrated in figure 10.7 is an obvious and well-used tactic for seeking fits to spectral peaks. The markers in lines a–e indicate calculated spectral transition energies for some systematic variation of the model parameters. The trace in f represents a rough sketch of an observed spectrum. Comparison of f with the calculated markers suggests that d provides the 'best fit' of those considered. This simple method is most effective. It involves little effort to draw diagrams of calculated eigenvalues from the program output and, in the search stage at least, they don't even need to be drawn very accurately. The eye can take in so much information at once (on the well-known principle that 'a picture is worth a thousand words') that one can quickly acquire a 'feel' for the progress of the work and hence decide more easily what to try next. It is worth mentioning such details, for it would be a pity if the theoretical, experimental and computational advances made in ligand-field analysis should stumble on the problem of searching a multivariable function. Again, while sophisticated methods do exist for doing this, it is doubtful if they are quicker or more informative than the simple sort of schemes outlined here.

One common circumstance with which the system in figure 10.7 can easily cope is that where some or all components of a higher-energy subset of transitions (e.g. $\rightarrow {}^3P$ components of d^8) experimentally lie obscured under a charge–transfer tail. Then it is a simple matter to glance at a figure like 10.7 and merely check that the relevant calculated bands all lie higher in energy than some particular value. This example introduces

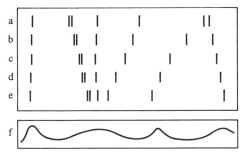

Fig. 10.7. Marker plots provide an obvious but very convenient way
of comparing observed and calculated spectral transition energies.

another simple tactic which is often useful by temporarily reducing the
number of significant parameters. Thus, if it can be established that certain
observed bands correspond with transitions to components of, say ^3P or
^4P in d^8 or d^7 systems, respectively, we can work with a sensible value
for the Racah B parameter held while lower-energy bands are fitted by
variation of the various AOM parameters. It is common to find that there
is relatively little correlation between interelectron repulsion and AOM
parameters, so that a small number of *subsequent* calculations in which
B is optimized, together with the $\{e_\lambda\}$, is usually sufficient to complete the
fit. Again, it is important not to involve bias in this process, which is after
all only a means to the end. We mentioned that this search aid can be
used *if* some aspects of the spectral assignment have already been settled.
This is an important matter and in analysis can be pursued with greater
confidence if all alternative assignments are considered honestly without
regard to prior expectation; the discussion sections of references [5] and
[6] provide examples of this approach. Altogether it is clear that, since
many spectra are not assigned by experiment, the process of assignment
often takes place concurrently with the location of a fit. In this respect,
the sort of information output with the eigenvectors, derived from the
projection operator (10.7), can be helpful; for example, when a weak
spectral band is to be assigned as a nominally, orbitally forbidden
transition.

Once a fit has been found, the process of refinement can be carried out
virtually automatically and a mapping scheme similar to that described
for magnetic properties can be employed: CAMMAG includes facilities
for this. One final remark on the question of refinement is that if the
basis is enlarged at the later stages – for example, from $^4F + {}^4P$ to
the complete d^7 manifold – some properties may be found to be more
sensitive than expected. An example of this occurs in the analysis[7] of

[Co(quinoline)Br$_3$]$^-$, reviewed in chapter 12, where the inclusion of the spin-doublet states affected the susceptibilities very little but the calculated g values quite significantly.

10.4 Alternative parametrization schemes

Much of Part III of this book is given to extolling the virtues of the angular overlap model, not only because of its efficacy to be made plain in chapter 12, but also because it has now been given a firm foundation within ligand-field theory and quantum chemistry,[8-10] as we shall see in the next chapter. However, there can be problems associated with the degree of parametrization involved in the AOM, and §9.8 hopefully provided an honourable account of these difficulties. It is especially in the case of higher-symmetry systems that the older, globally based parametrization schemes might be more appropriate. A particularly useful variation here is the parametrization of global, orbital energies.

This scheme has been successfully used with molecules possessing exact or approximate D_{4h} or D_{2h} symmetry, for example. Instead of parametrizing diagonal orbital energies within each *local* metal–ligand frame, we do this for the molecule as a whole within the *global* frame. In D_{2h} symmetry, for example, we use the ligand-field parameter set $\varepsilon(d_{xy})$, $\varepsilon(d_{yz})$, $\varepsilon(d_{zx})$, $\varepsilon(d_{x^2-y^2})$, $\varepsilon(d_{z^2})$ and $\langle d_{z^2}|V|d_{x^2-y^2}\rangle$: these correspond to orbitals transforming, respectively, as b_{1g}, b_{3g}, b_{2g}, a_g and a_g – and hence the need for the off-diagonal term connecting d_{z^2} and $d_{x^2-y^2}$ orbitals. Now the scheme is especially useful for planar molecules – like the Co(mesityl)$_2$(PPhEt$_2$)$_2$ discussed in §1.4 – in which the energy of the $d_{x^2-y^2}$ (in-plane) orbital can be expected to be high; and, more significantly, very different from that of the d_{z^2}. Under these circumstances, the off-diagonal term might not be too important and $\varepsilon(d_{x^2-y^2})$ probably need not be varied too much from some large value. Altogether, the number of important variables in the system might be reduced thereby to three – being the energy *differences* between the d_{xy}, d_{yz}, d_{zx} and d_{z^2} orbitals. As discussed earlier in §10.3.2, if the global orbital parametrization scheme is used as a path towards an AOM analysis (provided the latter is valid so far as the degree of parametrization is concerned), this approximate and temporary reduction in the degrees of freedom can be most helpful and ultimately without bias. But even if used in its own right, the approach can provide a fair approximation for minimal effort. Above all – and this is the main advantage of this model *within alternative global parametrization schemes* – the approach maximizes the opportunity for relating the system parameters to local bonding features. Depending, of course, upon the orientation of the ligands in such a planar molecules – but bearing in mind that this is frequently of a special nature, perhaps by virtue of chelation – energies of the d_{xz} and d_{yz} orbitals might be related uniquely to the π-bonding character of ligands placed along the x and y axes respectively. So this parametrization scheme can provide the user with some 'feel' for the ligand-field potential. This can be spoilt to some extent if the off-diagonal matrix element required for fit is comparable in magnitude with $\varepsilon(d_{z^2})$ or $\varepsilon(d_{x^2-y^2})$ but even then some contact with chemical thinking can be maintained. As a technical matter, the way in which CAMMAG operates the facility of using this scheme is simply to overwrite much of the produce of SETUP and establish the global multipole coefficients corresponding to the parameters above by direct substitution into table 9.2. It is apparent that the orbital-energy parametrization scheme is simply a special subset of one in which all 14 independent matrix elements of a real, global basis are directly parametrized. Although little chemical insight is likely to be gained by using that scheme, it is possible that alternative subsets, perhaps ones quantified with respect to a three-fold axis, might be useful: if so, table 9.1 will provide the path to their implementation.

While we prefer the scheme just described to other, more traditional global parametrization schemes, it is nevertheless appropriate to comment briefly upon these alternative approaches. The best known, of course, are those relating to molecules possessing strict four- or three-fold symmetry; for example, D_{4h} and D_{3d}. The first involve the parameter set Dq, Dt, Ds, referring to V_{oct}, fourth-, and second-order multipole contributions respectively. These last two are parametrized with $D\tau$ and $D\sigma$ in the case of systems with three-fold symmetry. Both schemes have been discussed extensively elsewhere.[111] They appear to offer little over a simple parametrization of harmonics appearing in a multipole analysis and fail to provide means for comparison of the ligand fields of molecules of each symmetry type, let alone with those of less symmetrical species. There is no obvious connection with generalized notions of chemical bonding and so the user is left with no 'feel' for the relevance of the results of a given analysis or even whether the conclusions are sensible. The CAMMAG system includes the option of calculating ligand-field properties in terms of these parameters, merely so that published reports may be checked. In recent years, Donini, Hollebone & Lever[112] have constructed a more comprehensive symmetry-based scheme in which a whole array of parameters are defined with respect to subgroup chains. The so-called 'normalized spherical harmonic' (NSH) approach[112] succeeds in its attempt to provide a means of parametrizing the ligand fields in molecules with a wide variety of lower symmetry. However, like its more limited antecedents, it still fails to make comprehensible contact with the world of chemical bonding. In fairness to the method, which undoubtedly involved considerable effort in its construction, it would have provided the only *ligand-field* model to be applicable to low-symmetry systems, had the AOM not been shown[8-10] to be a genuine ligand-field scheme also: but the AOM is far more comprehensible and chemically relevant. We do not, therefore, pursue the NSH method further.

10.5 Alternative bases

Throughout discussions of the tensor-operator method in chapter 8 and of its application to the AOM in chapter 9, we have exclusively used a function basis defined in terms of J, M_J quantization within the free-ion, Russell–Saunders coupling scheme. This choice was determined by the fact, which we demonstrate in the next chapter, that we *can* build ligand-field calculations upon a 'spherical' or free-ion type basis and, that being so, the coupling theory appropriate for all parts of (III.1) is conveniently available within the J, M_J basis. Others have preferred to work within a strong-field coupling scheme using a basis of octahedral functions, $(t_{2g})^n(e_g)^m$. Now we recognize that any computations performed within the *complete d^n* basis of a transition-metal complex will give identical results whatever the chosen basis – J, M_J or $(t_{2g})^n(e_g)^n$, for example – provided that identical radial characteristics, and hence parameters, are associated with all d wavefunctions. Some workers prefer to admit the possibility that this assumption may be invalid, that the radial extension of t_{2g} and e_g orbitals, for instance, may differ from one another, in which case the strong-field basis might seem more appealing. In response to this we note firstly that the undoubted formation of metal-ligand molecular orbitals with mixing coefficients varying from d orbital to d orbital is *not* forgotten nor unduly idealized within a ligand-field formalism based upon a 'spherical' d-orbital basis perturbed by an effective operator called the ligand-field potential: this is discussed at length in the following chapter. Secondly, even if a strong-field basis were to be preferred, it is not at all clear why a basis derived for *octahedral* symmetry should be particularly useful for systems with geometries which are not even distantly related to that point group. Once again, it must be stressed that, from the beginning, we have sought a system that was equally applicable to molecules of any symmetry.

For pressing reasons of economy we must frequently work with somewhat restricted basis sets and the manner in which a complete d^n configuration basis is truncated becomes an important issue. Consider, for example, wanting to compute ligand-field properties for a low-spin d^5 complex. The full d^5 configuration basis is 252-fold degenerate and so the construction and diagonalization of a complete basis would require a great deal of time and computer store: remember that such diagonalizations are carried out hundreds or thousands of times throughout the course of an analysis. In many low-spin d^5 systems, components of the formal $^2T_{2g}$ term of O_h parentage form a manifold of energetically low-lying levels well separated from others. For many purposes a calculation performed within such a basis might be adequate, in the earlier stages of an analysis at least. Were we to diagonalize the free-ion ^2I term, from which this $^2T_{2g}$ term arises in the weak-field limit, by providing SETUP with the cards

BASE 2 5
CONF 2I

we would simply find that *any* choice of ligand-field and spin-orbit coupling parameters would yield the one result of a 26-fold degenerate level. This follows, of course, from the fact that we have one-electron operators acting within a single term of d^n configuration which is its own hole equivalent. In order to calculate any splitting of the ^2I term in a weak-field basis we must include other free-ion terms, especially spin doublets (recognizing that $^2T_{2g}(t_{2g})^5$ is built from several weak-field, spin-doublet terms which give rise to $^2T_{2g}$ terms) arising from this configuration and we have no generally reliable method available for picking out which of these are likely to contribute most importantly. First-order splitting *is* observed, however, within a strong-field $(t_{2g})^n(e_g)^m$ basis – here t_{2g}^5 – and the calculation involves a matrix of order only six. However, if we wish to improve the accuracy of such a calculation by inclusion of other strong-field configurations, we are again somewhat unsure which to include; and, once more, this problem is exacerbated if the molecular geometry is not compatible with O_h or its subgroups. Recently, McMeeking[13] has developed and programmed a scheme which recognizes and solves these problems in basis-set truncation procedures in a powerful way.

McMeeking's approach[13] aims to greatly reduce computational effort in ligand-field calculations by truncation of a d^n or f^n configurational basis in conformity with the real molecular *geometry* and mean ligand-field *strength* and hence by incurring the least penalty in terms of accuracy:

though a product of the computer age, it draws heavily upon the application of group theory to ideal molecular symmetries. Its use is *not*, however, restricted to symmetric molecules. The approach involves two forms, referred to as 'J blocking' and 'L blocking', acting within bases spanned by $|J, M_J\rangle$ or $|LSM_LM_S\rangle$ eigenkets respectively. The central aim in both forms is the reduction in the size of matrix which must be diagonalized in, say, the RUN segment of CAMMAG. The problem is so acute in $d^{4(6)}$ and d^5 systems, involving bases of 210 and 252 functions respectively, that diagonalization of such large, complex matrices can be insufficiently accurate for the subsequent satisfactory calculation of magnetic properties; particularly if these depend rather sensitively upon small zero-field splittings, for example. In any case, calculations of these magnitudes are too expensive to repeat more than a few times. McMeeking's methods establish sensible ways for basis truncation: naturally they involve an 'overheads' cost in setting up, so that a new form of SETUP takes considerably longer than before, but this is recouped many times over within the RUN computations of the final analysis: and, as usual of course, SETUP is executed only once while RUN is performed perhaps hundreds of times. We begin by reviewing the 'J-blocking' scheme.

Suppose we wish to compute ligand-field properties for an octahedrally coordinated metal ion but one whose detailed geometry possesses less strict symmetry perhaps than O_h: in short, that octahedral symmetry represents a more-or-less good description of the molecular geometry. Note immediately that the ultimate calculation will not be so restricted. Let the original free-ion basis of $|J, M_J\rangle$ eigenkets form a set $\{\phi\}$. We project from $\{\phi\}$, linear combinations $\{\chi\}$ which transform as the irreducible representations of O_h (or of the double group O'_h in odd-electron systems). For example, the spin-orbitals $\{\phi\}$ arising from the d^6 configuration transform as $\{\chi\}$ belonging to $5^1A_{1g} + 2^1A_{2g} + 5^1E_g + 4^1T_{1g} + 7^1T_{2g}$, $+ 1^3A_{1g} + 2^3A_{2g} + 3^3E_g + 7^3T_{1g} + 5^3T_{2g}, + 1^5E_g + 1^5T_{2g}$ of the O_h 'idealizing' group. On regrouping the basis functions $\{\chi\}$ according to spin and spatial symmetry speciation, we then need to diagonalize under the ligand-field potential, V_{oct}, a blocked matrix of the form shown in figure 10.8, rather than a single matrix of size 210×210 for the complete basis $\{\phi\}$ of d^6. In addition to the symmetry blocking made possible by reference to spin and space representations, appropriate definition of the projection operators which form the $\{\chi\}$ from the $\{\phi\}$ can be made to separate matrices referring to different spatial components of the higher-dimensional representations. For obvious reasons of scale, all submatrices in figure 10.5 are not labelled completely, but the idea may be inferred

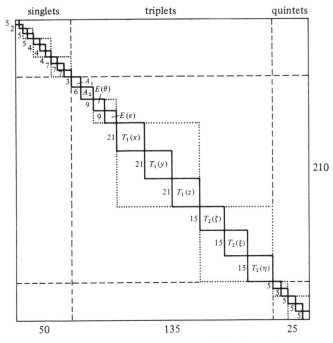

Fig. 10.8. The J blocking of the complete d^6 basis in O_h symmetry. All elements outside blocks marked by solid lines are identically zero.

from the example of the seven $^3T_{1g}$ terms. Thus, appropriate projection will separate spatial components transforming as either x or y or z in T_{1g}: there are seven separate $^3T_{1g}(x)$ terms for example, and hence 21 functions of each type. Similar subblocking of T_{2g} terms into ζ, ξ and η parts, or of E_g terms into θ and ε parts, complete the structure.

If we have written the complete ligand-field potential for the real complex in the form

$$V = aV_{\text{oct}} + b_1(Y_q^k)_1 + b_2(Y_q^k)_2 + \dots, \qquad (10.16)$$

then, in addition to the matrix in figure 10.8 for the O_h-adapted d^6 basis under V_{oct}, we must construct a further n matrices, where n is the number of b coefficients in (10.16). Although these matrices are constructed for the same basis $\{\chi\}$ they will not, of course, be blocked.

Consider now that we can identify a set of interelectron repulsion and octahedral ligand-field parameters $\{F_2, F_4, \Delta_{\text{oct}}\}$ which may be taken as an appropriate 'median' set for the system under study. Using the appropriately grouped basis set $\{\chi\}$, we construct matrices for each of these parameters, sum them and diagonalize. *Basis-set truncation* is now

implemented by selecting the N lowest-lying eigenvectors produced by
this process. The number N must be determined on grounds of economy
and acceptable accuracy; and it would surely correspond with a suitable
gap in the eigenvalue spectrum. Each member of the truncated set $\{\psi\}$
transforms as one or other representation of the idealizing group – here
O_h – and corresponds to a linear combination of members of the set $\{\chi\}$
transforming similarly. By defining a truncated basis in this way, we have
optimized our choice to the particular molecular complex in hand, both
with respect to detailed geometry and to 'ligand-field strength'. For
example, by choosing interelectron repulsion parameters as very small
compared with the ligand-field parameters (here $F_2, F_4 \ll \Delta_{oct}$), we would
establish a basis which corresponds closely with a 'strong-field' $(t_{2g})^n(e_g)^m$
basis: but note that such a choice is only one of an infinity of choices
possible within this scheme.

The process of obtaining eigenfunctions corresponding to the param-
eters $\{F_2, F_4, \Delta_{oct}\}$, of which the $\{\psi\}$ are the N members of lowest
energy, is one of performing a unitary transformation of the complete
matrix that leaves it in its diagonal form. Let us include in that
transformation the, essentially trivial, step of reordering the diagonalized
form so as to collect together the N functions $\{\psi\}$. We now apply a similar
unitary and reordering transformation on each of the matrices referring
to all parts of (III.1) – the interelectron repulsion, spin-orbit coupling and
all terms of the total ligand-field potential (10.16). At this stage, therefore,
we have constructed a set of matrices, similar to those in (10.3), one per
system parameter, and each matrix is constructed with respect to a
truncated basis which has itself been optimized to a median ligand-field
strength and geometry. Subsequently, within a barely modified[††] version
of the RUN segment of CAMMAG, each of these matrices is multiplied
by an appropriate parameter value, as in (10.3), the ligand-field multipole
multipliers ($a, b_1, \ldots, b_n$ in (10.16)) being derived as usual[††] from an AOM
parameter set: the matrices are then summed and finally diagonalized in
the normal way. Although the basis was truncated with respect to O_h
symmetry-adapted functions in the first place, the true geometry, together
with spin-orbit coupling, is recognized in the complete RUN computation.

[††] The form of (10.16) has been chosen to simplify the present discussion. If, instead
of V_{oct}, we use the explicit components $Y_0^4 + \sqrt{(\tfrac{5}{14})}(Y_4^4 + Y_{-4}^4)$, no modification
whatever is required in the RUN segment of CAMMAG, as the process summarized
in (10.3) is identical. However, certain off-diagonal terms occur for the separate
components of V_{oct} which cancel in the complete V_{oct}. One can adopt the tactic of
ignoring these off-diagonal terms, knowing that they ultimately vanish in the total.
We do not review this sort of detail further in the present outline.

Considerable latitude exists in this truncation process, not only with respect to the size (N) of the basis set $\{\psi\}$ ultimately chosen, but also with respect to the idealizing point group initially selected. At one extreme we may envisage selecting the point group C_1 as an 'idealizing' group for a totally unsymmetrical molecule. In this case, the whole process reduces to one of merely selecting the energetically lowest N levels belonging to the *actual* system calculated with a median set of parameters. Such an approach could certainly be expected to provide an ultimately optimized, truncated basis for subsequent calculations. The objection to this procedure is not that it is time consuming, for in the new SETUP segment it would be executed only once, of course. The difficulty is rather one of numerical accuracy: as mentioned above, diagonalizations of the large, complex matrices that would result (for the d^4–d^6 configurations, anyway) can certainly be insufficiently accurate for our purposes. In any case, a highly optimized form of truncated basis set of this form is not really necessary: recall that for many purposes the 'weak-field' truncation of, say, d^2 to $^3F + {}^3P$ is quite adequate and yet, in the spirit of our present discussion, this cannot be considered as an especially well-optimized basis. In general, therefore, a truncation procedure based upon some moderately symmetrical point group – though one chosen with some regard to the approximate molecular geometry – should be satisfactory.

If a given investigation is concerned with spectral transition energies rather than with magnetic susceptibilities and g values, even greater savings can be obtained using McMeeking's second approach.[13] In a so-called 'L-blocking' scheme, we work with an initial basis $\{\phi\}$ quantized within the LSM_LM_S scheme. The most obvious and important economy in computation which may be effected in this method derives from the spin factorization which is now possible: thus if spin-orbit coupling is not included in the initial blocking process, we need only consider one member of each spin multiplet – say those with $M_S = +3/2$ for spin-quartet levels – for, under ligand-field and interelectron repulsion perturbations, all members of the spin multiplet behave identically. Let us consider the example of the d^3 configuration, giving rise to the free-ion terms $^4F + {}^4P + 2{}^2D + {}^2F + {}^2G + {}^2H$. In octahedral symmetry these terms split as: $F \to A_{2g} + T_{2g} + T_{1g}, P \to T_{1g}, D \to T_{2g} + E_g, G \to A_{1g} + E_g + T_{1g} + T_{2g}$, and $H \to E_g + 2T_{1g} + T_{2g}$. Instead of the 120×120 matrix formally required for the complete d^3 manifold without symmetry adaptation, we obtain the blocked system shown in figure 10.9. The small dimensions of these blocks arises from the restriction to a single M_S value, as discussed, but also because the absence of spin-orbit coupling means we may construct

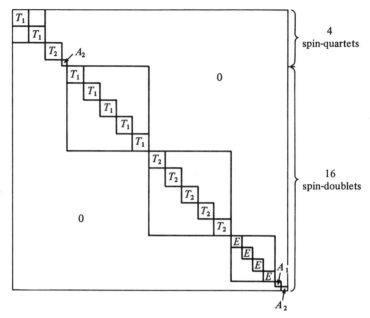

Fig. 10.9. The L blocking of the complete d^3 basis in O_h symmetry. Only elements outside the major subblocks are necessarily zero.

projection operators producing these $\{\chi\}$ from the LS-quantized $\{\phi\}$ such that each spatial component of an orbitally degenerate term transforms identically. Thus the 5×5 block for $^2T_{2g}$ terms involves, say, a basis of $\{T_{2g}(\xi), M_S = +\tfrac{1}{2}\}$ for each repeating $^2T_{2g}$ term: off-diagonal elements connect these functions amongst themselves, of course.

Basis truncation is achieved as for the 'J-blocking' scheme. While the method is most economic when spin-orbit coupling is completely neglected – as might be suitable for some spectral analyses – it can used even when the spin-orbit effect is included, as in the 'J-blocking' scheme. Of course, when spin-orbit coupling *is* subsequently included in the RUN segment, we cannot remove all but one M_S function set from the basis. On the other hand, nor is there any need to expand a matrix such as that in figure 10.9 to include all M_S values, for we merely need to incorporate the extra functions *after* basis truncation by applying an identical unitary transformation to them as that which defined the diagonalization of the matrix with the representative M_S value. While all this is straightforward in principle, it is clear that considerable 'book-keeping' is involved in its implementation within a computer program: the same is true, of course, for the *J*-blocking scheme. Nevertheless, once established – as it is – the scheme is powerful and simply applied.

The relative merits of the *J*- and *L*-blocking schemes are reasonably clear. If calculations without spin-orbit coupling are all that are required – and hence those restricted to some spectral studies – the *L*-blocked system clearly leads to the diagonalization of smaller matrices. If spin-orbit coupling is included, the *J*-blocked system may often be the more economical because spin-orbit coupling matrices generally do not block off in the LSM_LM_S quantization scheme while lower-symmetry, ligand-field matrices can be made to block off in the *J*-blocking procedure by choosing the actual molecular symmetry (when there is any) as the idealizing group. When applying McMeeking's techniques to the larger transition-metal configuration bases (d^4–d^6), let alone to lanthanide systems, it will generally be worthwhile experimenting with both blocking schemes, perhaps using more than one idealizing group, in order to go part way, at least, towards maximizing the efficiency of the whole process.

References

[1] 'CAMMAG', a FORTRAN program by D.A. Cruse, J.E. Davies, J.H. Harding, M. Gerloch, D.J. Mackey & R.F. McMeeking.

[2] Nielson, C.W. & Koster, G.F., *Spectroscopic Coefficients for p^n, d^n and f^n Configurations*, MIT Press, Cambridge, Mass., 1963.

[3] Johnston, D.L., Rohrbaugh, W.L. & Horrocks, W. DeW. Jr, *Inorg. Chem.*, **10** 547 (1971).

[4] Blake, A.B., *J. Chem. Soc. Dalton Trans.*, p. 1041 (1981).

[5] Gerloch, M. & Hanton, L.R., *Inorg. Chem.*, **20**, 1046 (1981).

[6] Gerloch, M., Morgenstern-Baderau, I. & Audière, J.P., *Inorg. Chem.*, **18**, 3220 (1979).

[7] Gerloch, M. & Hanton, L.R., *Inorg. Chem.*, **19**, 1692 (1980).

[8] Woolley, R.G., *Mol. Phys.*, **42**, 703 (1981).

[9] Gerloch, M., Harding, J.H. & Woolley, R.G., *Structure and Bonding*, **46**, 1 (1981).

[10] Gerloch, M. & Woolley, R.G., *Prog. Inorg. Chem.*, **31**, 371 (1983).

[11] Gerloch, M. & Slade, R.C., *Ligand-Field Parameters*, Cambridge University Press, 1973.

[12] Donini, J.C., Hollebone, B.R. & Lever, A.P.B., *Prog. Inorg. Chem.*, **22**, 225 (1977).

[13] McMeeking, R.F., personal communication (to be published.)

11

The nature of ligand-field theory and of the angular overlap model

---

11.1 Introduction

After a history of some 50 years, ligand-field theory remains a vital procedure in the study and interpretation of the spectral and magnetic properties of transition-metal and lanthanide complexes. Yet the nature of the many parameters spawned by ligand-field models often appears unclear since the relationship between the theory and underlying fundamental principles is seldom apparent. Our purpose in this chapter is to investigate this relationship and so provide a viewpoint from which ligand-field theory may be seen to emerge naturally from the main body of quantum chemistry. In so doing we hope to define the basic assumptions of the theory more clearly than hitherto, to interpret the parameters employed, and to discover the bounds of applicability of the model: we follow closely the developments of Woolley;[1] Gerloch, Harding & Woolley;[2] and of Gerloch & Woolley.[3]

One assumption above all gave rise, via the early exploitation of the techniques of group theory, to the various qualitative generalities which established the success of ligand-field theory. It is that the electronic levels which determine the spectral and magnetic properties of metal complexes derive from a simple, atomic-like d^n configuration in the transition block, or f^n in the lanthanides. Predictions of the spin and orbital degeneracy of the ground states of complexes classified according to their geometry and d configuration, as evidenced by their bulk paramagnetism and anisotropy, have been totally successful. In later years, the identification of broad bands in d–d spectra as spin-allowed transitions confirmed electronic spectroscopy as the other major experimental technique by which the qualitative, group-theoretical predictions of ligand-field theory

444

could be verified. Ballhausen[4] has reviewed these pioneering developments within an historical context.

In recent years, most work concerned with ligand-field theory, especially of d-block complexes, has been performed by the chemist rather than the physicist in capacity of *user* rather than phenomenologist. Of course, to some extent the theory has been continuously checked in that, as calculations have become increasingly sophisticated and experiments more detailed, the ability of ligand-field models to reproduce experimental data has been demonstrated repeatedly. The emphasis in so many latter-day studies in this area, however, has been upon the system rather than on the phenomenon. The chemist wishes to use ligand-field theory, in conjunction with optical spectroscopy and the measurement of paramagnetism, as a tool with which to probe the nature of the bonding in specific molecules. As we saw in chapter 1, his realization of this ambition has fallen well below the expectation which derived from the early, largely group-theoretical achievements in the prediction of coordination number, stereochemistry or even something of the thermodynamics of heats of hydration, for example. In solving the problems of less-than-cubic molecular symmetry and spin-orbit coupling as revealed in optical term splittings, polarization ratios, or anisotropy of paramagnetic susceptibilities and esr g values, sight of the ultimate purpose of the effort had, until recently, been lost. The quantities deduced from these specialized studies frequently reflected little of obvious chemical interest and have not been understood by the more general chemist. It was inevitable, perhaps, that interest in ligand-field theory waned during the past fifteen years. However, as argued in our review in Part I and as will be demonstrated by the detailed examples in Part IV, in its latest form, ligand-field theory genuinely does offer means by which the bonding in transition-metal complexes may be assessed in some detail and in a way not easily accessible by other means.

The central feature of the new approach stems from a recognition of the powerful and ubiquitous chemical concept of the *functional group*. Earlier, symmetry-based ligand-field models failed explicitly to recognize obvious local features of the electronic potential. Their parameters are *global* and so inevitably unable to provide any commentary upon *local* interactions of individual ligands with the central metal. By contrast, the angular overlap model[5-7] of Schäffer and Jørgensen not only establishes close links between ligand-field phenomena and chemical bonds but, as we saw in outline in chapter 1 and in somewhat more detail in chapter 9, provides an approach which may be exploited equally well for unsymmetrical molecules of virtually any geometry. The AOM was introduced as a molecular-orbital approach and it is probably true that a somewhat specious correlation with broader bonding concepts is most apparent in that formulation. We shall argue, however, that the original *spirit* of the AOM, as a scheme to apportion contributions to spectral splitting patterns amongst discrete metal–ligand interactions, takes the form of a 'localized potential model', as in chapter 9, falling entirely within ligand-field theory: as such, this 'ligand-field version' of the AOM must be distinguished from the 'molecular-orbital version' currently presented in various teaching texts,[8–10] and this is something we consider in detail later in this chapter. Meanwhile it is possible to view the AOM *as it is actually used* in ligand-field analysis as divided into two parts. The first concerns the sort of manipulations described in chapter 9, which depend on the idea of the cellular division of the non-spherical potential into separate ligand contributions, such that the ligand potential is diagonal with respect to the metal orbitals when quantized in a coordinate frame reflecting the local metal–ligand pseudosymmetry. The second part involves interpretation of the best-fit e_λ parameters by association of e_σ with σ bonding between metal and ligand, e_π with π bonding and their signs with ligand donor–acceptor function. Considered merely as two stages of a *modus operandi*, the subsequent interpretations do not affect the mechanical part of defining and obtaining values for the various AOM parameters: as discussed in the preceding chapter, it is certainly best to adopt this view in the analytical process, if only to avoid unnecessary bias. Nevertheless, as we shall see in the second half of the present chapter, the definition and interpretation of the AOM parameters are intimately related, though not in the axiomatic way originally proposed. It is, of course, of the utmost chemical interest whether the AOM parameters are related to conventional notions of molecular-orbital theory and whether the simple overlap criteria described in chapter 1 can be defended within ligand-field theory.

Enquiries of this kind must be taken together with similar concerns regarding all operators appearing in the total effective Hamiltonian (III.1) that we use to describe the ligand-field situation. Some problems which come to mind are more apparent than real, however. Consider the question of the calculation of magnetic properties which requires descriptions of both energies and wave-functions after perturbation by the effective Hamiltonian (III.1). It is customary to use Stevens' orbital reduction factor k in the magnetic moment operator $\mu_\alpha \equiv (kl_\alpha + 2s_\alpha); \alpha = x, y, z$, supposedly because the wavefunctions are no longer characterized by an integral l value. As the k factor was originally introduced,[11] the formation of molecular orbitals from metal and ligand functions was responsible for a reduction in the expectation value of l_z. Indeed, Orgel had described[12] the k coefficient as an 'electron delocalization factor', although the relationship between its numerical value and 'covalency' has since been shown[13] to be very complicated. Now the esr spectoscopist often prefers to fit his g-value data without the use of the effective operator kl_z, parametrizing his model instead by explicit molecular-orbital mixing coefficients. It is not unusual to see ligand-orbital, mixing coefficients as large as 0.3 or more from such analyses. The question *apparently* arises, therefore, whether the 'impurity' of the d-orbital basis, evidenced by either of these two approaches, invalidates the use of the transformation matrices in the AOM (or other ligand-field models) which, as we saw in §9.4, are functions of the angular parts of *pure d* (or f) orbitals. Recall here, Schäffer's so-called 'perturbation' formulation[7] of the AOM in §1.11, the first assumption (A) of which states that admixtures of ligand functions into the metal orbital basis are ignored. If such admixtures are to be considered negligible, then we must point again to the large magnitude of the mixing coefficients often established by esr studies.

A similar situation arises in connection with interelectron repulsion parameters and the nephelauxetic effect. Jørgensen in particular has been much concerned[14,15] with the desire to recognize and establish differential orbital expansion by the introduction of the nephelauxetic parameters $\beta_{33}, \beta_{35}, \beta_{55}$. Consistently lower values for β_{33} associated with the e_g orbitals of octahedral configurations relative to β_{55} associated with t_{2g}, have been taken as evidence for the greater nephelauxetic effect of σ orbitals relative to π. In turn, these effects, which derive from experimental spectral features, are considered to indicate differing radial properties amongst the d-orbital basis. Again, if this is so, we might ask if it is satisfactory to use the simple, 'pure' rotation matrices of §9.4 in the remainder of the ligand-field calculation.

We are questioning the use of a pure d- (or f-) orbital basis in ligand-field calculations, therefore. The matter is obviously crucial as all our ligand-field procedures, including the AOM, depend upon that basis. As will become increasingly clear, the definitions of ligand-field basis functions and of the ligand-field effective potential go hand in hand: they are *interdependent*. Looking ahead, we remark that no problem about the use of pure d (or f) basis really exists, for, *as a matter of formal manipulation*, we can always express the ligand field as projected onto some convenient basis. By suitably transforming the Hamiltonian and all other relevant operators, we occasion no approximation whatever by formulating the

ligand-field problem within a pure *l* basis. As to the problem of the MO
mixing coefficients referred to above, or the 'anisotropic' nephelauxetic
parameters, there is again no formal difficulty here for those quantities
refer to different models with different bases.[2,15] They are just examples
of the common mistake of attempting to compare quantities which are
not of the same kind. The chemical conclusions drawn from any one of
these models may well have some validity but the connections between
the different approaches are not obvious.

11.2 Effective Hamiltonians

The ligand-field regime is characterized by an exclusive concern with a
set of low-lying excitations in a complex, which are considered mostly
responsible for the magnetic behaviour and the weak spectral features we
call *d–d* (of *f–f*) transitions. This latter description points up, in line
with the accuracy of Laporte's rule in ligand-field spectroscopy, the notion
that these 'ligand-field states' are dominated by orbital contributions of
metal-ion *d*- (or *f*-) electron parentage. Conventional ligand-field theory
is concerned with an explicit treatment of just the metal *d* (*f*) electrons
and, while we can hope that such analysis will reflect broader bonding
issues, it is worth emphasizing that ligand-field models have never been
intended to provide a comprehensive theory of bonding in these complexes.
Ligand-field theory is only concerned, therefore, with the eigensolutions
of the effective operator, the ligand-field Hamiltonian,

$$\mathcal{H}_{LF} = T + \sum_{i<j}^{N} U(i,j) + \sum_{i}^{N} V_{LF}(\mathbf{r}_i) + \zeta \sum_{i}^{N} \mathbf{l}_i \cdot \mathbf{s}_i$$

$$= T + \mathcal{H}_{LF}', \tag{11.1}$$

in which the eigenvalues span a range in energy of only a few electron
volts above the ground state: *T* is the electronic kinetic energy and the
last three terms, referring to the two-electron interelectron repulsion and
to the one-electron ligand-field potential and spin-orbit coupling effects,
respectively, are those described within (III.1). We are thus given the form
of the final equations and our task is twofold; namely to manipulate the
many-body theory of the full many-electron Hamiltonian *H* into this form,
and to provide a justification for the usual chemical bonding interpretation
of the ligand-field parameters.

Compare here the formalisms of ligand-field theory and π-electron
theory[16] in organic systems. In the Hückel approach, for example,
attention is focussed upon the π electrons of the system while the σ

framework is recognized only as providing an 'external' potential field via variable 'Coulomb integrals'. The corresponding 'external' field to which metal orbitals are subject in the ligand-field situation is that represented by the effective Hamiltonian $\mathcal{H}_{LF}$ of (11.1). Both ligand-field and π-electron theories are examples of a formalism which partitions the complete many-electron theory into a consideration of separate subsystems referring to different groups of electrons whose separate roles may be distinguished physically – at least approximately. In computational chemistry, the ultimate reason for dividing up the problem in this way is to simplify the task of calculating the effects of electron correlation. The notional separation of a large system into small subsystems with physically recognizable parts, followed by final corrections for correlation, seems to offer computational chemists maximal scope for a fruitful combination of physical intuition, mathematical elegance and practicable computing methods.[17] We begin by clarifying the meaning of an *effective* operator as one which acts within a strictly limited basis.

The eigenvectors Ψ of some full, many-electron Schrödinger equation,

$$H\Psi = E\Psi, \tag{11.2}$$

may be considered as built from a complete, orthonormal set of functions $\{\Phi_k\}$, so that, for a given eigensolution, we have

$$\Psi = \sum_k c_k \Phi_k, \tag{11.3}$$

where

$$c_k = \langle \Phi_k | \Psi \rangle. \tag{11.4}$$

Writing

$$H_{kl} = \langle \Phi_k | H | \Phi_l \rangle \tag{11.5}$$

for an element of the matrix of H with respect to the (column) vector $\mathbf{c} = \{c_k\}$, we may construct the usual matrix representation of (11.2):

$$\mathbf{Hc} = E\mathbf{c}. \tag{11.6}$$

This is, of course, a system of linear equations in the unknown coefficients c_k. Let us partition[18] the basis Φ, *quite arbitrarily*, into two parts, Φ_a and Φ_b; and similarly divide up the matrix equation (11.6) as:

$$\mathbf{H}_{aa}\mathbf{c}_a + \mathbf{H}_{ab}\mathbf{c}_b = E\mathbf{c}_a, \tag{11.7}$$

$$\mathbf{H}_{ba}\mathbf{c}_a + \mathbf{H}_{bb}\mathbf{c}_b = E\mathbf{c}_b. \tag{11.8}$$

Note that, in general, both (11.7) and (11.8) describe systems of equations so that $\mathbf{H}_{aa}$ and $\mathbf{H}_{bb}$ represent diagonal blocks of the matrix $\mathbf{H}$ of (11.6), with $\mathbf{H}_{ab}$ and $\mathbf{H}_{ba}$ corresponding to off-diagonal blocks. Now, solving for $\mathbf{c}_b$ from (11.8) gives,

$$\mathbf{c}_b = (E \cdot \mathbf{1}_{bb} - \mathbf{H}_{bb})^{-1} \mathbf{H}_{ba} \mathbf{c}_a, \tag{11.9}$$

which, on substitution into (11.7) yields

$$\mathbf{H}_{aa} \mathbf{c}_a + \mathbf{H}_{ab} (E \cdot \mathbf{1}_{bb} - \mathbf{H}_{bb})^{-1} \mathbf{H}_{ba} \mathbf{c}_a = E \mathbf{c}_a. \tag{11.10}$$

This takes the form

$$\bar{\mathbf{H}}_{aa} \mathbf{c}_a = E \mathbf{c}_a \tag{11.11}$$

if we define the matrix $\bar{\mathbf{H}}_{aa}$ as

$$\bar{\mathbf{H}}_{aa} = \mathbf{H}_{aa} + \mathbf{H}_{ab} (E \cdot \mathbf{1}_{bb} - \mathbf{H}_{bb})^{-1} \mathbf{H}_{ba}, \tag{11.12}$$

provided that the inverse $(E\mathbf{1}_{bb} - \mathbf{H}_{bb})^{-1}$ is defined. $\bar{\mathbf{H}}_{aa}$ has eigensolutions determined by the usual secular equation

$$|\bar{\mathbf{H}}_{aa} - E \cdot \mathbf{1}_{aa}| = 0. \tag{11.13}$$

The significance of this result is that, by formal manipulation of the full, many-electron Schrödinger equation (11.2) in its matrix representation (11.6), we can obtain some of the *same* eigenvalues from a determinantal equation whose dimensionality is restricted to that of the subspace spanned by the functions Φ_a. The influence of the subspace b is fully described by the second term in (11.12) which accounts for the interactions $a \rightarrow b \rightarrow b \rightarrow a$. As a pointer to later discussion, note that a partitioning of the basis, such that the magnitude of $\mathbf{H}_{ab}$ is small compared with $\mathbf{H}_{aa}$ or $\mathbf{H}_{bb}$, would correspond to the eigenvalues Ψ of (11.2) being quite well described by either Φ_a or Φ_b alone: in such circumstances, when $\bar{\mathbf{H}}_{aa}$ differs only a little from $\mathbf{H}_{aa}$, we recognize a minimal 'coupling' between the two subsets Φ_a and Φ_b.

The partitioning technique can be expressed in terms of operators rather than matrices. We define a projection operator P_a onto the subspace a (of dimension N_a, say) spanned by the subbasis Φ_a, and a projector Q_b onto the orthogonal complement:

$$P_a = \sum_i^{N_a} |\Phi_i\rangle\langle\Phi_i| \tag{11.14}$$

$$Q_b = 1 - P_a. \tag{11.15}$$

Now if (11.11) is taken as the matrix representation of

$$(\bar{\mathscr{H}} - E)|\varphi\rangle = 0, \tag{11.16}$$

$\mathscr{H}$ must be defined, in concordance with (11.12), as

$$\mathscr{H} = H + \Delta\bar{H}(E) \qquad (11.17)$$

with

$$\Delta\bar{H}(E) = HQ_b(E \cdot Q_b - Q_bHQ_b)^{-1}Q_bH. \qquad (11.18)$$

Equations (11.17) and (11.18) are an *exact*, albeit highly formal, reorganization of the many-electron Hamiltonian in (11.2) which enable us to work with a strictly limited finite basis. Thus the *effective operator* $\mathscr{H}$ in (11.16) acts upon functions $\{\varphi\}$ spanning the subspace a; that is, each eigenvector of $\mathscr{H}$ takes the form,

$$\varphi = \sum_i^{N_a} c_i\Phi_i \qquad (11.19)$$

and yet, at the same time, the eigen*values* of $\mathscr{H}$ are *exactly* the same as (N_a of) those of the 'true', many-electron Hamiltonian H. Hence the definition of $\mathscr{H}$ and the *interdependent* nature of the effective operator and the basis upon which it acts are expressed by the equation,

$$E_n = \frac{\langle \Psi_n|H|\Psi_n \rangle}{\langle \Psi_n|\Psi_n \rangle} = \frac{\langle \varphi_n|\mathscr{H}|\varphi_n \rangle}{\langle \varphi_n|\varphi_n \rangle}, \qquad (11.20)$$

for all eigenfunctions spanned by the subspace a of dimension N_a.

The effective Hamiltonian so constructed necessarily involves the energy-dependent term $\Delta\bar{H}(E)$ of (11.18) through which the complementary subspace b is *implicitly* treated. Explicitly, this means that the effective Hamiltonian $\mathscr{H}$ varies from eigensolution to eigensolution. However, in the approximate model we call ligand-field theory, the same effective operator – $\mathscr{H}_{LF}$ of (11.1) – is taken to act *throughout* the restricted subspace spanned by a metal orbital basis with appropriate l quantum number. If $\mathscr{H}_{LF}$ varies significantly in practice, that is noticeably within the accuracies of the various experimental techniques we use in ligand-field studies, we should be obliged to use a different set of ligand-field parameters for each (many-electron) state of the system. Under these circumstances, ligand-field theory would cease to be useful. To emphasize this point with an extreme example: we could envisage constructing a 'ligand-field model' for, say, purely organic molecules in which the eigenproblem is projected onto a d-orbital basis. While this would be totally proper, in the sense that (11.16) is an exact representation of (11.2), the procedure would be ridiculous and worthless by virtue of an extreme energy dependence of the effective Hamiltonian: the ensuing large variation of the system parameters from eigensolution to eigensolution would rob the formalism of any unifying efficacy. That ligand-field theory survives as a workable model within the transition-metal and lanthanide blocks apparently means that the energy dependence of $\mathscr{H}$ is reasonably slight with respect to d- (or f-) type basis. Undoubtedly this is due in considerable part to the localized nature of the metal basis orbitals within such complexes and so the accuracy of Laporte's rule, for example, may be regarded as part of the definition of the *ligand-field regime*.

An exact treatment of the effective Hamiltonian would not be significantly easier than the direct solution of (11.2) and so we seek to define a subspace of practical size which allows for reasonable approximation to the additional term $\Delta \bar{H}(E)$ and, at the same time, does so in a manner which minimizes the energy dependence of the term so that the ligand-field formalism may ultimately be *recovered*. In terms of the partitioning equations (11.7) and (11.8), we look for a basis which is optimally decoupled from its complementary subbasis and we might guess that this should reflect a chemistry-based belief that the participation of transition metal d orbitals in coordination bonding is relatively slight. We approach this search by reference to an alternative partitioning procedure which, though of an ultimately approximate nature, makes possible an intelligent choice of basis by rather more explicit consideration of the problem of electron correlation effects within multi-configuration SCF-type computations.

11.3 Group product functions

We review here some aspects of so-called, group product functions,[17] used in an approach by which a total system may be mentally separated into smaller subsystems comprising various physically recognizable parts of a molecule. The scheme is useful in computational studies when electron correlation within any subsystem may be important but much less so between two different subsystems. Thus, instead of considering the full, many-electron problem, we deal, in effect, with the electrons a few at a time. Although describing some more general features of the approach in this section, we shall nevertheless specialize our treatment straightaway to the ligand-field problem.

Suppose that a metal complex contains n electrons which we divide into two subsets M and L and so put $n = N_M + N_L$. We write a trial wavefunction $\phi_k^{N_M}$ to be used in (11.2) in the so-called, group product form:

$$\phi_k^{N_M} = C\mathscr{A}^n[\Psi_{Mm}(\mathbf{x}_1, \mathbf{x}_2, \ldots, \mathbf{x}_{N_M})\Psi_{Ll}(\mathbf{x}_{N_M} + 1, \ldots, \mathbf{x}_n)], \quad (11.21)$$

where C is a normalization constant and $\mathscr{A}^n$ makes this binary product antisymmetric for all permutations of the n electrons. Ψ_{Mm} and Ψ_{Ll} are normalized, antisymmetric wavefunctions for the groups of electrons M and L (ultimately to be associated with metal basis and the 'rest') with quantum numbers m and l, respectively: m and l here refer to general quantum numbers, of course, and not simply to angular momenta. The index k picks out the combination (m, l). The $\{x_i\}$ are combined space $\{\mathbf{r}_i\}$ and spin $\{\mathbf{s}_i\}$ coordinates. Notice how group product functions may be regarded as generalizations of the usual antisymmetrized spin-orbital products, the individual factors (Ψ_{Mm} and Ψ_{Ll}) now describing whole groups of electrons rather than a single electron. In the limit of zero interaction between the groups M and L, the product (11.21) could in principle give an exact description of the state of the n electrons, provided that the component functions Ψ_{Mm} and Ψ_{Ll} were already constructed as exact functions for the separate groups of electrons.

The characterization of group product functions as generalized analogues of conventional, determinantal spin–orbital products is only completed by the imposition of a strict orthogonality requirement, more stringent than that imposed on ordinary spin-orbitals. The so-called, *strong orthogonality* condition[17] is written,

$$\int \Psi^*_{Mm}(\mathbf{x}_1, \mathbf{x}_i, \mathbf{x}_j, \ldots)\Psi_{Ll}(\mathbf{x}_1, \mathbf{x}_p, \mathbf{x}_q, \ldots)d\mathbf{x}_1 \equiv 0, \quad M \neq L, \quad (11.22)$$

meaning that the result of integrating over *any* one variable (here $\mathbf{x}_1$), common to two different group functions, must vanish identically for all values of the other variables $\mathbf{x}_i, \mathbf{x}_j, \ldots, \mathbf{x}_p, \mathbf{x}_q$. Contrast the usual orthogonality condition for simple spin-orbitals,

$$\int \Phi^*_a(\mathbf{x}_1, \mathbf{x}_2, \ldots, \mathbf{x}_n)\Phi_b(\mathbf{x}_1, \mathbf{x}_2, \ldots, \mathbf{x}_n)d\mathbf{x}_1 d\mathbf{x}_2 \ldots d\mathbf{x}_n = \delta_{ab}. \quad (11.23)$$

The strong orthogonality condition (11.22) merely serves to keep the different groups M and L really separate: thus, if we make any determinant Ψ_{Mm} from the first N_M of our total of n spin-orbitals, and Ψ_{Ll} from the remaining N_L, then Ψ_{Mm} and Ψ_{Ll} are strong orthogonal. Given this definition of strong orthogonality, we may proceed to outline an energy-calculation scheme using group product functions that parallel the form of ordinary Hartree–Fock theory with spin-orbitals.

In the fixed nucleus approximation, the n-electron Hamiltonian H is written as a sum of one- and two-electron operators,

$$H = \sum_{i=1}^{n} H_N(i) + \sum_{i<j}^{n} 1/r_{ij}, \quad (11.24)$$

where

$$H_N(i) = -\tfrac{1}{2}\nabla_i^2 - \sum_\sigma Z_\sigma/r_{\sigma i} + h_i^{\text{rel}} \tag{11.25}$$

in which Z_σ is the charge on nucleus σ (in units of $+e$) and $r_{\sigma i}$ is the distance of electron i from nucleus σ: we include in $h^{\text{rel}}(i)$ all additional relativistic terms like the mass–velocity and Darwin corrections of §7.4 which may be important in heavy elements, and the spin-orbit coupling potential which is always important when considering transition-metal magnetism.

The energy of a single determinantal, many-electron wavefunction $\Psi \equiv |\psi_1\psi_2,\ldots,\psi_p\psi_q|$ formed from the spin–orbitals $\{\psi\}$, under the Hamiltonian (11.24) is

$$
\begin{aligned}
E &= \langle \Psi|H|\Psi\rangle \\
&= \sum_p \langle \psi_p|H_N(p)|\psi_p\rangle + \sum_{p<q} [\langle \psi_p\psi_q|1/r_{pq}|\psi_p\psi_q\rangle \\
&\quad - \langle \psi_p\psi_q|1/r_{pq}|\psi_q\psi_p\rangle|] \\
&\equiv \sum_p H_N(pp) + \sum_{p<q} [J^{pq}(pp,qq) - K^{pq}(pp,qq)],
\end{aligned}
\tag{11.26}
$$

where p,q run over all occupied spin-orbitals. In the self-consistent-field method, each electron is deemed to move in the effective field provided by the nuclei and all other electrons and the best estimate of the ground state energy is obtained by variation of the spin-orbitals $\{\psi\}$ subject to an orthonormality constraint leading to a minimum total energy. The corresponding expression for the energy of a single generalized product function is

$$E = \sum_P H^P(pp) + \sum_{P<Q} [J^{PQ}(pp,qq) - K^{PQ}(pp,qq)], \tag{11.27}$$

provided the groups $\{P,Q,\ldots\}$ are strong orthogonal. Note here that the first term in (11.27) includes both one- and two-electron operators of (11.24) acting *within* the electrons of each separate group. If the electron groups have been well chosen according to recognizable physical parts or attributes of the system, the first term may already provide a good account of electron-correlation effects within each group and the second term of (11.27) is then left to describe the (relatively weak) correlation effects between groups (P and Q).

Specializing the discussion to the case of the *two*-group product function (11.21), the energy in (11.27) becomes

$$E = H^M(mm) + H^L(ll) + J^{ML}(mm,ll) - K^{ML}(mm,ll). \tag{11.28}$$

We can envisage optimizing this energy by a variational procedure, requiring that E be stationary for first-order variations of Ψ_{Mm} and Ψ_{Ll} subject to the orthonormality of the two functions. Since each variation makes its own first-order change in E we need only consider one group: we choose $\Psi_{Mm} \to \Psi_{Mm} + \delta\Psi_{Mm}$ where $\delta\Psi_{Mm}$ is a variation constructed from the M group orbitals alone so that Ψ_{Mm} will automatically remain strong orthogonal to Ψ_{Ll}. It is then useful to write the energy in 'separable' form:

$$E = \langle \phi_k^{N_M} |H|\phi_k^{N_M} \rangle = E_{Mm}^{\text{eff}} + E_{Ll}. \tag{11.29}$$

Thus,

$$E_{Mm}^{\text{eff}} = H^M(mm) + J^{ML}(mm, ll) - K^{ML}(mm, ll) \tag{11.30}$$

and

$$E_{Ll} = H^L(ll), \tag{11.31}$$

so that E_{Ll} is the energy the L-group electrons would have *by themselves* in the field of the nuclei, while E_{Mm} is the energy of the M-group electrons in the effective field of the electrons of group L. Variation of the M-group wavefunctions affects all parts of E_{Mm} but leaves E_{Ll} quite unchanged. The stationary value conditions are therefore,

$$\frac{\partial E_{Mm}}{\partial \Psi_{Mm}} = 0, \quad \langle \Psi_{Mm}|\Psi_{Mm} \rangle = 1. \tag{11.32}$$

An exactly similar criterion obtained by interchanging the roles of M and L, and of m and l, applies to the optimum L-group wavefunctions. For computational purposes, the power of the group product function approach arises by replacing, in effect, an n-electron calculation by a succession of smaller calculations on systems of fewer electrons (here, two such referring to the M and L groups) in which each is optimized in the effective field of the rest. This is a well-defined optimization problem which may be solved iteratively, starting from some reasonable guess for the group functions, and its solution already provides a good treatment of electron correlation.[17] We shall return to the optimization of these functions in a little more detail in §11.5: for the moment, we collect together the definitions in (11.29) *et seq.* and (11.24) *et seq.* and write,

$$E_{Mm} = \int \Psi_{Mm}^* \mathcal{H}_M \Psi_{Mm} d\tau_M, \tag{11.33}$$

$$E_{Ll} = \int \Psi_{Ll}^* \mathcal{H}_L \Psi_{Ll} d\tau_L, \tag{11.34}$$

where

$$\mathscr{H}_M = \sum_{\kappa=1}^{N_M} \mathscr{H}(\kappa)_{\text{core}} + \tfrac{1}{2} \sum_{\kappa,\lambda=1}^{N_M} 1/r_{\kappa,\lambda}, \qquad (11.35)$$

$$\mathscr{H}_L = \sum_{\mu=N_M+1}^{n} H_N(\mu) + \tfrac{1}{2} \sum_{\mu,\nu=N_M+1}^{n} 1/r_{\mu\nu}, \qquad (11.36)$$

in which

$$\mathscr{H}(\kappa)_{\text{core}} = H_N(\kappa) + J_L^l(\kappa) - K_L^l(\kappa). \qquad (11.37)$$

The Coulomb and exchange operators, J_L^l and K_L^l respectively, are defined within the group function basis by the formulae,

$$J_L^l \psi(\mathbf{x}_1) = \int d\mathbf{x}_2 \, 1/r_{12} \, \rho_1^L(ll:\mathbf{x}_2\mathbf{x}_2)\psi(\mathbf{x}_1) \qquad (11.38)$$

$$K_L^l \psi(\mathbf{x}_1) = \int d\mathbf{x}_2 \, 1/r_{12} \, \rho_1^L(ll:\mathbf{x}_1\mathbf{x}_2)\psi(\mathbf{x}_2) \qquad (11.39)$$

where $\rho_1^L(ll:\mathbf{x}_1\mathbf{x}_2)$ is a general matrix element of the first-order, reduced density operator[17] constructed for the many-electron function Ψ_{Ll}. Thus J_L^l operates on $\Psi(\mathbf{x}_1)$ simply by multiplying it by the value of the potential at point $\mathbf{x}_1$ arising from electrons distributed according to the density function for the collective L groups in state l: and K_L^l is an integral operator as in standard Hartree–Fock theory. Both J_L^l and K_L^l are 'Coulomb' and 'exchange' operators for an electron in the effective field due to the L group of electrons. Thus, in the optimization of E_{Mm}^{eff}, the L-group electrons have been formally eliminated by absorption into the effective Hamiltonian $\mathscr{H}_M$.

11.4 The basis orbitals

We now consider the orthonormal set of spin-orbitals we shall use in the determinantal expansion of the many-electron functions $\{\Psi_{Jj}\}$ for the groups M and L. First we suppose that we have a set of k orbitals describing the one-electron states in the metal atom: these will be orthonormal solutions of a Schrödinger equation for a spherically symmetric potential $V_c(\mathbf{r})$ which may be thought of as the average potential about the metal atom in a complex which an electron experiences:[2]

$$[-\tfrac{1}{2}\nabla^2 + (\varepsilon_i - V_c(\mathbf{r}))]\psi_i(\mathbf{r}) = 0, \quad i = 1,\ldots,k \qquad (11.40)$$

$$\langle \psi_i | \psi_j \rangle = \delta_{ij}, \quad 1 \leqslant i,j \leqslant k, \qquad (11.41)$$

$$\psi_i(\mathbf{r}) = \psi_{nL}(r)C(\hat{r}), \quad L \equiv l, m_l. \qquad \text{††}(11.42)$$

---

†† We again use the notation that r and $\hat{r}$ are the radial and angular coordinates of the vector $\mathbf{r}$.

Equivalently, (11.42) may be written

$$\psi_i(\mathbf{r}) = R_{nl}(r)\, Y_m^l(\theta, \phi) \tag{11.43}$$

where, as usual,

$$Y_m^l(\theta, \phi) = \left(\frac{4\pi}{2l+1}\right) C_{lm}. \tag{11.44}$$

With only minor modifications, the spin-independent relativistic corrections can be included in (11.40) and so be absorbed into the definition of the basis orbitals $\{\psi_i\}$: the spin-orbit potential, however, will be treated explicitly in the effective Hamiltonian.

In the ligand-field problem the five degenerate valence orbitals with angular momentum quantum number $l = 2$ obtained from (11.40) comprise the orbitals used to describe the M set of electrons; that is, for the transition-metal block – we use the seven with $l = 3$ for the lanthanides, of course. Note that these are *not* the same as the free-atom or free-ion functions, because of the presence of the mean potential $V_c(\mathbf{r})$ in (11.40): they do, of course, share the same *angular* properties. All the other orbitals, referring to the metal-core electrons and the higher-valence levels will be placed in the L set and the remaining one-electron orbitals in the L set describe the ligand electrons. As our purpose here is to investigate the structure of ligand-field theory rather than to perform explicit numerical computations, the choice of ligand orbitals need not concern us in detail: we simply imagine having enough of them to describe the polarization and interaction of the free ligands as they are brought together on complex formation and that they are orthonormal and orthogonal to the metal orbitals $\{\psi_i\}$. This latter requirement can always be achieved, for example, by Schmidt orthogonalization. However, in order to facilitate the exploitation of chemical information, it is desirable to choose the L set as optimally localized orbitals according to any convenient criterion. Altogether therefore, we imagine having a set of, say N, localized spin-orbitals divided up into the sets M and L as:

$$\begin{pmatrix} M & \bigg| & L \\ \chi_1, \ldots, \chi_{10} & \bigg| & \chi_{11}, \ldots, \chi_N \end{pmatrix}.$$

In addition to this definition of basis *orbitals*, we must characterize the implicit group product function (11.21) further, with regard to the number of electrons taken in the M group. In seeking to reconstruct the form of ligand-field theory it is natural, therefore, to define this number as that given by the metal ion configuration: with the total number of electrons $n = N_d + N_L$ we have $1 \leqslant N_d \leqslant 9$, or for $n = N_f + N_L$, $1 \leqslant N_f \leqslant 13$.

11.5 A primitive ligand-field parametrization

The functions $\{\Psi_{Mm}\}$ of (11.21) are thus built from N_d-fold products of the ten spin-orbitals (or equivalent for lanthanides) whose spatial parts are defined by (11.40). They are to be optimized in the effective field provided by the ground state Ψ_{L0} of the L-group electrons which will normally be a non-degenerate spin-singlet. Similarly, the $\{\Psi_{Ll}\}$ are the optimized functions for the L group of electrons in the average field of the ground state of the d (f) electrons, Ψ_{M0}. We thereby generate a set of many-electron wavefunctions $\{\phi_k^{N_d}\}$ as in (11.21), which span a subspace a, of dimension N_d, and are eigenfunctions of the Hamiltonian $P_a H P_a$, which is the projection of H onto this subspace. The optimized functions $\{\phi_k^{N_d}\}$ are not, therefore, eigenfunctions of the full, n-electron Hamiltonian (11.2) as we note that $P_a H P_a$ neglects the additional energy-dependent term $\Delta\bar{H}(E)$ as the 'primitive ligand-field scheme',[2] reserving consideration of the complexities of the additional term to §11.6. Meanwhile, however, we note that the importance of $\Delta\bar{H}(E)$ will have been diminished by the favourable choice of the M basis in (11.40) and, indeed, that the functions $\{\phi_k^{N_d}\}$ provide a better basis than is immediately obvious. Thus, if we assume, as we may, that the optimization has been carried out completely, the $\{\phi_k^{N_d}\}$ will be highly accurate wavefunctions for the complex, except in one respect, Solving for the functions $\{\Psi_{Ll}\}$ under the assumption that the final number, N_d, of the M subset of electrons only provides an *average* potential field, means that the wavefunctions $\{\phi_k^{N_d}\}$ will not be able to describe effects arising from $M \leftrightarrow L$ interchanges: that is, the interactions between metal d electrons (M) and both metal valence (s, p) electrons and ligand electrons (collectively, L) are treated in a self-consistently, averaged way, Despite this reservation, the $\{\phi_k^{N_d}\}$ are, in one sense, good to first order in such configuration mixing, as the following argument shows.

Consider the effects of configuration interaction between the ground- and excited-state functions: these may differ in either one or two orbital excitations. For a single excitation in the M group, say, in which $m \to m'$ we have

$$\langle \phi_k^{N_d} | H | \phi_{k'}^{N_d} \rangle = H^M(mm') + J^{ML}(mm', ll) - K^{ML}(mm', ll)$$
$$= \langle \Psi_{Mm} | \mathscr{H}_M | \Psi_{Mm'} \rangle \equiv H_{\text{eff}}^M(mm'). \tag{11.45}$$

When we have orbital differences in both M and L groups, there results

$$\langle \phi_k^{N_d} | H | \phi_{k''}^{N_d} \rangle = J^{ML}(mm', ll') - K^{ML}(mm', ll') \tag{11.46}$$

and if the energy after configuration interaction is written

$$E = E^{(0)} + E^{(1)} + E^{(2)} + \dots, \tag{11.47}$$

with $E^{(0)}$ given by (11.28), the CI corrections are

$$E^{(1)} = - \sum_{m' \neq m} \frac{|H_{\text{eff}}^M(mm')|^2}{\Delta E(Mm \to Mm')} \qquad (11.48)$$

and

$$E^{(2)} = - \sum_{\substack{m' \neq m \\ l' \neq l}} \frac{|J^{ML}(mm', ll') - K^{ML}(mm', ll')|^2}{\Delta E(Mm \to Mm', Ll \to Ll')}. \qquad (11.49)$$

However, as the group functions were determined variationally, so that (11.32) was satisfied for arbitrary variations $\Psi_{Mm} \to \Psi_{Mm} + \delta\Psi_{Mm}$ within the subspace M, it follows that

$$H_{\text{eff}}^M(mm') = \langle \Psi_{Mm} | \mathcal{H}_M | \Psi_{Mm'} \rangle = 0 \qquad (11.50)$$

because the $\{\Psi_{Mm}\}$ form an orthogonal set of functions, diagonal with respect to $\mathcal{H}_M$. Therefore, all *single*-excitation contributions $E^{(1)}$ to the energy in a CI calculation vanish identically. In the standard Hartree–Fock theory, the corresponding result is known as Brillouin's theorem. A similar argument follows for single excitations in the L group, the functions $\{\Psi_{Ll}\}$ having been obtained by an analogous optimizing process. In general, therefore, we have

$$\langle \phi_k^{N_d} | H | \phi_{k'}^{N_d} \rangle = 0 \text{ iff } k = m, l \text{ and } \begin{cases} k = m, l' \\ \text{ or } \\ k' = m', l \end{cases} \qquad (11.51)$$

Now admixture of only the *singly* excited functions polarizes the charge density[17] and so, to this order of perturbation theory, polarization of each group by the presence of the other is fully accounted for by optimizing the one-configuration $\{\phi_k^{N_d}\}$ function. The double-excitation contribution $E^{(2)}$, though non-zero, might reasonably be expected to be relatively small in view of its being associated with the modification of the so-called pair function[17] which should itself be minimized if the M and L groups were chosen to reflect the relatively small interaction between the metal $d(f)$ basis and the rest (L). Thus the localized nature of the metal $d(f)$ basis, together with hindsight of the viability of ligand-field theory, both serve to illuminate the suitability of the $\{\phi_k^{N_d}\}$ as defined above, as a well-partitioned basis and so justify the projected operator $P_a H P_a$ as a good first approximation to $\mathcal{H}$ of (11.17).

We shall now imagine that the optimized function Ψ_{L0} for the L group of electrons has been obtained, and look in more detail at the variational calculation for the M-group electrons, in which the $\{\Psi_{Mm}\}$ are expanded in terms of Slater determinants built from the M subset of the orthonormal spin-orbital basis $\{\chi\}$, the ten spin-orbitals of d-orbital character,

$$\left\|\Omega_u\right\rangle = \mathscr{A}^{N_d}\left|\chi_1, \ldots, \chi_m\right\rangle, \tag{11.52}$$

$$\Psi_{Mm} = \sum_u C_{mu}^M \left\|\Omega_u\right\rangle. \tag{11.53}$$

As there are N_d electrons in the M set, there are at most $^{10}C_{N_d}$ terms in the expansion (11.53). The L-group electrons now enter $\mathscr{H}_M$ only through the matrix elements of the density operator derived from their ground-state wavefunction Ψ_{L0}, $\rho_1^L(00 : \mathbf{x}_1, \mathbf{x}_2) \equiv \rho^L(\mathbf{x}_1, \mathbf{x}_2)$ for short. The stationary value condition for E_{Mm}^{eff} can be transformed in the usual way into a matrix eigenvalue equation for the eigenvalues E_{Mm}^{eff} and the expansion coefficients $\{C_{mu}^M\}$ in (11.53), namely

$$(\mathbf{H}_M - E_{Mm}^{\text{eff}}\mathbf{1})C_m^M = 0. \tag{11.54}$$

This is precisely the normal diagonalization procedure in ligand-field analysis. The elements of the matrix $\mathbf{H}_M$ are

$$H_M(uu') = \left\langle \left\|\Omega_u\right\| \mathscr{H}_M \left\|\Omega_{u'}\right\| \right\rangle, \tag{11.55}$$

with

$$\mathscr{H}_M = \sum_{\kappa=1}^{N_d} H_N(\kappa) - \sum_{\kappa=1}^{N_d} J_L^0(\kappa) - \sum_{\kappa=1}^{N_d} K_L^0(\kappa) + \sum_{\kappa<\lambda}^{N_d} 1/r_{\kappa\lambda} \tag{11.56}$$

and

$$J_L^0\chi(\mathbf{x}_1) = \int d\mathbf{x}_2 \, 1/r_{12} \, \rho^L(\mathbf{x}_2, \mathbf{x}_2)\chi(\mathbf{x}_1) \tag{11.57}$$

$$K_L^0\chi(\mathbf{x}_1) = \int d\mathbf{x}_2 \, 1/r_{12} \, \rho^L(\mathbf{x}_1, \mathbf{x}_2)\chi(\mathbf{x}_2) \tag{11.58}$$

specifying the effect of the Coulomb and exchange operators on a spin-orbital χ. At the level of the primitive parametrization, equations (11.54)–(11.58) constitute an explicit *ab initio* statement of the ligand-field problem. In this order of approximation we identify $\mathscr{H}_M$, of (11.35) and (11.37), acting on the basis with spatial parts defined by (11.40), with the effective Hamiltonian $\mathscr{H}_{\text{LF}}$ of (11.1) acting on metal d- (f-) orbital basis. We are therefore in a position to comment upon the basis and parameters of $\mathscr{H}_{\text{LF}}$ against a quantum-chemistry background rather than from the more usual *ad hoc*, phenomenological stance.

The spherical parts of the operators in $\mathscr{H}_M$ define the mean one-electron potential $V_c(\mathbf{r})$ used to generate the basis orbitals $\{\psi_i\}$ and hence, when taken together, lead to a constant energy term in all of the diagonal matrix elements $H_M(uu)$: this constant energy shift is of no interest in ligand-field studies and may be discarded. As we have been able to justify the use of $\mathcal{H}_{LF}$ with metal basis orbitals with well-defined orbital angular momenta, the interelectron repulsion integrals reduce exactly to combinations of the usual Slater–Condon parameters $F_k(k = 0, 2, 4$ for d configurations) or, of course, to their combinations as Racah parameters A, B, C. Once again, the parameter A does not affect term splittings and, as discussed above, is absorbed in the orbital energy ε_d of (11.40). We are therefore left with the usual two interelectron repulsion parameters (for d configurations), F_2 and F_4, or B and C, related conventionally by

$$B = F_2 - 5F_4; \quad C = 35F_4. \tag{11.59}$$

These parameters will not generally have values equal to those in the corresponding free ions because the spin-orbitals $(\psi_1, \ldots, \psi_{10})$ are eigensolutions of the Schrödinger equation for the mean potential in the complex rather than in the free ion. The nephelauxetic effect is thus contained within this difference in mean potentials and, at a qualitative level at least, may be correlated with the well-known concept of 'central-field covalency'.[14,15]

While the spherical parts of the one-electron operators in $\mathcal{H}_M$ – the kinetic energy operator, the electron–nuclear attraction arising from the metal nucleus, and the spin-independent relativistic terms – are accounted for within the definition of the basis orbitals $\{\psi_i\}$, the spin-orbit coupling operator is dealt with in exactly the same way as for the corresponding free ion: thus, as in (7.106), we define the spin-orbit parameter ξ by

$$\xi = -\frac{e}{2m^2c^2}\left\langle \chi_d \left| \frac{1}{r}\frac{dV_c}{d\mathbf{r}} \right| \chi_d \right\rangle, \tag{11.60}$$

where V_c is again the mean potential in the complex. Justification for introducing the relativistic corrections through the spherical potential V_c lies in the fact that these corrections are important mostly near the metal nucleus, a region which to a very good approximation *can* be described by this spherical potential.

Finally, the remaining electron–nuclear attraction operators which involve ligand nuclei are collected together with the operators J_L^0 and K_L^0 to form the ligand-field potential V_L:

$$V_L(\mathbf{x}) = -\sum_Q^{\text{ligands}} \frac{Z_Q}{|\mathbf{r} - \mathbf{Q}|} + J_L^0(\mathbf{x}) - K_L^0(\mathbf{x}) \tag{11.61}$$

and, as discussed above, the spherical part of $V_L(\mathbf{x})$ can be discarded because it can be absorbed within the definition of the basis orbitals $\{\psi_i\}$. Note that $V_L(\mathbf{x})$ has been written as a function of combined space–spin coordinates $\mathbf{x}$ because, while the spin summations can be carried out in $J_L^0(\mathbf{x})$ before calculating matrix elements, $K_L^0(\mathbf{x})$ may connect spin-orbitals which are off-diagonal in the spin wavefunctions: however, in the particular case of the density matrix $\rho^L(\mathbf{x}_1, \mathbf{x}_2)$ arising from a wavefunction which is a spin singlet $(S = 0)$ one can show that K_L^0 must also be diagonal.[17] This leads to a useful simplification here since we can usually assume this property for Ψ_{L0}, and it means that $V_L(\mathbf{x})$ reduces to a (non-local) function of the space variable $\mathbf{r}$ only: we can therefore consistently parametrize the matrix elements for the whole potential, $\langle \chi_R | V_L(\mathbf{x}) | \chi_S \rangle$ without having to decompose them into different spin combinations for Coulomb and exchange potentials.

11.6 Parameter renormalization

In the 'primitive parametrization' scheme outlined above we restricted attention to the matrix elements of the true Hamiltonian H in the basis of group product states $\{\phi_k^{Na}\}$. We turn now to an examination of the matrix elements of the effective Hamiltonian $\mathcal{H}$ of (11.17), in the same basis, noting that a complete treatment of the energy-dependent correction term $\Delta\bar{H}(E)$ would lead to the *exact* eigenvalues of the full molecular Hamiltonian. We first consider some of the formal properties of the effective Hamiltonian $\mathcal{H}$ and then move on to a more detailed discussion of the assumptions which make a connection with the traditional ligand-field procedures possible. It must be recognized at the outset that the operator $\Delta\bar{H}(E)$ has a very complicated structure. We shall not, therefore, analyse $\Delta\bar{H}(E)$ in detail as a prelude to a quantitative computational study but rather adopt the modest aim of investigating, by formal means, the extent to which the reduction of $\Delta\bar{H}(E)$ to the form of $\mathcal{H}_{LF}$ can be achieved. Our expectation that the chemical and physical insights which suggested that the set of group product functions $\{\phi_k^{Na}\}$ would be a good set of trial functions implies, of course, that the effects of $\Delta\bar{H}(E)$ will be quantitatively small.

We shall first review a crude argument[2] which reveals that $\Delta\bar{H}(E)$ consists of a sum of 1-, 2-,..., n- particle operatiors, where n is again the total number of electrons in the complex. This already implies that the ligand-field model neglects some terms in the true Hamiltonian since our 'empirical', effective Hamiltonian $\mathcal{H}_{LF}$ of (11.1) can only connect states differing in no more than two spin-orbitals. The primitive parameters of §11.5 describe contributions to many-electron matrix elements $\langle u|H|u'\rangle$ arising from the one- and two-electron operators of $\mathscr{H}_M$. To the extent that we can identify orbital matrix elements in $\langle u|\Delta\bar{H}(E)|u'\rangle$ which occur in the same way as in $\langle u|\bar{H}|u'\rangle$, the two quantities can simply be summed and handled as a single so-called, 'renormalized' parameter.[2]

The most general n-electron wavefunction Φ_j can be written as a linear combination of the $\{\phi_k^{N_M}\}$,

$$\Phi_j = \sum_{k,N_M} C_{jk}^{N_M} \phi_k^{N_M}. \tag{11.62}$$

The set of functions $\{\phi_k^{N_M}\}$ of (11.21) which has some definite fixed value of $N_M \equiv N_d$ (for d configurations) thus defines a subspace, say a_{N_d}, of the n-electron function space, and if we define the complementary subspace $\bar{a}_{N_d}$ through a set of orthogonal functions $\{\Delta_i\}$ such that

$$\langle \phi_k^{N_d}|\Delta_i\rangle = 0, \quad \langle \Delta_i|\Delta_j\rangle = \delta_{ij}, \tag{11.63}$$

we can decompose a general wavefunction Φ_j into its components in the two subspaces:

$$\Phi_j = \sum_k a_{jk}\phi_k^{N_d} + \sum_i C_{ji}\Delta_i. \tag{11.64}$$

In view of (11.51), we can take $a_{jk} = \delta_{jk}$ without loss of generality. Since the functions $\{\phi_k^{N_d}\}$ are constructed as antisymmetrized products of states for the d-electron manifold and the *ground-state* wavefunction Ψ_{L0} for the L subset of electrons, there are functions in the set $\{\Delta_i\}$ describing all the excited states of the L electrons; that is, we can write these functions as antisymmetrized products of the $\{\Psi_{Mm}\}$ and $\{\Psi_{Ll}\}$, discussed in §§11.3–11.5, with $l \neq 0$. The remaining functions in the set $\{\Delta_i\}$ describe all partitions between the d electrons and the remaining electrons other than the choice $n = N_d + N_L$ already used in the construction of the subspace a_{N_d}: these wavefunctions therefore describe charge-transfer states involving exchanges of electrons between the d-electron subset and the ligands, and states involving interaction between the metal-core and valence s, p electrons and the metal d electrons.

The projection operator Q_b of (11.15) can be expressed in terms of the $\{\Delta_i\}$ as

$$Q_b = \sum_i |\Delta_i\rangle\langle\Delta_i| \equiv 1 - P_a \tag{11.65}$$

and $\Delta\bar{H}(E)$ of (11.18) then rewritten

$$\Delta\bar{H}(E) = \sum_{i,k} H|\Delta_i\rangle G(E)_{ik}\langle\Delta_k|H, \tag{11.66}$$

where

$$G(E)_{ik} = \left\langle \Delta_i \left| \frac{1}{E \cdot 1 - H} \right| \Delta_k \right\rangle \tag{11.67}$$

is the (i, k)th element of the resolvent, $G(E)$, of H in the subspace spanned by the functions $\{\Delta_i\}$. The complexities engendered by (11.66) can be illustrated by the following rough analysis.

Using the operator identity[18]

$$(A - B)^{-1} = A^{-1} + A^{-1} \sum_{p=1}^{\infty} (BA^{-1})^p, \tag{11.68}$$

we write the resolvent operator $G(E)$ in a formal power series expansion

$$G(E) = (E \cdot 1 - H)^{-1} \equiv [(E - E_0)1 - (H - E_0)]^{-1}$$

$$= \frac{1}{E - E_0} + \sum_{p=1}^{\infty} \frac{(H - E_0)^p}{(E - E_0)^{p+1}}, \tag{11.69}$$

so that

$$G(E)_{ik} = \frac{\delta_{ik}}{E - E_0} + \sum_{p=1}^{\infty} \frac{\langle \Delta_i | (H - E_0)^p | \Delta_k \rangle}{(E - E_0)^{p+1}}. \tag{11.70}$$

Neglecting the sum in (11.70) in the present rough illustration and using (11.63) and (11.65), we then obtain

$$\langle \phi_j^{N_d} | \Delta \bar{H}(E) | \phi_j^{N_d} \rangle \approx \frac{1}{E - E_0} \sum_i \langle \phi_j^{N_d} | H | \Delta_i \rangle \langle \Delta_i | H | \phi_j^{N_d} \rangle$$

$$= \frac{1}{E - E_0} \langle \phi_j^{N_d} | H(1 - P_a)H | \phi_j^{N_d} \rangle$$

$$= \frac{1}{E - E_0} \left[\langle \phi_j^{N_d} | H^2 | \phi_j^{N_d} \rangle - \sum_i \langle \phi_j^{N_d} | H | \phi_i^{N_d} \rangle \langle \phi_i^{N_d} | H | \phi_j^{N_d} \rangle \right]$$

$$= \frac{\langle \phi_j^{N_d} | H^2 | \phi_j^{N_d} \rangle - E_j^2}{(E - E_0)}. \tag{11.71}$$

The numerator in (11.71) is a measure of the decay rate of $\phi_j^{N_d}$ in the full Hamiltonian H, as will be demonstrated shortly: note for the moment, however, that if $\Delta \bar{H}(E) \to 0$, the $\{\phi_j^{N_d}\} \to$ eigenfunctions of H, which are, of course stationary states. Now the energy E_0 may be freely chosen, but should best take a value which minimizes the contribution from the sum in (11.70) to the matrix multiplication in (11.66): thereby E_0 acquires a dependence on the state $\phi_j^{N_d}$ involved; $E_0 = E_0^j(E)$. We also see that the term in H^2 introduces up to 4-electron terms, and if we had not neglected the sum in (11.70) we would have found operators involving 1-, 2-,..., n-electron variables arising from the higher powers of the Hamiltonian. Further, the energy denominator had a dependence on the particular state in question and hence we may expect that the renormalized ligand-field parameters which lead to the true eigenvalues will also acquire a dependence on the individual states. This simple analysis serves to amplify the broad assertions made about $\Delta\bar{H}(E)$ in §11.2.

We shall not pursue this approach, however, since the effective Hamiltonian $\mathscr{H}$ is only hermitian when considered as an operator acting at a fixed energy: this has the immediate consequence that two eigenvectors $\varphi_n, \varphi_{n'}$ belonging to distinct eigenvalues $E_n, E_{n'}$ need not be orthogonal; although they can be assumed to be linearly independent. This is important when we recall that the many-electron formulations of ligand-field theory have always assumed[19,20] that the ligand-field states are orthonormal eigenvectors of some hermitian ligand-field Hamiltonian as in (11.1). Des Cloizeaux[21] has shown that one can explicitly construct a similarity transformation of $\mathscr{H}$ which leaves the eigenvalue spectrum invariant but enforces hermiticity and therefore orthogonal eigenvectors for distinct eigenvalues:

$$\hat{\mathscr{H}} = \Lambda \mathscr{H} \Lambda^{-1} \equiv H + h, \quad \hat{\mathscr{H}} = \hat{\mathscr{H}}^\dagger \tag{11.72}$$

and

$$(\hat{\mathscr{H}} - E_n)|\phi_n> = 0, \quad \langle \phi_n | \phi_{n'} \rangle = \delta_{nn'}. \tag{11.73}$$

Hence the Schrödinger equation for the n-electron Hamiltonian, H, (11.2), can always be formally transformed into the eigenvalue problem (11.73) for an effective Hamiltonian $\hat{\mathscr{H}}$ acting in the subspace a spanned by a finite set of orthonormal vectors $\{\phi_n\}$: the ligand-field Hamiltonian $\mathscr{H}_{LF}$ of (11.1) must therefore be an approximation to $\hat{\mathscr{H}}$.

As discussed for $\mathscr{H}$ above, the hermitian operator $\hat{\mathscr{H}}$ can also be expected to have a complicated many-particle structure, due to the presence of the term h (11.72): it can be written formally as a sum of 1-, 2-,..., n-electron operators,

$$\hat{\mathscr{H}} = \sum_i \mathscr{H}_i + \frac{1}{2!} \sum_{i \neq j} \mathscr{H}_{ij} + \frac{1}{3!} \sum_{i \neq j \neq k} \mathscr{H}_{ijk} + \dots. \tag{11.74}$$

It can easily be shown that orbital matrix elements which are off-diagonal in p spin-orbital indices only have non-vanishing matrix elements with $\mathscr{H}_{1,2,\dots,p}$, $\mathscr{H}_{1,2,\dots,p,p+1,p+2}$: in particular, Freed[22] has demonstrated that the fully diagonal ($p = 0$) matrix elements of $\mathscr{H}_{123}, \mathscr{H}_{1234}, \ldots$ vanish identically so that these matrix elements describe simple one- and two-body interactions. No simple explicit expressions for (11.74) can be written down, although the two-electron operators in h, for example, can be shown to represent screened Coulomb interactions

$$\exp\{-r_{ij}/\lambda\}/r_{ij} \tag{11.75}$$

with λ of the order of the Thomas–Fermi screening length as would be expected on physical arguments: their scalar parts will contribute to both the Slater–Condon F_k integrals and to the Trees correction.[23,15,24]

Renormalization of the 'primitive' parameters is achieved by replacing $\mathscr{H}_M$ in (11.55) by $\hat{\mathscr{H}}_M$, so that we have the matrix elements in the basis of determinants $\{|\Omega_u|\}$ formed from the basis orbitals $\{\chi_i\}$,

$$H_M(uu') = \langle |\Omega_u| |\hat{\mathscr{H}}_M| |\Omega_u| \rangle \equiv \langle |\Omega_u| |\mathscr{H}_M + h_M| |\Omega_u| \rangle. \tag{11.76}$$

As in §11.5, explicit dependence upon the L set of electrons has been integrated out using the ground-state wavefunction Ψ_{L0}, so that the parameter renormalization arises through the term $h_M = \langle \Psi_{L0} |h| \Psi_{L0} \rangle_L$. Since h_M is a diagonal ($p = 0$) matrix element of h, only one- and two-body terms survive in the integration over $\Psi_{L0}^* \Psi_{L0}$. Accordingly, $\hat{\mathscr{H}}_M$ is still a functional of the density matrix $\rho_L(\mathbf{x}, \mathbf{x}')$ for the L set of electrons, as before: in other parts of $\hat{\mathscr{H}}_M$, however, there are 1-, 2-,..., N_d- body operators, of course.

The way in which renormalized orbital matrix elements become configuration dependent may be seen in the following way: the discussion is complementary to that following (11.71). We apply Slater's rules[25] to the determinantal matrix elements (11.76) and consider, for simplicity, a case in which the determinants u and u' differ in one spin-orbital only: that is, $u = Uu_1$ and $u' = Uu_1'$, where u_1, u_1' are spin-orbitals and U is a determinantal function for $(N_d - 1)$ electrons. The renormalized matrix elements, off-diagonal in one spin-orbital, then take the form,[22]

$$\langle u_1 |\hat{\mathscr{H}}_M(U)| u_1' \rangle \equiv \langle u_1 |\mathscr{H}_1| u_1' \rangle + \sum_{v \in U} \langle v u_1 |\mathscr{H}_{12}(1 - P_{12})| v u_1' \rangle$$

$$+ \sum_{v < v' \in U} \langle \mathscr{A}^3 [vv'u_1] |\mathscr{H}_{123}| \mathscr{A}^3 [vv'u_1'] \rangle + \ldots, \quad u_1 \neq u_1', \tag{11.77}$$

where P_{12} is the two-particle permutation operator, and we have indicated the dependence of the renormalized, true parameter, and operator, on the $(N_d - 1)$ electron configuration U which is not excited. Here we see, perhaps more explicitly than before, that in contradistinction to the primitive parametrization, we now require parameters for every distinct $(N_d - 1)$-

electron configuration which can be built from the d-orbital basis $\{\chi\}_M$. Similar remarks apply to matrix elements with more than one spin-orbital difference or, indeed, to the diagonal elements $(u = u_1)$ themselves. It is evident, therefore, that there can be no exact map from the effective Hamiltonian theory sketched above to the ligand-field procedures. The success of the conventional ligand-field treatment of transition-metal or lanthanide complexes depends upon the validity of an assumption of what may be termed, 'anonymous configuration parentage' for the renormalized parameters. A failure of this assumption in a given case will manifest itself either through an inability to obtain any reasonable parameter fit to experimental data, or possibly a fit with highly abnormal parameter values. Here, so to speak, lie the borders of the ligand-field regime. The assumption will certainly fail when any of the coefficients $\{C_{jk}\}$ in (11.62) become comparable with unity. In that case there is strong coupling between ligand-field states $\{\phi_k^{N_d}\}$ and at least one of the states $\{\Delta_j\}$, and hence significant corrections to the primitive eigenvalues will emerge from $\Delta\bar{H}(E)$. This situation would occur whenever there is ambiguity in the formal identification of a unique d^n (or f^n) configuration of the metal ion. Altogether, therefore, conventional ligand-field theory can be regarded as an approximate version of the full, effective Hamiltonian theory outlined here that uses a weighted configuration average of the renormalized ligand-field parameters.

The requirement for an averaging procedure suggests that, as far as ligand-field theory is concerned, little is to be gained by an elaborate and explicit, many-body analysis of (11.77). One can, however, formulate a simple criterion for deciding how good an approximation the set of states $\{\phi_k^{N_d}\}$ really are. We might reasonably suppose that the primitive scheme of §11.5 is within the bounds of what is possible in computational quantum chemistry, at least for simple transition-metal complexes, so the analysis presented below should allow a computational check on ligand-field theory once one has obtained the states $\{\phi_k^{N_d}\}$: in any event, the argument provides further insight into the numerator of (11.71). It is based on the observation by Haydock *et al.*,[26] that, given any Hamiltonian H and a trial wavefunction u_0, it is always possible to construct a 'chain' representation of the quantum mechanics of the system. By this is meant that we can construct a sequence of states $\{u_n\}$ such that they satisfy a three-term recurrence relation generated by H, starting from u_0,

$$Hu_n = a_n u_n + b_{n+1} u_{n+1} + b_{n-1} u_{n-1}, \qquad (11.78)$$

with $b_{-1} = 0$. One can interpret this equation by saying that the coefficients

$\{a_n, b_n\}$ give a matrix representation of H which is tridiagonal (Jacobi form), the $\{a_n\}$ being the diagonal elements and the $\{b_n\}$ the off-diagonal ones. Since any hermitian matrix may be brought into Jacobi form by a similarity transformation, it follows that all quantum-mechanical theories can be discussed in a chain representation, if convenient. Let us take H to be the full Hamiltonian for a transition-metal complex and $u_0 = \phi_k^{Nd}$ for some quantum number k, and consider the first step of the recurrence. We have

$$H|\phi_k^{Nd}\rangle = a_k|\phi_k^{Nd}\rangle + b_{1k}|u_{1k}\rangle, \tag{11.79}$$

where a_k, b_{1k} and $|u_{1k}\rangle$ are given by[26]

$$a_k = \langle \phi_k^{Nd}|H|\phi_k^{Nd}\rangle, \tag{11.80}$$

$$b_{1k}^2 = \langle (H - a_k)\phi_k^{Nd}|(H - a_k)\phi_k^{Nd}\rangle, \tag{11.81}$$

$$|u_{1k}\rangle = b_{1k}^{-1}(H - a_k)|\phi_k^{Nd}\rangle. \tag{11.82}$$

Thus a_k is the average energy of the Hamiltonian in the state $|\phi_k^{Nd}\rangle$ and if this state is the solution of the 'primitive' Hamiltonian $\mathcal{H}_M$, $a_k = E_k$: on the other hand, the coefficient b_{1k} is simply a normalization constant for the state $|u_{1k}\rangle$. Consider (11.79) from the point of view of the time-dependent Schrödinger equation: evidently if $|\phi_k^{Nd}\rangle$ were an eigenstate of H, $b_{1k} = 0$. Conversely, if $|\phi_k^{Nd}\rangle$ is some approximation to the true eigenstate, the magnitude of b_{1k} provides an estimate of how good the approximation is because it measures the rate at which $|\phi_k^{Nd}\rangle$ decays. Finally, (11.81) may be expanded and simplified to yield

$$b_{1k}^2 = \langle \phi_k^{Nd}|H^2|\phi_k^{Nd}\rangle - E_k^2, \tag{11.83}$$

which we recognize as the numerator of (11.71).

11.7 The orbital reduction factor

The renormalization theory of the effective Hamiltonian implied by the restriction to some subspace a of the full Hilbert space also imposes a requirement for renormalization of expectation values of other operators.[22] Suppose we have some operator $\hat{B}$ for which we require its expectation value in a state Φ_k of the full Schrödinger equation (11.2). In complete analogy with the effective Hamiltonian theory described above, we define an effective operator $\bar{B}$ by the requirement that its expectation value in a state $|k\rangle$ in the subspace a should equal the exact expectation value for the exact state with the same eigenvalue (see equation (11.20))

$$\langle \hat{B} \rangle_k = \frac{\langle \Phi_k | \hat{B} | \Phi_k \rangle}{\langle \Phi_k | \Phi_k \rangle} \equiv \frac{\langle \phi_k^{N_d} | \bar{B}_k | \phi_k^{N_d} \rangle}{\langle \phi_k^{N_d} | \phi_k^{N_d} \rangle}; \qquad (11.84)$$

thus $\phi_k^{N_d}$ is the eigenvector of (11.73) belonging to the same eigenvalue as Φ_k. Substituting (11.64) with $a_{jk} = \delta_{jk}$ into (11.84) gives

$$\langle \hat{B} \rangle_k = \{ \langle \phi_k^{N_d} | \hat{B} | \phi_k^{N_d} \rangle + \langle \phi_k^{N_d} | \hat{B} | \sum_i C_{ki} \Delta_i \rangle + \langle \sum_i C_{ki} \Delta_i | \hat{B} | \phi_k^{N_d} \rangle$$

$$+ \sum_i \sum_j C_{ki}^* C_{kj} \langle \Delta_i | \hat{B} | \Delta_j \rangle \} / \langle \phi_k^{N_d} + \sum_i C_{ki} \Delta_i | \phi_k^{N_d} + \sum_j C_{kj} \Delta_j \rangle$$

$$= \{ \langle \phi_k^{N_d} | \hat{B} + \sum_i C_{ki} \hat{B} | \Delta_i \rangle \langle \phi_k^{N_d} | + \sum_i C_{ki}^* | \phi_k^{N_d} \rangle \langle \Delta_i | \hat{B}$$

$$+ | \phi_k^{N_d} \rangle (\sum_i \sum_j C_{ki}^* C_{kj} \langle \Delta_i | \hat{B} | \Delta_j \rangle) \langle \phi_k^{N_d} | \phi_k^{N_d} \rangle \} / (1 + \sum_i | C_{ki} |^2).$$

$$(11.85)$$

whence

$$\bar{B}_k = D_k^{-1} \Big\{ \hat{B} + \sum_i C_{ki} \hat{B} | \Delta_i \rangle \langle \phi_k^{N_d} | + \sum_i C_{ki}^* | \phi_k^{N_d} \rangle \langle \Delta_i | \hat{B}$$

$$+ | \phi_k^{N_d} \rangle \Big(\sum_i \sum_j C_{ki}^* C_{kj} \langle \Delta_i | \hat{B} | \Delta_j \rangle \Big) \langle \phi_k^{N_d} | \Big\}, \qquad (11.86)$$

with

$$D_k = 1 + \sum_i | C_{ki} |^2 \geq 1. \qquad (11.87)$$

Now if the mixing between the two subspaces induced by $\hat{B}$ is small – and this will certainly by true for a perturbation by a magnetic field, for example – the additional terms in the numerator of (11.86) can be neglected provided the coefficients $\{C_{ki}\}$ are also small (corresponding to there being only small mixing of M and L sets under H), in which case we have the approximate result

$$\bar{B} \approx D_k^{-1} \hat{B}, \qquad (11.88)$$

where the denominator D_k is approximately independent of the operator $\hat{B}$. As with the renormalized Hamiltonian it is impractical to work with a different scaling factor for each state $|k\rangle$ and so some average $\bar{D}$ is generally assumed. It is this sort of argument which must be used to provide some justification, in the many-electron context, for the conventional modification of the orbital angular momentum operator $L = \sum_i l_i$ with Stevens' orbital reduction factor k in the magnetic moment operator,

$$\mu_\alpha = \beta_e(kl_\alpha + 2s_\alpha); \quad \alpha = x, y, z, \tag{11.89}$$

used when fitting esr and magnetic susceptibility data.[11,12,13] Empirically, k usually, if not invariably, takes values less than unity, but from the foregoing discussion it is clear that connections between k values and other aspects of the quantum-mechanical situation are complex and certainly not generally correlatable with molecular-orbital mixing coefficients or even the mean potential V_c of (11.40), which latter is some determinant of the effective spin-orbit coupling parameter, as in (11.60).

11.8 The ligand-field formalism

Let us review the position reached so far. The supposition that a transition-metal ion is subjected to a classical, electrostatic field of surrounding charges formed the basis of early crystal-field theory. The octahedral-field splitting parameter Dq was expressed as a product of the charge q on the surrounding ligands and D, a factor related to ion separation and the radial properties of the metal-ion d wavefunction. It soon became clear that the magnitude of Dq calculated from any reasonable estimate of the charge q was far too small:[15] even earlier, Pauling had raised qualitative objections to the simplistic basis of crystal-field theory, as we reviewed in chapter 1. *Ligand*-field theory effectively began when Van Vleck[27] pointed out that major contributions to the splitting parameter must arise from covalency and demonstrated, as in §1.1, how the formalism of crystal-field theory could be reconciled with Mulliken's molecular-orbital theory. While a qualitative understanding of the relative magnitudes of Δ_{oct} values – the spectrochemical series – has been available for many years, it is only relatively recently that extensive, all-electron computations of the SCF type with CI have been able to reproduce observed values of Δ_{oct} at all well; and then only for a limited number of simple complexes. Throughout all this time, however, the ligand-field formalism has been continuously successful, not least because of the presumption that a single d^n- (or f^n-) configurational basis appears apposite. Beyond the exploitation of symmetry, which has been tremendously effective, the characteristics of ligand-field theory as practised are (i) that we use a simple $l = 2$ basis (or $l = 3$ for lanthanides), (ii) that the perturbation of the metal ions by ligands may be represented by a one-electron operator, and (iii) that, within the d^n manifold, vector-coupling procedures may be employed because we find empirically that all many-electron states within that manifold may be characterized with a common set of parameters. Our task in the first part of this chapter has been to show how and why these qualities exist.

The expression of ligand-field theory within a d^n- (or f^n-) configurational basis has a parallel in how attention is focussed[16] in π-electron theory upon the π orbitals. In its simplest form – for example, the Hückel approach to delocalized systems in organic species – the σ framework is recognized only as providing an 'external' potential field via variable 'Coulomb integrals'. In neither case need the restriction to a limited function basis constitute any approximation, however. We have seen how, as a matter of formal manipulation, the full many-body problem may be projected on to any desired basis. What is then important, is whether such a basis, together with its co-defined effective Hamiltonian, has convenience in that it leads, under subsequent approximations, to a significant and comprehensible parametrization scheme. Empirically, we know the answer is affirmative in both π-electron and ligand-field theories.

Projection onto a limited basis, or subspace, entails a concomitant replacement of the full Hamiltonian with an effective operator, some parts of which are necessarily energy dependent: similarly, all other operators for the system must be replaced by appropriate, effective operators chosen so that their matrix elements in the projected basis coincide with the 'true' matrix elements in the basis of the exact energy eigenstates. The first stage of the treatment described above – the 'primitive parametrization' – explicitly neglects these energy-dependent terms. Within this level of approximation, however, all three characteristic attributes of ligand-field theory cited above emerge naturally. The d^n-configuration basis is chosen by definition, though it does not correspond to the principal shell of a given free ion. While the angular quality, $l = 2$, is defined explicitly, the radial parts might be thought of as made from various free-ion nd shells, corresponding to a basis definition in terms of a mean spherical potential experienced within the complex. The neglect of energy-dependent terms in the primitive scheme ensures that the ligand-field parameters will be common to all states formed by vector coupling of the basis functions.

Ligand-field theory is conventionally concerned only with splittings; that is, with the non-spherical parts of the perturbation operator (11.1). The absorption of all spherical terms within the definition of the d^n basis via the mean spherical potential has the immediate result that the parameters of interelectron repulsion and spin-orbit coupling will differ from the corresponding free-ion values. The notion of 'central-field covalency' within the nephelauxetic effect thus flows naturally, though implicitly, from the theory. However, 'symmetry-restricted covalency', relating to the idea of differential orbital expansion, does not. Within the primitive theory (and somewhat beyond, as we shall discuss) the logically self-consistent approach requires that interelectron repulsion effects be treated exactly as in the Slater–Condon–Shortley theory of free ions; this follows from the pure $l = 2$ nature of the basis (or course, the numerical values of the F_k parameters are now different). This is not to say that the theory denies the possibility of different radial properties for, say, t_{2g} and e_g orbitals. Within a molecular-orbital approach, it may be quite appropriate to model them in that way: in conventional ligand-field theory it is not. In passing, we note that an alternative definition of the basis functions referring to some 'mean *octahedral* potential' *could* have been adopted, but the complications this would have introduced, especially with respect to the ensuing nine or ten interelectron repulsion parameters,[14,15] would define a ligand-field model of unnecessary complexity and of diminished utility: in any case this approach does not correspond with the conventional formalism. The empirical finding,[14,15] that the nephelauxetic parameters $\beta_{33} < \beta_{35} < \beta_{55}$ for many complexes, in no way invalidates this assertion. Firstly, these β parameters are rather ill defined;[28,29] but most immediately, the *qualitative* trends between β values have been shown[15] merely to reflect a greater reduction in F_2 than in F_4 from the corresponding free-ion values. In short, these parameters furnish rather little evidence for differential orbital expansion. Further, even if such differentiation 'really' exists – by which we presumably mean 'within a conventional molecular-orbital scheme' – it must manifest itself within the numerical values of F_k and ligand-field parameters. While working within a d^n (f^n) basis, therefore, there is no inconsistency whatever in recognizing the non-spherical ligand field within V_{LF} of (11.1) at the same time as representing the interelectron repulsion – $U(i,j)$ of (11.1) – in an 'isotropic' or 'spherical' way, as for the usual free-ion Condon–Shortley theory.

In §11.6 we considered some aspects of the extra terms in the effective Hamiltonian which the primitive scheme neglected. A complete treatment of these terms would lead to exact solutions of the full many-body problem, but this is not practical. What has emerged, however, is that some part of these extra terms can be arranged to yield matrix elements which occur in exactly the same way as those of the primitive theory with which they may therefore be summed. Insofar as this 'renormalization' process may be carried out, the technical procedures of ligand-field theory – angular transformations, vector coupling and so on – remain valid. Our general experience of fitting spectra and magnetism, and our observation of the remarkable applicability of the Laporte rule, suggests that ligand-field theory remains useful throughout much of the transition-metal and lanthanide blocks. It should be remembered, however, that ligand-field parameters obtained by fitting spectral and magnetic properties are frequently defined to no better than 10–15% and this may be taken as a generous estimate of the typical fluctuations in the parameters caused by energy-dependent terms in the effective Hamiltonian. In principle, these terms require the parametrization to refer to each individual state in the complex rather than to the system as a whole, and ligand-field theory can only be implemented as a practical scheme when the 'true' parameters have relatively small fluctuations about the mean 'system parameters' actually used. Within the limitations set by this caveat, ligand-field theory may be viewed as a parametrized form of the quantum mechanics of the many-electron system, so far as states and properties falling within the usual compass of the ligand-field regime are concerned. There will, of course, be situations where the 'correction' terms in $\Delta \bar{H}(E)$ achieve such a significance related to the primitive scheme that ligand-field theory loses all utility: no doubt such will be the case, for example, when 'charge-transfer' states lie very close to the ground electronic state.

It is interesting to observe in the treatment described above how the concept of an orbital reduction factor in the magnetic moment operator does *not* emerge in the primitive theory. All wavefunctions at that level of approximation, both before and after diagonalization by the usual ligand-field perturbation (III.1), involve only parts with the original $l = 2$ characteristic. Stevens' orbital reduction factor k only emerges as the neglect of configuration interaction is taken into account via $\Delta \bar{H}(E)$ and a justification of the use of a single k factor relies again on the notion of some averaged, anonymous configuration interaction. This implies that simple relationships between orbital reduction factors, on the one hand, and nephelauxetic or reduction in spin-orbit coupling coefficients on the other, should not be expected. The conventional and simple molecular-orbital approaches normally used to introduce and explain orbital reduction factors have suggested,[12,13,30,31] *inter alia*, some close parallels between these three phenomena. It is a fact, however, that such relationships have seldom been established *empirically* and, indeed, are frequently refuted in practice. Some workers, especially within the esr field, have preferred to work with explicit molecular-orbital bases rather than to introduce Stevens' orbital reduction factor – usually in the interests of 'reality'. Such success as has been achieved by this stratagem relies completely upon a belief that a basic molecular-orbital theory offers an adequate approach to the problem and, explicitly, a better one than ligand-field theory. Once more, the philosophy appears to stem from an unspoken and naive belief that 'reality' means molecular orbitals. Molecular orbitals are a product of molecular-orbital theory and not of molecules: the relationship between a single-determinant, molecular-orbital theory and the full, many-body Schrödinger equation may, on occasion, be slight and, in any case, is less transparent or even delineated than that between ligand-field theory and the many-body problem.

The remaining one-electron terms in the effective Hamiltonian, the 'ligand-field potential', comprise the non-spherical parts of

$$V_{LF} = -\sum_Q V_Q + \tilde{J}_L - \tilde{K}_L, \tag{11.90}$$

in which we note, by comparison with the expression in (11.61) written within the primitive scheme, that some modifications (denoted by the tildes) have been made to allow for the renormalization described in §11.6. In the primitive scheme, we were able to present an *explicit* reduction of the matrix elements of the full Hamiltonian H to the ligand-field form by confining attention to a particular type of trial wavefunction set – $\{\phi_k^{Nd}\}$. The conclusion from §11.6 is that when we include, as best we can, the effects of $\Delta \bar{H}(E)$, we may approximately retain the same structure but with modified operators. Physically, these modified operators occur because the matrix elements of H involving the states $\{\Delta_i\}$ can be represented approximately in the effective operator $\mathscr{H}$ as screened Coulomb potentials between the charged particles, so that we must make the replacement,

$$|\mathbf{r}_i - \mathbf{r}_j|^{-1} \rightarrow U(|\mathbf{r}_i - \mathbf{r}_j|), \tag{11.91}$$

where U is something like (11.75), throughout the explicit equations given in the primitive scheme. This has the immediate consequence that the Slater–Condon parameters for the metal ion in the complex differ from the free-ion values because (i) the d-orbital basis is different in the two cases, and (ii) the screening between the d electrons is altered by the presence of the ligands, partly because the metal d orbitals hybridize with the ligand orbitals and partly because the metal s electrons become strongly involved in metal–ligand bonding. Effects (i) and (ii) are closely related, of course, in that the processes described by (ii) contribute to the mean potential which is used to construct the d-orbital basis $\{\psi_i\}$ as well as to the non-spherical parts of the potentials. With regard to V_{LF} in (11.90), in which V_Q is the electron–nuclear Coulomb potential at ligand Q, $\tilde{J}$ and $\tilde{K}$ are modified Coulomb and exchange operators acting on spin-orbitals $\chi(\mathbf{x})$ as follows:

$$\tilde{J}_L\chi(\mathbf{x}) = \int d\mathbf{x}'U(|\mathbf{r} - \mathbf{r}'|)\rho^L(\mathbf{x}',\mathbf{x}')\chi(\mathbf{x}), \qquad (11.92)$$

$$\tilde{K}_L\chi(\mathbf{x}) = \int d\mathbf{x}'U(|\mathbf{r} - \mathbf{r}'|)\rho^L(\mathbf{x},\mathbf{x}')\chi(\mathbf{x}'). \qquad (11.93)$$

We shall not specify further the screened Coulomb interaction $U(r_{12})$ which is taken to be some average of the true configuration-dependent operator, as discussed in §11.6: but we note its use, along with V_{LF} of (11.90), in the ligand-field Hamiltonian $\mathscr{H}'_{LF}$ of (11.1) and (III.1). With this background of the nature of the ligand-field potential (11.90) within the many-electron Hamiltonian theory, we aim now to explore the *structure* of V_{LF} within a local metal–ligand frame by developing a one-electron theory of the ligand-field potential: at the same time, we reconstruct and interpret the angular overlap model within ligand-field theory.

11.9 Ligand-field orbitals

As originally introduced,[33,5] the AOM was intended to provide a means of distinguishing the differing contributions of separate ligands within a complex to the splittings observed in electronic spectra; and so its original *purpose*, therefore, lay totally within ligand-field theory. The *manner* of its introduction, however, was as a derivative of the Wolfsberg–Helmholz method[34] which was currently in favour in computational inorganic chemistry. This has led by degrees to a view in which the AOM appears as a description of the angular dependence of Hückel molecular-orbital energies in which the ensuing Walsh-type diagrams are used to rationalize the thermodynamics and structure of both transition-metal and main-group compounds. This form of the AOM will be familiar from various standard teaching texts.[8−10] Thus, confusion has attached to the AOM from the beginning, the name AOM being used to entitle two very different models in chemistry. It is crucial to distinguish these two models and to set out a clear description of the AOM in ligand-field theory, at least, if only to prevent the doubts – and for some, odium – which attaches to one viewpoint, obscuring the real achievements of the other.

Being originally formulated as a derivative of the Wolfsberg–Helmholz model, it seemed from the outset that the AOM carried with it all the weaknesses of that model. The Wolfsberg–Helmholz scheme had fallen from favour, however, partly because of its common failure at that time to reproduce various experimental data but especially because the assumptions and quantities embodied in the model had not been clearly stated.[35] In the preface of his recent book,[36] Ballhausen writes: 'Unfortunately, the temptation is to elaborate an approximate theory and to introduce an increasing number of loosely defined "effects" in order to "explain" the movement of the parameters... there is little reason to expect that deeper insight can be gained in this way.' One can only agree, for the end-result of such ill-defined practice can scarcely be more than a parametrization which achieves only the uninteresting goal of reproducing the experimental data, yet is incapable of extension or further analysis. If chemical insight into ligand-field properties is to be gained, it is vital to base the treatment of the experimental data on the *best* theory that can be assembled: in particular, approximation schemes which were originally motivated essentially by the practical needs of numerical computation must be recognized as such and discarded if inappropriate to the physical situation in hand. Thus the Hartree–Fock approximation, let alone cruder schemes derived from it, may not be appropriate in discussions of d-electron systems because of the important effects that arise from electron correlation. This does *not* mean that one is obliged to resort to the most accurate *ab initio* computational schemes of quantum chemistry, for *qualitative* arguments and parametric models may certainly succeed in predicting the trends which would emerge from such calculations *provided* the approach is made with due intellectual rigour. A sacrifice of numerical rigour in the interest of practicality and economy is, by contrast, totally defensible. Too often, however, the adjective 'qualitative' is introduced to disguise or excuse what is really only an inadequate basis for discussing chemistry. Commonly, 'orbitals' are referred to without the blessing of prior definition or even of a relationship with the system under study. Such imprecise discussions, lacking any assessment of their significance and approximations, make for confusion rather than insight and are poor theory rather than qualitative theory.

In aiming to avoid such objections, we now study a careful argument[3] which justifies the chemical interpretation of the results which are presented in Part IV. We have shown above how ligand-field theory, considered as a configuration-interaction method, can be embedded in the general body of quantum chemistry. In order to make an interpretation of this theory based on chemical bonding ideas, however, we must recognize that bonding notions are based on *one-electron* theory. Our task, therefore, is to forge a link between the one- electron model describing the orbital interactions of the metal ion and its ligands, and the many-electron, ligand-field theory that is necessary for the calculation of the molecular electronic states to which the observables can be properly related. We make that connection at the 'output' end, so to speak; that is, through an analysis of the ligand-field *parameters* which are, by definition, the matrix elements of the ligand-field potential in a basis of metal d (f) orbitals. Our strategy is to construct a one-electron theory leading to a definition of these matrix elements which is determined by their use in ligand-field analyses.

We begin with some remarks about the conventional one-electron theory of electronic structure. This can be based on a Hamiltonian operator H of the form

$$H = T + U, \tag{11.94}$$

where T and U represent kinetic and potential energy terms:

$$T = -\frac{\hbar^2}{2m}\nabla^2 \equiv -\frac{\hbar^2}{2m}\left[\frac{\partial^2}{\partial x^2} + \frac{\partial^2}{\partial y^2} + \frac{\partial^2}{\partial z^2}\right], \tag{11.95}$$

$$U = U_{\text{Coulomb}} + U_{\text{exchange/correlation}}. \tag{11.96}$$

Some well-known approximations can take the form in (11.96) for potential energy: for example,

Hartree: $\qquad U \approx U_{\text{Coulomb}}.$ $\hfill (11.97a)$

Hartree–Fock: $\qquad U \approx U_{\text{Coulomb}} + U_{\text{exchange}}^{HF}.$ $\hfill (11.97b)$

$X\alpha$: $\qquad U \approx U_{\text{Coulomb}} + U_{\text{exchange correlation}}^{X\alpha}.$ $\hfill (11.97c)$

Naturally, eigenfunctions of these, and other one-electron Hamiltonians differ from one approximation to another. Further, these orbitals, and the one-electron Hamiltonians that define them, may be chosen as convenient. Subsequent connection between one-electron and many-electron theories must be correctly made, of course, but we may nevertheless regard orbitals and their defining Hamiltonians as artefacts within the total treatment of many-electron molecular *states*. The nature of our freedom here is not just that of constructing conventional LCAO molecular orbitals from atomic orbital sets of varying size, for example, or of using Gaussian functions rather than Slater-type orbitals: rather, as implied by the examples of (11.97), it includes the option of incorporating varying amounts of electron correlation within the chosen basis. Provided we *define* a one-electron Hamiltonian and proceed consistently therefrom, all will be well. Many misunderstandings of the AOM, however, derive either from the absence of such definitions or from comparisons between theoretical segments made with respect to differently defined bases.

Although the choice of orbital basis is free in the sense above, there is one respect in which a 'best' set of orbitals may be defined. It stems from the centrally important theorem of so-called density functional theory,[32,37] namely that there exists a well-defined correspondence between the one-electron orbitals and the many-electron *ground* state because the potential energy U has a unique functional dependence on the charge density in this state; $U = U[\rho]$. This result means that a *best* choice *exists* for U in (11.94) yielding eigenfunctions ϕ (orbitals) which, without further many-electron correlation terms, lead to the correct ground-state energy and to the correct ground-state electron distribution, obtained simply by summation of $\phi^*\phi$ over populated orbitals, which in turn self-consistently defines the form of U in the one-electron Hamiltonian. The 'best' orbitals, whose existence this theorem proves (but without providing a path by which they may be determined), incorporate in this sense all electron-correlation effects. Emphasizing: the point of the density functional theorem is that it proves that orbitals *can* be constructed, in principle at least – but that is sufficient for our examination of structure – from which the energy and electron distribution of the ground state of a system may be computed exactly without any further two-electron terms being considered explicitly. Note that, in common with Hartree–Fock theory or any other model, there is no reason to suppose that these same orbitals can simultaneously describe *excited*-state properties with the same accuracy. Nevertheless, it is not vain to hope that this kind of approximation should be adequate for the sort of low-lying excited states which characterize the ligand-field regime outlined at the beginning of §11.2. Further, 'adequate' should be viewed against the renormalized theory of §11.6 in recognition that we cannot contemplate using a different ligand-field potential for every distinct many-electron configuration and that some averaging over configurations is implicit in the ligand-field parametrization. Because of this averaging in the ligand-field formalism itself, there cannot be a precise mapping of the results of density functional theory of the ground state onto experimental ligand-field parameters: however, the *structure* of these parameters should be adequately revealed by such an approach.

We are not so much attempting to *derive* ligand-field theory from a defined starting point in quantum chemistry as to map one-electron theory onto well-known and empirically successful ligand-field procedures. So let us recall briefly what these procedures are. For transition-metal complexes, for example, we work with a basis of d orbitals whose radial character is never empirically defined. Ligand-field theory is applied to the interpretation of d–d spectra and other related properties of complex ions with some d^n ($1 \le n \le 9$) configuration. It is part and parcel of a ligand-field analysis, therefore, to consider so-called 'interelectron repulsion terms' *simultaneously* with, but *separately* from the ligand-field potential and spin-orbit coupling: this is expressed explicitly in the ligand-field Hamiltonian $\mathscr{H}'_{LF}$ of (III.1). Now insight into the AOM parametrization (or that of any other model of the ligand-field potential) requires the elaboration of a model in which the orbitals, $\{d_i\}$ say, and the potential V_{LF}, are well-defined quantities. In ligand-field theory the effects on the d electrons of interelectron repulsions are dealt with fully using the Condon–Shortley parametrization in terms of the quantities F_k, $k = 0, 2, 4$. Consequently, the ligand-field potential V_{LF} contains *no* d–d correlations of this kind. However, the one-electron Hamiltonian H in (11.94) does, of course, include this effect in a self-consistently averaged way through the dependence of the potential U on the d electrons' contribution to the electron density. Hence, if we are to make a *valid* connection between ligand-field theory and one-electron theory, it is clear that we must omit this part of the potential from the one-electron equations we use, by excluding the d-electron density. The inescapable conclusion of this argument, therefore, is that the *best* molecular-orbital theory (that is, that based on the potential U determined by the density functional theorem) does not correspond directly with ligand-field theory and, *a fortiori*, nor do the various approximation schemes like the Hartree–Fock or 'extended Hückel' models which have been popular in discussions of ligand-field theory. This implies, further, that such approximations cannot in principle reproduce the ligand-field parameters obtained from the experimental observables. So arguments which invoke approximate molecular-orbital schemes – like 'extended Hückel' theory – to rationalize trends in AOM parameters, even in a 'qualitative' sense, are ill founded: such success as may have been claimed doubtless arises from the fact that the 'theoretical' AOM parameters are never actually calculated numerically using the chosen molecular-orbital model.

Consider, for example, the use of a relationship that emerged from the original, historical introduction of the AOM. As a derivative of the Wolfsberg–Helmholz model, the AOM defined parameters $\{e_\lambda\}$ as proportional to the squares of the appropriate overlap integrals,[15]

$$e_\lambda = k(S_{ML}^\lambda{}^*)^2; \quad \lambda = \sigma, \pi, \delta, \tag{11.98}$$

where k is independent of λ. We know from chapter 9, for example, that the AOM parametrization scheme, as actually used in reproducing experimental ligand-field data, does not employ this, or analogous, relationships. However, it has been popular to use (11.98) in order to fix a relationship between the magnitudes of e_σ and e_π parameters in a given system when such a tactic appeared necessary in view of the paucity of experimental data (see §9.8). Empirically, we note that there are many instances where e_π/e_σ ratios vary widely from corresponding values calculated from (11.98) and, as we shall see in Part IV, the variations make sense when taken with broader concepts of chemical bonding. Theoretically, we remark that since the AOM e parameters refer to bonding in the complex as it is, rather than in some putative condition, we have no way of knowing what the appropriate overlap integrals would be. Even more graphically, consider the following paradox. The criterion (11.98) implies that e_δ values should be negligible and, indeed, so does 'common sense'. However, since the formula rightly belongs in molecular-orbital theory, it is pertinent to ask what ratios for e_δ/e_σ and e_δ/e_π would emerge from a detailed computational study within this framework. Recently, state energies in idealized complexes of the type $[\mathrm{MnF}_{6-i}\mathrm{Cl}_i]^{k-}$, ($i = 1\text{--}6$, $k = 2, 4$), and distorted octahedral $[\mathrm{MnF}_6]^{k-}$, ($k = 2, 4$), have been calculated within an SCF molecular-orbital scheme and the ensuing theoretical transition energies were subsequently parametrized within the AOM. The result of this study[38] was that e_δ values were found to be not at all negligible and, indeed, were fitted with values close to those for e_π. This might cause concern for two reasons: (i) that if δ interactions in the AOM ligand-field parametrization cannot be neglected, as is current practice, already heavily parametrized analyses will be rendered insoluble, and (ii) the numerical conclusion contradicts the simple, 'qualitative' expectations reached beforehand. If this situation were not rationalized satisfactorily, the AOM approach would lie open to the charge of being no more than a clever fitting procedure but one lacking any real chemical significance. The SCF–MO study is irrelevant, however. It is precisely the potential associated with the d-electron density itself that is responsible for the conclusion that e_δ parameters are important. But, again, in ligand-field theory, the one-electron prameters, $V_{ij} \equiv \langle d_i | V_{\mathrm{LF}} | d_j \rangle$, *explicitly exclude* any two-electron d–d correlation. Thus, conventional molecular-orbital theory does not give a qualitative account of AOM parameters in ligand-field analysis – it gives a wrong account. This does not undermine the validity of the AOM but simply means that we cannot look to conventional MO theory based on (11.97) for the basis of the AOM. Of course, as a quite separate matter, such MO theories can be used as bases for equivalent many-electron treatments of metal complexes.

Having identified the distinction between ligand-field and molecular-orbital procedures, let us pause to restate what is being attempted here. Our aim is two-fold – to justify the formalism of the AOM within ligand-field theory, and to provide a framework for the chemical interpretation of the model parameters: it is not to establish means for the *ab initio* calculation of molecular eigenvalues, states and properties. We are concerned with analysis rather than synthesis. Suppose, then, that we are appraised of a given system, perhaps in the form of the one-electron Hamiltonian H yielding 'best' orbitals as described earlier with reference to density functional theory. We begin our mapping procedure by dividing the system in a way which reflects how ligand-field theory focusses attention on metal d electrons (a similar argument follows throughout for the f electrons of lanthanides). We remove the d-electron density ρ_d from the total electron density ρ, which defines H, prior to the calculation of the potential energy, which we now call V: thus if

$$\rho' = \rho - \rho_d, \tag{11.99}$$

then

$$V = U[\rho'] \tag{11.100}$$

and we construct a modified one-electron Hamiltonian

$$\mathcal{H} = T + V, \tag{11.101}$$

instead of H in (11.94). Note that V and hence $\mathcal{H}$ still contain the potential responsible for the formation of the chemical bonds in the molecule. We shall *define* the eigenfunctions of this Hamiltonian, that is, solutions of the Schrödinger equation

$$(\mathcal{H} - E_n)\psi_n = 0, \tag{11.102}$$

as *ligand-field orbitals* (LFO). These one-electron wavefunctions are, of course, distinct from conventional molecular orbitals simply because MOs are defined by a Hamiltonian H like (11.94) which includes two-electron or correlation **d–d** terms, while the LFOs are defined by $\mathcal{H}$, which does not.

However, analogously to the partitioning of LCAO molecular orbitals into atomic or subgroup orbitals, we can express the molecular ligand-field orbitals as linear combinations of corresponding fragment basis functions $\{\phi_i\}$:

$$\psi_n = \sum_i c_{ni}\phi_i. \tag{11.103}$$

There are important differences, however. We shall divide the *basis* $\{\phi_i\}$ into two subsets – metal d orbitals and *all* others. The d functions are to take the role of the valence d orbitals appearing eventually within explicit ligand-field matrix elements while the second set includes *metal s* and *p* valence orbitals as well as all the ligand orbitals. Secondly, notice that metal d functions must appear in the basis (11.103) because the set $\{\phi_i\}$ is complete, even though the Hamiltonian $\mathscr{H}$ was constructed with d-electron *density* explicitly removed. Altogether, it is this construction which leads to the important characteristic of ligand-field theory in which the orbitals appearing in ligand-field matrix elements are not determined in any way by d–d electron correlation, such effects being dealt with separately and explicitly through the first term in (III.1). The reader will note the similarity between the present construction and division of the $\{\phi_i\}$, and that used in §11.4: indeed, the similarity grows as we complete the definition of the basis metal d orbitals we shall use.

In many-electron atomic theory we are able to make a natural definition of a mean spherical potential experienced by an electron in a transition-metal ion and to use that potential in a Schrödinger equation to define atomic $s, p, d, \ldots$ orbitals. An analogous definition is made in one-electron theory by taking the V of (11.100) and using that component $\langle V \rangle$ which is spherically symmetric with respect to the metal nucleus as the potential defining the metal orbitals:

$$\mathscr{H}^{(0)}\phi_i \equiv (T + \langle V \rangle)\phi_i = \varepsilon_i\phi_i, \tag{11.104a}$$

$$\phi_i(\mathbf{r}) = R_{nl}(r)Y^l_m(\theta,\varphi). \tag{11.104b}$$

The basis orbitals so defined are determined in part by the electronic details of the *molecule* in question and, without pressing the point too far, their construction may be considered as related to the definition[39] of (linear) muffin-tin orbitals in $X\alpha$-type calculations. The angular properties of the $\{\phi_i\}$ for transition-metal complexes are therefore exactly like those of atomic d functions but their radial forms, being partly determined by the metal-ion environment, differ and so the $\{\phi_i\}$ are *not* simply atomic $3d$ orbitals, for example. This construction gives a meaning to the notion of a 'd orbital' for the metal ion *in the complex* and, in a sense we discuss later, may be considered a 'best' choice: for the moment we note that the exact radial form of the d orbitals we use is not explicitly referred to within the formalism we discuss shortly, Finally, we note the relationship between $\mathscr{H}$ of (11.101) and $\mathscr{H}^{(0)}$ of (11.104a) is simply

$$\mathscr{H} = \mathscr{H}^{(0)} + \mathscr{H}^{(1)}, \tag{11.105}$$

where $\mathcal{H}^{(1)}$, the non-spherical part of $\mathcal{H}$, is defined by the non-spherical part of the potential energy:

$$\mathcal{H}^{(1)} = V - \langle V \rangle. \tag{11.106}$$

11.10 AOM parameters

Ligand-field parameters for transition-metal complexes are written as matrix elements of the *basis d* orbitals – not of LCAO molecular orbitals – under an *effective* operator, V_{LF}:

$$V_{ij} = \langle d_i | V_{LF} | d_j \rangle. \tag{11.107}$$

AOM parameters are similarly defined but refer to individual metal–ligand interactions and modes. Let us connect these definitions with the one-electron theory of the preceding section, first by consideration only of a simple complex involving a single ligand, M—L: the complexities introduced with increasing coordination number are considered in the following sections. The idea of bonding modes is achieved directly by reference to coincident coordinate axes on metal and ligand so that the interactions may be classified as usual by the local M—L pseudo-symmetry. This is conveniently illustrated by the diagrams in figure 11.1. These diagrams must **NOT** be confused with the rather similar ones used in the molecular-orbital theory of transition-metal complexes.[10,40] Taken with these accompanying remarks, these diagrams encapsulate the essential character of ligand-field theory and of the AOM version we champion.

The diagram depicts the formation of ligand-field orbitals (LFOs) as in (11.103) by combination of symmetry-matched basis orbitals. Those on the left of each diagram are the metal basis *d* orbitals of (11.104b), while those on the right involve suitable combinations of *all* other basis orbitals. If we refer to the metal *d* orbitals and 'the rest' (*r*), the LFOs are written

$$\psi = c_d \phi_d + c_r \phi_r: \tag{11.108}$$

'the rest' comprise both ligand *and* metal *s* and *p* orbitals. So the σ orbital, for example, on the right of the top diagrams involves a combination of ligand σ orbitals (AOs or MOs) with metal *s* and p_z functions obtained from (11.104b); the π_x orbital involves ligand π_x orbitals and metal p_x orbitals; and so on. The reader might ask whether it would be possible instead to hybridize the appropriate metal *d*, *s* and *p* orbitals to form a purely metal orbital, on the left of each diagram, which then interacts with a purely ligand orbital on the right. It is indeed possible, of course,

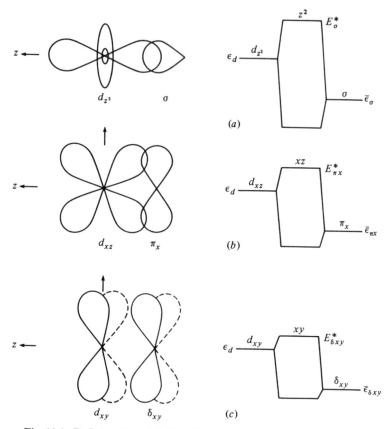

Fig. 11.1. Defining ligand-field orbitals (LFO). On the left are pure
metal orbitals whose angular parts are described *exactly* by $l = 2$: on
the right are represented *all* other orbitals (ligand and metal) which
may interact within the local M—L symmetry designation. LFO are
not the same entities used in conventional MO theory.

for that is part of the usual practice in conventional molecular-orbital
theory. However, the whole point of the construction we are describing
here is to *re*construct the procedures of ligand-field theory. Bearing in
mind the jargon of §11.2, we are 'projecting the bonding in transition-
metal complexes onto a *d*-orbital basis'. Notice straightaway that the
construction we describe, and which characterizes ligand-field procedures,
does not permit the identification of overlap integrals of the kind $\langle \phi_d | \phi_r \rangle$
between, say, 'd_{z^2}' and 'σ', with those of conventional molecular-orbital
theory. While the latter are calculable by reference to standard tabulations
of metal *d* and ligand *s*, *p* atomic or ionic orbitals, the former $\langle \phi_d | \phi_r \rangle$

are not, by virtue at *least* of our ignorance of the values taken by the mixing coefficients $\{c_d, c_r\}$ of (11.108) in any particular system. It is therefore immediately obvious that it is unreasonable to *expect* that relationships between *ligand-field* parameters shall be given simply by an expression like that in (11.98).

The second important element of the construction of figure 11.1 is that the resulting orbitals after interaction – the LFOs in the centre of the diagrams on the right – are eigenfunctions of $\mathscr{H}$ rather than H, and hence admit no possibility of two-electron d–d correlation, by construction. Therefore, the energies of the LFOs must be expressible as one-electron perturbations of the metal d orbitals; and so, by definition, the antibonding orbital energies E_λ^* in the diagrams are to be identified with eigenvalues from ligand-field d^1 calculations. There remains one further point, concerning conventions. Thus, it is customary[7,15] to identify the energy separation between the LFOs (as we now recognize them to be) and the metal d orbital – the antibonding shift – with the appropriate AOM e_λ parameters: $e_\lambda = (E_\lambda^* - \varepsilon_d)$. However, this definition is deficient in that the e_λ parameter so defined depends explicitly upon the *choice* of metal d orbital (and hence, for example, on the particular radial form defined by (11.104)), which is largely arbitrary: on the other hand, the energy E_λ^* of the LFO does not depend on the choice of basis, and it is better to make a definition of the *theoretical* AOM parameters which makes no reference to the d-orbital energy ε_d, and so identify e_λ with the energies E_λ^*. Experimentally, we can at best only determine $(n-1)$ of the n AOM parameters for the system and so extract useful values by referring all e_λ to e_δ, usually taken as zero: we return to this point in §11.14. Summarizing: *in the present case of a single-ligand complex*, we *define*

$$\text{LFO energy } E_\lambda^* \equiv \text{AOM parameter } e_\lambda. \tag{11.109}$$

By labelling the LFOs with the subscript of the parent d orbital, we may then write:

$$
\left.
\begin{aligned}
E_{z^2}^* &= \langle z^2 | \mathscr{H} | z^2 \rangle = \langle d_{z^2} | v_\sigma^l | d_{z^2} \rangle = e_\sigma, \\
E_{xz}^* &= \langle xz | \mathscr{H} | xz \rangle = \langle d_{xz} | v_{\pi x}^l | d_{xz} \rangle = e_{\pi x}, \\
E_{yz}^* &= \langle yz | \mathscr{H} | yz \rangle = \langle d_{yz} | v_{\pi y}^l | d_{yz} \rangle = e_{\pi y}. \\
E_{xy}^* &= \langle xy | \mathscr{H} | xy \rangle = \langle d_{xy} | v_{\delta xy}^l | d_{xy} \rangle = e_{\delta xy}, \\
E_{x^2-y^2}^* &= \langle x^2 - y^2 | \mathscr{H} | x^2 - y^2 \rangle = \langle d_{x^2-y^2} | v_{\delta x^2-y^2}^l | d_{x^2-y^2} \rangle \\
&= e_{\delta x^2-y^2}.
\end{aligned}
\right\} \tag{11.110}
$$

The point of the construction of $\mathscr{H}$ in (11.99)–(11.101) is now fully exposed: for by this division of the total one-electron Hamiltonian H, we have begun to provide an analysis in which ligand-field AOM parameters are identified with matrix elements of LFOs under $\mathscr{H}$, where both LFOs and $\mathscr{H}$ *are defined*. The construction of the Hamiltonian $\mathscr{H}$ (11.101), and of figure 11.1, leading to the definitions in (11.110), have been made to conform with two essential features of ligand-field theory: firstly, that the interelectron repulsion between d electrons is dealt with separately and explicitly through the Slater–Condon–Shortley parameters and, secondly, that the one-electron matrix elements we use are conventionally expressed, as on the right of (11.110), in terms of a d-orbital basis subjected to the perturbation of a ligand-field potential. As in the many-electron analysis earlier in this chapter, we have considered the metal–ligand interaction as projected onto a d-orbital basis. The adoption of LFOs as in (11.108) and the unfamiliar construction of figure 11.1 both deliberately emphasize the role that metal d orbitals play in the ligand-field *formalism*. This is *not* to say, however, that the metal–ligand interaction cannot be represented by more conventional molecular-orbital schemes. The conventional diagrams and those in figure 11.1 lead to precisely the same energy splittings for orbitals of predominantly d parentage, *provided* (i) that we recognize that the 'atomic' orbitals in the conventional diagrams cannot refer to the free metal atom and ligands since they must allow for the charge polarization which takes place on bond formation; that is, the diagrams represent the *final* situation in a self-consistent molecular-orbital calculation: and (ii), that this equivalence is only true for the d^1 case. The reader may care to review the opening paragraphs of this book: as described in §1.1, the molecular-orbital description of Δ_{oct} is a landmark in the history of ligand-field theory because of the way it rescued interpretations of level splittings from the sterility of the too-literal description offered by the old *crystal*-field theory. On the other hand, even for the d^1 case, Δ_{oct} of ligand-field theory may not be identified with the t_{2g}–e_g orbital splitting of the conventional MO diagram. Therefore, the commentary after figures 1.1 and 1.2 was made with care: those figures must represent the *final situation* in a self-consistent MO calculation and the various metal orbitals, for example, were labelled d, s, p, not so much to permit a general treatment of first-, second-, and third-row transition-metal complexes, but to indicate an ignorance of the exist radial functions of these orbitals in any actual system; and all that goes with the present description.

In systems involving configurations $d^n(1 < n < 9)$ the construction of the diagrams in figure 11.1 continues to be appropriate because of the definition of the Hamiltonian $\mathscr{H}$. However, as the more usual molecular-orbital diagrams refer to the Hamiltonian H, perhaps in the Hartree–Fock form, the equivalence of the two representations is lost in the presence of d–d correlations. For example, *no* amount of refinement of an LCAO–MO diagram for an ML_6 complex within conventional one-electron Hartree–Fock theory will yield the same value for the t_{2g}–e_g splitting as Δ_{oct} determined, via the construction of figure 11.1, from a ligand-field analysis. The difference, due to the inclusion or exclusion of d–d correlations in the Hamiltonian, can be very large indeed; for example, Watson & Freeman[41] calculated a value of $2815\,\text{cm}^{-1}$ for this energy separation in $[\text{NiF}_6]^{4-}$ ions using H and Hartree–Fock theory as compare with a value[15] of $7250\,\text{cm}^{-1}$ for the ligand-field *parameter* Δ_{oct}. Of course the inclusion of enough configuration-interaction via *ab initio* theory 'corrects' this situation, Richardson *et al.*,[42] for example, having computed a value of $7985\,\text{cm}^{-1}$ in this way. The point we emphasize repeatedly is that the ligand-field formalism explicitly separates d–d correlation effects from the rest and, in effect, parametrizes the true situation as it is and is therefore not subject to the particular approximations or simplifications of many computational models, however necessary such approximations may be in practice.

11.11 The ligand-field potential

By the construction of $\mathscr{H}$ and the LFOs we are in a position to define the local ligand-field potentials $\{v_\lambda^l\}$ of (11.110), being the equivalent operators to $\mathscr{H}$ but acting within the pure d-orbital basis rather than in that of these LFOs. Only when the $\{v_\lambda^l\}$ are *explicitly* defined may we hope to discuss what chemical information might be contained within the phenomenological AOM e parameters.

Writing the LFOs in the form (11.108) leads in the usual way to secular equations for the coefficients c_d and c_r,

$$(H_{dd} - E)c_d + (H_{dr} - ES_{dr})c_r = 0, \qquad (11.111a)$$

$$(H_{rd} - ES_{rd})c_d + (H_{rr} - E)\,c_r = 0, \qquad (11.111b)$$

where

$$H_{jk} = \langle \phi_j | \mathscr{H} | \phi_k \rangle; \quad S_{jk} = \langle \phi_j | \phi_k \rangle; \quad j,k \equiv d,r \qquad (11.111c)$$

being the matrix elements of $\mathscr{H}$ and the overlap integrals for the basis orbitals $\{\phi_i\}$ which we suppose are normalized to unity. The energies of the LFOs can be obtained from the secular determinant in the usual way.

It is more enlightening however, to solve these equations directly using a procedure[3] that will generalize to the many-orbital case we study later, and which the reader will recognize from our discussion in §11.2 of Löwdin's partitioning procedure.[18]

Provided that $E \neq H_{rr}$, we can solve (11.111b) for the coefficient c_r;

$$c_r = (H_{rd} - ES_{rd})(E - H_{rr})^{-1}c_d, \quad E \neq H_{rr} \tag{11.112}$$

and this may be substituted into (11.111a) to give

$$[(H_{dd} - E) + (H_{dr} - ES_{dr})(E - H_{rr})^{-1}(H_{rd} - ES_{rd})]c_d = 0, \tag{11.113}$$

a quadratic equation which would lead to two eigenvalues, E and E^*. Again, rather than solve (11.113) directly, let us keep the eigenvalue condition as an *implicit* equation for E_n, and rearrange it to the form

$$E = H_{dd} + (H_{dr} - ES_{dr})(E - H_{rr})^{-1}(H_{rd} - ES_{rd}). \tag{11.114}$$

Using (11.104a) and (11.104b), we have

$$\begin{aligned}
H_{dd} &= \langle \phi_d | \mathcal{H} | \phi_d \rangle = \langle \phi_d | \mathcal{H}^{(0)} + \mathcal{H}^{(1)} | \phi_d \rangle \\
&= \varepsilon_d + \langle \phi_d | \mathcal{H}^{(1)} | \phi_d \rangle = \varepsilon_d + H_{dd}^{(1)};
\end{aligned} \tag{11.115a}$$

$$\begin{aligned}
H_{rd} &= \langle \phi_r | \mathcal{H} | \phi_d \rangle = \langle \phi_r | \mathcal{H}^{(0)} + \mathcal{H}^{(1)} | \phi_d \rangle \\
&= \varepsilon_d S_{rd} + \langle \phi_r | \mathcal{H}^{(1)} | \phi_d \rangle = \varepsilon_d S_{rd} + H_{rd}^{(1)};
\end{aligned} \tag{11.115b}$$

and

$$H_{rr} = \langle \phi_r | \mathcal{H} | \phi_r \rangle = \bar{\varepsilon}_r, \tag{11.115c}$$

in which, since the $\{\phi_r\}$ have *not* been defined as eigenfunctions of $\mathcal{H}^{(1)}$ in the way that $\{\phi_d\}$ are of $\mathcal{H}^{(0)}$, the bar over $\bar{\varepsilon}_r$ denotes an expectation or mean value for the energy of the function ϕ_r. On substitution into (11.114), we get

$$(E - \varepsilon_d) = H_{dd}^{(1)} + \frac{|H_{dr}^{(1)} - (E - \varepsilon_d)S_{dr}|^2}{(E - \varepsilon_d) - (\bar{\varepsilon}_r - \varepsilon_d)}, \tag{11.116}$$

expressing energy shifts of the LFOs in terms of the matrix elements of the purely non-spherical potential $\mathcal{H}^{(1)}$.

It is usual in transition-metal complexes for the interaction between ϕ_d and ϕ_r to be rather heteropolar,[15,19] expressing the limited participation of metal d orbitals within the total bonding in the complex, so that $(\bar{\varepsilon}_r - \varepsilon_d)$ is large, and the overlap integral S_{dr} is small. It follows that there is a root of (11.116) for which $(E - \varepsilon_d)$ is small compared with $(\bar{\varepsilon}_r - \varepsilon_d)$ and that we may approximate this root by putting $E = \varepsilon_d$ on the right-hand side (this corresponding to the first step of an iterative solution):

$$E_M^* \approx (\varepsilon_d + H_{dd}^{(1)}) + \frac{|H_{dd}^{(1)}|^2}{\varepsilon_d - \bar{\varepsilon}_r}$$

$$= \left\langle \phi_d \left| \left\{ \mathscr{H} + \frac{\mathscr{H}^{(1)} |\phi_r\rangle \langle \phi_r| \mathscr{H}^{(1)}}{\varepsilon_d - \bar{\varepsilon}_r} \right\} \right| \phi_d \right\rangle. \qquad (11.117)$$

But the eigenvalues E_M^* are the AOM e_λ parameters of (11.109) and figure 11.1 and so we have obtained an expression for the matrix elements of the ligand-field potential v_λ^l in our chosen d-orbital basis. Since v_λ^l is an *effective* potential defined concomitantly with the d-orbital basis upon which it operates, we may formally write:

$$v_\lambda^l \approx \mathscr{H} + \frac{\mathscr{H}^{(1)} |\phi_\lambda\rangle \langle \phi_\lambda| \mathscr{H}^{(1)}}{\varepsilon_d - \bar{\varepsilon}_\lambda}, \qquad (11.118)$$

where λ labels the orbitals ϕ_λ built from all basis functions, other than the metal d set, of local λ symmetry. It is *not* claimed that (11.117) is an *exact* expression for the LFO eigenvalues E_M^* but, given the experimental accuracy with which such ligand-field quantities are generally determined, we should expect the expression (11.117) to be satisfactory.

Before (11.118) may be used to provide a basis for the chemical interpretation of AOM parameters, we shall have to specify the $\{\phi_\lambda\}$ orbitals in more detail than hitherto; and this is something we return to in the next section. Meanwhile, we can at least comment here on the structure of v_λ^l as displayed in (11.118). The potential consists of two distinct parts, the first simply being the one-electron Hamiltonian $\mathscr{H}$. As $\mathscr{H}$ carries the point-group symmetry of the molecule it gives rise to a 'crystal-field'-like splitting of the d orbitals: we shall refer to this term in v as the *static* contribution to the potential. Note, in passing, that the well-known, extended-Hückel, molecular-orbital version[8-10] of the AOM is deficient for transition-metal complexes, not only because it uses H of (11.94) rather than $\mathscr{H}$ of (11.101) but also because the matrix elements $H_{dd}^{(1)}$ in (11.115a) are taken to be the *same* for all d orbitals, being approximated by the d-orbital VSIPs. As we shall see later, there is no reason to believe that the static contribution for the various d orbitals are equal or negligible. The second term in (11.118) will be referred to as the *dynamic* or 'overlap' part of the ligand-field potential because it depends upon the ligand orbitals within ϕ_r and their mean energy $\bar{\varepsilon}_r = \langle \phi_r| \mathscr{H} |\phi_r\rangle$ relative to ε_d, as well as the potential $\mathscr{H}^{(1)}$. We make one further observation about (11.118) which refers back to the discussion immediately prior to (11.110): the argument used to obtain (11.117) from (11.116) is best justified with the d orbitals constructed in (11.104b) because this set of orbitals has absorbed the large, but uninteresting, spherical perturbation potential which accompanies the process of bond formation between the metal ion and the ligands. In a physically obvious way, the decomposition of $\mathscr{H}$ envisaged in (11.105a) leads to the *smallest* perturbation ($\mathscr{H}^{(1)}$) of a set of 'atomic-like' orbitals (eigenfunctions of $\mathscr{H}^{(0)}$) that can be made. It is only the effect of $\mathscr{H}^{(1)}$ that is of interest, of course. With a different choice of the radial part of the d orbitals, corresponding to some other division of $\mathscr{H}$, and hence not leading to $(E - \varepsilon_d)$ being small compared with $(\bar{\varepsilon}_r - \varepsilon_d)$, we would not generally be justified in stopping the iteration after the first stage, and therefore an expression more complicated than (11.117) would be required to express the LFO energy E_M^*. In this sense, the choice of d orbitals specified by (11.104b) is the 'best' possible because it leads directly to the most transparent form for the ligand-field potential.

It may seem that the theoretical justification for the AOM is complete and that the chemical significance of the parameters will follow once we have specified the ϕ_λ in (11.118) more fully. However, that potential was derived for the interaction of the metal ion and a *single* ligand. The usual procedure in the empirical application of the AOM is simply to sum similarly constructed potentials, one for each M—L interaction in the whole complex. The theoretical validity of this addition process is less obvious than might appear, however: we must be clear about what is being summed and that no double-counting is involved nor contributions omitted. The next section therefore concerns the other main characteristic of the AOM concerning additivity and the *local* nature of AOM parameters. Once that is secure, we may proceed to identify the chemical relevance of the AOM parameters.

11.12 The additivity principle

The localized potential model we call the AOM is based on the decomposition of the total ligand-field perturbation in a complex into parts which refer to individual metal–ligand interactions, and uses the local M—L pseudosymmetry of each ligand to further subdivide these contributions. We are thus concerned with a double summation – one over ligands, and one over bonding modes. The classification of bonding modes according to local M—L pseudosymmetry is summarized in figure 11.1 and equation (11.110), and is associated with the diagonal nature of (11.117). These ideas were embodied in the AOM as set out originally by Schäffer & Jørgensen.[5-7,31,33] The justification for additivity of ligand-field perturbations from *different* ligands, so natural and appealing an assumption, requires some care however. While it is obviously true that the total ligand-field 'potential' can be expressed as a sum of parts, the AOM additivity principle concerns the association of the ligand-field 'perturbation' with separate, real, chemical and structural moieties in a given complex. The distinction here between 'potential' and 'perturbation' must therefore be clarified: to see what is involved, let us write the total potential from a general, multiligand complex as a sum of contributions from N different sources,

$$V = \sum_{l=1}^{N} v^l. \tag{11.119}$$

Then the d-orbital matrix element (11.107) of the potential can be written as a sum of integrals over *all* space,

$$V_{ij} = \langle d_i | V | d_j \rangle = \sum_{l=1}^{N} \int d\tau \, d_i^*(\mathbf{r})(V d_j)(\mathbf{r}). \tag{11.120}$$

In the AOM we use the local pseudosymmetry associated with the source l to relate the terms in the sum to the AOM parameters $\{e_\lambda^l\}$. Now, in order to make a *valid* association between each ligand and the individual terms in the summation, it is apparent that the contributions to an integral like

$$\int d\tau \, d_\lambda^*(\mathbf{r})(v_\lambda^l d_\lambda)(\mathbf{r}) = e_\lambda^l \tag{11.121}$$

should *mostly originate from the volume of space associated with ligand l.* If this is not true, so that each term in the sum (11.120) contains contributions from parts of the molecular space *other* than that labelled by the index l, then this index serves as a label for the *sources* of the potential, but does not really separate their effects.

Consider instead another decomposition of the matrix elements V_{ij} in which the *integration* over all space is divided into a sum of contributions from a set of N disjoint regions of space (cells), each one associated with a source,

$$d\tau = \sum_{l=1}^{N} d\tau_l. \tag{11.122}$$

Then,

$$V_{ij} = \langle d_i | V | d_j \rangle = \sum_{l=1}^{N} \int d\tau_l d_i^*(\mathbf{r})(V d_j)(\mathbf{r}) \tag{11.123}$$

or

$$V_{ij} = \sum_{l}^{N} \int d\tau_l d_i^*(\mathbf{r})(v^l d_j)(\mathbf{r}) + \sum_{l \neq l'}^{N} \int d\tau_l d_i^*(\mathbf{r})(v^{l'} d_j)(\mathbf{r}) \tag{11.124}$$

after using (11.119) with (11.123). Although the left-hand sides of (11.120) and (11.124) are the same, the individual terms in the summations coincide only if the double sum with $l \neq l'$ is negligible in comparison with the sum over l in (11.124). *The additivity principle involved in the AOM is based on the decomposition* (11.122) *rather than* (11.119).

To see how this decomposition is implemented we first construct the total ligand-field potential V_{LF} for a complete multiligand complex, not by summing potentials like (11.118) which would lead us into tautology, but from the same point that we began our study of the simple M—L complex in §11.10. Recall that the LFO in (11.108) was already localized with respect to the (only) given ligand, so that for a σ interaction, for example, only one basis d orbital appears on the right-hand side of (11.108). As we do *not* wish to *presuppose* knowledge of how the total potential may be divided amongst the ligands, we must recognize that in a general multiligand complex each LFO ψ involves participation from all five d orbitals and all other basis functions, from the metal and *all* ligands:

$$\psi = \underbrace{c_1 \phi_1 + c_2 \phi_2 + \dots + c_5 \phi_5}_{\text{metal } d \text{ orbitals}} + \underbrace{c_6 \phi_6 + \dots}_{\text{metal } s, p} + \underbrace{c_7 \phi_7 + \dots}_{\text{metal } s, p} + \dots$$

$$+ \qquad\qquad + \qquad \text{etc.}$$

$$\underbrace{\text{ligand no. 1} \quad \text{ligand no. 2}}_{\text{orbital set } R} \tag{11.125}$$

Analogously to (11.108) we may still divide the basis into two parts, (d and R)

$$\psi = \boldsymbol{\phi}_d \mathbf{c}_d + \boldsymbol{\phi}_R \mathbf{c}_R, \tag{11.126}$$

where $\mathbf{c}_d$ is a column vector of coefficients of the d-basis, row vector $\boldsymbol{\phi}_d$, and $\mathbf{c}_R$ is a similar column vector of coefficients for the remaining functions we call 'the rest', $\boldsymbol{\phi}_R$: $\mathbf{c}_d$ and $\mathbf{c}_R$ are of dimensions 5 and, say, m. Secular equations, analogous to those in (11.111) but now appropriate to the interaction of the central metal d orbital with all other basis functions of the complete complex, take the form:

$$
\begin{aligned}
(H_{11} - E)c_1 + H_{12}c_2 + \ldots + H_{15}c_5 \quad &\bigg| \quad + (H_{16} - ES_{16})c_6 + \ldots + (H_{1n} - ES_{1n})c_n = 0 \\
\vdots \qquad\qquad\qquad &\bigg| \qquad\qquad \vdots \qquad\qquad\qquad \vdots \\
H_{51}c_1 + \ldots \qquad\qquad + (H_{55} - E)c_5 \quad &\bigg| \quad + (H_{56} - ES_{56})c_6 + \ldots \qquad\qquad = 0 \\
\text{---} \text{---} \text{---} \text{---} \text{---} \text{---} \text{---} &\bigg| \text{---} \text{---} \text{---} \text{---} \text{---} \text{---} \text{---} \text{---} \\
(H_{61} - ES_{61})c_1 + \ldots \qquad\qquad\quad &\bigg| \quad + (H_{66} - E)c_6 + \ldots \qquad\qquad\qquad = 0 \\
\vdots \qquad\qquad\qquad\qquad &\bigg| \qquad\qquad\qquad\qquad\qquad \vdots \\
(H_{n1} - ES_{n1})c_1 + \ldots \qquad\qquad &\bigg| \quad + H_{n6}c_6 + \ldots \qquad\qquad + (H_{nn} - E)c_n = 0
\end{aligned}
$$

$$(11.127)$$

where $n = m + 5$. Although not yet fully specified we can assume, without loss of generality, that the R set of basis functions are orthonormalized, as are the metal d functions: overlap integrals therefore appear only in the off-diagonal blocks. These secular equations are more tidily written in the form (see (11.8) and (11.9));

$$(\mathbf{H}_{dd} - E\mathbf{1}_{dd})\mathbf{c}_d + (\mathbf{H}_{dR} - ES_{dR})\mathbf{c}_R = 0, \qquad (11.128a)$$

$$(\mathbf{H}_{Rd} - ES_{Rd})\mathbf{c}_d + (\mathbf{H}_{RR} - E\mathbf{1}_{RR})\mathbf{c}_R = 0, \qquad (11.128b)$$

where

$\mathbf{H}_{dd}$ is a 5×5 matrix containing the matrix elements of $\mathcal{H}$ in the d-orbital basis,

$\mathbf{H}_{RR}$ is an $m \times m$ matrix containing all other matrix elements of $\mathcal{H}$ which do not involve the d orbitals,

$\mathbf{H}_{dR}$ and $\mathbf{S}_{dR}$ are rectangular $5 \times m$ matrices containing the matrix elements of $\mathcal{H}$ and overlap integrals connecting the two parts of the basis, and

$\mathbf{1}_{dd}$ and $\mathbf{1}_{RR}$ are unit matrices of dimensions 5 and m respectively.

We now repeat the previous arguments, equations (11.112)–(11.118), to express the eigenvalues E in terms of the matrix elements of an effective ligand-field potential for the whole complex in the d-orbital basis. First, we solve for $\mathbf{c}_R$ in (11.128b),

$$\mathbf{c}_R = (E\mathbf{1}_{RR} - \mathbf{H}_{RR})^{-1}(\mathbf{H}_{Rd} - ES_{Rd})\mathbf{c}_d, \qquad (11.129)$$

which is a well-defined expression provided that E is not equal to any of the eigenvalues of the matrix $\mathbf{H}_{RR}$, so that the inverse matrix $(E\mathbf{1}_{RR} - \mathbf{H}_{RR})^{-1}$, known as the *resolvent* of $\mathbf{H}_{RR}$, exists. Then, on substitution of (11.129) into (11.128a), we obtain the analogue of (11.113):

$$[(\mathbf{H}_{dd} - E\mathbf{1}_{dd}) + (\mathbf{H}_{dR} - ES_{dR})(E\mathbf{1}_{RR} - \mathbf{H}_{RR})^{-1}(\mathbf{H}_{Rd} - ES_{Rd})]\mathbf{c}_d = 0.$$

$$(11.130)$$

As before, we solve approximately for the eigenvalues E_d which lie close in energy to the d-orbital energy ε_d and eventually produce the matrix representation for the effective ligand-field potential,

$$\mathbf{V} = \mathbf{H}_{dd} + \mathbf{H}_{dR}^{(1)}[(\varepsilon_d \mathbf{1}_{RR} - \mathbf{H}_{RR})^{-1}]\mathbf{H}_{Rd}^{(1)}. \qquad (11.131)$$

Superficially, this expression for the global ligand-field potential $\mathbf{V}$ resembles that for the local, single-ligand, potential v (11.118) very closely indeed. However, in general, $\mathbf{V}$ is a *non-diagonal* matrix and is no longer related to the AOM e_λ parameters in the immediate way that v^l is in (11.118). Nevertheless, diagonalization of $\mathbf{V}$ still yields LFO eigenvalues $\{E^*\}$ and the corresponding eigenvectors $\mathbf{c}_d$ for the complete complex just as in the 'single-ligand' case (11.117).

Let us review the significance of (11.128) and (11.131). Equations (11.128) are simply the usual secular equations, like (11.111), but 'blocked' in such a way as to separate quantities relating to the purely metal d orbitals from the rest. The eigenvalues E_d and eigenvectors $\mathbf{c}_d$ of the ensuing matrix $\mathbf{V}$ are thus *expressed* in a *pure* d-orbital basis but this *does not imply* negligible 'mixing' or interaction between the d orbitals and the ligand (or other metal) orbitals because the matrix $\mathbf{V}$ contains all such effects 'folded-in' in the second term of (11.131). We have effectively projected the problem onto a pure d-orbital basis by concomitantly defining an effective ligand-field potential V_{LF} (through its matrix representation (11.131) as in §11.2). There remains considerable misapprehension of this 'd-orbital-purity' question in the inorganic chemistry literature and we emphasize, as in §11.2, that no approximations are involved in the choice of a pure d-orbital *basis*, *per se*. In principle we could think of returning to (11.129) and solving for the complementary set of eigenvectors $\mathbf{c}_R$ in order to reconstruct the complete eigenfunctions of the Hamiltonian $\mathscr{H}$. However, the practical application of ligand-field theory does not lead to the determination of the matrix $(E\mathbf{1}_{RR} - \mathbf{H}_{RR})^{-1}$ in (11.129) so that this second step is not possible. This is the information that is unavoidably lost by using the ligand-field formalism.

Having obtained the general form of the ligand-field potential V, we now investigate its decomposition in accordance with the additivity principle. This is conveniently dealt with in two parts since, as before, V consists of a static and a dynamic part which can be discussed separately.

(i) *Static matrix elements* $-\langle d_i|\mathscr{H}|d_j\rangle$. Apart from the orbital energy ε_d, the static matrix element contains the non-spherical part of the classical electrostatic contribution (the Hartree term) as well as the non-spherical parts of terms describing the exchange-correlation potential experienced by an electron in the average field of all the other electrons (excepting the d electrons, of course) reflecting the operation of the Pauli principle and including some other quantum-mechanical corrections to the mean Hartree potential. The static perturbation due to $\mathscr{H}$ can be broken down into contributions from different parts of space,

$$\mathscr{H} = \mathscr{H}^{(0)} + \mathscr{H}^{(1)} = \mathscr{H}^{(0)} + \sum_{l=1}^{N} \mathscr{H}^{(1)l}, \qquad (11.132)$$

simply because the potential experienced by an electron is *short-range* due to the dielectric screening[43] by the remaining electrons. This is true even though part of the exchange-correlation potential is a so-called 'non-local' potential.[††]

(*ii*) *Dynamic matrix elements.* We have said little so far about the orbitals we need other than the metal d functions. We have described how correct eigenvalues can be obtained, by projection onto the metal d-orbital basis, 'folding-in' the ϕ_R within the effective potential V of (11.131). These other functions cannot be kept implicit for ever, however, for if we are to understand the structure of the dynamic potential and hence provide a theoretical framework for the chemical interpretation of the ligand-field parameters it contributes to, we must consider the complementary orbital set *explicitly*. In principle we may use any set of linearly-independent orbitals as the basis, chosen according to whatever criteria seem appropriate to the problem in hand. Here, we need a basis which is *consistent*

[††] A 'local potential' involves an operator V which acts on an orbital *multiplicatively*, that is, the values of the potential and orbital at each point, $\mathbf{r}$ in space are simply multiplied together:

$$(V\phi)(\mathbf{r}) = V(\mathbf{r})\phi(\mathbf{r})$$

Other forms of operation by V are termed *non-local* as, for example, the Hartree–Fock exchange operator K which acts as

$$(K\phi)(\mathbf{r}) = \int d\mathbf{r}' K(\mathbf{r}, \mathbf{r}')\phi(\mathbf{r}');$$

in this case the value of the left-hand side at the point r requires the values of the orbital ϕ at *all* points in space. A 'local potential' is therefore a special case of this second type of expression, obtained when the integration is reduced to evaluation at a point by a Dirac delta function,

$$K(\mathbf{r}, \mathbf{r}') = V(\mathbf{r}')\delta(\mathbf{r} - \mathbf{r}').$$

Since the potential V in (11.120), for example, may contain both local and non-local contributions, we used the notation illustrated here. This technical use of the word 'local' is to be distinguished from our discussion of the *physical* localization of the ligand-field perturbation.

with our interpretations of AOM parameters as genuinely reflecting local electronic perturbations. To see what this involves we must look at a typical matrix element of the dynamic contribution to the potential (11.131) which, written in full,[††] is

$$\langle d_i | \mathcal{H}^{(1)} \sum_{\alpha' \alpha} | \phi_{Rd} \rangle \langle \phi_{Rd} | (\varepsilon_d \mathbf{1} - \mathcal{H})^{-1} | \phi_{R\alpha'} \rangle \langle \phi_{R\alpha'} | \mathcal{H}^{(1)} | d_j \rangle, \quad (11.133)$$

where the double summation is over all the orbitals in the R set, comprising metal s and p orbitals as well as all ligand functions as shown in (11.125).

We see from the explicit form of (11.133), and by reference to the original secular equations (11.127) that, in general, the resolvent operator $(E\mathbf{1} - \mathcal{H})^{-1}$ will couple orbitals associated with different ligands, so that the additivity principle we seek to construct requires that we can describe a set of orbitals associated with the spatially disjoint cells, in a natural way, such that $(E\mathbf{1} - \mathcal{H})^{-1}$ is approximately diagonal when evaluated at $E = \varepsilon_d$. Evidently the requirement that the resolvent should not cause *significant* coupling between different ligands implies that the ligands are in some sense 'independent': we must be very clear, however, about what this means. We refer to a spatial independence *after* complex formation rather than to a chemical independence of the various free ligands coming together to form the complex: after all, the modification of the properties of one ligand by others *in the complex* is the very stuff of chemistry. Thus it would *not* be correct, for example, to suppose that we can take the orbitals in the R set as molecular orbitals for the metal atom and each of its ligands *separately*: for such a basis the modification of ligand properties on complex formation *must* be described by off-diagonal matrix elements of $\mathcal{H}$, and hence of $(E\mathbf{1} - \mathcal{H})^{-1}$, connecting different ligands since, by definition, such effects were not incorporated in the basis $\{\phi_R\}$. On the contrary, the orbitals we seek must be related to the full Hamiltonian $\mathcal{H}$ for the *whole* complex – which contains information about charge redistribution accompanying bond formation as summarized by the electroneutrality principle – and yet be associated with the individual ligands in such a way that a cellular decomposition of the matrix elements (11.133) is valid. In other words, we require *a set of localized orbitals for the ligand as it finds itself in the transition-metal complex.* The

[††] The matrix V in (11.133) comprises matrix elements of the d basis: therefore the dummy indices ($R\alpha$ and $R\alpha'$) must *conform* in the matrix multiplication of the dynamic part. Note also that the matrix $[(\varepsilon_d \mathbf{1}_{RR} - \mathbf{H}_{RR})^{-1}]$ comprises matrix elements $\langle \phi_{R\alpha} | O | \phi_{R\alpha'} \rangle$, where $O \equiv [(\varepsilon_d \mathbf{1}_{RR} - \mathcal{H})^{-1}]$. Altogether, therefore a dynamic matrix element takes the form (11.133).

construction of such orbitals is the principal achievement of so-called 'local-orbital' or pseudopotential theories of the electronic structure of solids.[37]

Such a basis will lead to a nearly-diagonal resolvent provided that ε_d is not close to an eigenvalue ε_R of $\mathbf{H}_{RR}$ because this is the condition which justifies the neglect of 'small' (compared with $|\bar{\varepsilon}_R - \varepsilon_d|$) ligand–ligand interaction matrix elements. One of the summations over cells in (11.133) is then eliminated and so, using (11.132), we may cast the matrix elements of the dynamic potential in the form of a superposition of cellular contributions;

$$\sum_{l,\alpha=1}^{N} \frac{\langle d_i | \mathscr{H}^{(1)l} | \phi_{R\alpha} \rangle \langle \phi_{R\alpha} | \mathscr{H}^{(1)l} | d_j \rangle}{(\varepsilon_d - \bar{\varepsilon}_{R\alpha})}, \tag{11.134}$$

where $\bar{\varepsilon}_R = \langle \phi_{R\alpha} | \mathscr{H} | \phi_{R\alpha} \rangle$ is the mean energy of the orbital $\phi_{R\alpha}$ under the Hamiltonian $\mathscr{H}$. So we have discovered the way in which the AOM additivity principle is implemented and finds justification theoretically but, by the construction of the bond orbitals described above, *the angular overlap scheme achieves consistency only when AOM parameters refer to the various M—L groups as they occur in the complex*: the parameters reflect the M—L interactions rather than putative acceptor and donor functions of separated metal and ligand. The, seemingly obvious, idea that in ligand-field analysis, the 'd-orbital energies' provide probes of the total metal–ligand bonding thus appears *explicitly* within the theoretical framework of the AOM.

(iii) The AOM parametrization scheme. Now we saw in §9.3 that the key equation of the AOM as a localized-potential model,

$$V_{ij} = \sum_{l}^{N} \sum_{\lambda}^{n} R_{i\lambda}^{l\dagger} R_{\lambda j}^{l} e_{\lambda}^{l}, \quad \begin{cases} 1 \leq i,j,\lambda \leq n \\ 1 \leq l \leq N \end{cases} \tag{11.135}$$

followed more-or-less directly from the notion of a cellular potential: that is, that we write the total ligand-field potential V_{LF} as a superposition of local, 'non-overlapping' cellular potentials v^l,

$$V_{\mathrm{LF}} = \sum_{l=1}^{N} v^l. \tag{9.13}$$

The utility of the AOM as a basis for a parametrization scheme rests upon the observation that the matrices $\mathbf{R}^l$ of (9.15) and (11.135) are approximately determined by the molecular structure of the complex and so can be taken as known, *provided that an appropriate choice of cells l*

is made. The usual practice is to define a local coordinate system for each
ligand l in turn by taking a vector along the transition metal atom–ligand
donor-atom direction as the local polar axis z, together with x and y axes
oriented so as to best reflect the local metal–ligand pseudosymmetry:
the corresponding matrices $\mathbf{R}^l$ can then be calculated directly from the
geometry of the molecule as *rotation* matrices. This procedure is perfectly
satisfactory *provided* that the local pseudosymmetry about ligand l is
sufficient to make the potential v^l diagonal with respect to the d orbitals
$\{d^l\}$ in the local frame. While this is usually the case, we note two kinds
of circumstance when it need not be: the cases refer to a lack of diagonality
within the second summation of (11.135) on the one hand, and a problem
with the additivity principle on the other.

The first situation provides no difficulty in principle, and was discussed
in §9.9 in connection with the ligand-field effects of 'non-bonding', ligand
lone pairs of electrons, for example. All that is required in such cases of
'misdirected valence' (that is, when the local metal–ligand pseudosymmetry is too low to permit a ready and transparent classification of
bonding modes) is to forgo the requirement that the $\mathbf{e}^l$ matrix of (9.15)
be completely diagonal and to introduce a 'cross-parameter', like $e^l_{\pi\sigma}$,
into the local parametrization scheme. The procedure is formally correct
but involves some small cost with respect to the clarity of the chemical
interpretation of such quantities.

Rather more important is the situation in which the choice of cells, and
their associated rotation matrices just described, fails to satisfy the
AOM additivity principle. Recall that the ligand-field potential V_{LF} arises
from all the electron density about the metal ion with the exception of
the d electrons. Thus, even in a planar-coordinated complex there will
still be contributions to V_{LF} above and below the molecular plane. If these
contributions are lumped in with the ligands the matrices $\mathbf{R}^l$ are not
determined simply by the (planar) geometry of the molecular coordination
as described above, since the rotation matrices for the ligands will *not*
bring $\mathbf{v}^l$ into the diagonal form $\mathbf{e}^l$. Were we involved in a computational
scheme, this would not matter in the least, since any decomposition of
space into cells could be made and the matrices $\{\mathbf{R}^l\}$ corresponding to a
given set of cell potentials $\{v^l\}$ could be calculated. We emphasize again,
however, that this is not our aim; we seek to establish and justify the
parametrization scheme we call the ligand-field, angular overlap model,
in which the choice and number of cells used in the decomposition of the
total ligand-field potential V_{LF} is governed by a competition between two
factors. We obviously wish to have the minimum number of AOM

parameters $\{e_\lambda^l\}$ to be determined by fitting to experimental data using (11.135); on the other hand, there remains the requirement that each cell potential matrix $\mathbf{v}^l$ be diagonalized with a unitary matrix $\mathbf{R}^l$ determined *only* by the structural parameters of the molecule. This latter condition implies that the number of cells *may* need to be *greater* than the number of ligands, so that the sums in (11.133)–(11.135) may involve contributions from 'empty' cells; that is, cells not directly associated with ligands. For example, considering again the case of planar complexes, the correct procedure is to introduce two cells, enclosing the coordinationally void regions symmetrically disposed above and below the coordination plane, in addition to cells enclosing the N actual ligating species in that plane. For this arrangement of $(N + 2)$ cells it is reasonable to assume that the molecular geometry then determines all the matrices $\mathbf{R}^l$ in the usual way; there will, of course, be additional localized orbitals and AOM parameters associated with the 'empty cells'. The phenomenon of 'd–s' mixing, mentioned in §9.10, finds its natural rationalization in terms of this construction, and is a matter we return to in §11.15.

First, however, we discuss the chemical interpretation of the AOM parameters for the more usual circumstances defined by the equations (11.132), (11.134), (11.135) and (9.18). We condense the notation of these equations a little and write;

$$e_\lambda^l = \langle d_\lambda^l | \mathcal{H}^{(0)} + v_\lambda^l(\text{static}) + v_\lambda^l(\text{dynamic}) | d_\lambda^l \rangle$$
$$= \varepsilon_d + (e_\lambda^l)_{\text{static}} + (e_\lambda^l)_{\text{dynamic}}, \tag{11.136}$$

where

$$v^l(\text{static}) = \mathcal{H}^{(1)l} \quad (\equiv \hat{U}^l) \tag{11.137}$$

and

$$v^l(\text{dynamic}) = \sum_\alpha \frac{\mathcal{H}^{(1)l} | \phi_{R\alpha} \rangle \langle \phi_{R\alpha} | \mathcal{H}^{(1)l}}{\varepsilon_d - \bar{\varepsilon}_{R\alpha}}, \tag{11.138}$$

and we are now using the d orbitals $\{d_\lambda^l\}$ qualitized in the local coordinate frame of cell l. In the next two sections we analyse the two contributions $(e_\lambda^l)_{\text{static}}$ and $(e_\lambda^l)_{\text{dynamic}}$ separately. Before doing so, it is appropriate to make a comment about the relative magnitudes of the symmetry-allowed matrix element of $\mathcal{H}^{(1)}$ between d orbitals, involved in the static contribution, as compared with the corresponding matrix elements between a d orbital and a bond orbital, occurring in the dynamic part. Qualitatively, one expects the type of matrix element involving valence ligand orbitals to be larger than those involving only d orbitals. This is because $|r\phi_d(r)|^2$ is strongly localized about the metal nucleus, a region where the non-spherical potential is small: by contrast, the d-orbital–valence-ligand-orbital, matrix elements involve contributions from a much larger region of space which includes the bond overlap region where $\mathscr{H}^{(1)l}$ is expected to have its largest values. Hence, other things being equal, we expect the dynamic part of the potential to be more important than the static part.

11.13 The static contribution to AOM parameters

The static contribution to AOM e parameters refers to the second term in (11.136); namely, the integral $\langle d_i | \mathscr{H}^{(1)l} | d_j \rangle$ which will be written $\langle d_i | \hat{U}^l | d_j \rangle$ throughout the present section, following (11.137) in order to simplify notation. It is apparent that this 'static' integral has a form closely resembling the classical *crystal*-field matrix element. While it was agreed in the preceding paragraph that we might reasonably expect 'static' contributions generally to be of less importance than 'dynamic' ones, there are, nevertheless, two reasons why it is essential to analyse them in some detail. The first is that we need to derive some ground rules concerning the *relative* magnitudes for the static contributions to σ, π and δ AOM parameters so that we may know whether conclusions about the relationship between the dynamic contributions and generalized notions of chemical bonding may be upset significantly by the properties of the static contributions. The second, which is really a special case of the first enquiry, concerns the significance of the empirically determined zero, or near-zero, value for e_π in any given metal–ligand interaction. Thus, viewing the static contribution as a *quasi* crystal-field phenomenon leads one to suppose that a crystal-field potential which destabilizes a metal σ orbital ($d_{z^2}^l$) must also destabilize metal π (d_{xz}^l, d_{yz}^l) and δ ($d_{x^2y^2}^l, d_{xy}^l$) orbitals, though by decreasing amounts. At this stage, therefore, it seems possible that a significant static contribution to the e_σ value in, say, a metal—NH_3 interaction might inevitably be accompanied by significant (though smaller) static contributions to e_π (and e_δ): if so, the empirical association of vanishing e_π values with a lack of metal–ligand π bonding would be hard to justify.

Now the static potential in cell l, $\hat{U}^l$, may generally consist of a local and a non-local part,

$$\langle \mathbf{r} | \hat{U}^l | \mathbf{r}' \rangle = U_l(\mathbf{r})\delta(\mathbf{r} - \mathbf{r}') + \hat{U}(\mathbf{r}, \mathbf{r}'): \qquad (11.139)$$

however, since any non-local effects are short range, they can be taken with sufficient accuracy to vanish outside all l, recalling that the 'true' $\hat{U}_l$ is much more screened than the Hartree–Fock exchange operator (see equations (11.92) and (11.93)). We may now use the angular momentum techniques of chapter 8 to analyse the matrix elements of $\hat{U}^l$ in a d-orbital basis. The local part can be expanded in spherical harmonics about the metal atom as origin, while a similar expansion can be given for the non-local part using bipolar harmonics. If the unitary matrix $\mathbf{R}^l$ of (9.15) *et seq.* is taken as a Wigner rotation matrix, as in §8.2, the angular interactions can then be carried out trivially, and in both cases the diagonal matrix elements can be expanded as a sum of terms consisting of matrix elements involving the radial parts of the d orbitals and the $(J = 0, 2, 4; M = 0)$-components of the potential. Hence the whole potential $\hat{U}^l$ can be consistently parametrized in terms of these quantities.

The cellular decomposition of the *local* part of the potential is trivially exact and allows a complete correspondence between the geometric arrangement of the ligands in space and their additive contributions to V_{LF}. The decomposition of the non-local part, however, is not so obvious and is worth considering in some detail. It is convenient here to recall (11.93) for the screened operator, $\hat{K}_L$, a typical matrix element of which can be put in the form

$$\langle d_i|\hat{K}_L|d_j\rangle = \int d\mathbf{r}\int d\mathbf{r}'\, d_i^*(\mathbf{r})\rho^L(\mathbf{r},\mathbf{r}')U(|\mathbf{r}-\mathbf{r}'|)d_j(\mathbf{r}')$$

$$= \sum_{l=1}^{N}\int d\mathbf{r}\, d_i^*(\mathbf{r})v_{jL}^l(\mathbf{r}), \qquad (11.140)$$

where we have defined

$$v_{jL}^l = \int_{\mathrm{cell}\,l} d\mathbf{r}'\,\rho^L(\mathbf{r},\mathbf{r}')U(|\mathbf{r}-\mathbf{r}'|)d_j(\mathbf{r}') \qquad (11.141)$$

and the integration is restricted to cell l. However, the potential $v_{jL}^l(\mathbf{r})$ is defined over *all* space and, unless it is associated mainly with the volume of cell l, the identification between ligands and potential components is purely formal. That the partitioning *is* actually real, however, follows from the fact that (i) $\rho^L(\mathbf{r},\mathbf{r}')$ is a smoothly varying function of its arguments and (ii) the screened interaction $U(|\mathbf{r}-\mathbf{r}'|)$ has a *much shorter* range than a pure Coulomb potential so that $U(|\mathbf{r}-\mathbf{r}'|)\to 0$ quickly as $|\mathbf{r}-\mathbf{r}'|$ becomes large. Hence, unless $\mathbf{r}$ is approximately contained within cell l, v_{jL}^l must be extremely small: once again, therefore, we can take v_{jL}^l to be the contribution from cell l with only a small error occurring because v_{jL}^l is not strictly zero outside cell l.

Since the non-local potential $\hat{U}_l(\mathbf{r},\mathbf{r}')$ depends on two directions $(\hat{r},\hat{r}')$, it is convenient to make an expansion in bipolar harmonics, with the metal atom as origin. These harmonics were given in (8.125) as vector-coupled combinations of the spherical harmonics, constructed to give a resultant J, M as

$$B_M^J(jj') = \sum_{m,m'} \langle JM|jj'mm'\rangle C_m^j(\hat{r})C_{m'}^{j'}(\hat{r}') \qquad (11.142)$$

and normalized as in (8.126). Therefore, we can write

$$\tilde{U}_l(\mathbf{r},\mathbf{r}') = \frac{1}{16\pi^2}\sum_{\substack{j,j'\\J,M}}(2j+1)(2j'+1)U_l(r,r')_{JM}^{jj'}B_M^J(jj'), \qquad (11.143)$$

in which the non-angular multipliers $U_l(r,r')_{JM}^{jj'}$ are formally defined via the orthogonality relation (8.126) by

$$U_l(r,r')_{JM}^{jj'} = \int d\hat{r}\int d\hat{r}'\, U_l(\mathbf{r},\mathbf{r}')B_M^J(jj'). \qquad (11.144)$$

If we now express the $|d_k^l\rangle$ of (9.17) in polar form with respect to the metal–ligand l vector as Z axis.

$$|d_k^l\rangle = \phi_d(r)C_k^{(2)}(\theta,\varphi), \qquad (11.145)$$

where the modified spherical harmonics $C_k^{(2)}$ are defined in (8.115) as usual, the diagonal element of the non-local potential $\hat{U}_l$ can be written

$$\tilde{I}_{kk}^l = \langle d_k^l|\hat{U}_l|d_k^l\rangle$$

$$= \sum_{\substack{j,j'\\J,M}}\frac{(2j+1)(2j'+1)}{16\pi^2}\tilde{\varepsilon}_{jj'JM}^l\,\Omega_{jj'JM}^{kk}, \qquad (11.146)$$

where we define $\tilde{\varepsilon}^l$ as a 'potential parameter' by

$$\tilde{\varepsilon}_{jj'JM}^l = \int_0^\infty dr r^2\int_0^\infty dr'r'^2\phi_d^*(r)U_l(r,r')_{JM}^{jj'}\phi_d(r') \qquad (11.147)$$

and the angular quantity Ω by

$$\Omega_{jj'JM}^{kk} = \int d\hat{r}\int d\hat{r}'C_k^{(2)*}(\hat{r})B_M^J(jj')C_k^{(2)}(\hat{r}'). \qquad (11.148)$$

As we take bipolar harmonics in (11.142) constructed from modified spherical harmonics referred to the same Z axis used to define the $|d_k^l\rangle$ in (11.145), we may make use of the orthogonality relationship (8.126) and (11.142) to reduce (11.146), noting the survival of only one term in the inner summation over (j, j'): thus

$$\tilde{I}^l_{kk} = \sum_{JM} \tilde{\varepsilon}^l_{22JM} (-1)^k \langle 22k, -k | J, M \rangle. \qquad (11.149)$$

From the properties of Clebsch–Gordan coefficients, discussed in §8.3.1, we note that the non-zero terms in this sum are simply those for which $J = 0, 2, 4$ with $M = 0$ only.

For the *local* part of the potential, $U_l(\mathbf{r})$ in (11.139), we write

$$U_l(\mathbf{r}) = \sum_{JM} U^l_{JM}(r) C^J_M(\hat{r}) \qquad (11.150)$$

and hence

$$I'^l_{kk} = \sum_{JM} \langle \phi_d | U^l_{JM} | \phi_d \rangle \langle 2k | C^J_M | 2k \rangle, \qquad (11.151)$$

and, by analogy with the non-local part, define a radial matrix element as the potential parameter

$$\varepsilon'^l_{JM} = \langle \phi_d | U^l_{JM} | \phi_d \rangle = \int_0^\infty dr\, r^2 | \phi_d(r) |^2 U^l_{JM}(r). \qquad (11.152)$$

Now the angular part of (11.151), the so-called Gaunt coefficient $\langle 2k | C^J_M | 2k \rangle$ is simply proportional to the Clebsch–Gordan coefficient $\langle 22k, -k | JM \rangle$, of course, and so the local and non-local contributions to the total static, ligand-field matrix elements (11.139) have identical structures: accordingly we can combine the local and non-local radial matrix elements and work with collective potential coefficients ε,

$$\varepsilon^l_{JM} \equiv \tilde{\varepsilon}^l_{22JM} + \varepsilon'^l_{JM}. \qquad (11.153)$$

Finally, evaluation of the Clebsch–Gordan coefficients in (11.149) via the standard tables of $3-j$ coefficients, leads to the explicit forms for $I^l_{kk} \equiv \tilde{I}^l_{kk} + I'^l_{kk}$:

$$I^l_{00} = \varepsilon^l_{00} + \tfrac{2}{7}(\varepsilon^l_{20} + \varepsilon^l_{40}) \qquad (11.154)$$

$$I^l_{\pm 1, \pm 1} = \varepsilon^l_{00} + \tfrac{1}{7}(\varepsilon^l_{20} - \tfrac{4}{3}\varepsilon^l_{40}) \qquad (11.155)$$

$$I^l_{\pm 2, \pm 2} = \varepsilon^l_{00} - \tfrac{2}{7}(\varepsilon^l_{20} - \tfrac{1}{6}\varepsilon^l_{40}). \qquad (11.156)$$

Note that all of the foregoing analysis, following (11.142), is only appropriate for static contributions with *cylindrical* symmetry about the local metal–ligand axis; that is, for so-called 'linear ligators'. Therefore, (11.149), etc. make no distinction between $\pm k$ values, so yielding only the three different values for I^l_{kk} given in (11.154)–(11.156). While a more elaborate analysis would have been required in cases where the local metal–ligand pseudosymmetry approximates C_{2v} or less – leading to the non-zero values for $I^l_{\mp k, \pm k}$ and ultimately unequal contributions to $e_{\pi x}/e_{\pi y}$ and $e_{\delta xy}/e_{\delta x^2 y^2}$ – no further physical insight would emerge. In the present exploration of the relative magnitudes of the static contributions to the various e_λ parameters, we continue, therefore, with the simpler situation representative of linear ligation.

For the moment, let us adopt the 'renormalized' definition (9.35) of the AOM parameters and so write the static contributions to e_σ and e_π as

$$S^l_\sigma \equiv I^l_{00} - I^l_{\pm 2, \pm 2} = \tfrac{4}{7}\varepsilon^l_{20} + \tfrac{5}{21}\varepsilon^l_{40}, \tag{11.157}$$

$$S^l_\pi \equiv I^l_{\pm 1, \pm 1} - I^l_{\pm 2, \pm 2} = \tfrac{3}{7}\varepsilon^l_{20} - \tfrac{5}{21}\varepsilon^l_{40}. \tag{11.158}$$

The inverse forms of these equations,

$$\varepsilon^l_{20} = S^l_\sigma + S^l_\pi \tag{11.159}$$

$$\varepsilon^l_{40} = \tfrac{3}{5}(3S^l_\sigma - 4S^l_\pi) \tag{11.160}$$

establish a *formal* equivalence between the point-charge electrostatic model and the angular overlap model. These equations, however, are much more general than the crude electrostatic forms and retain their validity, with appropriately redefined potential parameters, when the non-local part of $\hat{U}^l$ is considered. Note, too, that one can make an analysis of this kind of the *whole* potential in cell l: the same equations are obtained, although the potential parameters now take a different physical significance since they must incorporate the effects arising from the dynamic part of the potential. Hence the formal equivalence with the electrostatic model survives (though only with the point-charge model in the case of linear ligators, of course): however, from the point of view of *chemical* interpretation, the separation of the potential into its static and dynamic parts is much more rewarding.

We now attempt to establish likely relationships between the magnitudes of static contributions to the various AOM e_λ parameters. Formal relationships are available immediately from the equations (11.157) and (11.158): since ε^l_{00} is discarded, the integrals $\{I^l_{kk}\}$, or equivalently the parameters S^l_σ, S^L_π, depend on only two potential parameters and we find:

$$I^l_{00} \left\{{\geq \atop <}\right\} I^l_{\pm 2, \pm 2} \quad \text{or} \quad S^l_\sigma \left\{{\geq \atop <}\right\} 0 \quad \text{when} \quad \varepsilon^l_{20} \left\{{\geq \atop <}\right\} -\tfrac{5}{12}\varepsilon^l_{40} \tag{11.161}$$

$$I^l_{\pm 1, \pm 1} \left\{{\geq \atop <}\right\} I^l_{\pm 2, \pm 2} \quad \text{or} \quad S^l_\pi \left\{{\geq \atop <}\right\} 0 \quad \text{when} \quad \varepsilon^l_{20} \left\{{\geq \atop <}\right\} \tfrac{5}{9}\varepsilon^l_{40}. \tag{11.162}$$

In order to use these equations in some *a priori* fashion, we require information about the potential parameters themselves, these depending in turn upon the potential components $\{U^l_{J0}; J = 0, 2, 4\}$. It is a difficult problem to reach rather specific conclusions about these quantities solely from a consideration of their general form. Our task is made easier by consideration first of the limiting case of a potential $U_l(\mathbf{r})$ which is concentrated along the metal–ligand-l bond axis. Disregarding the spherical component this may be idealized as

$$U_l(\mathbf{r}) = U^l(r)\delta(\theta - \theta_l)\delta(\phi - \phi_l), \qquad (11.163)$$

where $\{\theta_l, \phi_l\}$ are the polar angles of the metal–ligand l vector in the global coordinate frame. By expanding the angular delta function in spherical harmonics, it can be shown[44] that, for *any* radial function $U^l(r)$, a simple, fixed relationship exists between the components of different J, namely

$$\varepsilon^l_{JM} = (2J + 1)\delta_{M,0}\bar{U}^l, \qquad (11.164)$$

where

$$\bar{U}^l = \tfrac{1}{2}\int\limits_0^\infty dr\ r^2 |\phi_d(r)|^2\, U^l(r). \qquad (11.165)$$

On substitution of (11.164) into (11.154)–(11.156), *or* into (11.157), (11.158), we find

$$S^l_\sigma = 5\bar{U}^l, \quad S^l_\pi = 0, \quad S^l_\delta = 0. \qquad (11.166)$$

Note that this result, giving both S^l_π and S^l_δ, independently, exactly zero contributions to the AOM e_π and e_δ parameters, follows for *any* potential $U^l(r)$ which is concentrated along the line joining metal and ligand.

This result corresponds nicely with our physical view of the limiting case of σ-bond formation through overlap of linearly localized orbitals on the two centres with σ symmetry – a ball and stick model. Now the zero values of S^l_π and S^l_δ result from the directional property of the delta potential (11.163) and this points to an obvious defect in the point-charge (and radially directed, point-dipole) models of classical crystal-field theory.[15] There we take the potential $U_l(\mathbf{r})$ due to ligand l as proportional to $|\mathbf{r} - \mathbf{O}_l|^{-1}$ with *no restrictions* on $\mathbf{r}$, so that $U_l(\mathbf{r})$ is non-zero in the regions of space occupied by all the other ligands $l' \neq l$ and these regions make contributions to the matrix elements $\langle d^l_k | U_l(\mathbf{r}) | d^l_k \rangle$ for all ligands. The radial coefficients $U^l_{J0}(r)$ in (11.152) are simply $r^J_< / r^{J+1}_>$ according as $r \gtrless |\mathbf{O}_l|$, where $|\mathbf{O}_l|$ is interpreted as an effective bond length. Since $r_<^J/r_>^{J+1}$ is a positive, *decreasing* function of J we have

$$\varepsilon_{J0}^l > \varepsilon_{J'0}^l \quad \text{for} \quad J < J' \quad \text{with fixed} \quad |O_l|. \tag{11.167}$$

Hence, we can write

$$\varepsilon_{20}^l = x, \quad \varepsilon_{40}^l = x - \delta, \quad x \geq \delta > 0 \tag{11.168}$$

and so, from (11.157) and (11.158), we have

$$S_\sigma^l = \tfrac{17}{21}x - \tfrac{5}{21}\delta > 0 \tag{11.169}$$

$$S_\pi^l = \tfrac{4}{21}x + \tfrac{5}{21}\delta \tag{11.170}$$

and hence

$$\tfrac{4}{3} \leq S_\sigma^l/S_\pi^l \leq \tfrac{17}{4} \tag{11.171}$$

for the point-charge model: in this, the conventional point-charge, electrostatic model, therefore, σ and π contributions are necessarily of the same order of magnitude. The same is true for the radially directed, point-dipole model, for which the inequality (11.167) still holds. At the heart of the difference between (11.171) and (11.166) is the contrast between (11.167) and (11.164).

We make two observations about this use of the classical electrostatic models in the context of the angular overlap model. Firstly, the classical calculation of the radial coefficients ignores the fact that *in the AOM* the cell potential $\hat{U}_l$ vanishes outside of cell l, *by construction* using (11.122); that is, we recognize *explicitly* that $\mathbf{r}$ in $U_l(\mathbf{r})$ should be restricted to the space *occupied* by cell l. A convenient way to construct the cellular potentials $\hat{U}^l$ is by projection. We can define a cellular projection operator P_l such that

$$P_l(\mathbf{r}) = \begin{cases} 1 & \mathbf{r} \in \text{cell } l, \\ 0 & \text{otherwise} \end{cases} \tag{11.172}$$

$$\sum_{l=1}^{N} P_l = 1 \tag{11.173}$$

and

$$(P_l)^2 = P_l, \quad P_l^* = P_l. \tag{11.174}$$

With this notation, the projected potential

$$U_l(\mathbf{r}) = P_l(\mathbf{r}) \sum_{l'=1}^{N} |\mathbf{r} - \mathbf{O}_{l'}|^{-1} q_{l'} \tag{11.175}$$

would be the proper form for the point-charge model within the additivity formalism of the AOM. Further, the potential of an unscreened charge varies too slowly to be used as a basis for realistic representation of the ligand field set up by bonded ligands in real complexes. Both factors contribute to the unphysical behaviour shown by (11.171).

Bearing in mind our remarks at the beginning of this subsection, we turn now to enquire if the AOM cellular decomposition of the static potential in *realistic* systems more nearly resembles the result (11.166) for the delta potential or that of the incorrectly formulated, point-charge model (11.171). We shall assume that each cell is a cone, with apex at the metal nucleus, subtending a solid angle Ω_l ($0 \leq \phi \leq 2\pi$, $0 \leq \theta \leq \theta_l$ with polar axis along the metal–ligand l vector), so that $P_l(\mathbf{r})$ only depends on these angles. This simplifying assumption for geometry does not affect the final conclusion as the boundary shape of the cell is not important. We now employ the idea of the delta function model (11.163) to represent the *non-spherical* ligand-field potential, say $V(\mathbf{r})$, arising from (11.139), as a continuous linear superposition of delta function potentials using the identity

$$V(\mathbf{r}) = V(r, \hat{r}) = \int d\hat{r}' V(r, \hat{r}') \delta(\hat{r} - \hat{r}'). \tag{11.176}$$

The potential associated with cell l is then written as

$$U_l(\mathbf{r}) = \int d\hat{r}' P_l(\mathbf{r}') V(r, \hat{r}') \delta(\hat{r} - \hat{r}') \tag{11.177}$$

where the cellular projection operator $P_l(\mathbf{r}')$ restricts the integration domain to the solid angle Ω_l. Expanding the angular delta function in spherical harmonics, with the metal–ligand vector as polar axis, we again obtain radial coefficients as

$$U_{JM}^l(r) = \tfrac{1}{2}(2J + 1) \int d\hat{r}' P_l(\mathbf{r}') C_M^J(\hat{r}') V(r, \hat{r}'). \tag{11.178}$$

In passing, we note that (11.163) is recovered if we put $V(r, \hat{r}') = U_l(r)\delta(\hat{r}' - \hat{O}_l)$, since $C_M^J(0) = \delta_{M0}$. Now, proceeding as before, we calculate the potential parameters corresponding to (11.152), and substitute into (11.154)–(11.156) to obtain $\{I_{kk}\}$ integrals for the 'superposed delta model' as

$$I_{kk}^{\Sigma\delta} = \tfrac{1}{14} \int d\hat{r}' P_l(\hat{r}') \langle \phi_d | V(r, \hat{r}') | \phi_d \rangle D_{kk}(\hat{r}'), \tag{11.179}$$

where

$$D_{00}(\hat{r}') = 7 + 10C_0^{(2)}(\hat{r}') + 18C_0^{(4)}(\hat{r}'), \tag{11.180}$$

$$D_{\pm 1, \pm 1}(\hat{r}') = 7 + 5C_0^{(2)}(\hat{r}') - 12C_0^{(4)}(\hat{r}'), \tag{11.181}$$

$$D_{\pm 2, \pm 2}(\hat{r}') = 7 - 10C_0^{(2)}(\hat{r}') + 3C_0^{(4)}(\hat{r}'). \tag{11.182}$$

The $C_0^J(\hat{r}')$ are simply Legendre polynomials $P_J(\cos\theta)$: writing $x = \cos\theta$, the $\{D_{kk}\}$ reduce to

$$D_{kk}(x) = \tfrac{105}{2} d_k(x) \geq 0 \tag{11.183}$$

with

$$d_0(x) = \tfrac{1}{6}(1 - 3x^2)^2, \quad d_{\pm 1}(x) = x^2(1 - x^2), \quad d_{\pm 2}(x) = \tfrac{1}{4}(1 - x^2)^2 \tag{11.184}$$

whence:

$$I_{kk}^{\Sigma\delta} = \frac{15\pi}{2} \int\limits_1^1 dx\, \bar{V}(x) d_k(x), \tag{11.185}$$

where

$$\bar{V}(x) = \langle \phi_d | V(r, \hat{r}') | \phi_d \rangle. \tag{11.186}$$

As noted at the end of §11.12, the main contribution to the static part, that is, to the d-orbital radial matrix element $\langle \phi_d | V(r, \hat{r}') | \phi_d \rangle$ comes from the integral at r values less than the radius of the transition metal atom: thus, we envisage the potential decreasing from some positive value as the angle $\hat{r}'$ moves away from the metal–ligand vector $\hat{O}_l$. The presumption here is that the behaviour of the potential follows that of the electron density which is enhanced in the bonds and depleted elsewhere (as opposed to a spherical, atomic-like density) and that the Hartree potential outweighs the exchange-correlation potential. Note that we are *not* concerned with the anisotropic effects arising from the incomplete d^n configuration as revealed in some accurate X-ray crystallographic studies[45,46] because, as discussed several times already, the d-electron density is excluded from the potential $V(\mathbf{r})$. Now $V(\mathbf{r})$ is the non-spherical part of the potential, so that $\bar{V}(x)$ is a maximum along the metal–ligand vector, that is when $x = 1$, and decreases as x decreases towards zero. In the outer (that is, angularly distant) parts of the cell, the total potential may be less than the spherical average, so that $\bar{V}(x)$ may be negative for, say, $x \lesssim 3/4$. Using this suggested approximate behaviour for $\bar{V}(x)$, together with those of the $d_k(x)$ functions, shown in figure 11.2, it is easy to infer qualitatively the relative magnitudes of the integrals (11.185). First, it is immediately obvious that the integrals $I_{\pm 2, \pm 2}^{\Sigma\delta}$ are negligible

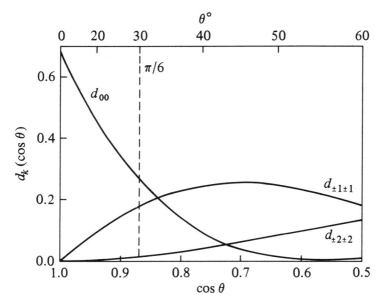

Fig. 11.2. Variation of the functions $d_k(\cos\theta)$ throughout a cell, $1 > \cos^{-1}\theta > 0.5$.

in comparison with those for $k = 0, \pm 1$, and so the static parameters S_σ^l, S_π^l are numerically close to the integrals $I_{00}, I_{\pm 1,\pm 1}$. That the σ parameter is likely to be much greater than the π parameter can be seen from the following argument. Suppose we make the simple model,

$$\overline{V}(x) = \begin{cases} V & 1 \geq x \geq \cos^{-1}(\pi/6) \\ 0 & x \leq \cos^{-1}(\pi/6) \end{cases} : \tag{11.187}$$

evaluation of (11.185) then yields the result,

$$I_{00} : I_{\pm 1,\pm 1} : I_{\pm 2,\pm 2} = 82 : 19 : 1. \tag{11.188}$$

Consideration of figure 11.2 suggests that I_{00} for a realistic $\overline{V}(x)$ may be reasonably approximated by this model (suppose that $\overline{V}(x)$ decreases like $d_0(x)$), but that the integrals with $k \neq 0$ are greatly overestimated so that the ratio in (11.188) gives a poor *lower* bound to (S_σ^l / S_π^l).

Altogether, therefore, while S_σ^l may make a significant contribution to the AOM e_σ^l parameter, we can expect static contributions to e_π^l to be unimportant. It then follows that the crucial determinant of e_π^l will be the dynamic contribution. Thus, in the notation of (11.136), we have

$$(e_\sigma^l)_{\text{static}} > (e_\pi^l)_{\text{static}} \gg (e_\delta^l)_{\text{static}} \tag{11.189}$$

and so it is only the matrix element of v^l(static) involving the d orbital of σ symmetry in the local coordinate frame which is likely to be of any importance. While both parts of (11.136) must be expected to contribute significantly to e^l_σ, contributions to e^l_π and e^l_δ effectively arise from v^l(dynamic) alone. It is to these dynamic contributions to AOM parameters that we now turn.

11.14 The dynamic contribution to AOM parameters

The dynamic contribution to the AOM parameter is written explicitly as

$$(e^l_\lambda)_{\text{dynamic}} = \sum_\alpha \frac{|\langle d^l_\lambda | \mathcal{H}^{(1)l} | \phi_{R\alpha} \rangle|^2}{\varepsilon_d - \bar{\varepsilon}_{R\alpha}}. \tag{11.190}$$

This contribution will be quantitatively significant if there are orbitals $\{\phi_{R\alpha}\}$ with energies $\bar{\varepsilon}_{R\alpha}$ 'close' to the d-orbital energy ε_d (that is, the order of magnitude for $|\varepsilon_d - \bar{\varepsilon}_{R\alpha}|$ must be eV rather than Rybergs – say $10\,000\,\text{cm}^{-1}$ rather than $100\,000\,\text{cm}^{-1}$), for which the matrix elements $\langle d^l_\lambda | \mathcal{H}^{(1)l} | \phi_{R\alpha} \rangle$ are not negligible. Essentially both requirements can only be satisfied by $\{\phi_{R\alpha}\}$ which are 'valence' orbitals. Remember that we suppose the $\{\phi_{R\alpha}\}$ to be constructed using local-orbital (pseudopotential) theory; a qualitative approximation to the $\{\phi_{R\alpha}\}$ is to regard them as molecular orbitals made up from metal s and p orbitals and suitable ligand orbitals of the ligand in cell l. The local pseudosymmetry about the metal–ligand bond is useful here as it provides a 'symmetry'-based classification of the fragment orbitals which may interact to give the $\{\phi_{R\alpha}\}$ and, equally, it gives approximate selection rules for the matrix elements $\langle d^l_\lambda | \mathcal{H}^{(1)l} | \phi_{R\alpha} \rangle$ since the pseudosymmetry is carried by $\mathcal{H}^{(1)l}$. This is the obvious and usually sufficient condition which makes the potential v^l(dynamic) diagonal with respect to the orbitals $\{d^l_\lambda\}$: we see here the 'equivalent' statement in the AOM within ligand-field theory of the 'overlap criterion' of the extended-Hückel version of the AOM. The many similarities between the construction of the diagrams in figure 11.1 and those of conventional molecular-orbital theory ensure the connection between available ligand valence orbitals and the orders of magnitude of AOM e parameters: the very important differences between these two approaches, however, deny the possibility of quantifying the ratios of the e_λ parameters by simple procedures based on the use of (11.98), not least because of the occurrence of $(e^l_\sigma)_{\text{static}}$ in the σ parameter.

Frequently, there will be only one significant term in the sum (11.190) for a given bonding mode λ: then the sign of $(e^l_\lambda)_{\text{dynamic}}$ is governed by the sign of the energy denominator $(\varepsilon_d - \bar{\varepsilon}_{R\alpha})$, since the numerator in (11.190) is necessarily positive. It is to be expected, however, that there will be situations where there is more than one significant contribution to the sum in (11.193), as, for example, when bonding and antibonding (metal(p) – ligand) orbitals of π symmetry are both energetically close enough to the d orbitals to be considered as 'valence orbitals' (antibonding orbitals of σ symmetry are expected to make $|\varepsilon_d - \bar{\varepsilon}_{R\alpha}|$ too large to be important): as will be argued empirically in §12.5 for complexes of imine-type ligands, the contributions to $(e_\pi^l)_{\mathrm{dynamic}}$ may well have opposite signs so that the AOM parameters measure a *net* effect. It is a matter of definition, therefore, that we characterize a ligand l engaged in a bonding mode λ as an *acceptor* if e_λ^l is negative $(\bar{\varepsilon}_{R\lambda} > \varepsilon_d)$ and as a *donor* if e_λ^l is positive $(\bar{\varepsilon}_{R\lambda} < \varepsilon_d)$: to the extent that frontier orbital arguments can be used to estimate the relative energy of $\bar{\varepsilon}_{R\lambda}$, these definitions are in accordance with the customary terminology derived from Lewis acidity–basicity, but we must emphasize that the present definitions are made with respect to the energy of the metal d orbitals defined for the complex.

The donor–acceptor properties of any one ligand l in a series of different complexes should not be expected to be constant, for since the orbitals $\{\phi_{R\alpha}\}$ associated with ligand l are, by construction, related to the molecule *as a whole* we may expect to see the operation of the electroneutrality principle reflected in the variation of the magnitudes of the AOM parameters throughout the series of complexes. This also means that the number and types of other groups in the complex are expected to enhance or diminish the donor–acceptor properties measured for the ligand l. The interpretation of such parameter variation must be carried out with care, however, particular attention being paid to variations in the associated metal–donor-atom, bond lengths. Variations in bond lengths are, of course, intimately connected with variations in the donor–acceptor functions of ligands in a given system and so are not likely to be overlooked; their effects are manifested primarily through the matrix elements $\langle d_\lambda^l | \mathscr{H}^{(1)l} | \phi_{R\alpha} \rangle$ in (11.190), in the sense, other things being equal, that bond-length shortening enhances the magnitude of the matrix element. We shall see how all these qualitative principles arise naturally in the analyses described throughout chapter 12.

The foregoing outlined how the general aspects of the chemical interpretation of AOM parameters emerge from the theoretical framework of the approach: we consider now some more detailed properties. Thus, for R set orbitals $\{\phi_{R\alpha}\}$ of σ symmetry it is generally the case that a bonding orbital formed from the metal s and p orbitals with suitable ligand orbitals leads to a smallest value of $(\varepsilon_d - \bar{\varepsilon}_{R\alpha})$ which is positive, corresponding to a bonding orbital lying below the metal d orbital. It is usual, therefore, to expect that $(e_\sigma^l)_{\text{dynamic}}$ reinforces the static σ contribution to give a large, positive e_σ^l parameter value. There appears to be no simple but correct argument which affords estimates of the relative magnitudes of the static and dynamic contributions to the AOM parameter: while our prejudice is that the dynamic part will generally predominate, we must allow the possibility of important static contributions, whose magnitudes, by virtue of the non-spherical nature of $\mathscr{H}^{(1)}$, might generally be expected to vary significantly with respect to their ultimate effect on the various global d orbitals (see discussion following (11.118)).

In any complex there are always bonding orbitals of local σ symmetry, and hence we can always expect non-negligible AOM e_σ parameters. On the other hand, for other symmetry modes, if there are no valence orbitals $\{\phi_{R\alpha}\}$ with appropriate symmetry to match a particular d orbital (d_π or d_δ in the local frame), we expect the corresponding matrix element $(e_\lambda^l)_{\text{dynamic}}$ to be negligible as the only contributing terms would have $(\varepsilon_d - \bar{\varepsilon}_{R\alpha})$ large and positive ($\sim$ Rydbergs) with small (< 0.1 eV) numerator matrix elements $\langle d_\lambda^l | \mathscr{H}^{(1)l} | \phi_{R\alpha} \rangle$. Herein lies the basis of our expectation that both $e_{\delta xy}$ and $e_{\delta x^2 - y^2}$ in virtually all complexes will be negligible: we have shown that static contributions to AOM π and δ parameters are expected to be tiny and, here, a similar conclusion for the dynamic contribution follows by virtue of the lack of ligand δ orbitals that could give rise to suitable $\{\phi_{R\alpha}\}$. It is ultimately the points of similarity between (say) Hückel molecular-orbital diagrams and those in figure 11.1 which are responsible for this justification – of negligible contributions to dynamic parts of e_δ parameters – appearing trivially obvious: it is the important differences between LCAO MOs and LFOs which ensures that a follow-up, numerical calculation would lead to the *same* conclusion. The negligibility of the δ parameters is essential to the practical utility of the AOM, as discussed in §9.8 where a formal discussion of the degree of parametrization under these circumstances was given. Recall, for example, the theoretical definition we adopted in (11.109): this form was chosen so as to remove from the theoretical AOM parameters any dependence on the detailed and arbitrary choice of the radial parts of the basis d orbitals in (11.104). On the other hand, we require these theoretical e_λ parameters to map onto those determined experimentally and which are not at all arbitrary. This can only be done up to the point of defining a theoretical energy origin, as in (9.34) in effect, so that the use of $\tilde{e}_\lambda$ parameters, as defined in (9.35) is implicit here. Clearly, if the arguments given immediately above had not been available, the restriction of AOM parameters to the *relative* quantities defined in (9.35) would circumscribe the approach rather severely: however, the neglect of e_δ parameters, or better their being fixed at zero, is defensible to a high degree and the practical application of the AOM approach along these lines described in §9.8 is valid.

The power of the AOM in ligand-field analysis arises because the magnitudes and signs of the empirical parameters may be related to specific electronic features of the chemical bonding interactions. Much of the chemical interest generated by the analyses to be described in chapter 12 centres of π interactions. A common finding is that ligands like amines are generally associated with $e_\pi \approx 0$ and in many cases this value is *forced* by 'best fits' to the experimental data. It is valuable to make an explicit interpretation of this result. Thus, if the free ligand has no valence π orbitals, we do not expect there to be a localized orbital $\phi_{R\pi}$ which can make any significant contribution to e_π^l: there may well be low-lying, core-like orbitals of π symmetry, of course, but because their $|\bar\varepsilon_{R\pi}|$ values are so large, these will produce negligible contributions; and these must be added to the very small, positive, static term, the resultant being a small, positive e_π^l parameter, perhaps $< 100\,\mathrm{cm}^{-1}$. On the other hand – and this is the point we wish to emphasize here – any reasonably large e^l parameter can be ascribed confidently to interactions between at least one valence orbital $\phi_{R\pi}$ and a d orbital of π symmetry in the l coordinate frame. A 'large' value here is one between several hundreds and a very few thousand wavenumbers, *which by chemical bonding standards* is still a *small* energy. Thus, we do not interpret a large e_π^l parameter as indicative of substantial d_π-orbital participation in metal–ligand bonding; rather, such a result tells us that the ligand l has associated with it a *valence* orbital $\phi_{R\pi}$ reflecting the bonding interaction between metal p_π orbitals and appropriate ligand π functions. Once more; *the d orbitals act as a probe to monitor the significant chemical bonding interactions*: for parameters of σ or π symmetry, these involve metal s and p orbitals with ligand orbitals and it is these functions which determine the orbitals $\phi_{R\alpha}$ with energies $\bar\varepsilon_{R\alpha}$ which occur in the potential v^l_{dynamic}.

11.15 The potential in empty cells

We shall discuss various aspects of the magnetism and spectra of planar, four-coordinated transition-metal ions in chapter 12. Included in that survey is some of the detailed evidence, from studies of a quite large number of such complexes, relating to an 'anomalous' value for the one-electron matrix element $\langle d_{z^2} | V_{LF} | d_{z^2} \rangle$. When fitting various ligand-field properties of these systems within the angular overlap model, it is found that $\langle d_{z^2} | V_{LF} | d_{z^2} \rangle$ must take a value some 5000–$6000\ cm^{-1}$ less than that accounted for by the optimal values for the AOM e_λ parameters in the molecule. At first sight, this represents a failure of the AOM and hence of the ligand-field model in general. Smith[47] has proposed an extension to the AOM based on the idea that this 'anomalous' effect originates in a d–s mixing process: he argues that an interaction between $3d_{z^2}$ and $4s$ metal orbitals, for example, is possible, as both transform as a_{1g} in planar D_{4h} symmetry, and leads to a lowering of the $3d_{z^2}$-orbital energy within the conventional (historic) AOM scheme. In this sense, Smith argues for an extended basis (that is, $d + s$) to account for these unusual empirical situations. Following our remarks in §11.12(iii), we present here a discussion to show that the phenomenon does *not* represent any failure of the AOM at all and that it *can* be described quite satisfactorily within the d-orbital basis employed within all other AOM studies: the role of higher-lying metal s functions is entirely supported, however. In essence, the following analysis brings the concept of d–s mixing into the 'localized potential model' in the same way that the business of the rest of the present chapter has been to provide a sound basis for, and rationalization of, the qualitative features of the AOM which derived historically from a simplistic, molecular-orbital approach.

Throughout our descriptions of the AOM so far, we have associated each cell uniquely with each metal–ligand perturbation; specifically, we have assigned N cells for a complex ML_N. We now examine the nature of the potential in an 'empty cell' corresponding to the vacant coordination sites in, say, planar complexes. While this may seem to be a somewhat strange idea, especially within the historical foundation of the AOM as a molecular-orbital model, we argue that a *consistent* application of the AOM actually *requires* the concept of empty or void cells in coordinationally sparse complexes. For the phenomenological reasons outlined above, we develop our discussion with planar complexes in mind, assigning the global, molecular Z axis to lie perpendicular to the coordination plane: we therefore consider two empty cells placed above and below this plane, on the Z axis. We note straightaway that the required lowering in energy of the matrix element $\langle d_{z^2} | V_{LF} | d_{z^2} \rangle$ is implemented if a *negative* e_σ parameter is associated with these empty cells. We can base a discussion of this parameter on either (11.90)–(11.93) or on the theory presented in §11.14: both approaches are of interest.

Firstly, in relation to (11.90) *et seq.*, we note that no contribution to the 'empty' cell potentials arises from ligand nuclei, of course, so we are left with contributions of the electron density above and below the molecular plane to

$$\tilde{J}_L(\mathbf{x}) - \tilde{K}_L(\mathbf{x}).$$ (11.191)

The electron–nuclear attraction term in (11.90) involving the metal nucleus is taken into account in the definition of the basis orbitals. If an empty cell along the Z axis, above or below the XY plane, occupies a volume Ω we wish to examine terms like

$$\tilde{J}_L(\mathbf{x})|d_{z^2}\rangle = \int_\Omega d\mathbf{x}' \rho^L(\mathbf{x}', \mathbf{x}') U(|\mathbf{x} - \mathbf{x}'|) d_{z^2}(\mathbf{x}),$$ (11.192)

$$\tilde{K}_L(\mathbf{x})|d_{z^2}\rangle = \int_\Omega d\mathbf{x}' \rho^L(\mathbf{x}, \mathbf{x}') U(|\mathbf{x} - \mathbf{x}'|) d_{z^2}(\mathbf{x}).$$ (11.193)

Now the density matrix $\rho^L(\mathbf{x}, \mathbf{x}')$ can be expressed as a bilinear combination of one-electron orbitals for the metal complex *excluding* the d orbitals, say

$$\rho^L(\mathbf{x}, \mathbf{x}') = \sum_\alpha \eta_\alpha \phi_\alpha^*(\mathbf{x}) \phi_\alpha(\mathbf{x}'),$$ (11.194)

where $\eta_\alpha = 2$ for occupied orbitals, and $\eta_\alpha = 0$ for unoccupied orbitals (recall that ρ^L describes a spin singlet). The natural orbitals $\{\phi_\alpha\}$, which may be thought of as molecular orbitals for the complex, can be classified according to the irreducible representations of D_{4h} and this means that amongst the $\{\phi_\alpha\}$ there are orbitals with significant metal s orbital character which transform in the same way as d_{z^2}. When we recall that $U(|\mathbf{r} - \mathbf{r}'|)$ is a short-range interaction, it becomes clear that it will be just these molecular orbitals which make the dominant contributions to the integrals (11.192) and (11.193). These molecular orbitals involve both metal s and ligand character but *locally* in the volume Ω they can be taken to be largely of metal s character. Now the Coulomb potential arising from the metal s electron is spherically symmetric and is therefore taken up in the definition of the metal orbitals. Hence the main contribution to the matrix element of the 'empty' cell potential is simply

$$\langle d_{z^2}| - \tilde{K}_L(\mathbf{x})|d_{z^2}\rangle_\Omega \simeq -2\int dr \int_\Omega d\mathbf{r}' d_{z^2}^*(\mathbf{r}) s(\mathbf{r}) s^*(\mathbf{r}') d_{z^2}(\mathbf{r}') U(|\mathbf{r} - \mathbf{r}'|).$$ (11.195)

The non-spherical part of this 'exchange' term contributes to e_σ and is *negative*: it is simply some fraction of the total atomic s–d exchange integral, which is typically ≈ 2 eV.

An alternative view of the same phenomenon but one couched in the language of §11.14 is to note that a negative e_σ parameter requires a dominant interaction between the d orbital of σ symmetry in the local, 'empty' cell and an orbital $\phi_{R\alpha}$ with a mean energy $\bar{\varepsilon}_{R\alpha}$ which is *more positive* than ε_d, so that $(\varepsilon_d - \bar{\varepsilon}_{R\alpha}) < 0$ determines the negative sign of $(e_\sigma^l)_{\text{dynamic}}$. Were the cell not coordinationally void, there would exist 'bonding' orbitals $\phi_{R\alpha}$ with mean energies more negative than ε_d and these orbitals would interact more strongly with the metal d orbitals than corresponding 'antibonding' orbitals because of better overlap. However, being 'empty', the only localized orbitals which can be associated with these cells, and which have similar energy to the pure metal d orbitals, are the metal s and p orbitals. These orbitals almost invariably have energies *above* the d-orbital energy (smaller ionization energies) and so make $(\varepsilon_d - \bar{\varepsilon}_{R\alpha})$ negative as required: further, the dominant contribution here is likely to derive from the metal s rather than the metal p orbital. So, taking the local and global z axis perpendicular to the coordination plane, coupling between the d_{z^2} and s orbitals of the metal ion results in a negative e_σ parameter value for the 'empty' cells; in turn, this negative sign is responsible for the 'anomalous' *depression* of the (global) d_{z^2} orbital energy, observed in planar coordinated complexes.

In summary, therefore, the empirical finding that the AOM parametrization of planar complexes requires an additional e_σ parameter which is large and negative is explained in this way and can certainly be described as 'd–s mixing'. It is *not*, however, a 'failure' of the AOM since the assumption that the matrix $\{e^l\}$ of AOM parameters for each ligand l is diagonal is intimately connected with the choice *and number* of cells for the cellular decomposition of the ligand-field potential. It is evident that the 'd–s mixing' mechanism is only important in situations where there are vacant coordination sites in an overall molecular symmetry which gives rise to mixing between valence molecular orbitals and the metal d_{z^2} orbital. In high-symmetry situations, the 'ligand-field potential' V_{LF} does indeed arise from the ligands and d–s mixing may vanish identically, by symmetry, as in, say, O_h chromophores. In complexes with somewhat lower symmetry than these, slight asymmetry will cause a term like (11.167) to make some small contribution which, being sequestered within the various ligand-generated e_λ parameters, is unlikely to be detected in practice. Although introduced originally in an *ad hoc* fashion and viewed as evidence for a failure of the AOM, the concept of 'd–s' mixing seen through the present analysis is not only defensible but actually required for the *consistency* of the AOM when applied to transition-metal complexes in which the ligands do not 'fill up' the coordination sphere of the central metal ion.

11.16 Δ_{oct} and 10Dq

We have not been concerned in this chapter to provide a structure for the *ab initio* computation of ligand-field parameters. Instead, we have attempted to discover a structure for the AOM within ligand-field theory which bears a *definable* relationship with the principles of the theory of electronic structure at large. Hence, given the procedures and definitions of the AOM as it is actually used to reproduce spectral and magnetic data, we have sought to establish with what quantum-mechanical operations and entities these correspond. Much early criticism and misunderstanding of the AOM within ligand-field theory could have been avoided if the various quantities involved had been formally and rigorously defined: indeed, a somewhat similar remark can be made of ligand-field theory in general. Hopefully, these questions have been adequately addressed in the present chapter. For the sake of completion, however, there remains one small area of the subject, with which Ballhausen[36] has also concerned himself recently, which might benefit from more careful definition: namely, the octahedral ligand splitting factors Δ_{oct} and 10Dq.

The definition of Δ_{oct} has been rather fully dealt with in §§11.9 and 11.10. The parameter Δ_{oct} corresponds to the energy difference between the one-electron matrix elements of t_{2g} and e_g symmetry in the octahedral environment:

$$\Delta_{oct} \equiv \langle e_g | \mathcal{H} | e_g \rangle - \langle t_{2g} | \mathcal{H} | t_{2g} \rangle \tag{11.196a}$$

$$= \langle d_{z^2} | V_{oct} | d_{z^2} \rangle - \langle d_{xy} | V_{oct} | d_{xy} \rangle \tag{11.196b}$$

$$\approx 3e_\sigma - 4e_\pi, \tag{11.196c}$$

where the $\approx$ sign has been included in (11.196c) in recognition of the 'extra' physical assumptions in the AOM concerned with the definitions of disjoint cells and the identity of the $\mathbf{R}^l$ of (11.135) with rotation matrices. As in (11.110), $\mathcal{H}$ is defined so as to exclude all d–d interactions. Insofar as ligand-field theory 'works'; that is, that the system under consideration lies within what we have called the 'ligand-field regime', a *state* energy difference like, say, $E(^4A_{2g}) - E(^4T_{2g})$ in an octahedral d^3 species will also be given by Δ_{oct}, following the concomitant validity of vector coupling and the notion of anonymous configuration parentage for a mean

ligand-field potential, as in §11.6. As emphasized repeatedly, these definitions of Δ_{oct} and of the e parameters in (11.196c) – which equation is
appropriate for linear ligators – refer specifically to ligand-field theory as
it is actually applied, and hence to the Hamiltonian $\mathcal{H}$ of (11.101). State
energy differences as calculated by conventional molecular-orbital theory,
using a Hamiltonian H such as (11.94) with (11.97), are not given by the
sorts of expressions we have studied here and the MO and LF theories
do not map onto each other. While this may be a problem for the
computational chemist, as discussed by Ballhausen,[36] no difficulties arise
within the present discipline.

As mentioned at the beginning of §11.8, $10Dq$ is a quantity first
introduced within *crystal*-field theory when the individual letters, D and
q, had separate significance. Now, crystal-field theory explicitly neglects
metal–ligand overlap and disallows the possibility of electron permutation
between metal and ligands: indeed, the ligands are regarded only as
providing a classical electrostatic medium. Let us follow Ballhausen,
however, and write a core operator

$$\hat{h}^{core}(i) = -\tfrac{1}{2}\nabla^2(i) - \sum_{\sigma} Z_{\sigma}/r_{\sigma i} + \sum_{\substack{core\,orbitals \\ k}} (2\hat{J}_{ki} - \hat{K}_{ki}), \quad (11.197)$$

expressing the kinetic energy of a transition-metal d electron (i) and its
interactions with all ligand nuclei (σ) and ligand electron cores (k). Then,
in the electrostatic, crystal-field formalism, we have

$$10Dq = \langle d_e|\hat{h}^{core}|d_e\rangle - \langle d_t|\hat{h}^{core}|d_t\rangle$$
$$\equiv \bar{h}^{core}(e_g) - \bar{h}^{core}(t_{2g}). \quad (11.198)$$

Furthermore, in the crystal-field scheme, a state energy splitting like
$E(^4A_{2g}) - E(^4T_{2g})$ for d^3 systems, for example, is also given exactly by
this same expression. At the same time, note that the d_e and d_t orbitals
are appropriate free-ion d orbitals and that the interelectron repulsion
energies for a d^n system ($1 < n < 9$) are *exactly* those determined by
Slater–Condon theory for the corresponding free ions. Altogether, therefore, $10Dq$ as a crystal-field octahedral splitting parameter is transferable
from d^n configuration to d^n configuration while simultaneously being
defined with respect to conventional metal d orbitals.

The ligand-field diagram, figure 1.1, with which this book began, is just
that: it is not a crystal-field diagram, nor it is a molecular-orbital diagram.
It is indeed deceptively simple but, once properly understood, really
provides a powerful summary of the nature of ligand-field theory.

References

[1] Woolley, R.G., *Mol. Phys.*, **42**, 703 (1981).

[2] Gerloch, M., Harding, J.H. & Woolley, R.G., *Structure and Bonding*, **46**, 1 (1981).

[3] Gerloch, M. & Woolley, R.G., *Prog. Inorg. Chem.*, **31**, 371 (1983).

[4] Ballhausen, C.J., *J. Chem. Ed.*, **56**, 215, 294, 357 (1979).

[5] Schäffer, C.E. & Jørgensen, C.K. *J. Inorg. Nucl. Chem.*, **8**, 143 (1958).

[6] Schäffer, C.E. & Jørgensen, C.K. *Mat. Fys. Medd.*, **34**, 13, (1965).

[7] Schäffer, C.E., *Structure and Bonding*, **5**, 68 (1968).

[8] Cotton, F.A. & Wilkinson, G., *Advanced Inorganic Chemistry*, 4th edn, Wiley-Interscience, New York, 1980.

[9] Purcell, K.F. & Kotz, J.C., *Inorganic Chemistry*, W.B. Saunders, Philadelphia, USA, 1977.

[10] Burdett, J.K., *Molecular Shapes*, Wiley-Interscience, New York, 1980.

[11] Stevens, K.W.H., *Proc. Roy. Soc.*, **A219**, 542 (1953).

[12] Orgel, L.E., *An Introduction to Transition Metal Chemistry*, Methuen, 1962.

[13] Gerloch, M. & Miller, J.R., *Prog. Inorg. Chem.*, **10**, 1 (1968).

[14] Jørgensen, C.K., *Prog. Inorg. Chem.*, **4**, 73 (1962).

[15] Gerloch, M. & Slade, R.C., *Ligand Field Parameters*, Cambridge University Press, London, 1973.

[16] Lykos, P.G. & Parr, R.G., *J. Chem. Phys.*, **24**, 1166 (1956).

[17] McWeeny, R. & Sutcliffe, B.T., *Methods of Molecular Quantum Mechanics* Academic Press, London, 1969.

[18] Löwdin P.O. *In : Perturbation Theory and its Application in Quantum Mechanics*, ed. C.H. Wilcox, pp. 255–94, Wiley, New York, 1966.

[19] Griffith, J.S., *Theory of Transition Metal Ions*, Cambridge University Press, 1961.

[20] Sugano, S., Tanabe, Y. & Kamimura, H., *Multiplets of Transition Metal Ions in Crystals*, Academic Press, London, 1970.

[21] Des Cloizeaux, J. *Nucl. Phys.*, **20**, 321 (1960).

[22] Freed, K.F., *Chem. Phys.*, 463 (1974).

[23] Trees, R.E., *Phys. Rev.*, **83**, 756 (1951); **84**, 1089 (1951).

[24] Judd, B.R., *Second Quantization and Atomic Spectroscopy*, Johns Hopkins Press, Baltimore, 1967.

[25] Slater, J.C., *Quantum Theory of Molecules and Solids*, vol. **1**, McGraw-Hill, New York, 1963.

[26] Haydock, R., Heine, V. & Kelly, M.J. *J. Phys.*, **C5**, 2845 (1972); **C8**, 2591 (1975).

[27] Van Vleck, J.H., *J. Chem. Phys.*, **3**, 803, 807 (1935).

[28] Ferguson, J. & Wood, D.L., *Aust. J. Chem.*, **23**, 861 (1970).

[29] Ferguson, J., *Prog. Inorg. Chem.*, **12**, 159 (1970).

[30] Owen, J. & Thornley, J.M., *Rep. Prog. Phys.*, **29**, 676 (1966).

[31] Jørgensen, C.K., *Modern Aspects of Ligand Field Theory*, North-Holland, Amsterdam, 1971.

[32] Hohenberg, P. & Kohn, W., *Phys. Rev.*, **B136**, 864 (1964).

[33] Jørgensen, C.K., Pappalardo, R. & Schmidtke, H.H., *J Chem. Phys.*, **39**, 1422 (1963).

[34] Wolfsberg, M. & Helmholz, L., *J. Chem. Phys.*, **20**, 837 (1952).

[35] Ballhausen, C.J. & Dahl, J.P., *Theor. Chim. Acta*, **34**, 169 (1974).

[36] Ballhausen, C.J., *Molecular Electronic Structure of Transition Metal Complexes*, McGraw-Hill, 1979.

[37] Heine, V., *Solid State Phys.*, **35**, 1 (1980).

[38] Tennyson, J. & Murrell, J.N., *J. Chem. Soc. Dalton Trans.*, p. 2395 (1980).

[39] Andersen, O.K. & Woolley, R.G., *Mol. Phys.*, **26**, 905 (1973).

[40] Gray, H.B., *Electrons and Chemical Bonding*, Benjamin, New York, 1964.

[41] Watson, R.E. & Freeman, A.J., *Phys. Rev.*, **134A**, 1526 (1964).

[42] Soules, T.F., Richardson, J.W. & Vaught, D.M., *Phys. Rev.*, **B3**, 2186 (1971).

[43] Ashcroft, N.W. & Mermin, N.D., *Solid State Phys.*, chap. 17, Holt, Rinehart and Winston, 1976.

[44] Brink, D.M. & Satchler, G.R., *Angular Momentum*, Clarendon Press, Oxford, 1968.

[45] Iwata, M. & Saito, Y., *Acta Cryst.*, **B29**, 822 (1973).

[46] Coppens, P., *Angew. Chem. Int. Ed.*, **16**, 32 (1977).

[47] Smith, D.W., *Inorg. Chim. Acta*, **22**, 107 (1977).

PART IV

━━━

THE SYNTHESIS

We turn at last to review recent analyses of the ligand-field properties of several transition-metal complexes. All have been successfully completed within the angular overlap model of the ligand field and, taken together, surely mark the beginning of a coherent approach to ligand-field analysis. The central theme throughout has been to exploit the methods described in this book to obtain semi-quantitative measures of the extents of σ and π bonding between metal and ligands, and generally to quantify our understanding of the electron distribution in complexes: we use the techniques of ligand-field theory to address *chemical* issues.

The analyses we describe have been successful because they derive from a broad-based attack. We have emphasized the desire and need to study *real* molecules unhindered by restrictions of d^n (or f^n) configuration, of coordination geometry or symmetry, or of the number and types of ligands. We have argued that the ligand-field version of the AOM offers the only parametric model capable of dealing with so general a problem: at the same time, that it does so in a manner which accords directly with the natural physics of these systems and in comprehensible, in its results at least, to the mainstream chemist. It is crucial to the whole enterprise, of course, that the theoretical structure upon which the contemporary synthesis is based be sound: this seminal issue formed the substance of chapter 11, in which were established the theoretical *bona fides* of ligand-field theory in general and of the AOM in particular. An important practical aspect of the use of the AOM concerns the high degree of parametrization frequently involved and this question was confronted in §9.8. As pointed out there, satisfactory analyses of heavily parametrized angular overlap models are often possible only if reference is made to a wider data base afforded by the simultaneous study of a variety of ligand-field properties. Measurements of both esr g values and single-crystal paramagnetic susceptibilities offer valuable complements to optical spectroscopy in this regard. In Part II, we investigated the theory and practice of susceptibility measurement and we note two relatively recent developments which were required by the current programme. One concerned the derivation of the generalized susceptibility equation (7.203) which is needed for unsymmetrical molecules; and the other was the construction of procedures for measuring susceptibilities of triclinic crystals: apparatus design has been an important factor in this experimental problem. In addition to important advances required and achieved in both theoretical and experimental areas of magnetochemistry and ligand-field theory, the efficacy of the whole analytical package relies on the existence of the sort of computing facility described in chapter 10. Studies like those we are about to review are sensible only if we can make the necessary calculations quickly and cheaply and are able to compare theory and experiment directly and with ease. In short, ligand-field analysis has become tractable by virtue of a *synthesis* of theoretical, experimental and computational techniques.

12

Ligand-field analyses

==

We now review some recent ligand-field analyses. Our main concern, of course, is to summarize the essential chemical conclusions which may be drawn from these studies, not only because of their intrinsic interest but because the success of the synthesis we have studied is made manifest that way rather than by attention being focussed on technicalities. However, so much of this book is concerned with the procedures by which such results are obtained that it would be a pity not to comment here and there on some of the more technical, and less chemical, aspects. It is, however, more natural to describe such issues as they arise rather than to collect them into a separate section. Accordingly, this review is organized such that analyses are assembled into groups on the basis of the chemical interest they engender: any supplementary points to be made of a predominantly technical nature, that is, pertaining to the analytical process itself, which are too long to be incorporated in the main text without obscuring the main issue, are presented under separate headings at the 'subdecimal' level. For the moment, let these analyses speak for themselves. In the final section, we make concluding remarks and provide an overview of the subject.

12.1 Pyridine, quinoline and biquinoline complexes

The *trans*octahedral complexes $M(pyridine)_4(NCS)_2$; $M = Co(II)$, $Fe(II)$ crystallize in the monoclinic system and are isostructural, with the centrosymmetric molecular geometry shown in figure 12.1. The planes of the pyridine ligands are oriented far from the perpendicular to the MN_4

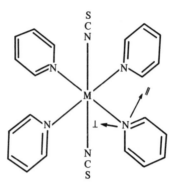

Fig. 12.1. Centrosymmetric molecules of $M(py)_4(NCS)_2$; $M = Co(II)$, Fe(II). AOM e_π parameters refer to directions parallel and perpendicular to the pyridine planes. The NCS groups are idealized.

plane and yet do not pack around the central metal in the common 'propeller' fashion: that would, of course, require the absence of the inversion centre at the metal atom. In this rather early study,[1] the ligand-field analyses were based on single-crystal, principal susceptibilities measured over the somewhat restricted temperature range 80–300 K. For the cobalt(II) complex, calculations employed the complete spin-quartet basis of the d^7 configuration $- ^4P + {}^4F$. The parameter set comprised the Racah B parameter for interelectron repulsion, the one-electron spin-orbit coupling coefficient ζ, and Stevens' orbital reduction factor k: the ligand field was parametrized within the AOM by $e_\sigma(py)$, $e_\sigma(NCS)$; $e_{\pi\parallel}(NCS)$ and $e_{\pi\perp}(NCS)$ (for π bonding with nodal planes perpendicular and parallel to the pyridine plane respectively); and by $e_{\pi av}(NCS)$ for a presumed average M—NCS π interaction – a simplification that was made to reduce the number of disposable parameters in view of the limited experimental data set.

All these parameters were varied independently over wide ranges until optimal agreement between observed and calculated crystal susceptibilities was achieved, corresponding approximately to a 2% discrepancy for all three principal crystal susceptibilities throughout the experimental temperature range. Several points emerged from the analysis. The first is typical of many such studies and serves as a warning. As described in chapters 4 and 10, the penultimate stage in fitting experimental crystal susceptibilities is the calculation of the corresponding molecular susceptibilities. In the present study, principal molecular susceptibilities corresponding to 'best fit' are found to lie along directions which bear no obvious relationship with bond vectors or bisectors. This arises from a small

π-bonding function of the pyridine groups, as we shall see, together with the non-special orientation of these groups. However, the point to emphasize is that there need be no obvious relationship between the orientation of molecular properties and geometrical features in the absence of *exact* symmetry and that magnetism often provides a very sensitive manifestation of this principle: the point was raised in §§1.4 and 1.5. No difficulty is made by this within the present analysis, indeed the contrary as we discuss in §12.1.1, but we take the occasion to emphasize again that it is important not to presume a knowledge of the orientation of paramagnetic susceptibilities or of g values from *approximate* geometrical symmetry.

As to the AOM parameter values affording good reproduction of the experimental paramagnetism; values of e_σ for both pyridine and thiocyanate ligands were not sensitively determined nor differentiated but are about 4000 cm^{-1}. We might expect that a rather more sensitive estimate of these values would emerge from a study of the optical spectra. Next, we find that $e_{\pi\parallel}(py) \approx 0$: it is clearly satisfying as well as immediately and non-technically obvious to observe a result indicating no M—L π interaction in the plane of the pyridine groups. Perpendicular to the pyridine ligands (that is, with a nodal plane *in* the plane of the ligand), some π interaction is evidenced by the value $e_{\pi\perp}(py) \sim 50\text{--}100$ cm^{-1}. The value is small but sensitively determined by the crystal magnetism: wholly unacceptable reproduction of the experimental properties are obtained for $e_{\pi\perp}(py)$ values of zero or ± 200 cm^{-1}, for example. The sensitivity with which the $e_{\pi\perp}(py)$ value was determined is associated with the orientation of the molecular susceptibilities and is discussed again in §12.1.1. In terms of the historical, molecular-orbital approach to the AOM, the small positive value determined for this parameter implies a small π-donor role for the pyridine with respect to the cobalt atom in this complex. However, in view of the theoretical analysis in chapter 11, we cannot tell whether so small a value merely represents a small 'static' contribution to the $e_{\pi\perp}(py)$ parameter. The only defensible conclusion we may draw from this value, therefore, is that any dynamic contribution from the metal–pyridine π bonding is of uncertain sign but slight.

An entirely analogous study[1] of the iron(II) complex, using the ^5D free-ion term basis, yielded similar results. This time, the principal molecular susceptibilities lay along different, though still uninformative, directions. The $e_\sigma(py)$, $e_\sigma(NCS)$, $e_{\pi\parallel}(py)$, $e_\pi(NCS)$ values are indistinguishable from those in the Co(II) complex and the $e_{\pi\perp}(py)$ lies in the range 60–130 cm^{-1}. It seems entirely reasonable that two such similar complexes, though possessing widely different magnetic susceptibilities, should be characterized by essentially identical AOM parameters. Here is an example of the transferability of parameter values we discussed in §9.8: we have already warned, and it will become increasingly clear, that such a feature is not general and should not be relied upon. Nevertheless, it is interesting to compare these results for $e_{\pi\perp}(py)$ with those obtained from spectral studies[2] of $[Cr(py)_4X_2]^+$; $X = F, Cl, Br$ and of[3-6] $[Ni(py)_4X_2]$; $X = Cl$, Br, I. By assuming that the $e_\lambda(X)$ parameters in this chromium(III) series are identical with those in $[Cr(NH_3)_4X_2]^+$, a value of about $-1000\ cm^{-1}$ was determined for $e_{\pi\perp}(py)$. Variable positive values for this parameter have been reported for the nickel(II) series. As Smith[65] has pointed out, pyridine is more likely to function as a π acceptor towards Ni(II) than Cr(III) and he cites one work of Barton & Slade[7] in which e_σ values for F range from 7200 to 9000 cm^{-1} and e_π from 1700 to 2500 cm^{-1} within a series of CrN_4X_2 systems, where the equatorial ligands are saturated amines. Smith makes the point, with which we must agree, that the presumed transferability of parameters for the halogens between the chromium series above might be inappropriate, so casting doubt on the negative $e_{\pi\perp}$ value for pyridine in those systems.

12.1.1 Orientation of the principal molecular susceptibilities in $M(py)_4(NCS)_2$; $M = Co(II)$ and $Fe(II)$

As an aside from the main theme, let us consider briefly the reasons for the unexpected (?) orientation of the molecular magnetic ellipsoids in these complexes. As defined by the donor atoms and ligand types, the co-ordination geometry of these molecules closely approximates tetragonally distorted octahedral MA_4B_2 or D_{4h} symmetry. On that basis, one might have expected that the molecular susceptibility tensor would be characterized, approximately at least, as axial with an essentially unique property nearly parallel to the SCN—M—NCS vector. Obviously the fact that the empirical results are so widely different from this expectation means that the relatively small departures from D_{4h} symmetry are important. We can discern three main characteristics of the lower symmetry: (i) that the M—NCS moieties are, as usual, not linear, (ii) that the pyridine planes are strongly inclined to planes normal to the equatorial MN_4 plane and in a centrosymmetric fashion which necessarily destroys the four-fold rotation symmetry, and (iii) that small displacements of even the donor atoms, no doubt due to both inter- and intramolecular forces, exist to lower the symmetry from the ideal. These various 'distortions' can manifest themselves within the ligand field approximately as follows. Small displacements of the donor atoms from the ideal will predominantly affect the metal–ligand σ bonding. The non-linearity of the M—NCS moieties might give rise to unequal values for $e_{\pi\parallel}(NCS)$ and $e_{\pi\perp}(NCS)$ and so will be detected within the π-bonding scheme. The orientation of the equatorial pyridine groups will similarly contribute to the π-bonding scheme, unless $e_{\pi\perp}(py)$ had been exactly zero (or equal to $e_{\pi\parallel}(py)$). Now consider the ground term of the d^7 system. At the level of O_h symmetry, which is satisfactory for the present discussion, the $^4T_{1g}$ ground term correlates with the strong-field O_h configuration $(t_{2g})^5(e_g)^2$. Similarly for the Fe(II) d^6 system, we are concerned with a $^5T_{2g}$ term arising from the configuration $(t_{2g})^4(e_g)^2$. Note that, in both cases, the orbital angular momentum of the T ground terms is associated with the incompletely filled t_{2g} submanifold. Therefore the magnetic anisotropy in the molecule arises predominantly from the raising of the degeneracy of the t_{2g} orbital manifold by 'distortions': and, of course, this is most directly associated, in the octahedral coordination, with metal–ligand π bonding. Now the final optimized fit was obtained using an average π-bonding parameter for the M—thiocynate interactions (in the interests of reducing the degree of parametrization) and, in any case, it was found that the calculated susceptibilities were not hypersensitive to anisotropy in $e_\pi(NCS)$ parameters. This means, then, that the final *calculated* orientations of the molecular magnetic susceptibilities owe nothing to the non-linearity of the M—NCS entity. In turn the determinant of these 'unusual' orientations is the anisotropic π bonding of the metal to the four pyridine groups. The large magnitude of the magnetic anisotropy in these molecules arises from the fact that the degeneracy of the t_{2g} orbitals is only raised a little. The sort of behaviour described in figure 1.6(*a*) and (*b*) is typical in this respect: maximum anisotropy occurs for distortions giving rise to splittings of the order of kT. The sensitivity with which the $e_{\pi\perp}(py)$ parameters were determined (*absolutely*, to better than $50\,cm^{-1}$) is associated with the same ligand-field condition. In essence, this arises because several components of the $^4T_{2g}$ or $^5T_{2g}$ terms, split by the low-symmetry fields plus spin-orbit coupling, are thermally populated which, together with large second-order Zeeman terms, means that the calculated temperature dependences of susceptibilities are sensitive functions of these distortions. Altogether, therefore, the magnetism is especially sensitive to the M—pyridine π bonding, at least so far as anisotropy and hence tensor orientation are concerned, and in this respect we can say that the paramagnetism 'samples' some parts of the electronic manifold more strongly than others. All these aspects are fully absorbed within the detail of our standard computations, of course. Problems can only arise if one tries to work in reverse by idealizing a molecular and electronic symmetry and hence to predict the orientation of the magnetic property. Of course, this may not always be a serious problem, as the sensitivity of magnetic properties to distortion we have reviewed depends strongly upon the configurations in question.

Leaving this more technical commentary, we next consider a recent study[8] involving the combined use of optical spectroscopy on single crystals in both polarized and unpolarized light; of single-crystal susceptibility measurements over a wide temperature range; and of both powder and single-crystal esr spectroscopy. The isomorphous[8, 9] complexes $[Bu_4N]^+[M(quinoline)Br_3]^-$; $M = Ni(II)$, $Co(II)$ crystallize in the triclinic space groups $P\bar{1}$ and involve tetrahedrally coordinated metal ions suffering very approximate three-fold distortion. The single-crystal susceptibilities were measured using the techniques described in §6.3. Maximum and minimum susceptibilities in each of eight different identified crystal planes of each complex were measured throughout the temperature range 20–300 K: details of this study were used to illustrate the method in §6.3.5. The results are illustrated in figure 12.2, showing principal sections of the molecular susceptibility ellipsoid at 25 K for both nickel(II) and cobalt(II) complexes. From the projected molecular geometries also presented in the figure, we observe that in both complexes a principal susceptibility lies within $ca.23°$ of the M—N vector but that while this is the largest susceptibility in the nickel complex, it is the smallest in the cobalt system: this difference merely reflects the different d^n configurations.

Transmission spectra for single crystals of these complexes at about 4 K, in figure 12.3, though not especially well resolved, show considerable structural detail. Bertini et al.[10] have reported a polarized crystal spectrum for the cobalt complex and, although the significance of the polarization may be unclear in these triclinic crystals, their observed peak maxima are included in the diagram in view of the resolution of two components of the absorption at about $7000\,cm^{-1}$.

AOM analysis for both complexes have been performed within the complete d^n manifolds for variation of the parameter sets: Racah B and C parameters; spin-orbit coupling ζ and orbital reduction factor k; and AOM parameters $e_\sigma(quin)$, $e_\sigma(Br)$, $e_{\pi\|}(quin)$, $e_{\pi\perp}(quin)$, $e_\pi(Br)$. The optimal fitting to ligand–field properties afforded by the parameter sets in table 12.1 are shown in figures 12.2 and 12.3. Fair reproduction of most of the many features in each optical spectrum is achieved and all details of the magnetic susceptibilities are reproduced very closely indeed: the 'best-fit' ellipsoid

[Ni (quinoline) Br$_3$]$^-$ [Co (quinoline) Br$_3$]$^-$

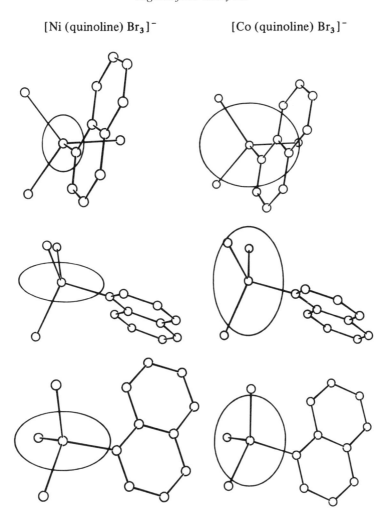

Fig. 12.2. Projections of complex ion geometry and susceptibility ellipsoids onto planes perpendicular to principal magnetic susceptibilities; on the left for [Ni(quinoline)Br$_3$]$^-$ and on the right for [Co(quinoline)Br$_3$]$^-$. Susceptibility ellipses are shown for values at 25 K and are virtually identical for observed and calculated results. The drawing scale for the susceptibilities of the cobalt complex is about five times smaller than that for the nickel one.

projections corresponding to the experimental ones shown in figure 12.2 overlay these within the thickness of the printed curves. Similarly, the observed[111] **g** tensor of the cobalt complex, characterized by the principal *g* values 1.61, 2.33, 6.51, is reproduced within 0.02 in each value and within 2° of the experimentally observed orientation (which latter, incidentally,

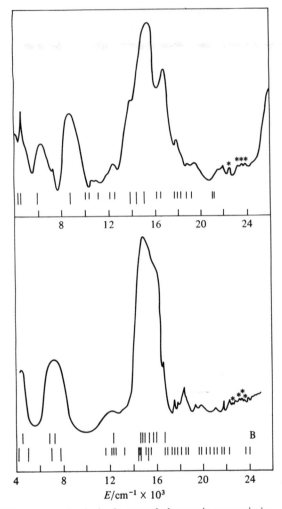

Fig. 12.3. Unpolarized, single-crystal electronic transmission spectra for [Ni(quinoline)Br$_3$]$^-$ (top) and [Co(quinoline)Br$_3$]$^-$ (bottom). Calculated 'best-fit' transition energies, corresponding to the para-meter sets in table 12.1, are represented by long markers for spin-allowed transitions and by short ones for spin-forbidden transitions. Completely or partly resolved peak maxima observed by Bertini *et al.*[10] for the cobalt complex are indicated at B. Asterisks mark presumed non-*d*–*d* transitions, being common to both complexes.

lies within 3% of the experimental and calculated orientation of the susceptibility ellipsoid). Altogether, we see that a wide variety of ligand-field properties can all be reproduced within a common ligand-field parameter set, for each of the two complexes. The chemically interesting result for

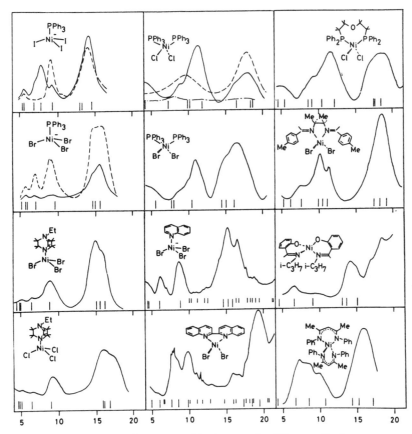

Fig. 12.4. Observed spectra (single-crystal, diffuse reflection or solution) and calculated transition energies (cm^{-1} × 10^3) for several nominally tetrahedrally coordinated complexes of nickel(II).

the present discussion is that these parameter sets identify a significant π-acid role for the quinoline ligands, for which $e_{\pi\perp}$ takes the value of about $-500\,\text{cm}^{-1}$.

Compare the π-acceptor role for the quinoline groups in these molecules on the one hand with the very small π (donor) role of the pyridines in the M(py)$_4$(NCS)$_2$ systems and, on the other, with an even greater π-acid role of biquinoline, established from a single-crystal, optical spectral study[12] of Ni(biquinoline)Br$_2$. The coordination geometry, spectrum and calculated band assignments for this tetrahedrally coordinated complex are shown in figure 12.4: optimized parameters are given in table 12.1 and include the value $e_{\pi\perp}$(biquinoline) as about $-1000\,\text{cm}^{-1}$. The trend in $e_{\pi\perp}$ values for pyridine, quinoline and biquinoline is qualitatively in agreement

Table 12.1. *AOM and Racah B parameters* (cm^{-1}) *for some tetrahedrally coordinated complexes of nickel(II) and cobalt(II)*

Complex	$e_\sigma(P)$	$e_\pi(P)$	$e_\sigma(hal)$	$e_\pi(hal)$	$e_\sigma(N)$	$e_{\pi\perp}(N)$	B	reference
[Co(quinoline)Br$_3$]$^-$	—	—	3000	450	3500	−500	670	8
[Ni(quinoline)Br$_3$]$^-$	—	—	3600	500	3600	−600	720	8, 12
Ni(biquinoline)Br$_2$	—	—	3500	850	4200	−1000	790	12
[Ni(PPh$_3$)Br$_3$]$^-$	5000	−1500	3000	700	—	—	620	15
[Ni(PPh$_3$)I$_3$]$^-$	6000	−1500	2000	600	—	—	490	15
Ni(PPh$_3$)$_2$Cl$_2$	(4500)	(−2500)	(4500)	(2000)	—	—	(550)	17
Ni(PPh$_3$)$_2$Br$_2$	(4000)	(−1500)	(4000)	(1500)	—	—	(550)	17
Co(PPh$_3$)$_2$Cl$_2$	(4000)	(−1000)	(3500)	(2000)	—	—	(575)	17
Co(PPh$_3$)$_2$Br$_2$	(3500)	(−1000)	(3500)	(1500)	—	—	(575)	17
Ni(P—O—P)Cl$_2$[a]	5000	−1500	3600	1500	—	—	550	12
Ni(P—O—O—P)I$_2$[b]	(6000)	(−1500)	(2500)	(500)	—	—	(330)	12

[a] See figure 12.4.
[b] See figure 12.6.
Values in parentheses are less sensitively determined.

with the tendency for ligands like biquinoline, bipyridyl or phenantholine, for example, to stabilize metals in their lower oxidation states. Bearing in mind our remarks in §11.14 concerning the significance of AOM parameters as probes of the metal–ligand bonding as it occurs in the bonded complex, we should, perhaps, read the above trend as; the biquinoline *has accepted* more π-electron density from the metal than has the quinoline or, in turn, than have the pyridines.

12.1.2 Enlargement of the basis in the analysis of [Co(quinoline) Br$_3$]$^-$

Of the many interesting details in the ligand-field analysis[8] of the [M(quinoline)Br$_3$]$^-$ complex ions, we consider just one, concerned with the relative sensitivities of calculated *g* values and susceptibilities to the chosen basis. As for the nickel complex, the AOM analysis began within a basis restricted to those terms of maximum spin multiplicity. Most of the ligand-field parameters were established more easily by reference to the electronic spectrum than the magnetic properties: indeed, reproduction of these latter tended to furnish a means of refinement more than search. Anyway, having completed that part of the analysis, an attempt was made to reproduce the spectral spin-forbidden bands: the same procedure had been carried out for the nickel complex also. As the weak, spin doublets of the d^7 system were not well resolved nor, in themselves, were any assignments at all apparent, the tactic adopted was to do no more than optimize their reproduction by variation of the Racah *C* parameter alone. Incidentally, this was achieved by using RUN input cards (see chapter 10) for F_2 and F_4, together with a LINK card, in such a way that F_2 and F_4 were varied while maintaining the relationship $B = F_2 - 5F_4 = \text{constant}$, as determined by fitting the spin-allowed bands. An order-of-magnitude fit for the spin doublets was readily achieved in this way, as indicated by the short markers in figure 12.3(*b*) but only by placing one calculated doublet as low as about $11\,500\,\text{cm}^{-1}$ above ground. The question then arose as to how large an effect on the calculated magnetic properties such 'close' doublets would have. It transpires that, so far as the susceptibilities are concerned, the effect is no more than about 0.5%. It is worth observing here that at least one example is known where the inclusion of doublets into the quartet manifold has a marked effect on susceptibilities.[13] While the effect is very small in the present complexes, that on the *g* values is not. On inclusion of the spin doublets, optimized to the spectrum as described, the calculated principal *g* values change by about 0.2, which is about ten times experimental error. It was found subsequently that relatively small changes in the e_π parameters allowed good fits with the esr experiment to be recovered, while having only a minor effect on the complete spectrum. The point of the present aside, therefore, is to relate how g-values and susceptibilities can (but do not need to) show significantly different sensitivities to the system parameters.

12.2 Phosphine complexes

Several studies of phosphine complexes have been made within the AOM, with results which bear strongly upon the π-bonding role of these ligands with transition metals. This subject has long been of interest – in connection with the *trans* effect, for example – although the extent of phosphine π bonding has been a controversial issue[14] in assessments made through indirect experimental measurements of infrared stretching frequencies, nmr coupling constants and X-ray-determined bond lengths.

Two monophosphine complexes, $[Ph_4As]^+[Ni(PPh_3)X_3]^-$; $X = Br$, I have been the subject of magnetic and spectral study.[15] Polarized, single-crystal spectra at various temperatures between 80 and 300 K have been reported for both complexes.[16] These and single-crystal paramagnetic susceptibilities for the bromo complex throughout the temperature range 20–300 K, and powder susceptibilities for the iodo one from 80 to 300 K formed the data set for the AOM analyses.[15] Fits to the spectral transitions are shown in figure 12.4. The optimal parameter sets affording these are given in table 12.1 and identify several interesting features of the metal-ligand bonding. Thus

(i) The phosphine groups act as strong σ donors and strong π acceptors, a conclusion which is based on comparison of the $e_\sigma(P)$ and $e_\pi(P)$ values, not only with corresponding values from the complexes already discussed and those which follow but also with the e_λ values observed for the halogens. In the present systems, all e_λ and B values are quite sensitively determined, by which is meant within about 10%. The idea that phosphine groups behave towards transition metals in a similar fashion to carbonyl groups, for example, is not new but the analyses summarized here present a clear indication of the separate σ-and π-bonding roles adopted by these ligands, and a semi-quantitative description of the synergic 'back-bonding' model.

(ii) The halogen ligands function as both σ and π donors, as is to be expected from their common occurrence in complexes with transition metals in higher oxidation states, for example, and we observe that both

σ- and π-donor roles are diminished somewhat as we replace bromines by iodines. The rather *modest* decrease in Δ_{oct} values throughout the spectrochemical series for the halogens arises from the conflicting trends from σ donation and π donation, together with a quantitative decrease in both contributions down the series.

(iii) Considering that the e_λ values of the halogens provide a measure of the donation of negative charge to the central metal, we note that three iodines, each with e_σ values of $2000\,cm^{-1}$, are associated with an $e_\sigma(P)$ value of $6000\,cm^{-1}$, as compared with one of only $5000\,cm^{-1}$ associated with the presence of three bromines, where $e_\sigma(Br)$ is $3000\,cm^{-1}$. These parameter values thus reflect the operation of the electroneutrality principle throughout the molecule as a whole and not just with respect to the synergic processes associated with a single M—L interaction. It is satisfying, too, to observe that the greater M—P σ interaction in the iodo complex is associated with a shorter bond length compared with that in the bromo molecule; 2.28 Å versus 2.32Å. The variation in $e_\sigma(P)$ values with other ligand substitutions, in these and other systems to be discussed, provides a clear example of the non-transferability of AOM parameters in detail and, at the same time, emphasizes the way in which AOM parameters can, directly and transparently, reflect features of chemical bonding.

An earlier AOM study[17] of phosphine complexes had been concerned with the species $M(PPh_3)_2X_2$; $M = Co(II)$, $Ni(II)$ and $X = Cl$, Br. One member of the series, $Ni(PPh_3)_2Cl_2$, is famed as the first tetrahedrally coordinated complex of nickel(II) to be synthesized.[18] The nickel complexes, at least, crystallize[17] in several different lattices but there is one monoclinic form common to all four molecules and this was the one studied in this analysis. Well resolved, low- and room-temperature polarized spectra have been reported for both cobalt complexes[19,20] and for the chloro nickel one[21]: a diffuse reflectance spectrum is available[22] for $Ni(PPh_3)_2Br_2$. Complete sets of single-crystal paramagnetic susceptibility data have been obtained for all four systems throughout the temperature range 20–300 K. Ligand-field analyses were characterized with the relatively small AOM parameter set of $e_\sigma(P)$, $e_\sigma(X)$, $e_\pi(P)$ and $e_\pi(X)$, as both types of ligand are linear ligators. The extensive experimental data set taken together with the small number of theoretical degrees of freedom was expected to provide a favourable situation for the determination of the system parameters. However, despite the excellent reproduction of the detailed susceptibility tensors; of the spectral transition energies; and of the spectral assignments (established by the application[19-21] of dipole selection rules to the polarized spectra), most parameter values were strongly correlated and somewhat wide regions of parameter space include equally satisfactory fits to these data. In general, it does not seem possible to predict those systems which should provide clear-cut results. In parenthesis, we note that this insensitivity of the fitting parameters discouraged our pursuing what otherwise might have been an interesting study: namely, to perform similar experimental and analytical work on $Ni(PPh_3)_2Br_2$ in its different crystalline forms. To find such similar molecular species in different lattices is not common and the effects of the different L—M—L angles in the coordination shells would have been very interesting: however.... Despite the spread in best-fit parameter values in the present study, some qualitative conclusions may be drawn firmly from the analyses. Thus,

(i) While representative best-fit parameter values are given in table 12.1, the parameter correlation referred to above is typified by that between e_π values for the phosphine and halogen ligands. Figure 12.5 shows a contour map for figures of merit, describing the quality of fit to the experimental data, as functions of $e_\pi(P)$ and $e_\pi(X)$, in the manner of figure 10.6. Very similar maps arise from all four complexes independently and for spectral and magnetic fits independently. The 'best-fit' e_λ values given in table 12.1 correspond to locations in the middle of these areas of good fit. Now there is clear and unambiguous evidence in that diagram that the phosphine ligands in these complexes function as strong π acceptors: certainly no satisfactory reproduction of either spectra or magnetism is possible for $e_\pi(P)$ near zero or positive. At the same time, the large negative values for $e_\pi(P)$ seem to be associated with large positive values for $e_\pi(X)$ and one would therefore conclude that the presence of two strong phosphine π acids augments the π-donor role of the halogen ligands. Again in terms of the electroneutrality principle, we would envisage the metal atoms as providing an electron delocalization pathway between the two types of ligand. It is especially significant that the effect is markedly greater in these bisphosphine complexes than in the monophosphine species described above, obviously reflecting the difference in the relative numbers of π acidic and basic ligands.

(ii) An additional manifestation of the proposed extended π-bonding network in these complexes is provided by the observed[17] nephelauxetic effects. The low Racah B values listed in table 12.1 are determined essentially by fitting the spectral components of $\rightarrow {}^3P$ or $\rightarrow {}^4P$ transitions. In effect the B values are established by the spectra, while AOM parameters

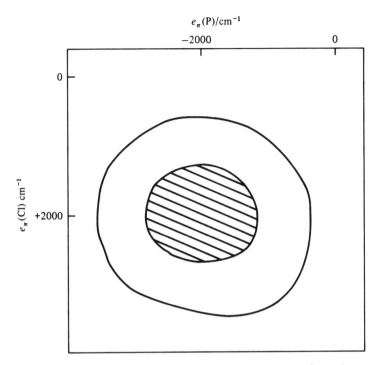

Fig. 12.5. Contour plot showing spread and correlation of e_π values yielding good reproduction of both single-crystal susceptibilities and electronic spectra of $Ni(PPh_3)Cl_2$. The hatched area represents about 1% disagreement with experiment and the annulus about 4%; agreement between theory and experiment outside both areas is poor.

derive from both spectra and magnetism: this is a nice illustration of the points raised in §10.3.3. The nephelauxetic ratios B/B_0 (where B_0 is the free-ion value) for these complexes are about 0.5–0.6. One can argue that the greater mean interelectron distances obtaining for spectral electrons in spatially extended π-bond networks, especially ones involving formally empty π^*-ligand functions, would be partly responsible for the large nephelauxetic effects observed. The same trend is also apparent from the B parameters in the monophosphine complexes. The even lower B value for the $[Ni(PPh_3)I_3]^-$ complex no doubt also reflects the softer nature of iodine ligands relative to bromine.

Some uncertainty about these conclusions inevitably attaches to this study in view of the correlation displayed by the fitting parameter values. It is surely dispelled, however, by the results of a subsequent analysis[12] of the spectrum of a chelating bisphosphine complex of nickel(II), which

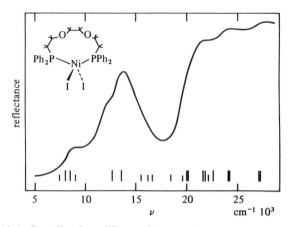

Fig. 12.6. Coordination, diffuse reflectance spectrum, and calculated transition energies for a 'tetrahedral', diamagnetic complex of nickel(II): long markers show spin-allowed singlets; and short markers, spin-forbidden triplets.

confirm each of the features just described. The geometry of the complex, first prepared by Sacconi & Greene[23] is shown in figure 12.4, together with the diffuse reflectance spectrum[24] and 'best-fit' calculated[12] transition energies. Both AOM and interelectron repulsion parameters were quite sensitively determined in the ligand-field analysis and the values shown in table 12.1 indicate an electron distribution in the complex which is very similar to that proposed for the $M(PPh_3)_2X_2$ species.

Estimates of phosphine π bonding have also been made from AOM analyses of nickel(II) and cobalt(II) complexes in their *low*-spin states. An interesting, though cursory, study of the distorted tetrahedral nickel(II) molecule shown in figure 12.6, has confirmed the general trend so far identified for phosphine complexes and at the same time has provided a basic understanding of the unusual diamagnetism in this system. The analysis[12] is based only upon the diffuse reflectance spectrum[25] at about 110 K and the calculated band energies are indicated in figure 12.6 by the longer markers for spin-allowed, spin-singlet transitions, and by the shorter ones for spin-forbidden triplets. A complete d^8 basis was used in the calculations, of course, and so two interelectron repulsion parameters had to be included in the parameter set. As the reflectance spectrum provides insufficient data with which to determine all the system parameters, the study was restricted to an investigation of whether the spectrum could be reproduced approximately (but without regard to uniqueness of fit) and simultaneously with a calculated spin-singlet diamagnetic ground term.

In line with free-ion ratios,[26,27] F_4/F_2 was fixed at 0.09 and the ground state could then be reproduced only with very low values of $B(=F_2 - 5F_4)$. Correlation between B and the AOM parameters is strong so that the parameter set given in table 12.1 is illustrative only. Nevertheless, two features are clearly established by this analysis: that this low-spin nickel(II) system is characterized by a very large nephelauxetic effect and that the phosphines function once more as strong σ donors and strong π acceptors.

Similar conclusions are reached, but with much greater confidence, in a study[28] of the low-spin, cobalt(II) complex shown in figure 1.8(*a*). The mesitylene groups in dimesitylbis(phenyldiethylphosphine)cobalt(II) lie in planes which are almost exactly perpendicular to the coordination plane.[29,30] As described in §1.4, the principal molecular g values[31] lie almost precisely parallel to the coordination bond vectors, defining a very approximately axial tensor with the quasi-unique axis parallel to P—Co—P: early analysis[31] had shown that the d_{z^2} orbital must lie much closer to the d_{yz} orbital than must the d_{xy} if the g-value pattern is to be reproduced. A more recent study,[28] using the techniques described in this book, has been based on single-crystal susceptibility data in the temperature range 20–300 K as well as the previous crystal esr work[31] and a crystal-transmission spectrum in the near-IR, visible and near-UV regions. The earlier analysis[31] had employed a basis of all spin-doublet, strong-field configurations which arise without electron occupation of the $d_{x^2-y^2}$ orbital (z is taken perpendicular to the coordination plane): this restriction was enforced by the contemporary availability of theoretical and computational power. The recent analysis,[28] on the other hand, employed the full 120-fold d^7 basis within the CAMMAG system,[32] and one technical point of interest in this analysis, therefore, was the question of how great an effect the enlarged function basis had on calculated quantities: at the same time, the final use of the *complete* d^7 basis removes any doubts concerning the effects of basis set truncation. However, as it turned out, at the particular (and atypical) location in polyparameter space defining 'best fit' the difference between calculations in the two bases was rather slight. The same was not true in the analyses[28] of some closely related analogues.

Parametrization of the ligand-field in the more complete analysis[28] was begun using the scheme described in §10.4. This was an especially useful tactic because of the relatively small number of ligand-field parameters required in the D_{2h} molecular symmetry, and also because subsequent 'interpretation' of the global orbital energies in terms of local AOM parameters follows naturally, and with minimum ambiguity, in this rather symmetrical molecule. In the first instance, off-diagonal global orbital matrix elements were ignored and, after a somewhat involved analysis, very good reproduction of susceptibilities, g values and the optical spectrum was achieved with the one-electron global orbital energies shown in figure 1.8(b). Now the ligand-field orbital energy of d_{yz} is taken as the energy reference because this orbital is not expected to interact with any ligand (for we presume zero π bonding with the mesityl groups parallel to their planes). The $d_{x^2-y^2}$ and d_{z^2} cobalt orbitals will interact with the ligand σ orbitals, d_{xy} with both phosphine and mesityl ($\perp$) π orbitals, and d_{xz} with phosphine π functions. The energy of the d_{xz} metal orbital theoretically corresponds to $2e_{\pi}(P)$; whence $e_{\pi}(P) \approx -3200\,\text{cm}^{-1}$. The large AOM parameters in this low-spin complex are accompanied by very large nephelauxetic effects such that B/B_0, for example, takes some value between 0.2 and 0.4 depending upon the estimate of the energy of the $d_{x^2-y^2}$ orbital ($18\,000\,\text{cm}^{-1}$ is a lower bound). Once again, therefore, this study furnishes further evidence of a strong π-acid role for phosphine ligands – even for this mixed alkyl/aryl phosphine – with the transition metal. It also provides qualitative comparison, at least, with the diamagnetic nickel(II) complex, described above, in that these low-spin species appear to be characterized by larger ligand-field parameters and much larger nephelauxetic effects than similar high-spin compounds.

12.2.1 Ligand-field parameters as properties of ground states

Ligand-field parameter values established by fitting to susceptibility or esr data tend to be associated more with the ground and a few low-lying excited states: by contrast, those parameters obtained by reproducing spectra are frequently thought to be more informative about higher-lying ligand-field states. It is not uncommon to summarize these observations by remarking that spectra are properties more of excited states while magnetism is a property essentially of the ground state. That such is a mistaken view is nicely illustrated by the very different values of both AOM e_{λ} values and interelectron repulsion F_k parameters obtained for complexes in high- and low-spin states as we see from a simple idealization.

Consider the following *gedanken* situation: that we may envisage the same complex in two separate circumstances corresponding to total occupancy either of a low-spin ground state or of a (nearby) excited high-spin state. We imagine that the parameter values for molecules occupying these two different states, separately and alone, can somehow be obtained and that they would show the typical sort of results we have reviewed for the real systems above; namely, that molecules in the low-spin state are characterized by significantly larger e_λ and smaller F_k values. Note, however, that parameter values for the system populating the low-spin state would derive from ligand-field properties referring to *all* ligand-field states – in principle, at least – and not just to the low-spin ones. Similarly, the 'high-spin parameter set' would also account for the complete ligand-field manifold – high and low spin. Hence both sets of parameter values, which are numerically very different, appear to describe the *same* set of molecular states. Actually the same apparent paradox would be evidenced from parameter sets referring to molecules in any two or more different states, whether they are of the same spin multiplicity or not: the present discussion concerns molecules in different spin states, essentially on the empirical grounds that herein lies the experimental demonstration of the effect.

Of course, the equilibrium nuclear configuration is different in the two spin states. Complexes in low-spin states are commonly observed to be more compact than those in high-spin states. No doubt there is rather more participation of the d orbitals in chemical bonding with the ligands in the low-spin situation. In what would have to be a self-consistent argument of interdependent cause and effect, the greater saturation of spins occasioned by the stronger bonding is associated with shorter coordination-bond lengths and larger ligand-field, AOM parameters: the greater nephelauxetic effects observed in these circumstances are to be viewed as a corollary of a less 'atomic-like' character of the d electrons as compared with the higher-spin, more weakly bonded environment. The paradox is not yet resolved by these observations, however, for it must be stressed that the various molecular vibronic states are uniquely defined: they do not change as different 'reference' states are populated – but then why should the ligand-field parameter sets?[††] The idealized energy diagram in figure 12.7 provides a focus for an interpretation.

Various energy surfaces are shown in the diagram as functions of a typical displacement coordinate. Smaller dimensions associated with the low-spin ground state are represented in the figure, but there is, of course,

---

[††] We are not concerned here with the point that experimental *data* obtained for molecules in high- and low-spin states will be different: for example, different magnetic susceptibilities and, because of selection rules, spectra of vastly different appearance. To point up this discussion, imagine (only in the context of this footnote) that all potential energy surfaces, like those in figure 12.7, are identical and have minima at a common nuclear configuration: then, for molecules populating different states, very different experimental data would be fitted within a ligand-field model (AOM or not) by the *same* parameter set.

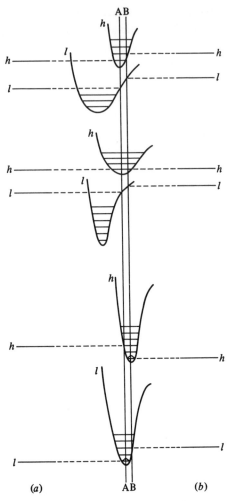

Fig. 12.7. Various high-spin (*h*) and low-spin (*l*) energy surfaces are shown as functions of a typical displacement coordinate.

no reason to suppose that all low-spin states share the same nuclear configuration. The vertical lines A–A and B–B mark the equilibrium configurations for molecules populating the ground and low-lying excited states. These lines have a two-fold significance. Most immediately, they illustrate the Franck–Condon principle for spectral transitions. As shown by the horizontal lines in (*a*) and (*b*), the transition energies, corresponding presumably most nearly to positions of band maxima, will clearly differ for molecules having these two different nuclear configurations. Manifestly, the sets of electronic transition energies are properties of the different populated, 'reference' states. Since these transition energies, together with any magnetic data, are then fitted by sets of ligand-field parameters, these parameter sets are similarly properties of the different reference states and so we identify the second significant attribute of the vertical markers in figure 12.7. We would, indeed, like to say that the two sets of parameter values (for high- and low-spin reference states) are *defined* for the appropriate equilibrium nuclear configurations by the theory we have expounded in chapter 11. In order to do this, however, we must decompose the preceding argument into two separate steps.

In any given ligand-field analysis, experimental data are reproduced by variation of parameters associated with terms appearing in the Hamiltonian $\mathcal{H}'_{LF}$ of (III.1). Only those terms within the so-called 'ligand-field potential', V_{LF}, bear any relation to the molecular geometry. Now within the conventional multipole expansion of V_{LF}, *all* geometrical details are sequestered within the expansion coefficients c_{kq}. Therefore, within this form of parametrization, different sets of parameter values obtained by 'best fit' to the different transition energies associated with the different 'reference' states, are defined *a fortiori* to be properties of those reference states. As described explicitly in chapters 9 and 10, however, the $\{e_\lambda\}$ of the AOM are linearly related to the $\{c_{kq}\}$ of the multipole expansion and so they too are *defined* as properties of the reference state. Within this last step, however, are contained the further, but conventional, features of the definition of AOM e_λ parameters: namely, that the $\{e_\lambda\}$ are *interpreted* by reference to the usual assertion that the unitary matrices **R** of (9.19) are rotation matrices relating metal and ligand coordinate frames and which render the local ligand-field matrices $\{\mathbf{v}^l\}$ diagonal. In short, the *formalism* of the model we have studied remains intact and separately describes parameter sets associated with different reference states. The present discussion extends the nature of the averaging implicit in the *interpretation* of AOM parameters, however. In §11.6, for example, we discussed the concept of 'anonymous configuration parentage' in recognition of the averaged nature of the ligand-field projection involving electronic states. Our present discussion shows that a further averaging is implicit in a process which identifies matrix elements of vibronic states under an electronic Hamiltonian with combinations of electronic orbital matrix elements like (11.107): this further averaging involves, effectively, the overlap integrals of the various nuclear wavefunctions implied by figure 12.7. Ultimately, therefore, it is a matter of definition and assertion that best-fit AOM e_λ parameters and interelectron F_k parameters are to be interpreted as properties of the reference ground state.

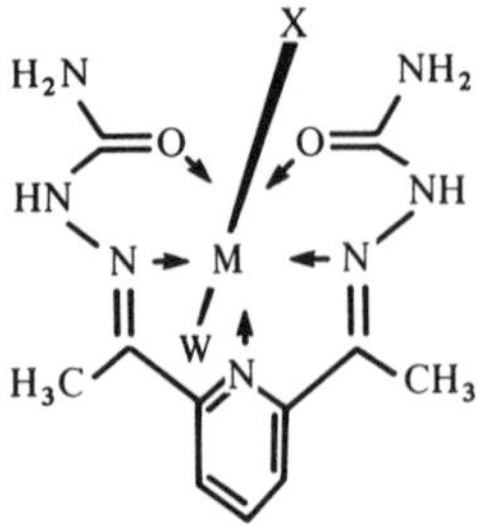

Fig. 12.8. Pentagonal bipyramidal, seven-coordination for the series $M(dapsc)WX^{n+}$; $W = H_2O$; $X = H_2O$ or Cl; $M = Fe(II)$, $Co(II)$, $Ni(II)$, $Cu(II)$.

12.3 Seven-coordinate complexes

The 'interactive' nature of AOM parameters is emphasized by some studies of seven-coordinate complexes having approximate pentagonal bipyramidal molecular geometry. We consider a series of molecules containing the, essentially planar, pentadentate ligand, 2,6-diacetylpyridinebis (semi-carbazone) – dapsc – shown in figure 12.8 X-ray structural analyses on a wide range of complexes $M(dapsc)WX^{n+}$; $M = Fe(II)$, $Co(II)$, $Ni(II)$, $Cu(II)$; $W = H_2O$ and $X = H_2O$ or Cl have been reported[34,35] by Palenik & Wester. AOM analyses[36,37] have been based upon single-crystal susceptibility measurements throughout the temperature range 20–300 K and unpolarized crystal transmission spectra for the four complexes $[Fe(dapsc)Cl.H_2O]^+$, $[Co(dapsc)Cl.H_2O]^+$, $[Co(dapsc)(H_2O)_2]^{2+}$ and $[Ni(dapsc)(H_2O)_2]^{2+}$, the choice of axial ligands being determined by the ability to prepare suitable, large crystals for the magnetic experiments. One would not expect these to be good systems for an AOM study. There are too many parameters: e_σ, $e_{\pi\|}$ and $e_{\pi\perp}$ ($\|$ and $\perp$ to the pentadentate plane) for each of the three different types of functional groups in the dapsc ligand – pyridine, imine and keto-type moieties – together with mean e_σ and e_π values for the diametrically opposite W and X ligands (assumed to be cylindrically symmetric, 'linear ligators', for simplicity). Further, the π functions of the axial ligands will overlap with the same metal orbitals (d_{xz} and d_{yz}, for z perpendicular to the dapsc plane) as do $\pi_\perp$ orbitals of the various dapsc functional groups. These and several other similar possible correlations amongst the whole set of AOM variables, together with the very approximate five-fold nature of the global geometry, were likely to frustrate any attempt to discriminate the individual ligating functions. In short, these systems were somewhat the reverse of the $M(PPh_3)_2X_2$ series in terms of promise; and yet it transpires that

Table 12.2. *AOM and Racah parameters* (cm^{-1}) *for* $[M^{II}LWX]^{n+}$ *complexes*,[36,37] $L = dapsc$, $W = H_2O$, $X = H_2O$ *or* Cl. *Values suffer correlation and are given as illustrative only – see text*

Parameter		FeLWCl$^+$	CoLWCl$^+$	CoLW$_2^{2+}$	NiLW$_2^{2+}$
N(py)	e_σ	4400	4500	4500	4800
	$e_{\pi\parallel}$	0	0	0	0
	$\lvert e_{\pi\perp}\rvert$	< 200	< 200	< 200	< 200
N(imine)	e_σ	3500	3500	3500	4200
	$e_{\pi\parallel}$	0	0	0	0
	$e_{\pi\perp}$	1000	1000	1000	1000
O(keto)	e_σ	2500	2500	2500	2000
	$e_{\pi\parallel}$	400	400	400	0
	$e_{\pi\perp}$	2000	2000	2000	0
mean of W, X	e_σ	4000	4000	4500	5200
	e_π	800	800	400	500
	B	?	800	800	850

considerable insight into these molecules, and into the AOM method itself, was to be gained from the present analyses.

The parameter values listed in table 12.2 represent typical 'best fit' sets: there are correlations between many of them which are detailed in the original work,[36,37] but the numbers describe the essential quality of the parameters in more than a subjective way. Most values for both cobalt(II) complexes are essentially identical and describe a situation in which unexceptional, 'normal' values are obtained for pyridine σ and π bonding, imine σ and π bonding and for the σ and π interactions with the axial water and chlorine ligands: even the greater π-donor character of the chlorine ligand relative to water is reflected in these data. Particular interest, however, attaches to the values established for the semi-carbazone keto groups. Firstly, we note that very small e_σ values for these functional groups obtain, despite the non-special M—O bond lengths shown in table 12.3. It appears that the neutrality of these oxygens bonded to —NH and —NH$_2$ moieties characterizes a low base strength, commensurate with an expected high acidity of the conjugate acid. Perhaps it should not be a surprise to find these keto groups acting as weak σ bases in comparison, say, with negatively-charged phenolic groups, for which an e_σ value of about 4000 cm^{-1} has been observed (see §12.6). Far more interesting are the large, positive $e_{\pi\perp}$ (keto) values found for these complexes. No doubt the considerable π-donor role of these keto groups derives in part from the adjacent trigonally hybridized nitrogen groups.

Table 12.3. *Mean bond lengths (Å) in* $[M^{II}LWX]^{n+}$ *complexes*[34,35]
$L = dapsc$, $W = H_2O$, $X = H_2O$ *or* Cl *(see figure* 12.8*)*

bond	$FeLWCl^+$	$CoLWCl^+$	$CoLW_2^{2+}$	$NiLW_2^{2+}$	$CuLW_2^{2+}$
M—N(py)	2.22	2.19	2.19	2.06	2.27
M—N(imine)	2.22	2.19	2.21	2.16	2.26
M—O(keto)	2.18	2.18	2.18	2.35	2.35
M—O(W)	2.15	2.14	2.15	2.07	1.92
M—Cl	2.51	2.48	—	—	—

It is also possible, however, that keto groups might generally function as good π donors because carbonyl groups act as good π acids. The π and π^* orbitals of the CO entity shown in figure 12.9 illustrate the greater electronegativity of oxygen relative to carbon. Bonded to a transition metal as a carbonyl group, CO acts as a strong π acid, augmented by the usual synergic process, but when part of a ketone group bonded via the oxygen atom, it functions as a π base: we might thus view the high π basicity of the keto group as a corollary of the high π acidity of the carbonyl group. It would, of course, be desirable to confirm this characteristic of the keto ligator in different systems.

On passing from the cobalt(II) complexes to the nickel(II) one, we note that the most striking change in the parameter values given in table 12.2 is the total loss of any keto π interaction. Only magnetic data were useful for the analysis of the iron(II) complex and the most that could be achieved here was to establish which of the two types of behaviour was characteristic – that is, requiring a large $e_{\pi\perp}$ (keto) value as in the cobalt species or a zero one as in the nickel. With other AOM parameter values held fixed near to those determined for the cobalt species. the principal magnetic susceptibilities of the iron(II) complex could be reproduced well, only for large $e_{\pi\perp}$ (keto) values. So, within the series of iron, cobalt and nickel systems in table 12.2, only the nickel one displayed a total lack of π interaction with the keto groups. This correlates immediately with the greater mean Ni—O distance compared with the other M—O bond lengths, as summarized in table 12.3. Since we expect bonding to be especially sensitive to bond-length variation, the explanation of the zero $e_{\pi\perp}$ (keto) value in the nickel complex seems ready enough. It is most interesting, however, to try and use the AOM analysis in this series to rationalize the bond-length changes responsible for the variation in the ligand-field parameters themselves.

Consider the ligand-field orbital energy diagram corresponding to the

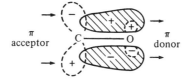

Fig. 12.9. π-acceptor functionality of carbonyls contrasted with π-donor property of ketones.

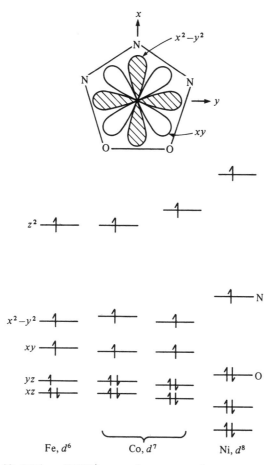

Fig. 12.10. M(dapsc)WX^{n+} complexes: one-electron, or orbital, energy diagrams calculated using typical 'best-fit' parameters from table 12.2 but without spin-orbit coupling. Orbitals are labelled with predominant parentage. Shading is for clarity and is not intended to reflect electron occupancy.

'best-fit' parameter sets given in table 12.2: these are constructed by performing calculations for the d^1 configuration with the parameter values in the table, but without spin-orbit coupling. In figure 12.10 these orbital energies are labelled according to their predominant cubic parentage (the speciation is reasonably 'pure' here, because of the fairly high molecular symmetry – C_{2v} tending to D_{5h}). Included in the figure is a diagram of the coordinate frame, together with sketches of the $d_{x^2-y^2}$ and d_{xy} orbitals in relation to the dapsc ligating groups. Note that the $d_{x^2-y^2}$ orbital is directed predominantly toward the three nitrogen donor atoms while the d_{xy} points mainly at the keto oxygen atoms: these orbitals have been labelled N and O accordingly in the orbital energy diagram. Now, as we pass from the d^7 cobalt system to the d^8 nickel one, the extra 'metal' electron is placed within the d_{xy} orbital which therefore acquires a decreased acceptor character. In our interpretations of e_λ parameters, we have so far concentrated on the *ligand* donor–acceptor function, but we must recall that these quantities describe the mutual interaction between metal and ligand in the complex, *as it is*; so we may now focus advantageously upon *metal* acceptor–donor character. The double-electron occupation of the d_{xy} orbital, directed more towards the keto groups, decreases the σ bonding and, in consequence, the oxygen atoms have moved further from the nickel atom. The most sensitive electronic feature of this bond lengthening is the metal-oxygen π bond weakening and the difference between the orbital energy diagram for the nickel complex and those for the d^6 and d^7 complexes derives from this effect. This viewpoint, in which we have interactively correlated metal configuration and detailed coordination geometry, invites extrapolation. Thus we predict that, following the addition of a further electron into the $d_{x^2-y^2}$ orbital on passing to the d^9, copper(II) system, we should observe similar lengthening of the Cu—N bonds also. Such is indeed the case, as we see from table 12.3. Moreover, the bond-length increases, measured by the X-ray studies,[34,35] are contrary to an expected contraction due to the increasing effective nuclear charge across the transition period. In fact, it appears that the in-plane bond-lengthening results in a closer approach of the axial ligands, also shown in table 12.3. A study[38] of the optical spectrum and esr g values in $[Cu(dapsc)(H_2O)_2]^{2+}$ confirms these ideas, describing a situation in which there is a significant enhancement of the ligand-field parameters for the axial water ligands as compared with the d^6–d^8 species, together with an essential lack of *any* π bonding with the dapsc ligand.

The chemical relevance of the AOM is demonstrated most dramatically in a study[39] of the spectrum and single-crystal paramagnetism of another

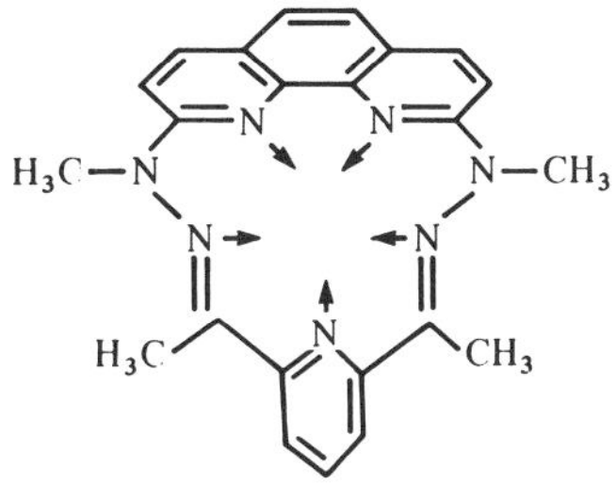

Fig. 12.11. The macrocycle tdmmb.

formally pentagonal bipyrimidal complex involving the macrocyclic ligand shown in figure 12.11. There are obvious similarities between the geometries of the donor set provided by this ligand (tdmmb) and by dapsc: instead of the two keto groups, we now have the two nitrogen ligators of a phenanthroline moiety. In its transition-metal complexes, the pentadentate macrocycle is slightly puckered from the plane[40,41] due to steric interaction of the methyl groups, but the effect is not important for the present discussion.

The analysis of the magnetic susceptibilities of $[Ni(tdmmb)(H_2O)_2]^{2+}$ began by recognizing the even closer approach to five-fold, ligand-field symmetry than in the corresponding dapsc complex. In the limit of exact five-fold symmetry, the orbital ordering could be expected to be $d_{z^2} > d_{xy}$, $d_{x^2-y^2} > d_{xz}$, d_{yz}, so that only two ligand-field parameters are required, corresponding to the transition energies $\varepsilon(d_{z^2} - d_{xz})$ and $\varepsilon(d_{xy} - d_{xz})$. Were this high-symmetry model successful in reproducing the experimental spectra and magnetism, virtually no useful chemical insight would emerge. As discussed in §9.8, this is a typical situation for many high-symmetry chromophores, but the high degree of parametrization required by an AOM approach cannot be tolerated in such circumstances. However, it transpires that no combination of values for these two transition energies, taken with a wide variety of Racah B values and spin-orbit coupling coefficients could reproduce the measured susceptibilities even qualitatively. Typically, calculated magnetic anisotropies are nearly an order of magnitude greater than observed. Analysis which recognizes the real, low symmetry of the complex is therefore both justified and necessary.

An AOM study[39] of this system had to incorporate various simplifying assumptions in view of the large number of ligand types and the sort of parameter correlations expected and observed in the dapsc series. Details of these approximations and their justification are not reproduced here but it is worth remarking that, despite the much-less-than-ideal subjects

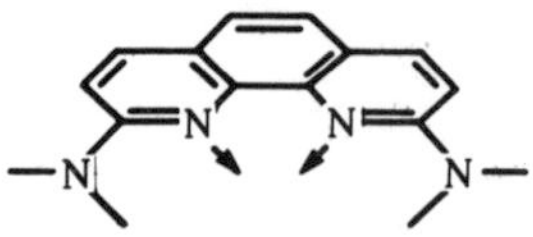

Fig. 12.12. The α, α′-diaminophenanthroline functions as a σ and π donor towards nickel(II) and cobalt(II).

of this and the preceding dapsc studies, the analyses have been tremendously valuable. The main conclusion of the present study, for example, is of considerable chemical interest. It is that the ligand-field perturbation is dominated by the phenanthroline moiety of the macrocycle. Not only does this group function as a strong σ donor towards the central metal, which was expected, but also as a strong π *donor*, which was not. Phenanthroline itself would surely act as a strong π acid, in line with the ability of this ligand to stabilize transition metals in their lower oxidation states. The pale yellow colour of the present macrocyclic complex is probably associated with the reversed π-bonding role of the phenanthroline group. The unusual π basicity of the group presumably arises from the presence of the adjacent, electron-rich, non-coordinating sp^2 nitrogen atoms, with the phenanthroline group itself providing a π delocalization pathway from these peripheral nitrogens towards the central nickel atom. So the effective functional group dominating the ligand-field properties in this complex appears to be the α, α′ diaminophenanthroline moiety shown in figure 12.12. It would be very interesting to see if synthetic work directed toward the replacement of the sp^2 nitrogen groups either by sp^3 nitrogens or by carbon atoms, for example, would yield systems in which the more usual π acidity of the phenanthroline moiety is restored. The transparent chemical relevance of the contemporary AOM approach to ligand-field phenomena is surely obvious from studies of this kind: at the same time, we are warned again of the possible non-transferability of ligand-field parameters between systems, for we are probing the electron distribution throughout the molecule as a whole.

12.4 Further examples of metal–ligand complementarity

The complexes $M(dabco^+)X_3$; $M = Co(II)$, $Ni(II)$; $X = Cl$, Br, where dabco$^+$ is the positively-charged, bicyclic tertiary amine shown in figure 12.13, provide examples[42,43] of trigonally distorted, tetrahedral complexes with nearly perfect C_{3v} symmetry. Once again, analyses[44] have been based upon complete single-crystal susceptibility measurements in the range 20 – 300 K, together with polarized and unpolarized optical spectroscopy.

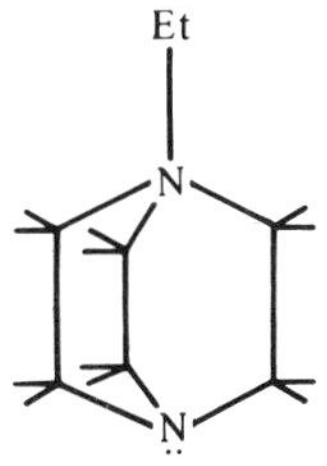

Fig. 12.13. The bicyclic, tertiary amine ion, dabconium[+].

Structure analyses[45] have been carried out on all four complexes and it was found that each complex crystallizes in a different lattice in two cases with two molecules in the asymmetric unit. In these latter cases, appropriate tensorial addition of the, presumed identical, molecular susceptibilities of each member of the crystallographically asymmetric unit was required and this, together with the very deliquescent nature of the $Ni(dabco^+)Cl_3$ crystals, provided some practical difficulties and necessitated some approximations. Nevertheless, values for all ligand-field parameters were quite sensitively determined in this study. The parameter set comprised $e_\sigma(dabco^+)$, $e_\sigma(X)$, $e_\pi(X)$, together with B, ζ and k as usual. It was presumed that the tertiary amines would not enter into π bonding with the metal atoms, for the high approximate molecular symmetry precluded the determination of a greater number of parameters than the set just described. Indeed, these analyses proved to be some of the simplest and most direct undertaken within the CAMMAG scheme and all the values given in table 12.4 have been well established. The most notable feature of the AOM parameters determined in this study is the considerably larger values for $e_\sigma(dabco^+)$ in the nickel(II) complexes than in the cobalt(II) ones; this despite an essential constancy for the e_σ and e_π parameters for the halogen ligands. These results may be understood along the lines pursued for the dapsc complexes in §12.3. The one-electron, ligand-field-orbital diagrams in figure 12.14 have been constructed using the best-fit AOM parameters listed in table 12.4. On passing from the d^7 cobalt species to the d^8 nickel ones, the extra 'metal' electron is placed in the $d_{xy}/d_{x^2-y^2}$ orbital pair (which are not exactly degenerate), where the z axis is defined to lie parallel to the M—vector. These orbitals are not expected to interact with the amine ligand – being of δ symmetry in the local ligand frame – but will contribute to the overall M—halogen bonding. The greater electron occupancy of the $d_{xy}/d_{x^2-y^2}$ orbitals in the d^8 system and their consequentially reduced acceptor properties might be expected to be manifest in decreased $e_\lambda(X)$ values in the nickel complexes. That no such decrease is observed,

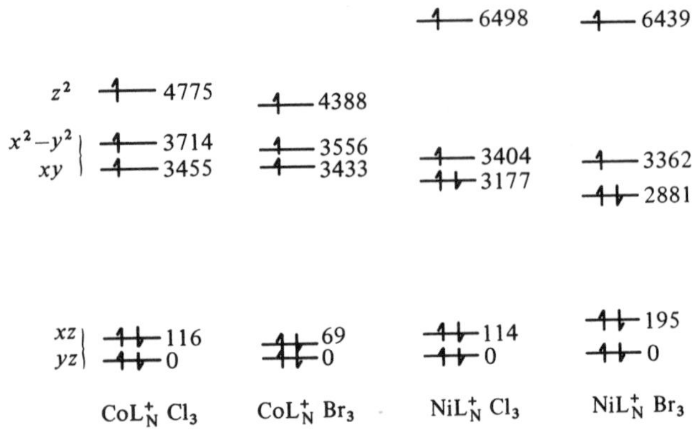

$$—\!\!\!\!—\ 6498 \qquad —\!\!\!\!—\ 6439$$

z^2 —$\uparrow$— 4775　　—$\uparrow$—4388

$\left.\begin{array}{l} x^2-y^2 \\ xy \end{array}\right|$ —$\uparrow$—3714　—$\uparrow$—3556　—$\uparrow$—3404　　—$\uparrow$—3362
　　　　　—$\uparrow$—3455　—$\uparrow$—3433　$\uparrow\downarrow$—3177
　　　　　　　　　　　　　　　　　　　　　　$\uparrow\downarrow$—2881

　　　　　　　　　　　　　　　　　　　$\uparrow\downarrow$—195
$\left.\begin{array}{l} xz \\ yz \end{array}\right|$ $\uparrow\downarrow$—116　$\uparrow\downarrow$—69　$\uparrow\downarrow$—114　$\uparrow\downarrow$—0
　　　　$\uparrow\downarrow$—0　　$\uparrow\downarrow$—0　$\uparrow\downarrow$—0

　　$CoL_N^+ Cl_3$　　$CoL_N^+ Br_3$　　$NiL_N^+ Cl_3$　　$NiL_N^+ Br_3$

Fig. 12.14. One-electron energies (cm^{-1}) for the series M(dabconium$^+$)X$_3$, corresponding to 'best-fit' parameters in table 12.4. Orbital purity corresponds fairly closely with a_1, e, e for D_{3d} symmetry. (L_N^+ = dabconium$^+$).

Table 12.4. *AOM and Racah parameters* (cm^{-1}) *for dabconium complexes*[44] M(dabco$^+$)X$_3$

Complex	e_σ(amine)	e_σ(hal)	e_π(hal)	B
Co dabco$^+$Cl$_3$	4250	3500	1100	740
Co dabco$^+$Br$_3$	4000	3500	1000	700
Ni dabco$^+$Cl$_3$	6100	3250	1000	760
Ni dabco$^+$Br$_3$	5900	3000	850	720

Table 12.5. *Dabconium complexes: mean bond lengths* (Å) *in coordination shells*[45]

Bond	Co dabco$^+$Cl$_3$	Co dabco$^+$Br$_3$	Ni dabco$^+$Cl$_3$	Ni dabco$^+$Br$_3$
mean M—N	2.09	2.13	2.04	2.04
mean M—X	2.26	2.38	2.24	2.38

presumably reflects the greater effective nuclear charge of the d^8 ion relative to d^7, together with the operation of the electroneutrality principle so far as the halogen ligands themselves are concerned. However, that same electroneutrality principle then requires the nickel–amine interaction to increase relative to that in the cobalt system, with the consequential increase in e_σ(dabco$^+$) values. This conclusion is supported by the coordinate-bond length given in table 12.5.

12.4.1 The magnetism of tetrahedrally coordinated nickel(II) complexes

The material described in the present review has been organized to emphasize the nature of the chemical information which derives from contemporary AOM studies. More traditional descriptions of the magnetic properties of the transition-metal complexes have tended to collate data according to ground state or metal d^n configuration. We consider now a division along these lines in order to make some rather general remarks about the magnetic properties of d^8, nickel(II) systems.

Years ago it was felt that a simple differentiation between tetrahedral and octahedral nickel(II) molecules would be possible by virtue of the formal T_1 ground term of the former conferring a large orbital contribution to, and temperature dependence of, the magnetic moment. It is now well known that such a diagnostic of d^8 coordination geometry is *empirically* unreliable. One reason for this, of course, is the partial quenching of the orbital moment on distortion from ideal T_d symmetry. Nevertheless, the frequently observed very low magnetic moments in 'tetrahedrally' coordinated nickel(II) systems – occasionally even lower than those observed for typical octahedral molecules – have been taken as indicative of 'gross' deviation from the tetrahedral ideal. Even allowing that 'gross' is a subjective epithet, it remains surprising that so large a degree of orbital quenching should be associated with unexceptional geometric distortions. Another empirical observation concerns the magnitudes of magnetic *anisotropies* commonly observed for distorted tetrahedral nickel(II) and cobalt(II) species. While such anisotropy can occur in, and indeed should be dominated by, first-order raising of ground-state degeneracy in the 'tetrahedral' d^8 species, only second-order mechanisms are available in the d^7 case. We expect, therefore, to observe generally rather larger magnetic anisotropies in the nickel(II) complexes than in roughly similar cobalt(II) species. It is not uncommon, however, to find little difference between the two cases or even a reversed ordering. Several factors may contribute to this: (i) the d^7 systems are intrinsically more paramagnetic – three unpaired electrons as compared with the two in the d^8 species: (ii) the geometric distortion may be so large as to reduce all anisotropy to a second-order role; (iii) somewhat as a special case of (ii) perhaps, large distortions, together with small Δ_{tet} values, will increase the significance of components of the excited $^4T_{2g}$ state in the d^7 case; and there are others. Altogether, however, the general behaviours of both average susceptibilities and magnetic anisotropies of 'tetrahedral' nickel(II) systems tend more towards those of ions with orbital singlet states than would be expected; and this requires explanation.

Table 12.6. *Spin-orbit coupling coefficients* (cm^{-1}) *for tetrahedrally coordinated nickel(II) and cobalt(II) complexes*

Complex	nickel system			cobalt system		
	ζ	ζ/ζ_0^a	reference	ζ	ζ/ζ_0	reference
$[M(PPh_3)Br_3]^-$	195	0.31	15			
$M(PPh_3)_2Cl_2$	350	0.56	17	500	0.97	17
$M(PPh_3)_2Br_2$	300	0.48	17	500	0.97	17
$[M(quinoline)Br_3]^-$	250	0.40	8	450	0.87	8
$M(dabco^+)Cl_3$	130	0.21	44	500	0.97	44
$M(dabco^+)Br_3$	120	0.19	44	500	0.97	44
$M(sal)^b$	350	0.56	48			
$[M Cl_4]^{2-}$	380	0.60	62	450	0.87	63
$[M Br_4]^{2-}$	260	0.41	64	470	0.91	63
$[M I_4]^{2-}$	15	0.02	46			

$^a \zeta_0 Ni(II) = 630 \, cm^{-1}$; $\zeta_0 Co(II) = 515 \, cm^{-1}$.
^{b} sal = N-isopropylsalicylaldiminato.

The reason lies in the greatly reduced values of the effective spin-orbit coupling parameter which has been demonstrated[15,44] to be a character-istic feature in tetrahedrally coordinated nickel(II) systems. Spin-orbit coefficients ζ determined from ten different molecules have been found to range from 15 to $380 \, cm^{-1}$, as compared with the free-ion value ζ_0 of $630 \, cm^{-1}$. The ζ/ζ_0 ratios are markedly less than corresponding ones (where known) for the analogous cobalt(II) complexes, as shown in table 12.6. Since orbital contribution, high temperature dependence, and magnetic anisotropy all decrease with decreasing spin-orbit coupling in species with formally, orbitally degenerate ground states, the unexpected magnetic properties of 'tetrahedral' nickel(II) complexes follow directly. There remains the question of why such low ζ values should characterize these systems. Two, possibly complementary, mechanisms have been proposed.[15,46]

One, after Collingwood et al.,[46] refers to the sign of the mixing coefficient of ligand functions within molecular orbitals of the formal T_1 ground term which, being negative, expresses a contribution to the effective ζ parameter from the ligands, which is opposed to that from the metal. This should be an especially significant process for heavier ligands with larger spin-orbit coefficients. The virtually vanishing value of $\zeta \sim 15 \, cm^{-1}$ in NiI_4^{2-} may well be due to this mechanism, although it must be cautioned that the empirical value was deduced[46] solely from the electronic spectra (absorption and mcd) and, unlike the other entries in table 12.6, may

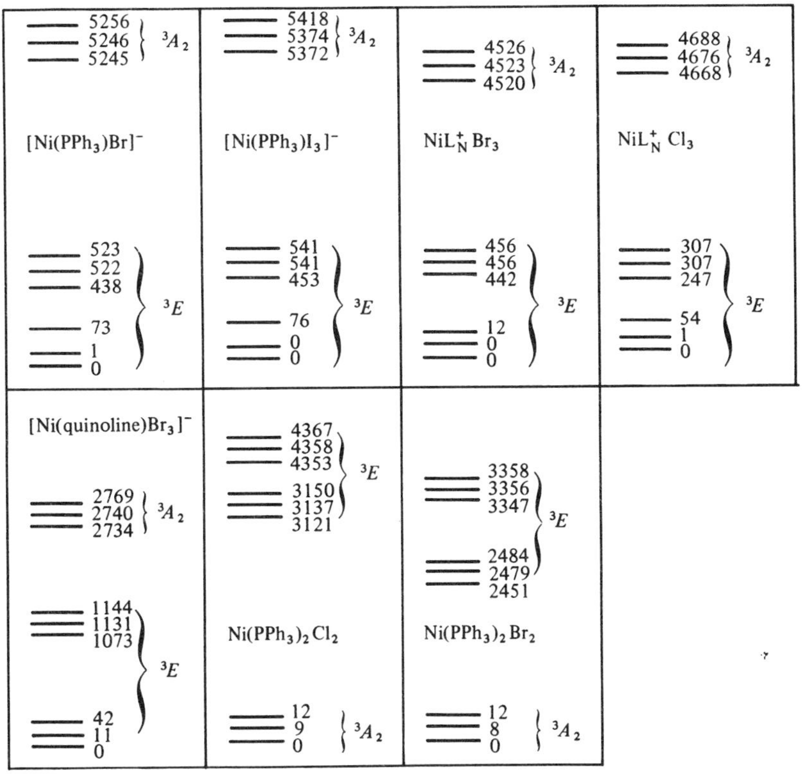

5256 5246 } ³A₂ 5245	5418 5374 } ³A₂ 5372	4526 4523 } ³A₂ 4520	4688 4676 } ³A₂ 4668
[Ni(PPh₃)Br]⁻	[Ni(PPh₃)I₃]⁻	NiL$_N^+$ Br₃	NiL$_N^+$ Cl₃
523 522 438 73 1 0 } ³E	541 541 453 76 0 0 } ³E	456 456 442 12 0 0 } ³E	307 307 247 54 1 0 } ³E
[Ni(quinoline)Br₃]⁻	4367 4358 4353 } ³E 3150 3137 3121	3358 3356 3347 } ³E	
2769 2740 } ³A₂ 2734 1144 1131 1073 42 11 0 } ³E	Ni(PPh₃)₂Cl₂ 12 9 } ³A₂ 0	2484 2479 2451 Ni(PPh₃)₂Br₂ 12 8 } ³A₂ 0	

Fig. 12.15. Splitting of the $^3T_1(T_d)$ ground term in some 'tetrahedral' nickel(II) complexes, calculated with the best-fit parameter sets given in tables 12.1 and 12.4. The top four complexes closely approach three-fold symmetry; the quinoline complex only very approximately so; and the last two possess two-fold symmetry. (L_N^+ = dabconium⁺).

not simultaneously account for ground-state properties. However, this mechanism is unlikely to account for the small, though larger, ζ values – typically $200 \, \text{cm}^{-1}$ – observed for tetrahedral nickel(II) complexes with lighter donors like chlorine, nitrogen or oxygen. It has been suggested,[15] therefore, that a second mechanism for the reduction may arise from the Ham effect[47] which relates to a quenching of orbital angular momentum within the dynamic Jahn–Teller regime. State energies, corresponding to 'best fit', are shown in figure 12.15 for the formal $^3T_1$ ground terms of several of the nickel(II) complexes listed in table 12.6. Those for the first four complexes (top row in diagram) describe only slightly split, orbital–doublet ground terms: these molecules conform fairly well to C_{3v} symmetry. The molecular geometries for the *bis*-phosphine complexes approximate C_{2v} and the state energy diagrams describe orbital-singlet ground terms. The lower-symmetry quinoline complex is characterized by a large splitting of an orbital-doublet ground term, the overall $^3T_1$ manifold splitting appearing intermediate between the first four and last two complexes in the group. Now the Ham effect refers to the quenching of orbital angular momentum and so one might expect the first four systems to provide better candidates for the operation of this effect than the last three: the data in table 12.6 may just support this. Nevertheless, if the Ham effect is the main cause of the generally much reduced ζ values in *all* these d^8 complexes, it appears to be operating in second order effectively as well as in first, in that the $^3T_1$ components span widths of up to $5500\,\mathrm{cm}^{-1}$. Another problem derives from the fact that no anomalously large thermal factors were observed in the X-ray analysis of these systems, as might have been expected for systems undergoing a dynamic Jahn–Teller coupling.

The picture is not entirely clear, therefore, but it is worth reiterating that the ligand-field formalism described theoretically in chapter 11 and whose applications are described in this, appears generally successful at the, say, 10% level, which is all that can be expected or, indeed, demonstrated. This suggests, at least, that it is unnecessary to contemplate restructuring the theory to parametrize separately 'minority' and 'majority' electron spins, even if one could tolerate the degree of parametrization. Finally we can summarize: the successful application of ligand-field methods empirically carries over to 'tetrahedrally' coordinated d^8 nickel(II) species but with concomitant, large reduction in ζ/ζ_0 ratios.

12.5 Imine-type ligands

Examples in this section illustrate the variability of the π-bonding function of imine groups and identify an interesting comparison which can be made between AOM parameters on the one hand and interelectron repulsion parameters on the other. We compare studies of the three complexes of 'tetrahedrally' coordinated nickel(II) shown in figure 12.16.

Analysis[48] of the first of these, the Schiff-base complex in figure 12.16(a) was based upon complete crystal-susceptibility data in the temperature range 20–300 K, together with a diffuse reflectance spectrum (shown in figure 12.4), while those[12] of the diimine and iminoamine complexes in figures 12.16(b) and (c) rely on solution spectra only. In the present section, we consider the values, in table 12.7, for $e_{\pi\perp}(\mathrm{N})$, being $+900$, -250 and $+200\,\mathrm{cm}^{-1}$ respectively (the last value is less well determined to lie

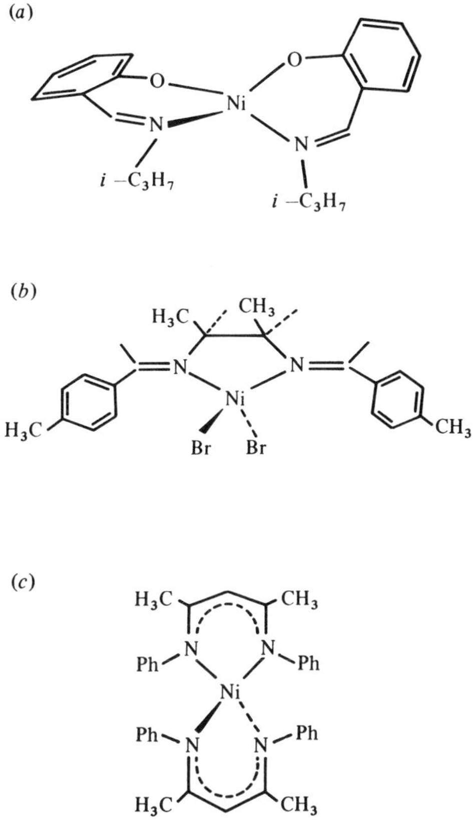

(a)

(b)

(c)

Fig. 12.16. Three tetrahedrally-coordinated complexes of nickel(II) involving imine type ligators.

anywhere between -50 and $+300\,\mathrm{cm}^{-1}$). In particular, let us concentrate upon the moderate π-*donor* function of the Schiff-base imine versus the weaker π-*acceptor* function of the diimine ligators. Manifestly, there is no question of parameter transferability in these systems and we are led to identify the relevant difference between these two ligand groups. It is, surely, that adjacent to the imine group in the Schiff base is the electron-rich phenolic oxygen atom and we anticipate that the π-electron drift in the diimine system is reversed by the presence of this group, as illustrated in figure 12.17.

The small positive value found for $e_{\pi\perp}(N)$ in the iminoamine complex in figure 12.16(c) presumably reflects the type-name of the ligand: thus we may write the canonical structures in figure 12.18 and anticipate that the electron drift indicated by the dotted arrows will be less facile than

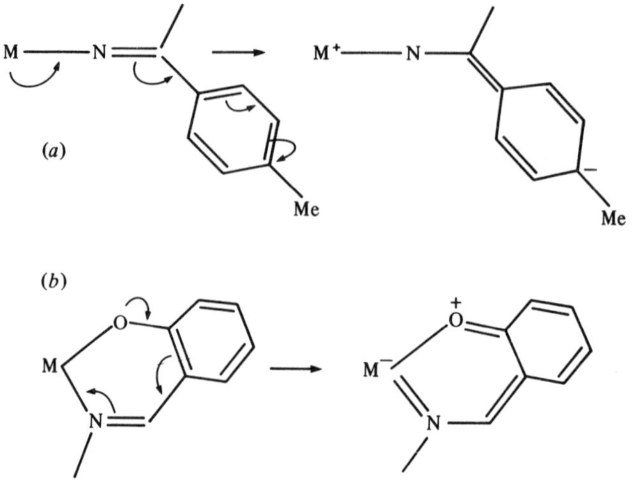

The synthesis

Fig. 12.17. (a) π-acceptor function of the conjugated imine group: (b) π-donor function of the imine moiety within the 'sal' group.

that shown in figure 12.17(a) by virtue of the 'initial' polarization of the chelate π delocalization and, particularly, in view of the marked non-coplanarity[49] of the phenyl and iminoamine moieties.

In all three complexes, we observe (table 12.7) quite substantial nephelauxetic reduction in the B values despite the variable and small $e_{\pi\perp}(N)$ values: contrast these Racah B values with those observed in quinoline or dabconium amine complexes, for example. Recall, however, that an AOM e_{λ} parameter provides a measure of the *net* λ bonding in the M—L interaction. Specifically, consider the change from the diimine to the salicylidene complex as one involving variations in *both* ligand π and π^* functions. We presume that the interaction of the π^* function of the diimine just prevails over that of the filled molecular orbital, resulting in the small, negative e_{π} value observed. Electron drift from the oxygen of the salicylidene system to the nitrogen raises the energy of the N π orbital to better match the energy of the metal orbital and simultaneously decreases the interaction with the antibonding, empty π^* molecular orbital. Whether this latter effect is to be viewed as the raising of the energy of the π^* orbital or the attenuation of the M—L π^* overlap is not important in the schematic representation of figure 12.19. The point to emphasize here is that the e_{π} AOM parameter measures the *net* M—N π bonding in these systems. The connection with the language of frontier-orbital theory, in which the π and π^* orbitals in the present description might

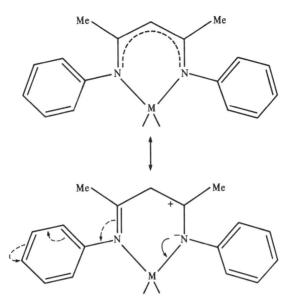

Fig. 12.18. The putative π-acceptor role of the imine group in this iminoamine system is not realized; probably due to the non-coplanarity of the ligand: the phenyl rings make an angle of about $56°$ with respect to the chelate moiety.

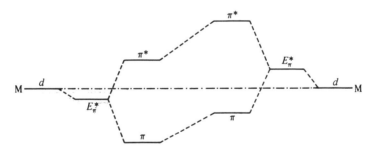

Fig. 12.19. Illustrating how the AOM measures *net* metal – ligand interactions. On the left is represented the net negative e_π parameter for the diimine complex (figure 12.16(b)) and, on the right, for the positive e_π value for the salicylaldiminato complex (figure 12.16(a)).

be labelled HOMO and LUMO has been remarked upon in §11.14. However, note that, whether the spectral electrons reside in orbitals of ligand π or π^* parentage or both, we may expect them to suffer a somewhat reduced interelectron repulsion in these spatially more extended orbitals. Therefore, while the AOM energy parameters e_π provide a measure of *net*

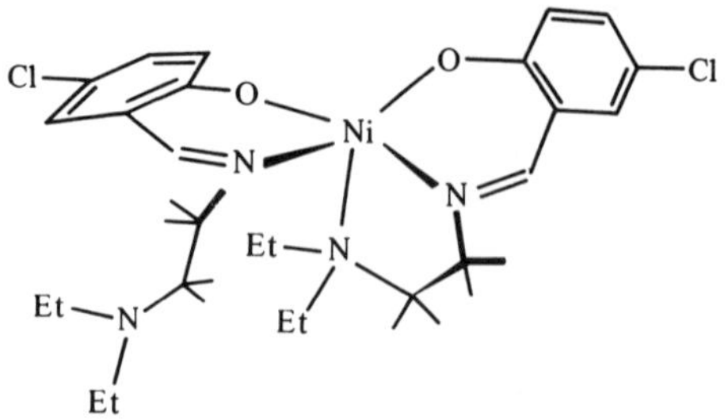

Fig. 12.20. Bis(5-chloro-N-β-diethylaminoethylsalicylaldiminato) nickel(II): approximate square-pyramidal geometry.

π bonding, the nephelauxetic effect describes, in part, the *gross* extent of π bonding.

12.6 Ligand fields from lone pairs

Results from the analysis[48] of the salicylidene Schiff-base complex in figure 12.16(*a*) may be compared with those for the five-coordinate analogue[50] shown in figure 12.20. The molecules differ in two respects: the isopropyl substituents on the imine nitrogens of the tetrahedrally coordinated molecule have been replaced by tertiary amines, one of which coordinates to the nickel atom to form the approximate, square-pyramidal structure shown (note that the amine group does *not* occupy the apical site, due to steric constraints of the ligand as a whole); secondly, the aromatic rings have chlorine substituent groups in positions para to the phenolic oxygens. Best-fit parameter values for both molecules are compared in table 12.7 and are based upon AOM analyses of single-crystal magnetic susceptibilities over the temperature range 20–300 K, together with diffuse reflectance spectra. In most respects, similar ligators in the two molecules are characterized by similar AOM parameter values. Thus e_σ values for imine and phenoline oxygen groups are all about 4000 cm^{-1}; the imine groups function as π donors perpendicular to the 'sal' planes and are not involved in any π interaction with the metal *in* the planes of the chelates. In the five-coordinate complex, the $e_\sigma(\text{amine})$ value is much less than that of the imines, but this merely reflects the much longer Ni—amine bond of about 2.20 Å as compared with 1.98 Å for the Ni—imine groups.[51] The most interesting differences between the two systems are associated with the π bonding AOM parameters for the phenolic oxygen groups.

Firstly, the value of $e_{\pi\perp}(\text{O})$ is about twice as large in the five-coordinate

Table 12.7. *AOM and Racah parameters* (cm⁻¹) *for some imine-type complexes of nickel*(II)

Complex: see fig:	$e_\sigma(N)^a$	$e_{\sigma\perp}(N)^a$	$e_\sigma(O)$	$e_{\pi\perp}(O)$	$e_{\pi\parallel}(O)$	$e_\sigma(Br)$	$e_\pi(Br)$	$e_\sigma(N_T)^b$	B	reference
12.16(a)	(4000)	+900	(4000)	800	600	—	—	—	660	48
12.20	(4000)	+900	(4000)	1500	600	—	—	2500	?	50
12.16(b)	5200	(−250)	—	—	—	3500	800	—	630	12
12.16(c)	4000	(+200)	—	—	—	—	—	—	640	12

[a] refers to imine or iminoamine ligators.
[b] refers to tertiary amine ligand.
$e_{\pi\parallel}(N)$ and $e_\pi(N_T)$ values demonstrated or assumed zero.
Values in parentheses are less sensitively determined.

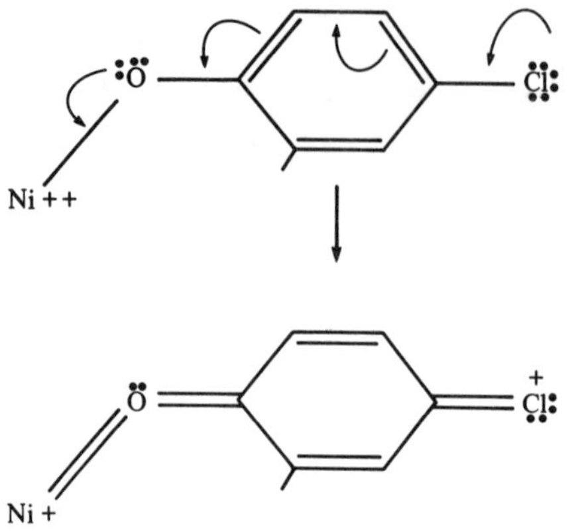

Fig. 12.21. Enhancement of the π-donor role of the phenolic oxygen ligator by mesomeric coupling with the para substituent chlorine atom.

complex as in the four-coordinate one. We ascribe this to the mesomeric donation from the para chlorine, as in figure 12.21. It would be interesting to undertake similar ligand-field analyses on a variety of related systems in view of the organic chemist's classical view of the relative mesomeric functions of ortho, meta and para substituents on the aromatic ring. It is worth noting that when the dissimilarity between the AOM parameter sets for these two salicylidene complexes was first recorded and rationalized,[50] it was the first time that this sort of argument (involving the participation of distant atoms or groups) common in other chemical disciplines, was applied directly to the interpretation of the magnetic susceptibilities of a central atom; and we remarked that this was an illustration of how 'magnetochemistry is about chemistry'.

Secondly, we consider π bonding *in* the plane of the 'sal' groups. While $e_{\pi\|}$(imine) values are essentially zero for each complex, $e_{\pi\|}$(O) values take on substantial, positive numbers in both molecules. Although incorrectly parametrized in these early analyses, we consider these results to reflect the absence and presence, for imine and oxygen groups respectively, of lone pairs of electrons in the plane of the 'sal' moieties. While the imine nitrogens possess only one formal lone pair – in the $p_\perp$ orbital – the oxygen atoms have two; one in the $p_\perp$ orbital (π) and one in the sp^2 hybrid, directed in the plane of the 'sal' chelate, about 120° away from the Ni—O vector. It is this latter lone pair which provides the potential represented, rather inadequately, by the $e_{\pi\|}$(O) parameter. Although the sp^2 lone pair lies displaced about a direction pointing strongly away from the metal, it must be remembered that lone pairs are laterally much more diffuse than

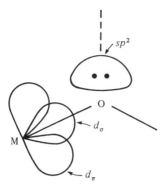

Fig. 12.22. The AOM 'cross-parameter' $e_{\sigma\pi}$ is used to account for the interaction of the 'misdirected' lone pair with both metal σ and π d orbitals.

bonding pairs (Gillespie–Nyholm), an idea amply supported by detailed calculations.[52] The inadequacy of the early analyses in representing the potential of the sp^2 lone pair as of $\pi_{\|}$ symmetry is made plain by the sketch in figure 12.22. The 'misdirected' valence of the lone pair is such as to allow interaction with *both* metal σ and $\pi_{\|}$ orbitals. So the AOM energy matrix in the local M—O frame is no longer diagonal and, if that plane is labelled xz, we may define an off-diagonal parameter, $e_{\pi\sigma} = \langle d^l_{z^2} | v^l | d^l_{xz} \rangle$, as discussed in §§9.9 and 11.12(iii). The $e_{\pi\|}(O)$ parameters of table 12.7 should be augmented, therefore, by $e_{\pi\sigma}(O)$. This refinement had not been developed at the time of the foregoing analyses,[48,50] but unpublished calculations since then suggest that values for $e_{\pi\sigma}(O)$ of similar magnitudes to the $e_{\pi\|}(O)$ in the table are approximately equally satisfactory in reproducing the experimental data. Again, application of the contemporary synthesis of ligand-field techniques might well be rewarding as a probe of metal–lone pair interactions including, for example, those which may be involved with metal–water coordination: we note that what initially appeared as a problem (in-plane π bonding without the apparent availability of ligand orbitals), turned out to be a sensible reflection, within ligand-field theory, of real electron distributions in bonds.

12.7 Planar molecules – the anomalous energy of the d_{z^2} orbital

In figure 12.6 were summarized the approximate ligand-field orbital energies in trans-dimesityl bis (diethylphenylphosophine) cobalt(II), deduced[28] from detailed studies of single-crystal esr and optical spectra and of paramagnetic susceptibilities. In idealized D_{2h} symmetry, to which this low-spin, cobalt(II) coordination geometry approximates, the energies of the $d_{x^2-y^2}$ and d_{z^2} orbitals are given[33] by $3\bar{e}_\sigma$ and $\bar{e}_\sigma$, respectively, where $\bar{e}_\sigma$ means the average e_σ parameter for the phosphine and mesitylene ligands. The analysis provided a lower bound for the energy of the $d_{x^2-y^2}$ orbital of 18 000 cm^{-1} (relative to an origin defined with no asymmetric ligand field and to that of the d_{yz} orbital which is totally unaffected by the ligand field, as described in §12.2); hence $\bar{e}_\sigma \geq 6000$ cm^{-1}. Accordingly, the d_{z^2} orbital is found to lie about 5500 cm^{-1} lower than predicted by theory; or, at least, by the theory contemporaneously available. At that time, it appeared that the system provided an example of a failure of the AOM (and of any other form of ligand-field theory) in that only four out of five orbital energies (or three out of four transition energies) could be properly accounted for.

The idea that the energy of the d_{z^2} orbital is anomalous was similarly reached in a study of planar nickel (II) diamine complexes by Hitchman & Bremner.[53] The same conclusion has also been drawn from an analysis[54] of the ligand-field properties of Gillespite, a mineral containing planar iron(II) chromophores. In this case, it was considered that the energy of the d_{z^2} orbital was anomalous, partly because a compromise 'fit' to the AOM yielded *very* large e_π values for the Fe—O interactions: using the relationship (11.98), it was found[54] that the proportionality constant for the π interaction was much larger than for the σ interaction and it is upon this that these workers asserted that the value of e_π was unacceptably high. These authors report a similar analysis, of the isomorphous copper(II) analogue, 'Egyptian blue', and reach similar conclusions. Finally, Smith[55] has considered the application of the AOM to a series of chlorocuprates, arguing, as have others in the more recent studies, that satisfactory fits to the optical spectra could not be obtained without special treatment for the d_{z^2} orbital.

The mechanism proposed by Smith[55] is that the $3d_{z^2}$ orbital is depressed energetically by hybridization with the $4s$ orbital, both orbitals transforming as the totally symmetric representation in this point group. Independent evidence that some degree of d–s mixing takes place in planar complexes, of low-spin cobalt(II) at least, derives from two esr studies of the isotropic hyperfine coupling with the metal nucleus. The sign of the Fermi contact term is positive if the $3d$ electron spin couples with the metal nuclear spin via the contracted $3s$ orbital, but negative if the process takes place via the more expanded $4s$ orbital. The effect is represented in figure 12.23. Exchange correlation is such as to cause like spins to attract,

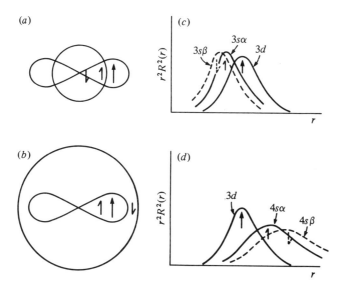

Fig. 12.23. Schematic representations of spin-polarization effects. The unpaired spin in the $3d$ orbital 'attracts' parallel spin-orbitals more than antiparallel ones so that the Fermi contact terms take opposite signs for inner ($3d$) and outer ($4d$) orbital participation.

so that the amplitude of the $s\alpha$ spin-orbitals maximizes nearer to the $3d\alpha$ function than does the $s\beta$ orbital; the figure shows the $3s\alpha$ spin-orbital polarized further out from the nucleus than the $3s\beta$, while the reverse situation obtains for the $4s$ spin-orbitals. This phenomenon may be quantified to yield estimates for the degree of $3d$–$4s$ mixing. Using a simple molecular-orbital approach, Rockenbauer *et al.*[56] have estimated $4s$ mixing coefficients of about 0.1–0.2 throughout a series of about 12 different planar cobalt(II) complexes. Even before that study, the analysis[31] of the esr isotropic hyperfine coupling in the mesityl phosphine complex with which we began this section indicated a similar degree of d–s mixing within the formalism of the model used.

Altogether, we can be confident of the basic phenomenon: that, in planar complexes of several transition-metal ions (certainly of Fe(II), Co(II), Ni(II), Cu(II)), the otherwise highly successful angular overlap model apparently fails to account for the anomalously low energies of the d_{z^2} orbital relative to those of other members of the d set, the anomaly being of the order of 4000–6000 cm^{-1}. That this phenomenon *is* explicable within a properly formulated scheme and that, far from being a 'failure' of the model, it is actually required in planar-coordinated species for reasons of theoretical consistency, has been demonstrated in §§11.12(iii) and 11.15.

12.8 Conclusions

Ligand-field theory defines only one small area of interest within chemistry at large: it has, nevertheless, provided a focus for considerable attention by physical, theoretical, and especially inorganic chemists over the years, partly because of the opportunity it has provided for the exercise of various group-theoretical skills but ultimately, and much more importantly, because of its early promise to provide a probe of the chemical bonding and electron distribution in transition-metal complexes. As described in chapter 1, however, refinements of the theory, and experimental practice, of the ligand-field theory of single-centre systems have tended towards introspection within the subject, sight of the ultimate chemical purpose of the discipline often having been lost. The angular overlap model, emerging against a backdrop of ideas from McClure,[57] Yamatera,[58] Schmidtke and others,[59,60] was first formulated by Jørgensen and Schäffer as a means by which electronic spectral splittings might be analysed in terms of local, chemical functional groups. Considering the ubiquitous nature of localized chemical functionality, it is surprising that it has taken so long for such an obviously attractive idea to be generally known and to be made to work. One may discern two main reasons for this slow emergence. One stems from an historical restriction in the application of the AOM to systems with relatively high molecular symmetry for which it is ill-suited and, even when such has not been the case, we find AOM parameter sets being underdetermined by a common reference to only one type of ligand-field property – usually electronic spectroscopy; and the other arises from explicit, written or spoken, doubts concerning the theoretical standing and veracity of the angular overlap model itself. Some have taken on board the original expression of the approach, seemingly forgetting its intentions within ligand-field theory, to the extent that the formulation of the model likely to reach the largest audience by way of several contemporary teaching texts is one which purports to identify the AOM as a molecular-orbital scheme closely related to the Hückel model. This increasingly popular version of the AOM is used ever less for the interpretation of $d-d$ spectra and other ligand-field phenomena, and increasingly for the rationalization of thermodynamic and structural features both inside and outside the d block. That model, deriving certainly from the first formulations of the AOM but equally undoubtedly denying the original purposes of the approach, appears to be a prime target for those who despair of theoretical models which rest on undefined or ill-defined premises. The purpose of the extensive discussions in chapter 11 was to reconstruct a model with the original spirit of the AOM, but whose form lies totally within the scope of ligand-field theory and whose precepts are explicitly defined. The cellular model we have created[61] makes contact with the original AOM in two ways: it adopts the same aims of spatially partitioning the ligand field and so reflecting chemical functionality and hence its application involves the use of the same AOM equation (9.20); and, although the e_λ parameters are interpreted differently in the two models, many of the qualitative assertions of the old AOM are supported by the defined, new scheme. So while some may prefer to rename the structure we have expounded in this book – the 'local potential model' perhaps – the similarities it shares with the original AOM suggest that a retitling is unnecessary. Throughout this book, therefore, we refer to the AOM, intending the ligand-field version at all times unless otherwise stated.

Hopefully it is clear from the review of the recent applications of the AOM to ligand-field analysis, described in this chapter, that the original aims, not only of the model but of ligand-field theory in general, are at last being fulfilled. The analyses allow us to comment directly and semi-quantitatively on the bonding in transition-metal complexes, to monitor the contributions made by individual ligands, and to separate local σ- and π-bonding features. Some results of these studies serve to confirm familiar views of the electron distribution in complexes – views which have arisen gradually from a wide spectrum of chemical notions and hints: others have been surprising enough to rekindle excitement in a long-established subject and suggest fresh appraisals of the chemical functionality of particular ligands and even to provoke further synthetic effort. These analyses have succeeded at the practical level often because the problem of 'overparametrization' has been solved by experimental recourse to the larger data base afforded by study of a variety of ligand-field properties rather than just one; and in this connection, measurements of electronic and esr spectra, and paramagnetic susceptibilities of single crystals have been useful, and often essential. The pragmatic success of the contemporary synthesis of ligand-field techniques derives in no small measure from the existence of a comprehensive computing package designed to minimize the frustrations and tedium of a wide coverage of parameter space that is essential within fitting processes. In short, in ligand-field analysis, the AOM works. There remains the question of why it should.

We were not concerned, in chapter 11, to provide for *ab initio* computation of ligand-field parameters. We attempted, rather, to discover a structure for the AOM within ligand-field theory which bears a *definable* relationship with the principles of the theory of electronic structure at large. Given the procedures and definitions of the AOM as it is actually used to reproduce experimental data, we sought to establish with what defined quantum-mechanical operations and entities these correspond: in this we followed the philosophy adopted in our analysis of the ligand-field formalism itself. Altogether it has been possible to provide a sound basis for the formalism and parameters of the AOM as used in ligand-field analysis. However, the internal structure of the model that is required to do this set it apart from conventional molecular-orbital theory. This should not be surprising really, when we recall that ligand-field theory is *not* a molecular-orbital model to replace the old crystal-field theory, but this point has been insufficiently emphasized in the inorganic chemistry literature. The failure to remember that ligand-field theory involves *projection* onto basis metal functions, explicitly defined to exclude valence–valence correlation (d–d or f–f, as appropriate), by the concomitant construction of an effective operator called the ligand-field potential, has often led to the false assertion that an explicit neglect of ligand functions within the molecular orbitals of d (f) parentage defines a fundamental limitation or approximation of the approach. There exists in this sort of criticism, a naive belief that the truth resides only in molecular-orbital theory. Now of course, a properly defined version of molecular-orbital theory can certainly provide for the computation of various observations of transition-metal complexes – the so-called, 'ligand-field properties'. In practice, however, it is found extremely difficult and costly to complete such calculations – which require configuration interaction – with satisfactory accuracy. Ligand-field theory focusses attention onto one part of the eigenvalue spectrum – that associated with the ground and low-lying states differing amongst themselves almost solely with respect to their d-orbital components – and furthermore expresses the situation in a parametric form: thus ligand-field theory has strictly limited aims and application, but ones which have been shown to be theoretically well based. By contrast, various approximate molecular-orbital schemes, employing ill-defined quantities and procedures, have been represented as providing a 'more realistic' view of the system under study. In that the aims of molecular-orbital models, good and bad, are more ambitious than those of ligand-field theory, the claim for greater realism may seem persuasive; but, in terms of scientific, definable achievements, the limited, but sound, results of ligand-field models present an altogether more serious accomplishment than the whole body of answers – however numerate – provided by ill-based, 'qualitative' molecular-orbital methods.

Viewed against the background of established ligand-field models, it seems obvious that the AOM parameters should monitor features of chemical.bonding in complexes as they actually are, rather than as properties of the notional reactants from which the molecule was formed. It follows that the exigencies of the electroneutrality principle will have already been met so far as possible and that, insofar that synergic, 'back-bonding' and other processes will have taken place to implement this, the AOM parameters must reflect the final result. Such was the conclusion of the theoretical analysis in chapter 11 and such is the stance adopted consistently throughout our interpretations of the analytical results described in this chapter. It is not surprising that simplistic notions of the ratios of e_λ parameters being expressed in terms of ratios of corresponding overlap integrals squared are indefensible by reference to either theory or experiment. The practices of fixing ratios of AOM parameters by this relationship (11.98) and of transfering parameter values from one complex to another as means of ameliorating a paucity of experimental data, arise explicitly or implicitly from the molecular-orbital version of the AOM, rarely find empirical support, and are to be discouraged. We have described a model which is soundly based and, in bringing to ligand-field theory the notion of the functional group, has finally provided the chemical relevance so long claimed for the subject. It would be a great pity if the pejorative influence of the molecular-orbital-based model should rob us of a promising, well-founded facility. The contemporary synthesis of experimental, theoretical and computational techniques has surely seen to it that ligand-field analysis has come of age.

References

[1] Gerloch, M., McMeeking, R.F. & White, A.M., *J. Chem. Soc. Dalton Trans.*, p. 2452 (1975).
[2] Glerup, J., Monsted, O. & Schäffer, C.E., *Inorg. Chem.*, **15**, 1399 (1976).
[3] Lever, A.P.B., *Coord. Chem. Rev.*, **3**, 119 (1968).
[4] Rowley. D.A. & Drago, R.S., *Inorg. Chem.*, **7**, 795 (1968).
[5] Hitchman, M.A., *Inorg. Chem.*, **11**, 2387 (1972).
[6] Schreiver, A.F. & Hansen, D.J., *Inorg. Chem.*, **12**, 2037 (1973).
[7] Barton, T.J. & Slade, R.C., *J. Chem. Soc. Dalton Trans.*, p. 650 (1975).
[8] Gerloch, M. & Hanton, L.R., *Inorg. Chem.*, **19**, 1692 (1980).
[9] Horrocks, W.D. Jr, Templeton, D.H. & Zalkin, A., *Inorg. Chem.*, **7**, 2303 (1968).
[10] Bertini, I., Gatteschi, D. & Mori, F., *Inorg. Chim. Acta*, **7**, 717 (1973).
[11] Bencini, A. & Gatteschi, D., *Inorg. Chem.*, **16**, 2141 (1977).
[12] Gerloch, M., Hanton, L.R. & Manning, M.R., *Inorg. Chim. Acta*, **48**, 205 (1981).
[13] Mackey, D.J., Evans, S.V. & McMeeking, R.F., *J. Chem. Soc. Dalton Trans.*, p. 160 (1978).

[14] Mason, R. & Meek, D.W., *Angew. Chem. Int. Ed.*, **17**, 183 (1978).
[15] Gerloch, M. & Hanton, L.R., *Inorg. Chem.*, **20**, 1046 (1981).
[16] Bertini, I., Gatteschi, D. & Mani, F., *Inorg. Chem.*, **11**, 2464 (1972).
[17] Davies, J.E., Gerloch, M. & Phillips, D.J., *J. Chem. Soc. Dalton Trans.*, p. 1836 (1979).
[18] Venanzi, L.M., *J. Chem. Soc.*, p. 719 (1958).
[19] Simo, C. & Holt, S., *Inorg. Chem.*, **7**, 2655 (1968).
[20] Tomlinson, A.A.G., Bellitto, C., Piovesana, O. & Furlani, C., *J. Chem. Soc. Dalton Trans.*, p. 350 (1972).
[21] Fereday, R.J., Hathaway, B.J. & Dudley, R.J., *J. Chem. Soc. (A)*, p. 571 (1970).
[22] Huddesman, K.D., Ph.D. Thesis, London University, 1977.
[23] Greene, P.T. & Sacconi, L., *J. Chem. Soc. (A)*, p. 866 (1970)
[24] Sacconi, L. & Gelsomini, J., *Inorg. Chem.*, **7**, 291 (1968).
[25] Sacconi, L. & Dapporto, P., *J. Amer. Chem. Soc.*, **92**, 4133 (1970).
[26] Ferguson, J., *Prog. Inorg. Chem.*, **12**, 159 (1970).
[27] Ferguson, J. & Wood, D.L., *Aust. J. Chem.*, **23**, 861 (1970).
[28] Falvello, L.R. & Gerloch, M., *Inorg. Chem.*, **19**, 472 (1980).
[29] Owsten, P.G. & Rowe, J.M., *J. Chem. Soc.*, p. 3411 (1963).
[30] Falvello, L.R. & Gerloch. M., *Acta Cryst.*, **B35**, 2547 (1979).
[31] Bentley, R.B., Mabbs, F.E., Smail, W.R., Gerloch, M. & Lewis, J., *J. Chem. Soc. (A)*, p. 3003 (1970).
[32] 'CAMMAG', a FORTRAN computer program by D.A. Cruse, J.E. Davies, J.H. Harding, M. Gerloch, D.J. Mackey and R.F. McMeeking.
[33] Gerloch, M. & Slade, R.C., *Ligand-Field Parameters*, Cambridge University Press, 1973.
[34] Palenik, G.J., & Wester, D.W., *Inorg. Chem.*, **17**, 864 (1978).
[35] Wester, D.W. & Palenik, G.J., *J. Amer. Chem. Soc.*, **95**, 6505 (1973).
[36] Gerloch, M., Morgenstern-Baderau, I. & Audière, J.P., *Inorg. Chem.*, **18**, 3220 (1979).
[37] Gerloch, M. & Morgenstern-Baderau, I., *Inorg. Chem.*, **18**, 3225 (1979).
[38] Gerloch, M. & Morgenstern-Baderau, I., in press.
[39] Gerloch, M. & Hanton, L.R., *Inorg. Chim. Acta*, **49**, 37 (1981).
[40] Bishop, M.M., Lewis, J., O'Donoghue, T.D., Raithby, P.R. & Ramsden, J.N., *J. Chem. Soc. Chem. Comm.*, p. 828 (1978).
[41] Hanton, L.R. & Raithby, P.R., *Acta Cryst.*, **B36**, 1489 (1980).
[42] Garrett, B.B., Goedken, V.L. & Quagliano, J.V., *J. Amer. Chem. Soc.*, **92**, 489 (1970).
[43] Quagliano, J.V., Banerjee, A.K., Goedken, V.L. & Vallarino, L.M., *J. Amer. Chem. Soc.*, **92**, 482 (1970).
[44] Gerloch, M. & Manning, M.R., *Inorg. Chem.*, **20**, 1051 (1981).
[45] Gerloch, M., Manning, M.R. & Raithby, P.R., in preparation.
[46] Collingwood, J.E., Day, P. & Denning, R.G., *J. Chem. Soc. Faraday Trans. II.* **69**, 591 (1973).
[47] Ham, F.S., *Phys. Rev.*, **138A**, 1727 (1965).
[48] Cruse, D.A. & Gerloch, M., *J. Chem. Soc. Dalton Trans.*, p. 152 (1977).
[49] Healy, P.C., Bendall, M.R., Doddrell, D.M., Skelton, B.W. & White, A.H., *Aust. J. Chem.*, **37**, 727 (1979).
[50] Cruse, D.A. & Gerloch, M., *J. Chem. Soc. Dalton Trans.* p. 1613 (1977).
[51] Orioli, P.L., DiVaira, M. & Sacconi, L., *J. Amer. Chem. Soc.*, **88**, 4383 (1966).
[52] Steiner, E. *The Determination and Interpretation of Molecular Wavefunctions*, Cambridge University Press, 1976.
[53] Hitchman, M.A. & Bremner, J.B., *Inorg. Chim. Acta*, **27**, 261 (1978).
[54] Mackey, D.J., McMeeking, R.F. & Hitchman, M.A., *J. Chem. Soc. Dalton Trans.*, p. 299 (1979).

[55] Smith, D.W., *Inorg. Chim. Acta*, **22**, 107 (1977).

[56] Rockenbauer, A.R., Budó-Zákionyi, E. & Sinándi, L.J., *J. Chem. Soc. Dalton Trans.*, p. 1729 (1975).

[57] McClure D.S., In: Advances in the Chemistry of Coordination Compounds, S. Kischner (ed.), Macmillan, New York, p. 498, 1961.

[58] Yamatera, H., *Bull. Chem. Soc. Japan*, **31**, 371 (1983).

[59] Jørgensen, C.K., Pappalardo, R. & Schmidtke, H.H., *J. Chem. Phys.*, **39**, 1422 (1963).

[60] Jørgensen, C.K., *Modern Aspects of Ligand-Field Theory*, North-Holland, Amsterdam, 1971.

[61] Gerloch, M. & Woolley, R.G., *Prog. Inorg. Chem.*, **31**, 371 (1983).

[62] Gerloch, M. & Slade, R.C., *J. Chem. Soc. A*, p. 1022 (1969).

[63] Gerloch, M., Lewis, J. & Rickards, R., *J. Chem. Soc. Dalton*

[64] Figgis, B.N., Lewis, J., Mabbs, F.E. & Webb, G.A., *J. Chem. Soc. A*, p. 1411 (1966).

[65] Smith, D.W., *Structure and Bonding*, **35**, 87 (1978).

Appendix A

Molecular geometry in the crystal

═══

We summarize here, one or two useful procedures for handling geometry calculations in molecules and crystals.

(a) An X-ray or neutron diffraction analysis provides a list of atomic fractional coordinates together with a unit cell. Given $(x/a, y/b, z/c)$ for a location in the cell $(a, b, c, \alpha, \beta, \gamma)$, we can obtain the real coordinates (X, Y, Z) in the orthogonalized frame $(a, c^* \wedge a, c^*)$ using the transformation

$$\begin{pmatrix} X \\ Y \\ Z \end{pmatrix} = \mathbf{U} \begin{pmatrix} x/a \\ y/b \\ z/c \end{pmatrix},$$

where

$$\mathbf{U} = \begin{pmatrix} a & b\cos\gamma & c\cos\beta \\ 0 & b\sin\gamma & c(\cos\alpha - \cos\beta\cos\gamma)/\sin\gamma \\ 0 & 0 & V/ab\sin\gamma \end{pmatrix} \qquad (A.1)$$

and the cell volume V is given by

$$V = abc(1 - \cos^2\alpha - \cos^2\beta - \cos^2\gamma + 2\cos\alpha\cos\beta\cos\gamma)^{1/2}. \quad (A.2)$$

(If fractional coordinates are required in the orthogonalized cell, the columns of $\mathbf{U}$ should be divided by a, b and c, respectively.)

All further geometry calculations are most easily performed in terms of these orthogonalized coordinates and with respect to the orthogonalized unit cell.

(b) To calculate the direction cosines of the vector joining P_1 to P_2 with respect to the orthogonal cell, we proceed in two steps.

(i) Subtract the coordinates of P_2 from P_1, effectively placing P_2 at the origin O.

(ii) Compute the distance $OP_3 = P_1 - P_2$ and divide the coordinates of P_3 by OP_3 to obtain the coordinates of the end point of a unit vector, centred at the origin and parallel to $P_1 - P_2$. These co-ordinates are the required direction cosines of $P_1 - P_2$ with respect to the orthogonal cell.

(c) The SETUP part of CAMMAG defines a global reference frame with a card like

$$\text{XREF} \quad 1 \quad 2 \quad 3$$

which requires the global z axis to lie parallel to the vector joining atoms 1 and 2 in a preceding list, and the y axis to lie perpendicular to the plane defined by atoms 1, 2 and 3. This is implemented as follows:

(i) construct the unit vectors 1–2 and 1–3 using the procedures in (b) above. Then the coordinates of the unit vector parallel to 1–2 and centred at the origin provide the elements $M(3, 1)$, $M(3, 2)$, $M(3, 3)$ *of the required orientation matrix* M directly, by identity. (Since we are dealing with unit vectors, normalization is achieved automatically.)

(ii) The second row of M – that is, the y axis – is provided by taking the vector product of the two unit vectors 1–2 and 1–3.

(iii) Finally, the top row (x axis), is constructed by taking the vector product of rows two and three.

(d) Ligand reference frames are constructed in a similar manner to that for XREF. The relationship between a ligand frame and XREF is determined subsequently by simple unitary transformation (axis rotation) as all frames at this stage are referred to an *orthogonal* cell.

Appendix B

Euler angles and direction cosines

Equation (A–1) on p. 148 of *Operator Methods in Ligand Field Theory*, by H. Watanabe (Prentice-Hall, 1966) provides for the transformation of coordinates (x, y, z) referred to an original frame to (x', y', z') referred to a final frame:

$$\begin{pmatrix} x' \\ y' \\ z' \end{pmatrix} = \begin{pmatrix} c_\alpha c_\beta c_\gamma - s_\alpha s_\gamma & s_\alpha c_\beta c_\gamma + c_\alpha s_\gamma & - s_\beta c_\gamma \\ - c_\alpha c_\beta s_\gamma - s_\alpha c_\gamma & - s_\alpha c_\beta s_\gamma + c_\alpha c_\gamma & s_\beta s_\gamma \\ c_\alpha s_\beta & s_\alpha s_\beta & c_\beta \end{pmatrix} \begin{pmatrix} x \\ y \\ z \end{pmatrix}, \qquad \text{(B.1)}$$

where $c_\alpha \equiv \cos \alpha$, $s_\beta \equiv \sin \beta$, etc. and α, β, γ are the Euler angles defined in §8.1. Using subscripted Greek letters for direction cosines as usual, the equivalent transformation is written

$$\begin{pmatrix} x' \\ y' \\ z' \end{pmatrix} = \begin{pmatrix} \alpha_1 & \beta_1 & \gamma_1 \\ \alpha_2 & \beta_2 & \gamma_2 \\ \alpha_3 & \beta_3 & \gamma_3 \end{pmatrix} \begin{pmatrix} x \\ y \\ z \end{pmatrix}. \qquad \text{(B.2)}$$

Comparison of the matrices in (B.1) and (B.2) provides a means of determining the Euler angles that are equivalent to a given set of direction cosines, the latter probably deriving from some process like that described in Appendix A.

The most obvious route is (i) $c_\beta = \gamma_3$ and hence β is known, (ii) $s_\beta s_\gamma = \gamma_2$ and so now γ is known, and (iii) $s_\alpha s_\beta = \beta_3$ leading to α. However, this procedure will not work if $\beta = 0$ for then $s_\beta = 0$ and α and γ are indeterminate: the same is effectively true, in terms of computational accuracy, if $\beta \approx 0$. The problem cannot be solved by taking ratios of $\beta_3/\alpha_3 = \tan \alpha$, or of $\gamma_2/\gamma_1 = - \tan \gamma$ because the condition $\gamma_3 \to 1$, means that $\alpha_3, \beta_3, \gamma_1, \gamma_2 \rightarrow 0$ and indeterminacy or ill-conditioned arithmetic arise once again. Instead, when $1 - |\gamma_3| < 0.01$ say, Euler angles α and γ are determined without the use of the last row or column of (B.1) and (B.2). Thus,

$$\beta_1 - \alpha_2 = s_\alpha c_\gamma(1 + c_\beta) + c_\alpha s_\gamma(1 + c_\beta) = (1 + c_\beta)\sin(\alpha + \gamma) \qquad \text{(B.3)}$$

and, similarly,

$$\alpha_1 + \beta_2 = (1 + c_\beta)\cos(\alpha + \gamma). \qquad \text{(B.4)}$$

Having already obtained c_β from γ_3, these two equations furnish a unique solution of the sum $(\alpha + \gamma)$ in the range $0-2\pi$. Then we also have

$$\beta_1 + \alpha_2 = (c_\beta - 1)\sin(\alpha - \gamma), \qquad \text{(B.5)}$$

$$\alpha_1 - \beta_2 = (c_\beta - 1)\cos(\alpha - \gamma), \qquad \text{(B.6)}$$

and hence a unique solution for $0 \le (\alpha - \gamma) < 2\pi$: values for α and γ follow trivially.

Appendix C
3–j and 6–j symbols

The following FORTRAN function subroutines accept REAL arguments and provide REAL values for Wigner's 3–j and 6–j coefficients. WIG3J codes equations (8.53) and (8.56) to yield $\begin{pmatrix} a & b & c \\ x & y & z \end{pmatrix}$. WIG6J codes equation (3–7) on p. 57 of *Operator Techniques in Atomic Spectroscopy*, by B.R. Judd, (McGraw-Hill, 1963), to produce $\begin{Bmatrix} a & b & c \\ d & e & f \end{Bmatrix}$. WIG6J calls function DEL. Selection rules are built into both WIG3J and WIG6J. (Source: D.J. Mackey and R.F. McMeeking in CAMMAG).

Function WIG3J

```
FUNCTION WIG3J (A,B,C,X,Y,Z)
DIMENSION D(6),K(9)
Q = 0.001
WIG3J = 0.0
HA = ABS(X+Y+Z)
HB = A+B−C+0.02
HC = B+C−A+0.02
HD = C+A−B+0.02
HE = A+X+0.02
HF = A−X+0.02
HG = B+Y+0.02
HH = B−Y+0.02
HI = C+Z+0.02
HJ = C−Z+0.02
HK = A+B+C+1.02
   IF((HA.GT.Q). OR.(HB.LT.Q).OR.(HC.LT.Q).OR.(HD.LT.Q).OR.(HE.LT.Q)
2.OR.(HF.LT.Q).OR.(HG.LT.Q).OR.(HH.LT.Q).OR.(HI.LT.Q).OR.(HJ.LT.Q)
3.OR.(HK.LT.Q))   GO TO 13
   SN = 1.0
   RN = 1.0
   SUM = 0.0
   V = 0.0
   SM = A+B+C
   GO TO 2
 1 CONTINUE
   V = V+1.0
   IF (V.GT.SM) GO TO 10
 2 CONTINUE
   PN = 1.0
   D(1) = A+B−C−V+0.02
   D(2) = A−X−V+0.02
   D(3) = B+Y−V+0.02
   D(4) = C−B+X+V+0.02
   D(5) = C−A−Y+V+0.02
   D(6) = V+0.02
   IF((D(1).LT.Q).OR.(D(2).LT.Q).OR.(D(3).LT.Q).OR.(D(4).LT.Q).OR.
2(D(5).LT.Q).OR.(D(6).LT.Q))   GO TO 1
   CONTINUE
   DO 3 M = 1,6
   N = IFIX(D(M))
   IF (N.EQ.0) GO TO 14
   DO 4 I = 1,N
   PN = PN*FLOAT(I)
 4 CONTINUE
14 CONTINUE
 3 CONTINUE
   PX = 0.5*(V+2.02)
   FACT = −1.0
   IF((PX−AINT (PX)−0.3).LT.0.0)   FACT = 1.0
   SUM = SUM + FACT/PN
   GO TO 1
10 CONTINUE
   K(1) = HD
   K(2) = HC
   K(3) = HB
   K(4) = HE
   K(5) = HF
   K(6) = HG
   K(7) = HH
   K(8) = HI
   K(9) = HJ
   K(10) = HK
   DO 5 M = 1,9
   N = K (M)
   IF (N.EQ.0)   GO TO 16
   DO 6 I = 1,N
   RN = RN*FLOAT(I)
 6 CONTINUE
16 CONTINUE
 5 CONTINUE
   DO 7 I = 1,K10
   SN = SN*FLOAT(I)
 7 CONTINUE
   QX = 0.5*(A−B−Z+100.02)
   PHF = −1.0
   IF((QX−AINT(QX)−0.3).LT.0.0)   PHF = 1.0
   WIG3J = PHF*SUM*SQRT(RN/SN)
13 CONTINUE
   RETURN
   END
```

Function WIG6J

```
      FUNCTION WIG6J (A,B,C,D,E,F)
      DIMENSION G(7)
      Q = 0.001
      WIG6J = 0.0
      HA = A+B−C+0.02
      HB = B+C−A+0.02
      HC = C+A−B+0.02
      HD = C+D−E+0.02
      HE = D+E−C+0.02
      HF = E+C−D+0.02
      HG = A+E−F+0.02
      HI = E+F−A+0.02
      HJ = F+A−E+0.02
      HK = D+B−F+0.02
      HL = B+F−D+0.02
      HM = F+D−B+0.02
      IF((HA.LT.Q).OR.(HB.LT.Q).OR.(HC.LT.Q).OR.(HD.LT.Q).OR.(HE.LT.Q)
     2.OR.(HF.LT.Q).OR.(HG.LT.Q).OR.(HI.LT.Q).OR.(HJ.LT.Q).OR.(HK.LT.Q)
     3.OR.(HL.LT.Q).OR.(HM.LT.Q))    GO TO 13
      Z = 0.0
      SM = A+B+C+D+E+F
      SUM = 0.0
      GO TO 2
    1 CONTINUE
      Z = Z+1.0
      IF(Z.GT.SM) GO TO 10
    2 CONTINUE
      PN = 1.0
      QN = 1.0
      G(1) = Z−A−B−C+0.02
      G(2) = Z−A−E−F+0.02
      G(3) = Z−D−B−F+0.02
      G(4) = Z−D−E−C+0.02
      G(5) = A+B+D+E−Z+0.02
      G(6) = B+C+E+F−Z+0.02
      G(7) = C+A+F+D−Z+0.02
      IF((G(1).LT.Q).OR.(G(2).LT.Q).OR.(G(3).LT.Q).OR.(G(4).LT.Q).OR.
     2(G(5).LT.Q).OR.(G(6).LT.Q).OR.(G(7).LT.Q))    GO TO 1
      CONTINUE
      DO 3 M = 1,7
      N = IFIX(G(M))
      IF (N.EQ.0) GO TO 14
      DO 4 I = 1,N
      PN = PN*FLOAT(I)
    4 CONTINUE
   14 CONTINUE
    3 CONTINUE
      JR = Z+1.02
      DO 17 J = 1,JR
      QN = QN*FLOAT(J)
   17 CONTINUE
      PX = 0.5*(Z+2.02)
      FACT = −1.0
      IF(PX−AINT(PX)−0.3).LT.0.0)    FACT = 1.0
      SUM = SUM+FACT*QN/PN
      GO TO 1
   10 CONTINUE
      WIG6J = DEL(A,B,C)*DEL(A,E,F)*DEL(D,B,F)*DEL(D,E,C)*SUM
   13 CONTINUE
      RETURN
      END
      FUNCTION DEL(U,V,W)
C     FUNCTION IS REQUIRED IN THE CALCULATION ON THE WIGNER 6−J SYMBOL
      DIMENSION J(3)
      SP = 1.0
      PP = 1.0
      J(1) = U+V−W+0.02
      J(2) = U−V+W+0.02
      J(3) = V+W−U+0.02
      J4 = U+V+W+1.02
      DO 1 M = 1,3
      N = J(M)
      IF (N.EQ.0) GO TO 2
      DO 3 I = 1,N
      SP = SP*FLOAT(I)
    3 CONTINUE
    2 CONTINUE
    1 CONTINUE
      DO 4 K = 1,J4
      PP = PP*FLOAT(K)
    4 CONTINUE
      DEL = SQRT(SP/PP)
      RETURN
      END
```

Appendix D

AOM transformations and f electrons

W. Urland has published a thorough analysis of the ligand-field potential for f electrons in the angular overlap model (*Chem. Phys.*, **14** 393 (1976)). He defines real f-electron wavefunctions as

$$
\left.\begin{aligned}
|\sigma\rangle &= \sqrt{(7/4\pi)}r^{-3}\tfrac{1}{2}(2z^2 - 3x^2 - 3y^2)z \\
|\pi s\rangle &= \sqrt{(7/4\pi)}r^{-3}\tfrac{1}{4}\sqrt{6}(4z^2 - x^2 - y^2)y \\
|\pi c\rangle &= \sqrt{(7/4\pi)}r^{-3}\tfrac{1}{4}\sqrt{6}(4z^2 - x^2 - y^2)x \\
|\delta s\rangle &= \sqrt{(7/4\pi)}r^{-3}\tfrac{1}{2}\sqrt{(15)}z(2xy) \\
|\delta c\rangle &= \sqrt{(7/4\pi)}r^{-3}\tfrac{1}{2}\sqrt{(15)}z(x^2 - y^2) \\
|\varphi s\rangle &= \sqrt{(7/4\pi)}r^{-3}\tfrac{1}{4}\sqrt{(10)}(3x^2y - y^3) \\
|\varphi c\rangle &= \sqrt{(7/4\pi)}r^{-3}\tfrac{1}{4}\sqrt{(10)}(x^3 - 3xy^2)
\end{aligned}\right\} \tag{D.1}
$$

and provides unitary rotation matrices – $\mathbf{R}^l$ of (9.15) – in terms of direction cosines or of Euler angles: but, in the latter case only, the rotation is specialized to $D(\alpha, \beta, 0)$, appropriate for linear ligators. More important, from the point of view of the procedures in §9.6, is his table 3 which provides the equivalent of our table 9.2 but for the real f-orbital basis (D.1). We abstract relevant parts of this to construct the table D below, appropriate for performing calculations as in §9.6 but restricted to σ and π interactions, in line with the philosophy expounded in §9.8.

Table D. *Relationships between c_{kq} and e parameters for σ and π bonding in an f-orbital basis within one local M—L interaction. (All c_{kq} are real; only non-zero c_{kq} are listed.)*

$$c_{00} = \tfrac{2}{7}\sqrt{\pi}(e_{\sigma} + e_{\pi x} + e_{\pi y})$$

$$c_{20} = \tfrac{2}{7}\sqrt{(5\pi)}e_{\sigma} + \tfrac{3}{14}\sqrt{(5\pi)}(e_{\pi x} + e_{\pi y})$$

$$c_{22} = c_{2,-2} = \sqrt{\left(\frac{30\pi}{14}\right)}(e_{\pi x} - e_{\pi y})$$

$$c_{40} = \frac{\sqrt{\pi}}{7}(6e_{\sigma} + e_{\pi x} + e_{\pi y})$$

$$c_{42} = c_{4,-2} = \frac{\sqrt{(16\pi)}}{7}(e_{\pi x} - e_{\pi y})$$

$$c_{60} = \tfrac{2}{7}\sqrt{(13\pi)}e_{\sigma} - \tfrac{3}{14}\sqrt{(13\pi)}(e_{\pi x} + e_{\pi y})$$

$$c_{62} = c_{6,-2} = \frac{1}{2}\sqrt{\left(\frac{39\pi}{35}\right)}(e_{\pi x} - e_{\pi y})$$

Expressions for 'misdirected valency' involving $e_{\pi\sigma}$ in the local frame give rise to the terms c_{21}, c_{41} and c_{61} as may be deduced from Urland's table 3.

Appendix E
Real ligand-field potentials

Here we show that the ligand-field potential for a system possessing two-fold rotation symmetry with respect to the standard,[††] polar y axis is necessarily real.

(a) Let asymmetric reference moiety (1) establish Euler angles (α, β, γ) with respect to some global frame. Consider moiety (2) obtained by rotation of (1) by π about global y axis, and defining equivalent Euler angles $(\alpha', \beta', \gamma')$. Then, for (1)

$$D(\alpha\beta\gamma) = e^{i\gamma j_z} e^{i\beta j_y} e^{i\alpha j_z} \qquad (E.1)$$

and, for (2)

$$
\begin{aligned}
D(\alpha'\beta'\gamma') &= e^{i\gamma j_z} e^{i\beta j_y} e^{i\alpha j_z} e^{i\pi j_y} \\
&= e^{i\gamma j_z} e^{i\beta j_y} e^{i\pi j_y} [e^{-i\pi j_y} e^{i\alpha j_z} e^{i\pi j_y}] \\
&= e^{i\gamma j_z} e^{i(\pi + \beta) j_y} e^{-i\alpha j_z} \\
&= D(-\alpha, \pi + \beta, \gamma). \qquad (E.2)
\end{aligned}
$$

As a check, refer to (B.1) and (B.2) relating Euler angles and direction cosines. Under the transformation $\alpha \to -\alpha$, $\beta \to \pi + \beta$, $\gamma \to \gamma$ the matrix of direction cosines changes signs according to

$$
\mathbf{M} \to \begin{pmatrix} - & + & - \\ - & + & - \\ - & + & - \end{pmatrix}
$$

[††] The polar angle θ is measured from z, while ϕ is measured in the xy plane from x. The polar coordinate scheme thus defines a standard y axis.

582

and this conforms with the relationship between two systems related by a diad parallel to y.

(b) Consider transforming a ligand-field multipole expansion for moiety (1) into the global frame, as in (9.31):

$$c_{kq}^G = \sum_{q'=-k}^{k} \mathscr{D}_{q'q}^k(\alpha\beta\gamma)c_{kq'}^{(1)} \tag{E.3}$$

where, from (10.1), k is even and $q' = 0, \pm 2$. Recall that

$$\mathscr{D}_{q'q}^k(\alpha\beta\gamma) = e^{i(\alpha q + \gamma q')}d_{q'q}^k(\beta), \tag{8.24}$$

$$d_{q'q}^k(\beta) = (-1)^{k+q}d_{q'-q}^k(\pi+\beta), \tag{E.4}$$

$$d_{q'q}^k(\beta) = (-1)^{q'-q}d_{qq'}^k(\beta) = d_{q-q'}^k(\beta), \tag{E.5}$$

(E.4) and (E.5) arising from (8.33). Then,

(i) For c_{kq}^G; $q = 0$.

$$c_{k0}^G = \mathscr{D}_{00}^k(\alpha\beta\gamma)c_{k0}^{(1)} + \mathscr{D}_{20}^k(\alpha\beta\gamma)c_{k2}^{(1)} + \mathscr{D}_{-20}^k(\alpha\beta\gamma)c_{k-2}^{(1)} \tag{E.6}$$

The first term is real, by (8.37). For the others,

$$\mathscr{D}_{20}^k(\alpha\beta\gamma) = e^{2i\gamma}d_{20}^k(\beta)c_{k2}^{(1)},$$

$$\mathscr{D}_{-20}^k(\alpha\beta\gamma) = e^{-2i\gamma}d_{-20}^k(\beta)c_{k-2}^{(1)},$$

but, from (10.1), $c_{k2}^{(1)} = c_{k-2}^{(1)}$; and from (E.5),

$$d_{-20}^k(\beta) \equiv d_{-2-0}^k(\beta) = (-1)^2 d_{20}^k(\beta) = d_{20}^k(\beta).$$

Therefore,

$$\mathscr{D}_{20}^k(\alpha\beta\gamma)c_{k2}^{(1)} + \mathscr{D}_{-20}^k(\alpha\beta\gamma)c_{k-2}^{(1)} = d_{20}^k(\beta)c_{k2}^{(1)},$$

which is real. Altogether, therefore, all c_{k0}^G are *real*.

(ii) For c_{kq}^G; $q \neq 0$.

$$c_{kq}^G = \mathscr{D}_{0q}^k(\alpha\beta\gamma)c_{k0}^{(1)} + \mathscr{D}_{2q}^k(\alpha\beta\gamma)c_{k2}^{(1)} + \mathscr{D}_{-2q}^k(\alpha\beta\gamma)c_{k-2}^{(1)}$$

$$\equiv A \qquad\quad + B \qquad\quad + C, \quad \text{say.} \tag{E.7}$$

We have,

$$\left.\begin{array}{l} A = e^{i\alpha q}d_{0q}^k(\beta)c_{k0}^{(1)}, \\[4pt] B = e^{i(\alpha q + 2\gamma)}d_{2q}^k(\beta)c_{k2}^{(1)}, \\[4pt] C = e^{i(\alpha q - 2\gamma)}d_{-2q}^k(\beta)c_{k-2}^{(1)}, \end{array}\right\} \tag{E.8}$$

each of which is generally complex. Now, for every such contribution from moiety (1) with Euler angles $(\alpha\beta\gamma)$, we have an equivalent contribution from moiety (2) with Euler angles $(-\alpha, \pi+\beta, \gamma)$. (E.8) for these takes the form

$$
\left.
\begin{aligned}
A' &= e^{-i\alpha q}d^k_{0q}(\pi+\beta)c^{(2)}_{k0}, \\
B' &= e^{i(-\alpha q+2\gamma)}d^k_{2q}(\pi+\beta)c^{(2)}_{k2}, \\
C' &= e^{i(-\alpha q-2\gamma)}d^k_{-2q}(\pi+\beta)c^{(2)}_{k-2}.
\end{aligned}
\right\}
\tag{E.9}
$$

Again, from (10.1) $c^{(1)}_{k2}=c^{(1)}_{k-2}$ and $c^{(1)}=c^{(2)}$ for equivalent moieties: using (E.4) and (E.5), we have

$$
d^k_{0q}(\beta)=(-1)^{k+q}d^k_{0-q}(\pi+\beta)=(-1)^k d^k_{0q}(\pi+\beta)=d^k_{0q}(\pi+\beta)
$$

and

$$
d^k_{-2q}(\pi+\beta)=(-1)^{k-q}d^k_{-2-q}(\beta)=(-1)^{k-q}d^k_{2q}(\beta)(-1)^{2-q}=d^k_{2q}(\beta).
$$

Hence, corresponding to

$$
B+C=e^{i(\alpha q+2\gamma)}d^k_{2q}(\beta)c^{(1)}_{k2}+e^{i(\alpha q-2\gamma)}d^k_{-2q}(\beta)c^{(1)}_{k-2}
$$

for moiety (1), we have

$$
C'+B'=e^{-i(\alpha q+2\gamma)}d^k_{2q}(\beta)c^{(2)}_{k2}+e^{-i(\alpha q-2\gamma)}d^k_{-2q}(\beta)c^{(2)}_{k-2}.
$$

Altogether, summing contributions from moieties (1) and (2), we find

$$
c^G_{kq}(1+2)=2\cos\alpha q\, d^k_{0q}(\beta)c_{k0}+2\cos(\alpha q+2\gamma)d^k_{2q}(\beta)c_{k2}
$$
$$
+2\cos(\alpha q-2\gamma)d^k_{-2q}(\beta)c_{k-2},
$$

which, because all $d(\beta)$ are real (from (8.26)), is *real*. Hence, for all systems in the AOM scheme, possessing (at least) a diad parallel to the global y axis, the resulting global potential, expressed as a multipole expansion, involves only *real* coefficients.

(c) The same result is true for any modelling of the ligand field. Thus, consider the potential in the global frame due to moiety (1):

$$
V(1)=\sum_{k,q}c_{kq}Y^k_q, \quad k \text{ even.}
$$

That for moiety (2), related to (1) by rotation by π about y is given by applying the transformation

$$
Y^k_q(2)=\sum_{q'=-k}^{k}\mathscr{D}^k_{q'q}(0\pi0)Y^k_{q'}(1).
$$

But

$$
\mathscr{D}^k_{q'q}(0\pi0)=d^k_{q'q}(\pi)=(-1)^{k-q}\delta_{q,-q'},
$$

from (8.38), and hence

$$
=(-1)^{q'}\delta_{q,-q'}
$$

as k is even. Therefore,

$$Y_q^k(2) \Rightarrow (-1)^{q'} Y_{q'}^k(1).$$

In particular:

$$Y_0^k(2) \Rightarrow Y_0^k(1),$$

for q odd:

$$Y_{\pm q}^k(2) \Rightarrow -Y_{\pm q}^k(1),$$

for q even:

$$Y_{\pm q}^k(2) \Rightarrow Y_{\pm q}^k(1),$$

and hence

$$V(1) + V(2) = 2\sum_k c_{k0} Y_0^k + \sum_k \sum_q^{\text{odd}} c_{kq}(Y_{-q}^k - Y_q^k) + \sum_k \sum_q^{\text{even}} c_{kq}(Y_{-q}^k + Y_q^k)$$

and it is simple to verify that each term in this expression is *real*.

QED

Appendix F

Frame-independent discrepancy index

Here we establish a measure of the discrepancy between complete observed and calculated tensors, to be used in constructing 'fitting maps' as in §10.3.1. Let $\mathbf{A}$ and $\mathbf{B}$ respectively represent the observed and calculated susceptibility tensor χ. We construct the difference tensor $\mathbf{C}$ whose elements are given by

$$C_{ij} = A_{ij} - B_{ij}. \tag{F.1}$$

Of course, we cannot base a discrepancy index on the sum of such elements, because of cancellations, but it might seem possible to do so using the sum of their moduli. It is not difficult to show, however, that such a sum is not invariant with respect to a change of reference frame and is therefore unsuitable as a basis for comparisons. On the other hand, the sum of the squares of these difference elements, $\sum_{ij} C_{ij}^2$, is independent of the reference frame, as we now show.

Let $\mathbf{C}'$ represent the difference tensor expressed in some other frame related to the reference frame by rotation: thus

$$\mathbf{C} = \mathbf{U}^{-1} \mathbf{C}' \mathbf{U}, \tag{F.2}$$

where $\mathbf{U}$ is a unitary matrix. Define $\mathbf{D}' = \mathbf{C}'^2$ and the transform

$$\mathbf{D} = \mathbf{U}^{-1} \mathbf{D}' \mathbf{U}. \tag{F.3}$$

Then

$$\mathbf{D} = \mathbf{U}^{-1} \mathbf{C}'^2 \mathbf{U} = \mathbf{U}^{-1} \mathbf{C}' \mathbf{U} \mathbf{U}^{-1} \mathbf{C}' \mathbf{U} = \mathbf{C}^2. \tag{F.4}$$

Now the trace of a symmetric matrix is invariant under a unitary transformation, so we have

$$\mathrm{Tr}(\mathbf{C}^2) = \mathrm{Tr}(\mathbf{D}) = \mathrm{Tr}(\mathbf{D}') = \mathrm{Tr}(\mathbf{C}'^2) \tag{F.5}$$

586

and hence

$$\sum_i \sum_j C_{ij}^2 = \sum_i \sum_j C_{ij}'^2,$$ (F.6)

which is the required result.

The discrepancy index used for mapping is formed by taking the ratio of that sum to the square of the trace of the observed tensor, so establishing a suitable proportionality and, finally, we use

$$R = 100 - S \sum_T \left[\sum_{i,j} C_{ij}^2(T) \middle/ \left(\sum_i A_{ii} \right)^2 \right]:$$ (F.7)

S is a scale factor and the outer sum is over the various sampling temperatures as in (10.14). A similar discrepancy index is used for comparing $\mathbf{g}^2$ tensors. Note that the $\mathbf{g}^2$ tensor is symmetric (while the $\mathbf{g}$ tensor may not be – see §7.7) and so the quantities C_{ij}^2 we now use correspond to the squares of the differences between elements of $\mathbf{g}^2$(obs) and $\mathbf{g}^2$(calc).

Index